AF361952

Novel and Emerging Strategies for Sustainable Mine Tailings and Acid Mine Drainage Management

Novel and Emerging Strategies for Sustainable Mine Tailings and Acid Mine Drainage Management

Editors

Carlito Tabelin
Kyoungkeun Yoo
Jining Li

MDPI • Basel • Beijing • Wuhan • Barcelona • Belgrade • Manchester • Tokyo • Cluj • Tianjin

Editors

Carlito Tabelin	Kyoungkeun Yoo	Jining Li
School of Minerals and Energy	Department of Energy and	School of Environment
Resources Engineering	Resources Engineering	Nanjing Normal University
University of New South Wales	Korea Maritime and Ocean	Nanjing
Sydney	University	China
Australia	Busan	
	Korea, South	

Editorial Office
MDPI
St. Alban-Anlage 66
4052 Basel, Switzerland

This is a reprint of articles from the Special Issue published online in the open access journal *Minerals* (ISSN 2075-163X) (available at: www.mdpi.com/journal/minerals/special_issues/ NESSMTAMDM).

For citation purposes, cite each article independently as indicated on the article page online and as indicated below:

LastName, A.A.; LastName, B.B.; LastName, C.C. Article Title. *Journal Name* **Year**, *Volume Number*, Page Range.

ISBN 978-3-0365-2747-5 (Hbk)
ISBN 978-3-0365-2746-8 (PDF)

Contents

About the Editors

Carlito Tabelin

Dr. Carlito Baltazar Tabelin is an environmental geochemist competent in the fields of inorganic contaminant geochemistry, rock-soil-water interactions, geochemical and reactive transport modelling, electrochemistry, mine waste management, prevention/control of acid mine/rock drainage (AMD/ARD), and remediation of contaminated sites. He earned his Ph.D. in Field Engineering for Environment from Hokkaido University, Japan in 2011. After his graduation, he was hired as a Postdoctoral researcher by an endowed program at Hokkaido University funded by Mitsubishi Materials Corporation before being promoted to the position of specially appointed assistant professor in 2013. In 2014, he joined the Laboratory of Mineral Processing and Resources Recycling at Hokkaido University as a tenure track Assistant Professor. He was awarded two research grants from the prestigious Japan Society for the Promotion of Science (JSPS) as principal investigator to develop advanced pyrite passivation techniques to limit AMD formation. From these projects, two passivation techniques that take advantage of how pyrite dissolves in nature as a targeting mechanism were developed/improved: (1) galvanic microencapsulation (GME), and (2) titanium- and aluminum-based carrier-microencapsulation (CME). He moved to Australia in 2019 and joined the School of Minerals and Energy Resources Engineering of UNSW Sydney as the lecturer of mine waste management. His current research projects include finding ways to add value to mine wastes for their sustainable management, AMD/ARD prevention and control, remediation of contaminated sites, and recovery of valuable metals from electronic and industrial wastes.

Kyoungkeun Yoo

Dr. Kyoungkeun Yoo is a Professor at Korea Maritime and Ocean University, Korea. After Kyoungkeun received his bachelor and master degrees from Hanyang University, Seoul, Korea, he completed his PhD in the Mineral Processing and Resources Recycling Laboratory at Hokkaido University, Japan. Prior to joining Korea Maritime and Ocean University, he served as a senior researcher at the Korea Institute of Geoscience and Mineral Resources (KIGAM). His research interest is the recovery or removal of metals from mineral, deep-seabed resources, urban or industrial wastes, wastewater, and contaminated soil/sediment using mineral processing and hydrometallurgical technologies. Currently, he is working on the recycling of lithium-ion batteries (LIBs), where he focuses on the continuous leaching of black powder obtained from LIBs.

Jining Li

Dr. Jining Li is an associate professor working at the School of Environment of Nanjing Normal University. His research interests mainly focus on the remediation of heavy metal(loid)s (e.g., lead, zinc, cadmium, arsenic, antimony, chromium, etc.) contaminated soils or solids wastes (e.g., tailings and smelting slags), and the long-term effectiveness evaluation of the remediation. Now he is doing some relevant researches mainly funded by the National Natural Science Foundation of China (42007111) and the Key Research and Development Program of China (2020YFF0218304).

Editorial

Editorial for Special Issue "Novel and Emerging Strategies for Sustainable Mine Tailings and Acid Mine Drainage Management"

Carlito Baltazar Tabelin [1,*], Kyoungkeun Yoo [2] and Jining Li [3]

1   School of Minerals and Energy Resources Engineering, University of New South Wales, Sydney, NSW 2052, Australia
2   Department of Energy and Resources Engineering, Korea Maritime and Ocean University, Busan 49112, Korea; kyoo@kmou.ac.kr
3   School of Environment, Nanjing Normal University, Nanjing 210023, China; ljn006@126.com
*   Correspondence: c.tabelin@unsw.edu.au

Citation: Tabelin, C.B.; Yoo, K.; Li, J. Editorial for Special Issue "Novel and Emerging Strategies for Sustainable Mine Tailings and Acid Mine Drainage Management". *Minerals* **2021**, *11*, 902. https://doi.org/10.3390/min11080902

Received: 16 August 2021
Accepted: 17 August 2021
Published: 20 August 2021

Publisher's Note: MDPI stays neutral with regard to jurisdictional claims in published maps and institutional affiliations.

Climate change is one of the most pressing problems facing humanity this century. In a recent report by the Intergovernmental Panel on Climate Change (IPCC), carbon dioxide (CO_2), nitrous oxide (N_2O) and methane (CH_4) were found to have reached annual averages of 410, 332 and 1866 ppm in 2019, respectively, leading to the global surface temperature increasing by 0.84–1.10 °C compared to measured values about a century ago [1]. Notably, the IPCC report showed significantly higher atmospheric CO_2 concentration in 2019 than historical values in the last two million years [1].

Spearheaded by the United Nations (UN) as enshrined in its sustainable development goals (UN-SDG 13 "Climate Action"), governments, companies and the research community have banded together to develop low-carbon alternatives to fossil-fuel-based technologies that dominate the transportation and heat/electricity generation sectors [2]. These alternatives include electric-based vehicles (EVs), clean storage and renewable energy technologies such solar, wind, geothermal and hydroelectric, along with secondary electricity storage media, such as lithium-ion batteries and flow-battery cells. There is, however, a big catch to all this. Low-carbon technologies are more metal, mineral and material intensive than conventional fossil-fuel-based technologies. Electric cars, for example, require up to 11 times more copper than conventional cars [3]. In addition to copper, the World Bank Group has identified aluminum, chromium, cobalt, graphite, indium, iron, lead, lithium, manganese, molybdenum, neodymium, nickel, silver, titanium, vanadium and zinc as critical elements/materials for the clean energy transition to succeed [4]. Thus, the successful transition from fossil-fuel-based to low-carbon technologies would require more extensive mining to maintain the stable supply of raw materials, such as metals and minerals, in the next 30–50 years.

The expansion of mining and mineral processing operations would mean more mining-related wastes, such as tailings, waste rocks and acid mine drainage (AMD), which are notorious for their devastating and long-term destructive impacts on the environment. Tailings are waste materials generated during the processing of ores, which are typically disposed of in tailings storage facilities (TSFs). The tonnage of tailings generated by a mine depends on the deposit. For example, large porphyry copper mining operations, such as those of Escondida and Antamina, could generate 50–150 million tons of tailings per year [3]. Waste rocks or overburden are the rocks, soils and sediments removed for access and exploitation of a deposit and could amount to about three times the tonnage of ore extracted in open-pit mining operations. Finally, AMD, the effluent generated in mine sites due to the oxidation of sulfide minerals like pyrite (FeS_2), is acidic and contains strictly regulated contaminants not only destructive to the surrounding ecosystem but also hazardous to human health. Moreover, sustainable AMD treatment is difficult because of the complex and site-specific nature of AMD and its long-term generation, which could persist for several centuries or even a few millennia [5]. Sustainable management of these three types of mine wastes remains a huge challenge for the resource sector. Most techniques and strategies employed by the sector are outdated and/or ineffective, so alternative novel ways to manage these wastes are needed, including improvements to existing waste management strategies employed in mine sites around the world [6,7].

In this Special Issue, two promising sustainable strategies for mine waste management, namely repurposing/reprocessing (i.e., valorization) and cost reduction, were explored. Longos et al. [8] repurposed nickel laterite mine waste and industrial wastes, such as coal fly ash, into geopolymers with unconfined compressive strengths around 20 MPa. Meanwhile, Wu et al. [9] demonstrated the potential use of tailings as specialty building materials resistant to corrosion when exposed to saline environments. Waste minimization via repurposing of tailings for construction and/or backfill materials is also gaining traction in the resource industry [10–12]. As high-grade ores are depleted, tailings from historic mine sites could be reprocessed to recover residual valuable metals and minerals [13]. Jung et al. [14], for example, recovered residual scheelite ($CaWO_4$) from tungsten mine tailings by flotation through reagent regime optimization. Meanwhile, Lazo and Lazo [15] utilized native plant species in Chile, such as *Oxalis gigantea*, *Cistanthe grandiflora*, *Puya berteroniana* and *Solidago chilensis*, to rehabilitate mine tailings and found that these plants could sequester molybdenum, copper and lead from the wastes and concentrate them in plant tissues. These native plants show promise for phytomining of mine wastes in the future.

For cost reduction and improved sustainability of AMD treatment, articles in this Special Issue investigated AMD volume reduction, utilization of cheaper alternative neutralizing materials and sequestration of valuable materials during treatment. Yamaguchi et al. [16] modeled up to 30% reduction of AMD generation from underground mine workings of a closed mine in Japan when mine waste was implemented for backfilling. Meanwhile, the use of cheaper, locally available materials like dolomite [17], and alkaline wastes, such as red mud [18,19], calcareous mine waste rocks [20] and demolition wastes [21], were effective alternatives to commercially available neutralizers [22].

From the perspectives of a circular economy and resource conservation, AMDs can be considered as future resources because they contain considerable amounts of valuable and critical metals. Aghaei et al. [23] demonstrated >95% copper and lead recovery from simulated mine/processing wastewater using aluminum-based bimetallic materials. Bimetallic materials, also known as bimetals, bimetallic particles and bimetallic catalysts, are promising because their action relies on reduction and galvanic interactions that specifically target redox-active elements, such as copper, nickel, gold, silver and most heavy metals [24]. The only drawback of this approach is the expensive reagent-grade aluminum powder needed for synthesis, but the idea of using aluminum wastes or scraps as raw material is recently gaining traction [24]. Another promising sustainable approach for AMD management based on circular economy concepts is the use of microbial fuel cells (MFCs) to simultaneously recover target metals and generate electricity. Ai et al. [25] demonstrated the treatment of AMD using MFCs and the recovery of copper in the cathode as elemental copper. MFCs could be a promising approach, especially in regions where the mine site is located close to cities with an abundant supply of organic-rich wastewaters.

Mine wastes exposed to the environment for a long time can also act as a natural laboratory where microorganisms evolve and develop resistance to high acidity and heavy metal concentrations, which can be isolated, cultured, grown and applied to improve existing bioleaching technologies. Chen et al. [26] studied AMD from Clarabelle Mill, Canada, and successfully isolated and identified new species of *Acidithiobacillus ferridurans* that can be used for the bioleaching of low-grade porphyry copper ores.

The Special Issue is completed with articles on the critical roles of redox conditions and galvanic interactions in the weathering of sulfide-bearing mine wastes when disposed of on land and under the sea [27], the importance of hydrogeochemical conditions on waste rock weathering [28] and the potential impacts of climate change on AMD formation in abandoned and closed mines [29].

The adoption of clean storage and renewable energy technologies is a classic example of the proverbial "double-edged sword". The benefits of transitioning to low-carbon technologies to mitigate climate change should be balanced with the socio-environmental impacts of extensive mining required to supply critical metals and materials. The collection of research and review articles in this Special Issue shows that the impacts of mining can be mitigated and managed more sustainably using circular economy concepts and the development of novel techniques to reprocess, repurpose and decontaminate mine waste streams, such as tailings, waste rocks and AMD.

Author Contributions: Writing—original draft preparation, C.B.T.; writing—review and editing, K.Y. and J.L. All authors have read and agreed to the published version of the manuscript.

Funding: This research received no external funding.

Conflicts of Interest: The authors declare no conflict of interest.

References

1. IPCC 2021 Summary for Policymakers. *Climate Change 2021: The Physical Science Basis. Contribution of Working Group I to the Sixth Assessment Report of the Intergovernmental Panel on Climate Change*; Masson-Delmotte, V., Zhai, P., Pirani, A., Connors, S.L., Péan, C., Berger, S., Caud, N., Chen, Y., Goldfarb, L., Gomis, M.I., et al., Eds.; Cambridge University Press: Cambridge, UK, 2021; in press.
2. Tabelin, C.B.; Dallas, J.; Casanova, S.; Pelech, T.; Bournival, G.; Saydam, S.; Canbulat, I. Towards a low-carbon society: A review of lithium resource availability, challenges and innovations in mining, extraction and recycling, and future perspectives. *Miner. Eng.* **2021**, *163*, 106743. [CrossRef]
3. Tabelin, C.B.; Park, I.; Phengsaart, T.; Jeon, S.; Villacorte-Tabelin, M.; Alonzo, D.; Yoo, K.; Ito, M.; Hiroyoshi, N. Copper and critical metals production from porphyry ores and E-wastes: A review of resources availability, processing/recycling challenges, socio-environmental aspects, and sustainability issues. *Resour. Conserv. Recy.* **2021**, *170*, 105610. [CrossRef]
4. WBG (World Bank Group). Minerals for Climate Action: The Mineral Intensity of the Clean Energy Transition. 2020. Available online: http://pubdocs.worldbank.org/en/961711588875536384/Minerals-for-Climate-Action-The-Mineral-Intensity-of-the-Clean-Energy-Transition.pdf (accessed on 30 September 2020).
5. Park, I.; Tabelin, C.B.; Jeon, S.; Li, X.; Seno, K.; Ito, M.; Hiroyoshi, N. A review of recent strategies for acid mine drainage prevention and mine tailings recycling. *Chemosphere* **2019**, *219*, 588–606. [CrossRef] [PubMed]
6. Igarashi, T.; Salgado, P.H.; Uchiyama, H.; Miyamae, H.; Iyatomi, N.; Hashimoto, K.; Tabelin, C.B. The two-step neutralization ferrite-formation process for sustainable acid mine drainage treatment: Removal of copper, zinc and arsenic, and the influence of coexisting ions on ferritization. *Sci. Total Environ.* **2020**, *715*, 136877. [CrossRef] [PubMed]
7. Tabelin, C.B.; Corpuz, R.D.; Igarashi, T.; Villacorte-Tabelin, M.; Alorro, R.D.; Yoo, K.; Raval, S.; Ito, M.; Hiroyoshi, N. Acid mine drainage formation and arsenic mobility under strongly acidic conditions: Importance of soluble phases, iron oxyhydroxides/oxides and nature of oxidation layer on pyrite. *J. Hazard. Mater.* **2020**, *399*, 122844. [CrossRef]
8. Longos, A.; Tigue, A.A.; Dollente, I.J.; Malenab, R.A.; Bernardo-Arugay, I.; Hinode, H.; Kurniawan, W.; Promentilla, M.A. Optimization of the mix formulation of geopolymer using nickel-laterite mine waste and coal fly ash. *Minerals* **2020**, *10*, 1144. [CrossRef]
9. Wu, J.; Li, J.; Rao, F.; Yin, W. Characterization of slag reprocessing tailings-based geopolymers in marine environment. *Minerals* **2020**, *10*, 832. [CrossRef]
10. Opiso, E.M.; Tabelin, C.B.; Maestre, C.V.; Aseniero, J.P.J.; Park, I.; Villacorte-Tabelin, M. Synthesis and characterization of coal fly ash and palm oil fuel ash modified artisanal and small-scale gold mine (ASGM) tailings based geopolymer using sugar mill lime sludge as Ca-based activator. *Heliyon* **2021**, *7*, e06654. [CrossRef]
11. Aseniero, J.P.J.; Opiso, E.M.; Banda, M.H.T.; Tabelin, C.B. Potential utilization of artisanal gold-mine tailings as geopolymeric source material: Preliminary investigation. *SN Appl. Sci.* **2019**, *1*, 35. [CrossRef]
12. Cao, S.; Yilmaz, E.; Song, W. Evaluation of viscosity, strength and microstructural properties of cemented tailings backfill. *Minerals* **2018**, *8*, 352. [CrossRef]
13. Lam, E.J.; Carle, R.; González, R.; Montofré, Í.L.; Veloso, E.A.; Bernardo, A.; Cánovas, M.; Álvarez, F.A. A Methodology Based on Magnetic Susceptibility to Characterize Copper Mine Tailings. *Minerals* **2020**, *10*, 939. [CrossRef]
14. Jung, M.Y.; Park, J.H.; Yoo, K. Effects of Ferrous Sulfate Addition on the Selective Flotation of Scheelite over Calcite and Fluorite. *Minerals* **2020**, *10*, 864. [CrossRef]
15. Lazo, P.; Lazo, A. Assessment of Native and Endemic Chilean Plants for Removal of Cu, Mo and Pb from Mine Tailings. *Minerals* **2020**, *10*, 1020. [CrossRef]
16. Yamaguchi, K.; Tomiyama, S.; Igarashi, T.; Yamagata, S.; Ebato, M.; Sakoda, M. Effects of Backfilling Excavated Underground Space on Reducing Acid Mine Drainage in an Abandoned Mine. *Minerals* **2020**, *10*, 777. [CrossRef]

17. Tangviroon, P.; Noto, K.; Igarashi, T.; Kawashima, T.; Ito, M.; Sato, T.; Mufalo, W.; Chirwa, M.; Nyambe, I.; Nakata, H.; et al. Immobilization of lead and zinc leached from mining residual materials in Kabwe, Zambia: Possibility of Chemical Immobilization by Dolomite, Calcined Dolomite, and Magnesium Oxide. *Minerals* **2020**, *10*, 763. [CrossRef]
18. Keller, V.; Stopić, S.; Xakalashe, B.; Ma, Y.; Ndlovu, S.; Mwewa, B.; Simate, G.S.; Friedrich, B. Effectiveness of Fly Ash and Red Mud as Strategies for Sustainable Acid Mine Drainage Management. *Minerals* **2020**, *10*, 707. [CrossRef]
19. Ran, Z.; Pan, Y.; Liu, W. Co-Disposal of Coal Gangue and Red Mud for Prevention of Acid Mine Drainage Generation from Self-Heating Gangue Dumps. *Minerals* **2020**, *10*, 1081. [CrossRef]
20. Retka, J.; Rzepa, G.; Bajda, T.; Drewniak, L. The Use of Mining Waste Materials for the Treatment of Acid and Alkaline Mine Wastewater. *Minerals* **2020**, *10*, 1061. [CrossRef]
21. Turingan, C.O.A.; Singson, G.B.; Melchor, B.T.; Alorro, R.D.; Beltran, A.B.; Orbecido, A.H. Evaluation of Efficiencies of Locally Available Neutralizing Agents for Passive Treatment of Acid Mine Drainage. *Minerals* **2020**, *10*, 845. [CrossRef]
22. Bortnikova, S.; Gaskova, O.; Yurkevich, N.; Saeva, O.; Abrosimova, N. Chemical treatment of highly toxic acid mine drainage at a gold mining site in Southwestern Siberia, Russia. *Minerals* **2020**, *10*, 867. [CrossRef]
23. Aghaei, E.; Wang, Z.; Tadesse, B.; Tabelin, C.B.; Quadir, Z.; Alorro, R.D. Performance Evaluation of Fe-Al Bimetallic Particles for the Removal of Potentially Toxic Elements from Combined Acid Mine Drainage-Effluents from Refractory Gold Ore Processing. *Minerals* **2021**, *11*, 590. [CrossRef]
24. Tabelin, C.B.; Resabal, V.J.T.; Park, I.; Villanueva, M.G.B.; Choi, S.; Ebio, R.; Cabural, P.J.; Villacorte-Tabelin, M.; Orbecido, A.; Alorro, R.D.; et al. Repurposing of aluminum scrap into magnetic Al0/ZVI bimetallic materials: Two-stage mechanical-chemical synthesis and characterization of products. *J. Clean. Prod.* **2021**, *317*, 128285. [CrossRef]
25. Ai, C.; Yan, Z.; Hou, S.; Zheng, X.; Zeng, Z.; Amanze, C.; Dai, Z.; Chai, L.; Qiu, G.; Zeng, W. Effective treatment of acid mine drainage with microbial fuel cells: An emphasis on typical energy substrates. *Minerals* **2020**, *10*, 443. [CrossRef]
26. Chen, J.; Liu, Y.; Diep, P.; Mahadevan, R. Genomic analysis of a newly isolated Acidithiobacillus ferridurans JAGS strain reveals its adaptation to acid mine drainage. *Minerals* **2021**, *11*, 74. [CrossRef]
27. Mun, Y.; Strmić Palinkaš, S.; Kullerud, K. The Role of Mineral Assemblages in the Environmental Impact of Cu-Sulfide Deposits: A Case Study from Norway. *Minerals* **2021**, *11*, 627. [CrossRef]
28. Vriens, B.; Plante, B.; Seigneur, N.; Jamieson, H. Mine waste rock: Insights for sustainable hydrogeochemical management. *Minerals* **2020**, *10*, 728. [CrossRef]
29. Tokoro, C.; Fukaki, K.; Kadokura, M.; Fuchida, S. Forecast of AMD quantity by a series tank model in three stages: Case studies in two closed Japanese mines. *Minerals* **2020**, *10*, 430. [CrossRef]

 *minerals*

Review

Mine Waste Rock: Insights for Sustainable Hydrogeochemical Management

Bas Vriens [1,*] , **Benoît Plante** [2] , **Nicolas Seigneur** [3] and **Heather Jamieson** [1]

1   Department of Geological Sciences & Geological Engineering, Queen's University, 36 Union St W,
    Kingston, ON K7N1A1, Canada; jamieson@queensu.ca
2   Institut de Recherche en Mines et en Environnement, Université du Québec en Abitibi-Témiscamingue,
    445 Boulevard de l'Université, Rouyn-Noranda, QC J9X 5E4, Canada; Benoit.Plante@uqat.ca
3   MINES ParisTech, PSL University, Centre de Géosciences, 35 rue St Honoré, 77300 Fontainebleau, France;
    nicolas.seigneur@mines-paristech.fr
*   Correspondence: bas.vriens@queensu.ca

Received: 7 July 2020; Accepted: 17 August 2020; Published: 19 August 2020

Abstract: Mismanagement of mine waste rock can mobilize acidity, metal (loid)s, and other contaminants, and thereby negatively affect downstream environments. Hence, strategic long-term planning is required to prevent and mitigate deleterious environmental impacts. Technical frameworks to support waste-rock management have existed for decades and typically combine static and kinetic testing, field-scale experiments, and sometimes reactive-transport models. Yet, the design and implementation of robust long-term solutions remains challenging to date, due to site-specificity in the generated waste rock and local weathering conditions, physicochemical heterogeneity in large-scale systems, and the intricate coupling between chemical kinetics and mass- and heat-transfer processes. This work reviews recent advances in our understanding of the hydrogeochemical behavior of mine waste rock, including improved laboratory testing procedures, innovative analytical techniques, multi-scale field investigations, and reactive-transport modeling. Remaining knowledge-gaps pertaining to the processes involved in mine waste weathering and their parameterization are identified. Practical and sustainable waste-rock management decisions can to a large extent be informed by evidence-based simplification of complex waste-rock systems and through targeted quantification of a limited number of physicochemical parameters. Future research on the key (bio)geochemical processes and transport dynamics in waste-rock piles is essential to further optimize management and minimize potential negative environmental impacts.

Keywords: mine waste; drainage; water quality; geochemistry; hydrogeology; modelling

---

1. Introduction

1.1. A global Environmental Perspective on Mine Wastes

Mineral resource extraction and ore processing operations around the world produce significant amounts of waste, predominantly wastewater and non-profitable solid residues, i.e., waste rock and tailings. While technical innovations in mining and processing techniques have increased the overall efficiency of extraction, the great majority of raw material moved to access mineral ores is still discarded as waste (e.g., up to 99% for precious metals). As a result, several gigatons of mine waste are produced around the world each year [1,2], and this number is expected to grow as increasingly lower-grade and larger-scale deposits are being mined to keep up with exponentially growing global demand for mineral resources.

In the absence of economic incentive or technical ability to re-utilize and re-valorize mine wastes [3–6], they are often stored on-site indefinitely. The impacts of these mine wastes, even when properly managed in engineered storage facilities, can be wide-ranging and include ecological, hydrological, geotechnical, climatic, and environmental aspects pertaining to the quality of natural habitat, i.e., the atmosphere, ground- and surface-water, and soils. High-profile mine waste catastrophes are well-known by the public (e.g., tailings dam failures in Canada [7], Hungary or Brazil [8]), but deterioration of environmental quality from mining waste more often occurs gradually [9,10] and can even linger unnoticed for years or become apparent until decades after mine closure. An example thereof is acid rock drainage (ARD); the weathering of sulfidic mine waste that leads to acidic drainage with high metal concentrations [1,11] (the term acid mine drainage is increasingly substituted by acid rock drainage to indicate that acidic drainage can originate from sources other than mines). Acid rock drainage is an environmental problem of global scale and deterioration of water quality from acid rock drainage which may persist for decades to millennia.

To minimize detrimental environmental impacts from mine wastes, mine sites are required to abide by legislative environmental quality standards during the entire mining cycle, i.e., from exploration and development to decades past closure [12,13]. To this end, effective long-term waste management strategies need to be developed [14,15]. Wastewater quantity and quality predictions are a critical component of these strategies, as some form of treatment is usually required before wastewater can be released to receiving downstream environments, and most wastewater treatment requirements are long-term. Drainage from on-site mine waste critically contributes to wastewater quantity and quality, and the processes underlying mine waste weathering and drainage in practical, industry-relevant settings thus need to be quantitatively understood.

Prediction of drainage quantity and quality from mine wastes requires knowledge on the local geology and weathering climate, as well as the mine waste's mineralogy, geochemistry, hydrogeology, et cetera. While local weathering conditions and the geology at a mining site are typically well-known, the properties of waste materials relevant to environmental management are usually unknown at early mine development stages and can hardly be determined a priori from theoretical calculations or by extrapolation of laboratory test results [13]. There can be a significant lag time between waste placement and onset of ARD, the composition of waste varies from mine site to mine site and can be highly heterogeneous even within sites, especially in complex geological settings. Therefore, the prediction of mine wastewater quality and quantity poses a major challenge for scientists and practitioners at sites around the world.

1.2. Waste Rock as Unique Class of Mine Waste

Mine waste rock and tailings are typically the two major waste types at mine sites, regardless of the mineral commodity (e.g., coal or base/precious metals), deposit type or extraction method (surface or open-pit mining versus in-situ or underground mining) [14]. Waste rock—distinct from other overburden spoil—consists of excavated low-grade bedrock that has been transported away to access profitable ore and is typically composed of relatively coarse, granular broken rock in the size range of sands to boulders. In turn, tailings are a composite slurry of (process) water and finely ground residuals that remain after ore comminution and beneficiation, which can contain secondary precipitates and processing reagents such as blasting agents or extraction chemicals [16]. For many types of ores, waste rock can behave in a geochemically contrasting way when compared to tailings, due to the following critical differences between tailings and waste rock:

1. The finer-grained nature of tailings materials compared to coarser-grained waste rock may yield elevated exposed mineral surface area (which can, depending on the mineralogy, increase geochemical reaction rates), whereas the wider particle size range and textural properties of waste rock give rise to quite unique (non-uniform) hydrodynamic behavior, and,

2. Storage practices for waste rock and tailings materials create distinct conditions that alter the controls of certain geochemical processes and physical transport mechanisms. Namely, waste rock is mostly placed in tall stockpiles that are porous, hydraulically unsaturated, and therefore relatively exposed to atmospheric conditions (i.e., mostly oxic environments) [16]. In contrast,

tailings slurries are often pumped into tailings ponds, where particulates settle under limited ambient exposure (i.e., fully saturated, inundated tailings that can exhibit sub-oxic, reducing conditions [17], although tailings may also be stored as backfills or dry stacks).

At the same time, similar minerals, geochemical reactions, and physical transport processes can occur in waste rock and tailings: the conceptual hydrogeochemical model is often comparable for waste rock and tailings [2]. The dimensions of industrial-scale waste-rock piles and tailings facilities are also comparable (i.e., hundreds of tons of material): once in place, both materials are prohibitively expensive to move and therefore are practically stored indefinitely. In this review, we will discuss a conceptual hydrogeochemical framework that largely applies to both types of mine waste, but, in our discussion, place emphasis on waste rock.

1.3. Scope of This Review

In recent years, the understanding of processes controlling mine waste dynamics has critically increased due to, e.g., breakthroughs in mineral characterization at nanoscales and in-situ geophysical characterization to long-term field studies and applications of big data and mechanistic numerical models. This work presents an overview of the key hydrogeochemical and physical processes relevant to water quantity and quality from mine waste rock and ultimately sustainable wastewater management. Novel insights into the geochemical and mineralogical characteristics of waste rock, the in-situ assessment of mass- and heat transfer processes in large waste-rock piles and emerging applications of advanced reactive-transport modeling are presented. Previous reviews on select aspects of mine wastes are listed in Table 1. Certain processes discussed in this work also receive attention in the context of hydrometallurgy and bioleaching/biomining, where often the opposite objective is targeted, i.e., acceleration of mineral dissolution to mobilize and recover valuable metals (e.g., in heap leaching [4,18–20]) rather than stabilizing wastes to prevent mobilization (e.g., through covers and barriers [4,21–23]). This review is structured as follows: major geochemical mobilization reactions (i.e., oxidation and dissolution) and relevant attenuation processes (i.e., sorption and secondary mineral formation) are introduced in Section 2. Section 3 outlines relevant physical transport processes in waste-rock piles and their parameterization in heterogeneous systems. In Section 4, we discuss the couplings between physical and chemical processes and upscaling phenomena, as well as the role of numerical modelling in resolving such couplings for practical long-term predictions.

Table 1. Reviews on various aspects of mine waste rock: weathering mechanisms, microbial interactions, characterization of physicochemical (bulk) properties, management, and reclamation.

Title:	Year	Reference
Molecular (bio-)oxidation mechanisms		
A review: Pyrite oxidation mechanisms and acid mine drainage prevention	1995	[24]
Leaching mechanisms of oxyanionic metalloid and metal species in alkaline solid wastes: a review	2008	[25]
The mechanisms of pyrite oxidation and leaching: A fundamental perspective	2010	[26]
A review of the structure, and fundamental mechanisms and kinetics of the leaching of chalcopyrite	2013	[27]
Principles of sulfide oxidation and acid rock drainage	2016	[28]

Table 1. *Cont.*

Title:	Year	Reference
Bioleaching: metal solubilization by microorganisms	1997	[29]
Geomicrobiology of sulfide mineral oxidation	1997	[30]
Heavy metal mining using microbes	2002	[31]
Microbial communities in acid mine drainage	2003	[32]
The microbiology of acidic mine waters	2003	[33]
The bioleaching of sulphide minerals with emphasis on copper sulphides—A review	2006	[34]
The microbiology of biomining: development and optimization of mineral-oxidizing microbial consortia	2007	[35]
Heap bioleaching of chalcopyrite: a review	2008	[20]
Biomining-biotechnologies for extracting and recovering metals from ores and waste materials	2014	[36]
Microbial ecology and evolution in the acid mine drainage model system	2016	[37]
Recent progress in biohydrometallurgy and microbial characterization	2018	[38]
Mine waste characterization and treatment techniques		
The environmental impact of mine wastes—roles of microorganisms and their significance in treatment of mine wastes	1996	[39]
Acid mine drainage remediation options: a review	2005	[22]
Acid mine drainage (AMD): causes, treatment, and case studies	2006	[40]
Passive treatment of acid mine drainage in bioreactors using sulfate-reducing bacteria: critical review and research needs	2007	[41]
Geochemistry and mineralogy of solid mine waste: essential knowledge for predicting environmental impact	2011	[15]
Remediation of acid mine drainage-impacted water	2015	[42]
Mineralogical characterization of mine wastes	2015	[43]
Characteristics and environmental aspects of slag: a review	2015	[44]
A critical review of acid rock drainage prediction methods and practices	2015	[45]
Acid rock drainage prediction: a critical review	2017	[46]
Acid mine drainage: prevention, treatment options, and resource recovery: A review	2017	[47]
Environmental indicators in metal mining	2017	[48]
Environmentally sustainable acid mine drainage remediation: research developments with a focus on waste/by-products	2018	[49]
A review of recent strategies for acid mine drainage prevention and mine tailings recycling	2019	[23]
Waste rock management		
The geochemistry of acid mine drainage	2003	[2]
Sustainable mining practices	2005	[50]
Mine wastes: past, present, future	2011	[1]
Hydrogeochemical processes governing the origin, transport, and fate of major and trace elements from mine wastes and mineralized rock to surface waters	2011	[51]
Management of sulfide-bearing waste, a challenge for the mining industry	2012	[52]
Acid mine drainage: challenges and opportunities	2014	[53]
Waste-rock hydrogeology and geochemistry	2015	[16]
Hydrogeochemistry and microbiology of mine drainage: an update	2015	[54]
Geochemical and mineralogical aspects of sulfide mine tailings	2015	[17]
Mine waste characterization, management, and remediation	2015	[14]
The mine of the future—even more sustainable	2017	[55]
Mining waste and its sustainable management: advances in worldwide research	2018	[56]
Guidance for the integrated use of hydrological, geochemical, and isotopic tools in mining operations	2020	[13]

2. Geochemical Processes in Mine Waste Rock

The exposure of previously buried geological material to atmospheric conditions triggers oxidation reactions and allows for water to infiltrate and percolate the waste rock (i.e., drainage) which can transfer dissolved solutes into the environment. In this section, major geochemical processes that affect drainage chemistry are discussed. On the broadest level, mine waste drainage pH is determined by the balance between acid-producing and acid-neutralizing (buffering) reactions, whereas solute concentrations are controlled by their waste-rock grade and solubility [16,51,54,57]. In addition, attenuation processes such as adsorption and secondary mineral formation [2] may reduce the mobility of solutes through (temporary) internal retention in the waste rock, as discussed below.

2.1. Acid-Producing Reactions

2.1.1. Metal-Sulfide Mine Waste

The majority of exploited chalcophile-metal ore deposits (e.g., Cu-, Zn-, Pb-bearing ores) are sulfide-type. While many mined chalcophile-metal deposits have undergone significant oxidative alteration of the ore zones prior to mining, sulfide minerals can remain abundant in the generated waste. Of the sulfide minerals in waste rock, pyrite (FeS_2) is typically the most abundant. Other major phases include chalcopyrite ($FeCuS_2$), covellite (CuS), sphalerite (ZnS), pyrrhotite (FeS), arsenopyrite ($FeAsS$), and galena (PbS). Because of their economic and environmental importance, the crystal structures, chemical compositions, physical properties, and phase relations of major sulfides have been well-established and previously reviewed [58]. The oxidative dissolution of a generalized metal sulfide (Me_xS_y) is illustrated in Figure 1a: this reaction consumes water and oxygen and produces protons, metal and sulfate ions, and heat.

The reaction pathway for FeS_2 oxidation has been studied extensively [26,59–61] and occurs via an electrochemical mechanism wherein electron exchange takes place on specific surface sites of the mineral [62]. Its overall congruent dissolution can be simplified by the equation:

$$FeS_2 + \frac{7}{2}O_2 + H_2O \rightarrow Fe^{2+} + 2H^+ + 2SO_4^{2-} \tag{1}$$

Under oxic conditions, dissolved ferrous Fe^{2+} rapidly oxidizes into ferric Fe^{3+} which can act as an additional oxidant:

$$Fe^{2+} + \frac{1}{4}O_2 + H^+ \rightarrow Fe^{3+} + \frac{1}{2}H_2O \tag{2}$$

$$FeS_2 + 14Fe^{3+} + 8H_2O \rightarrow 15Fe^{2+} + 16H^+ + 2SO_4^{2-} \tag{3}$$

Due to the capability of Fe^{3+} ions to break metal sulfide bonds more effectively than protons [59], the Fe^{3+} reaction speeds up FeS_2 oxidation by orders of magnitude and forms an important (auto)catalytic feedback (Equation (3) versus Equation (1); Figure 1a). Because FeS_2 is naturally abundant in many chalcophile metal ores, Fe^{3+} also serves as an oxidant to other metal sulfides, e.g.,:

$$CuFeS_2 + 16Fe^{3+} + 8H_2O \rightarrow Cu^{2+} + 17Fe^{2+} + 16H^+ + 2SO_4^{2-} \tag{4}$$

$$ZnS + 8Fe^{3+} + 4H_2O \rightarrow Zn^{2+} + 8Fe^{2+} + 8H^+ + SO_4^{2-} \tag{5}$$

$$PbS + 8Fe^{3+} + 4H_2O \rightarrow Pb^{2+} + 8Fe^{2+} + 8H^+ + SO_4^{2-} \tag{6}$$

The kinetics of abiotic sulfide oxidation by Fe^{3+} allow for much faster sulfide oxidation than by oxygen alone, so that the above equations for sulfide oxidation with Fe^{3+} are more commonly used than those with oxygen, e.g.,:

$$CuFeS_2 + 4O_2 \rightarrow Cu^{2+} + Fe^{2+} + 2SO_4^{2-} \tag{7}$$

$$ZnS + 2O_2 \rightarrow Zn^{2+} + SO_4^{2-} \tag{8}$$

$$PbS + 2O_2 \rightarrow Pb^{2+} + SO_4^{2-} \tag{9}$$

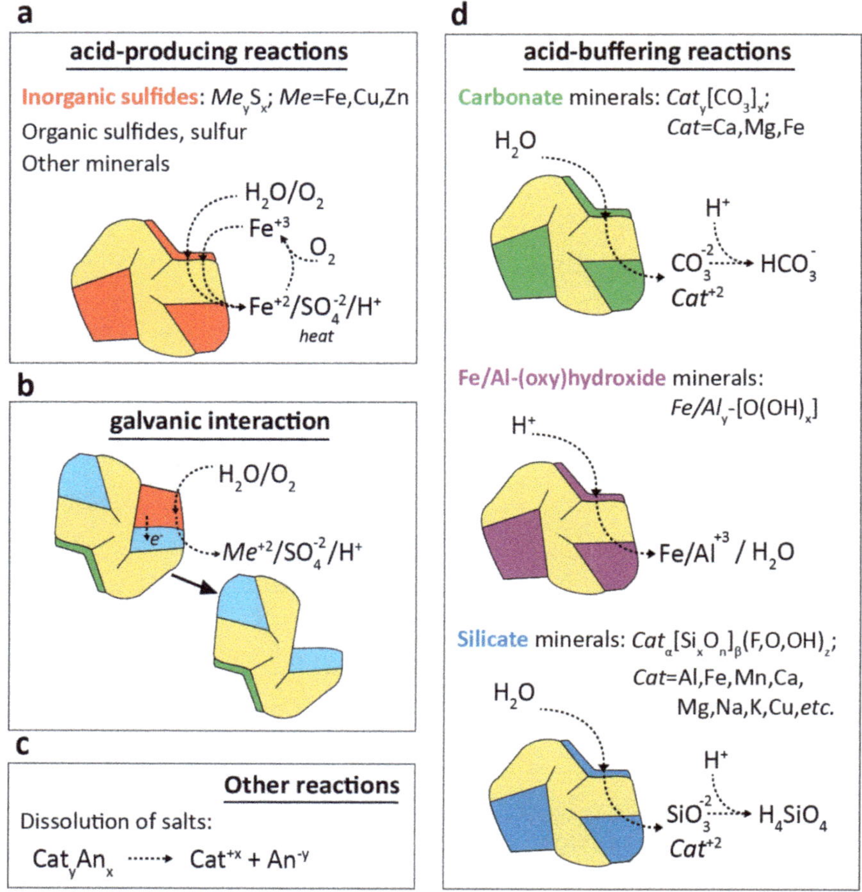

Figure 1. Schematic grouping of key geochemical reactions (not balanced) that control mine waste-rock drainage quality. The oxidative dissolution of sulfide minerals is the major acid-producing reaction (**a**), which has been extensively studied for pyrite and other major sulfides. Galvanic interactions (**b**) can promote the preferential dissolution of sulfide minerals (Section 2.5). Additional reactions (**c**) can introduce metals and other solutes into the waste rock drainage without directly affecting drainage pH. The dissolution of carbonates, Fe/Al-(oxy)hydroxides, and silicates (simplified, generalized stoichiometries; (**d**) consumes protons and thereby perform a net-buffering action.

Finally, ferrous Fe^{3+} is very insoluble under oxic and near-neutral conditions and readily precipitates as Fe^{3+}-(oxy)hydroxide to release additional acidity:

$$Fe^{3+} + 3H_2O \rightarrow Fe(OH)_3 + 3\,H^+ \tag{10}$$

The overall reaction equation for pyrite oxidation (combining Equations (2), (3), and (10)) is therefore often summarized as:

$$FeS_2 + \frac{15}{4}O_2 + \frac{7}{2}H_2O \rightarrow Fe(OH)_3 + 2H_2SO_4 \tag{11}$$

Despite the catalytic effect of Fe^{3+} on sulfide oxidation, the rates of the abiotic reactions described above are relatively slow, especially at low pH [16,63]. Sulfide oxidation reactions are catalyzed by a variety of Fe- and S-oxidizing bacteria that metabolically tap into the energy released during Fe and S oxidation [64]. Major chemolithotrophic bacteria involved in sulfide oxidation reactions include both acidophilic and neutrophilic iron and sulfide oxidizing species, e.g., ferrooxidans and thiooxidans species in the *Thiobacillus*, *Leptospirillum*, and *Ferrobacillus* genera [33,65,66]. The role of microbial sulfide mineral oxidation in controlling mine waste weathering rates has been well-established and the geomicrobiology of mine wastes previously reviewed [30,33,67,68] (Table 1).

Because oxidation of sulfidic minerals involves the transfer of many electrons from each sulfur atom to an aqueous oxidant, various intermediate sulfur species can exist (i.e., polysulfides, thiosulfate, and elemental sulfur) [59,69–71]. For major sulfides, (a)biotic oxidation pathways, intermediate reaction products, and corresponding kinetics have been characterized in a variety of settings, i.e., under ambient conditions versus at elevated temperatures to optimize leaching [27,72–79]. Various kinetic models exist for pyrite oxidation [2], but the oxidation mechanisms and kinetics for less-abundant sulfides remain comparatively underexplored (e.g., molybdenite (MoS_2) [72,80]). The exact reaction mechanisms and kinetics of sulfide oxidation vary with weathering conditions (e.g., as shown for pyrite [75,81–83]). Yet, (hydro)metallurgical and geochemical studies often reveal similar overall rate dependencies (e.g., on dissolved oxygen and Fe concentrations or mineral surface area) and may deploy comparable kinetic models [84–86].

Metal sulfide minerals are rarely pure, especially in waste rock: e.g., Fe, Cu, or As, may be readily substituted by a range of (trace) metals in solid-solutions: As, Co, and Ni can be present in pyrite with up to several weight percent and sulfosalts typically contain a range of metalloids including As, Sb, Bi, and Se. Mineralogical impurities have a demonstrated effect on abiotic weathering kinetics [87], and even though microorganisms in mine waste appear quite tolerant to otherwise harmful metals [88], impurities may also affect biotic oxidation rates through inhibitive effects or shifts in community structure and functional diversity [89]. Finally, select other non-sulfur minerals may also oxidize to produce acidity, e.g., selenide, arsenide, telluride, and antimonide minerals. Due to their natural scarcity (typical abundances are < ‰ of that of S), these metalloids tend to occur at trace levels in sulfides rather than as distinct phases [90] and are usually not relevant to the overall acid-producing capacity.

2.1.2. Coal Mine Waste

In addition to metal-sulfides waste rock, coal mine wastes can also be a major source of ARD [91–97]. Sulfur can occur in coal in three main forms: (i) inorganic sulfides, including authigenic minerals such as detrital pyrite, (ii) organically bound sulfur, including mercaptan, (di)sulfide and heterocyclic compounds, and (iii) inorganic sulfates, particularly in weathered coals [98,99]. Because of the relative abundance of inorganic sulfide compared to other forms of sulfur in coal waste rock, acidity production is typically attributed to pyritic material, even though pyrite concentrations in coal vary regionally [100–102]. While undesirable for the prevention of coal waste drainage acidification, oxidative de-pyritization (or desulfurization) as a major coal ore beneficiation objective has been extensively studied [103,104], including pathways [105] and kinetics [106–108].

An important difference between metal ore wastes versus coal wastes is that metal ore waste typically contains negligible organic carbon, whereas coal wastes can have several wt% organic carbon [93,98]. This organic material can, in the aqueous phase, subsequently ligate aqueous solutes and mineral surfaces and thereby alter the mobility of metals in coal waste-rock drainage, depending on the nature of the ligands and minerals (see Section 2.4.1). Organic matter may oxidize to induce oxygen depletion and alter the Fe^{2+}/Fe^{3+} equilibrium (Equation (2)) as well as those of other redox-sensitive elements (e.g., Mn, As, S) [109]. Subsequent shifts in the microbial community structure to favor Fe and S reducers [110,111] may lead to net alkali generation (i.e., acid-neutralization) and in fact, certain acid–rock drainage prevention or remediation strategies rely on the purposed addition of organic matter to induce oxygen depletion [112–114].

2.2. Acid-Buffering Reactions

When present and abundant in waste rock, the dissolution of carbonate minerals (e.g., calcite, dolomite, ankerite), Al- and Fe-(oxy)hydroxides (e.g., gibbsite, ferrihydrite, respectively) and silicate minerals (e.g., feldspars, chlorites, smectites, micas, and amphiboles) consumes protons and introduces alkalinity that offsets acidity produced by sulfide oxidation [2,16]. The following (simplified) equations illustrate these reactions:

$$CaCO_3 + H^+ \rightarrow Ca^{2+} + HCO_3^- \tag{12}$$

$$[Fe/Al]OOH + 3H^+ \rightarrow [Fe/Al]^{3+} + 2H_2O \tag{13}$$

$$CaSiO_3 + 2H^+ + H_2O \rightarrow Ca^{2+} + H_4SiO_4 \tag{14}$$

$$CaAl_2Si_2O_8 + 8H^+ \rightarrow Ca^{2+} + 2Al^{3+} + 2H_4SiO_4 \tag{15}$$

The acid-neutralizing dissolution reactions of carbonates, oxides and silicates accelerate at decreasing pH, in accordance with their solubility, creating a so-called buffering sequence for acid–rock drainage [16,115]: carbonates buffer acidity until they are typically depleted at pH < 6, Al- and Fe-(oxy)hydroxides dissolve to buffer acidity at pH < 4–5 and pH < 3–4, respectively, and the dissolution of aluminosilicates only significantly contributes to acid-buffering capacity at pH < 3. Carbonate dissolution neutralizes protons through the carbonate equilibrium, but dissolution of Fe-rich carbonate (e.g., siderite ($FeCO_3$)) also introduces dissolved ferrous Fe that upon oxidation and hydrolysis (Equation (10)) releases surplus protons, reducing the net-neutralizing action [2,46]. Acid-buffering minerals dissolve simultaneously alongside progressing sulfide oxidation when they are naturally co-located in the waste-rock matrix [115]. In addition, they may also be intentionally mixed, blended, or added to sulfide-rich waste material as an active or passive acid–rock drainage remediation strategy [116–118].

A wealth of information is available on the molecular mechanisms, kinetics, and pH-dependencies of dissolution reactions of carbonate minerals [119–121], Fe-(oxy)hydroxides [122,123], and silicate minerals commonly encountered in waste rock [124–126]. The kinetics of carbonate dissolution are typically rapid compared to sulfide oxidation (i.e., far from equilibrium [127]), whereas the dissolution of silicates is orders-of-magnitude slower [2]. This introduces important ramifications as to the timescales required for effective acid-neutralization, for instsance when transport times are fast or experimental durations short (e.g., static tests; Section 2.6.1). In addition to consuming protons, dissolution of carbonate, oxide, or aluminosilicate minerals introduces additional solutes to a leachate, including Ca, Mg, Mn, Al, and Fe, and possibly metal impurities if present at considerable levels [128,129]. Mobilization of such solutes and its effect on drainage quality must be considered if these neutralizing materials are to be employed to prevent or remediate acid drainage [13]. Similar to sulfide oxidation, mineralogical heterogeneities (Section 2.5) strongly affect the efficacy of acid-neutralizing reactions. In contrast to sulfide-dominated acid-production, acid-neutralization can be governed by many minerals: prediction of the extent and timing of neutralization therefore requires characterization of the complete mineral assemblage.

2.3. The Geochemistry of Neutral Drainage

While the environmental impacts of ARD have received extensive attention, neutral to alkaline drainage can also compromise water quality: neutral rock drainage (NRD), also referred to as contaminated neutral drainage (CND; [129,130]) or metal leaching (ML; [131,132]), is the mobilization of hardness (mostly Ca and Mg), major ions such as sulfate and chloride, metal(oid)s, and other contaminants that are mobile under near-neutral, non-acidic conditions. Most metallic cations show higher solubility in acidic conditions, such as Cu, Ni, Zn, Co, and Mn [133], but some can remain in elevated concentrations at near-neutral values, mostly because their oxyhydroxides precipitate at pH between 6 and 10. Oxyanionic elements such as As, Se, and Mo exhibit elevated mobility at near-neutral pH, which has been demonstrated in a variety of hornfels, carbonate [126], and sedimentary [134] waste-rock types. Cations and oxyanions generally show inverse behavior in mine drainage in the sense that cations tend to adsorb or precipitate in near neutral and alkaline conditions, while oxyanions tend to do more so under (slightly) acidic conditions (see Section 2.4.1). Despite a growing awareness of its potential environmental impacts, there is currently no standardized definition of neutral mine drainage [132]. Table 2 lists examples of reported NRD cases, illustrating how neutral drainage composition can be widely different across sites (slightly acidic versus alkaline, and negligible versus elevated concentrations of sulfate, metal(loid)s, and other (non-metallic) contaminants). NRD from waste rock can arise in different situations:

(i)   Sulfide oxidation in the presence of sufficient acid-buffering or weathering of non-acid generating minerals (i.e., low-sulfide waste rock such as carbonates and silicates [135]), dissolution of salts [136];
(ii) Insufficient treatment of ARD (e.g., abandoned mine sites using passive ARD treatment), where the pH is successfully increased to near-neutral but certain contaminants remain present at elevated concentrations;
(iii) Within reclaimed ARD-generating mine wastes, where the rate of acid generation is decreased to levels that can be buffered by neutralizing minerals, but still allows for the leaching of metals.

Contaminant loads in neutral drainage can be controlled by their occurrence in sulfidic minerals as well as carbonate or silicate phases, or through the dissolution of (secondary) salts and oxides (Figure 1). For instance, elevated As levels in high-alkalinity drainage have been mostly attributed to the oxidation of arsenopyrite or As-bearing pyrite that had been neutralized [137,138] but may also originate from natively As-rich carbonate phases in the absence of Fe-oxides [128]. With the increased interest in the development of rare earth element (REE) mines, there is an increasing interest in REE geochemistry and ecotoxicity [139–141]. REE-bearing minerals can be found in many different geological settings (although not necessarily at economic concentrations), and REEs can be found as trace contaminants in other minerals. Therefore, they might represent an accessory contaminant in mine drainage from non-REE operations. Indeed, they are found in many ARD cases [142,143] but most REE deposits are within geological settings that are not prone to ARD generation, such as pegmatites, carbonatites, and peralkaline igneous deposits, as well as within placers and clays ([144] and references therein). Upon their release in near-neutral mine waters, REE concentrations are mostly controlled by secondary precipitation and sorption phenomena (Section 2.4) [145–148]. However, significant knowledge gaps remain as to the fate of REE released from mine wastes.

Table 2. Examples of neutral rock drainage (NRD) chemistries reported at different mine sites around the world.

	Antamina, Peru [128]	Hitura, Finland [149,150]	Lac Tio, Canada [130]	Beaver Brook, Canada [151]	Greens Creek, United States [152]	Giant Mine, Canada [153]
pH	6.5–8.5	6.1–7.0	6.5–7.5	5.7–8.6	6.5–8.5	6.7
Ni (mg/L)	N/R	0.2–14.3	0.1–8.8	N/R	0–1	0.029
Zn (mg/L)	0.1–80	25–660	N/R	N/R	0–150	0.027
Mn (mg/L)	0.001–0.2	4.7–8.9	N/R	N/R	0–35	0.446
Co (mg/L)	N/R	0.05–7.2	N/R	N/R	N/R	<0.007
As (mg/L)	0.001–1.0	N/R	N/R	0–2.3	0–0.03	4060
Se (mg/L)	0.001–0.2	N/R	N/R	N/R	0–0.02	<0.03
Sb (mg/L)	0.001–0.2	N/R	N/R	0–26	0–0.06	11.9
Mo (mg/L)	0.0–1.0	N/R	N/R	N/R	0–0.02	0.07
SO_4 (g/L)	0.1–2	2.1–5.2	0.1–3.5	0.075–0.9	2–8	0.5
Ca (mg/L)	50–600	200–450	10–70	9–231	400–800	313

N/R: not reported.

2.4. Attenuation Processes

Attenuation processes can retain solutes in the waste-rock matrix and thereby influence drainage quality [51,115,154]. Solute-specific attenuation through adsorption or secondary mineral formation can lead to apparent discrepancies between the elemental composition of bulk waste rock and observed drainage loads. A quantitative assessment of relevant attenuation mechanisms, informed by robust mineralogical analyses and/or geochemical equilibrium modeling, can help predict such discrepancies and optimize waste rock management.

2.4.1. Adsorption

Geochemical adsorption refers to the reversible attachment of aqueous solutes to mineral surfaces (schematic in Figure 2a). While adsorption is highly surface- and solute-specific, the general relevance of adsorption for mine waste drainage quality has been widely acknowledged [2]. Field and laboratory waste-rock studies have demonstrated that sorption can be a dominant attenuation mechanism for various mine waste-relevant solutes, from metal cations [155] to metal(oid) oxyanions [128].

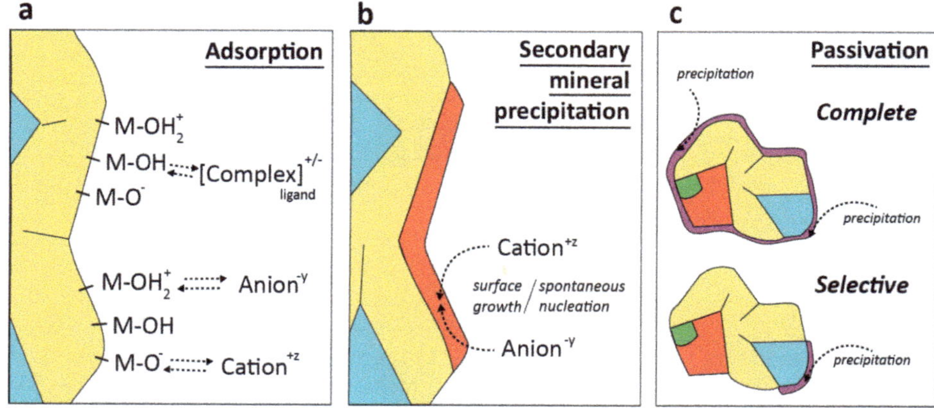

Figure 2. Schematic of key geochemical attenuation processes in mine waste rock: adsorption (**a**) and secondary mineral formation (**b**). Through adsorption, (hydrated) ionic or ligated solutes and complexes adsorb to mineral surface groups (-M-(x)) through covalent bonding or electrostatic interaction. Secondary minerals form on mineral surfaces or spontaneously in solution through (co-)precipitation of cationic and anionic aqueous solutes. The occlusion of mineral surfaces by secondary mineral precipitation is referred to as passivation or armoring (**c**), discussed in Section 2.5.

Adsorption can involve relatively strong covalent binding (inner-sphere) or weaker electrostatic attraction between aqueous ions and oppositely charged mineral surfaces (outer-sphere). In both cases, adsorption is governed by the solute's aqueous speciation as well as characteristics of the mineral surface (e.g., pH-dependent surface charge). The aqueous speciation chemistry for major inorganic ions in mine waste drainage has been previously reviewed (e.g., for Cu [156] and Zn [157]). Because most solutes present in mine drainage have an aqueous speciation that is strongly dependent on the solution chemistry, and because the surface characteristics of minerals present at a site may equally vary with the drainage type (i.e., acidic versus neutral versus alkaline), the many geochemical reactions underlying adsorption typically require modeling to be resolved and quantified. Geochemical equilibrium models, informed by thermodynamic and kinetic parameters for these reactions, are frequently used to study and predict net-attenuating effects under practice-relevant conditions [132]. However, even with thermodynamic databases supplying equilibrium or stability constants for aqueous complexation and hydrolysis reactions, the complete aqueous speciation of mine drainage may be challenging to predict, in particular for complex solutions that seasonally vary [115]. Evolving solution chemistries can induce desorption reactions, whereby previously adsorbed solutes may be re-introduced into the drainage [2]. Even though Lewis acid–base interactions and aqueous complexation reactions for major solutes are part of established geochemical equilibrium models, relevant aqueous ligands and potential competitive solutes need to be experimentally determined to inform such models. This may be especially challenging for e.g., coal waste-rock drainage [158,159] where a variety of organic acids can occur. Furthermore, significant temperature variations can occur within waste-rock piles (Section 3.3) and the ionic strengths of waste rock leachates can be unusually high, so that these aspects must be considered in aqueous speciation modeling.

The modes of surface binding are documented for major waste-rock solutes and detailed adsorption models exist for various mineral surfaces (e.g., multi-layer models, charge-distribution multi-site complexation model [160]). Yet, the application of quantitative sorption models in drainage predictions is scarce. This may be because the sorption properties of less-abundant solutes remain poorly studied (e.g., Mo [161]) and stability constants deduced in laboratory studies may not be representative for variable field conditions. Furthermore, the data required to inform sorption models, i.e., mineral surface characteristics such as sorption-site density or solution chemistry, is typically unavailable (or semi-quantitative at best) for large, heterogeneous waste-rock piles. Finally, the surface properties of waste rock may evolve as weathering progresses, e.g., through precipitation of amorphous secondary Fe-oxide coatings with elevated sorption capacity. As a result, adsorption modeling for waste rock often relies on estimated parameters, adoption of synthetic idealized phases (e.g., hydrous ferric oxide [128]), or extrapolation of generalized behavior (the Irving-Williams series) in stable drainage types. Adsorption is often considered a precursor for secondary mineral formation (Section 2.4.2) but the contribution of adsorption versus that of secondary minerals to overall attenuation remains to be quantitatively resolved.

2.4.2. Secondary Mineral Formation

Secondary minerals in waste rock are considered those minerals that form after the waste material is disposed (Figure 2b), i.e., distinct from primary minerals and oxidation products native to the excavated (ore) material. Major secondary mineral phases in mine waste have been reviewed previously [2] and include (oxy)hydroxides, (hydroxy)sulfates, (hydroxy)carbonates, and phosphates, or arsenates, depending on the waste-rock primary mineralogy. Specific secondary phases encountered at many mine sites include Fe-oxides (e.g., ferrihydrite, goethite, and lepidocrocite), the Fe-sulfates jarosite and melanterite, as well as gypsum ($CaSO_4$) [2,54,115].

Secondary mineral precipitates are typically composed of different polymorphs of variable hydration: water-soluble salts may initially precipitate as hydrated or hydroxy-phases but recrystallize into more stable forms as the precipitates age, accelerated by evaporation and drying of the waste rock. The physical and hydrological conditions of waste-rock piles (size, particle size, hydraulic retention time; discussed in Section 3) have a strong impact on the geochemical water-rock interaction and therefore occurrence of secondary minerals. Secondary minerals often occur as distinct rims or coatings on weathered host particles, but their occurrence is not necessarily related to the waste-rock composition in their direct vicinity: mixing of different drainage types within heterogeneous waste-rock piles [162] (Section 3.1) or generally rapid infiltration rates compared to slow precipitation rates may facilitate secondary mineral precipitation seemingly unrelated to the local waste-rock composition [163]. While certain secondary phases readily precipitate (e.g., Fe-(hydroxy)oxides [2,16]), others may take decades to precipitate and crystalize to detectable levels [115]. The bulk mineralogy and drainage pH of waste-rock piles can indicate whether certain secondary mineral classes are likely to occur (e.g., secondary phosphates and carbonates are unstable at drainage pH< 4 [164]; secondary phases such as scorodite may be stable even under strongly acidic conditions [165]. Thus, secondary minerals typically exhibit a widely different stability under oxic versus anoxic or acidic versus neutral drainage conditions [2]. Their controls on long-term drainage quality may thus be difficult to quantify for heterogeneous waste-rock piles with seasonally and spatially variable drainage signatures. For instance, re-dissolution of secondary minerals under gradually acidifying drainage conditions or due to the reductive dissolution of Fe-oxides, have both been invoked to explain spikes in loading rates on the timescales of years [115].

Secondary minerals are often qualitatively inferred from chemical equilibrium modeling (i.e., mineral saturation indices) rather than unequivocally identified analytically [115,165]. Yet, mineral stability constants obtained in controlled laboratory experiments may not apply under field conditions, and secondary mineral occurrence is dictated by their precipitation kinetics more than by geochemical saturation (e.g., slower crystal growth rates occur under low degrees of saturation whereas high precipitation rates may be sustained by spontaneous nucleation at high oversaturation). The identification, let alone quantification, of secondary minerals in waste rock, e.g., with X-Ray diffraction (XRD), remains challenging for dispersed and poorly crystalline phases. Yet, even secondary phases with low overall bulk abundance can be substantial attenuating sinks: e.g., the scarce but rapidly precipitating wulfenite ($PbMoO_4$) has been shown to present an important control on Pb and Mo drainage levels [161,166,167]. For successful drainage management, it is thus critical that the secondary mineral assemblage is quantified and monitored over time.

2.5. Mineral Reactivity

The generation of acidic and metal-bearing drainage from waste rock is the result of mineral-water interaction, and thus the nature of the fluid and the minerals themselves. The relative resistance to oxidation of common sulfide minerals in mine waste has been discussed: based mostly on field observations, pyrrhotite is considered the most reactive, followed by galena, sphalerite, bornite, pentlandite, arsenopyrite, marcasite, pyrite, and chalcopyrite [45,168,169]. The reactivity of potentially neutralizing minerals also influences mine drainage quality, e.g., calcite is considered to be 180 times more reactive than the most reactive silicate (wollastonite). In addition to its mineral composition (or overall sulfide grade), overall waste-rock reactivity is determined by additional petrographic and mineralogical factors that include crystallinity and morphology [24,170,171], surface defects and heterogeneities, and reduced liberation (i.e., reduced exposure to oxygen and fluids) caused by inclusion in primary minerals (occlusion) or secondary mineral precipitation (*passivation* or *armoring*; Figure 2c) [163,172–175].

Small-scale mineral features of waste rock are typically unknown in practice, since determining them requires more advanced analytical techniques that can be costly to apply to sufficient samples for a realistic assessment [45]. These factors, however, may explain why predictive laboratory testing does not always match field results (Section 2.6.1) [129,176]. For instance, occluded (unliberated) sulfides in silicates are less likely to oxidize, and both chemical and mineralogical acid–base accounting tests may thus overestimate potential acid generation, especially waste that has been exposed over times such that the exposed sulfides have oxidized. Further, sulfide oxidation may be slowed by the formation of secondary mineral rims of Fe-(oxyhydr)oxides, sulfates, or even secondary carbonates such as smithsonite, whereas sulfides that do not contain iron, such as stibnite, may oxidize more rapidly because they do not develop Fe-oxide rims [151]. Such passivation may thus occur selectively on specific minerals but also throughout the waste-rock matrix [177], and extensive precipitation can even reduce pore space and thereby alter the hydraulic properties of waste rock (Section 4). Deliberate sulfide passivation has also been proposed as an active acid drainage prevention technique [178,179], e.g., using metal-organic complex formation and passivation of sulfides [180,181]. Finally, the contact of two sulfidic phases in waste rock can result in electron exchange (galvanic interaction) that causes one phase to corrode more rapidly than the other (Figure 1b). For instance, the leaching of chalcopyrite can be enhanced through association with pyrite [182]. Galvanic interactions can substantially increase the leaching of one or both of the minerals that constitute the galvanic cell: galvanically promoted dissolution has been reported in laboratory and field studies with mine waste rock [163,183–185]. Even though mineral occlusion, passivation, and association are known to critically determine waste-rock reactivity [115,186], and techniques such as electron microscopy and automated mineralogy (e.g., MLA, QEMSCAN) now allow for such parameters to be quantified with increasing ease, few waste-rock studies and drainage prediction models have given consideration to mineralogical and petrographic aspects (e.g., through refinement of kinetic rates) [187–190], given the cost of a representative assessment at full scale (Sections 2.6 and 3.4).

In general, the grain size of primary minerals is inversely correlated with reactivity, and thus finely-crushed tailings are expected to be more reactive in terms of sulfide oxidation and carbonate dissolution than coarse waste rock fragments. On the other hand, larger pore spaces and unsaturation of waste rock may result in a higher ingress of oxygen and periodic wetting and drying of mineral surfaces, which may increase reactivity (Section 3.2). The nature of the water interacting with the waste rock also affects reactivity. Oxygen-rich, slightly acidic meteoric water interacting with recently blasted waste rock will encounter freshly broken minerals surface, and possibly highly reactive mineral dust from blasting. Porewater reaching the lower portions of a waste-rock pile may be oxygen depleted, and Fe^{3+} may act as an oxidant under acid conditions in the absence of dissolved oxygen (Equation (3)). Under anoxic conditions at the bottom of a pile, or if the waste rock is submerged, Fe-(oxyhydr)oxides are susceptible to reductive dissolution, resulting in release of elements of concern that were attenuated by adsorption or co-precipitation. Waste rock submerged in a marine environment is unlikely to generate acidic drainage, given the strong buffering capacity of seawater, although the availability of ligands may increase the solubility of some metals [191].

2.6. Characterization of Bulk Waste-Rock Reactivity

While the geochemical reactions in weathering waste rock are universal on a molecular level, there is typically large variability in the weathering conditions and waste-rock composition and grain sizes across mine sites. As a result, sulfide oxidation and bulk weathering rates reported from laboratory and field experiments vary by orders-of-magnitude across sites (Table 3). Because it is virtually impossible to resolve all molecular-scale mineralogical heterogeneity, waste-rock reactivity or weathering rates are typically presented through representative bulk parameters such as drainage loads (i.e., net sulfate leaching) or oxygen consumption or heat production rates [192–194]. Similarly, the estimation of the acid-producing versus acid-neutralizing nature of waste-rock material is often based on bulk laboratory tests (static testing or acid–base accounting (ABA)) rather than (or complementary to) microscale mineralogical analyses.

Table 3. Selection of waste-rock weathering rates reported for mine sites around the world, including field studies, laboratory experiments, and numerical modeling.

Reported Rate [Varying Units]	Waste-Rock Type	Mine [Main Ore Product]	Method of Estimation	Rate in g S per kg Waste Rock (Bulk) per Year *	Reference
		Laboratory studies			
$5 \pm 1 \times 10^{-7}$ [mol O_2 m^{-3} s^{-1}]	up to 1.5% sulfides	Aitik, Sweden [Cu]	Oxygen consumption	0.3	[195,196]
up to 7×10^{-8} [mol O_2 kg^{-1} s^{-1}]	up to 6 wt% py	Doyon, Canada [Au]	Oxygen consumption	up to 40	[192]
6 to 60 [mg SO_4 kg^{-1} wk^{-1}]	<0.5 wt% S	Cluff Lake, Canada [U]	Sulfate mass-loading	0.3 to 3	[183]
1×10^{-12} to 4×10^{-11} [kg O_2 kg^{-1} s^{-1}]	0.6 – 1.4% S	Duluth Complex, USA [Cu, Ni]	Drainage loading	1.8 to 52	[197]
		Field experiments			
3×10^{-9} to 1×10^{-7} [kg O_2 m^{-3} s^{-1}]	Reactive (>3% S)	Antamina, Peru [Cu, Zn]	Oxygen consumption	0.1 to 3.4	[198]
6×10^{-11} to 4×10^{-10} [kg O_2 m^{-3} s^{-1}]	Unreactive (<0.5% S)	Antamina, Peru [Cu, Zn]	Oxygen consumption	0.002 to 0.01	[198]
up to 3×10^{-3} (±87%) [kg S kg^{-1} yr^{-1}]	Reactive (1.6% S)	Antamina, Peru [Cu, Zn]	Sulfate mass-loading	3	[115]
up to 4×10^{-4} (±20%) [kg S kg^{-1} yr^{-1}]	Unreactive (0.5% S)	Antamina, Peru [Cu, Zn]	Sulfate mass-loading	0.4	[115]
1×10^{-7} [kg S m^{-3} s^{-1}]	Reactive (>10% S)	Antamina, Peru [Cu, Zn]	Heat production	1.8	[194]
2×10^{-7} [kg O_2 m^{-3} s^{-1}]	Reactive (>10% S)	Antamina, Peru [Cu, Zn]	Oxygen consumption	6.7	[194]
0.05 to 0.3 [g S kg^{-1} yr^{-1}]	Mixed (0.5–1.6%S)	Antamina, Peru [Cu, Zn]	Heat production	0.05 to 0.3	[199]
1×10^{-8} to 1×10^{-10} [kg O_2 m^{-3} s^{-1}]	0.6 vol% Sulfides	Aitik, Sweden [Cu]	Oxygen consumption	0.002 to 0.2	[200,201]
1×10^{-9} to 1×10^{-10} [mol O_2 kg^{-1} s^{-1}]	Up to 6 wt% py	Doyon, Canada [Au]	Heat and oxygen profiles	0.58 to 5.8	[192]
3 to 100 [mg SO_4 kg^{-1} wk^{-1}]	<0.5 wt% S	Cluff Lake, Canada [U]	Sulfate mass-loading	0.05 to 1.7	[193]
7 to 70 [mg SO_4 kg^{-1} wk^{-1}]	<0.5 wt% S	Cluff Lake, Canada [U]	Oxygen consumption	0.12 to 1.2	[193]
8×10^{-8} to 2×10^{-7} [kg Py m^{-2} s^{-1}]	Mixed (~3 wt% S)	Rum Jungle, Australia [U]	Thermal profiles	0.15 to 0.36	[202]
		Numerical modelling			
0.004–0.4 [kg O_2 m^{-3} yr^{-1}]	Mixed (6–0.1% S)	Doyon, Canada [Au]	Simulated	0.004 to 0.4	[203]
0.02 [kg Py m^{-3} yr^{-1}]	Mixed (0.1% Py)	Doyon, Canada [Au]	Simulated	0.006	[204,205]
0.15 [kg Py m^{-3} yr^{-1}]	Mixed (0.05% Py)	Nordhalde, Germany [U]	Simulated	0.04	[204,205]
$5 \pm 1 \times 10^{-7}$ [mol O_2 m^{-3} s^{-1}]	up to 1.5% sulfides	Aitik, Sweden [Cu]	Calibrated to measurements	0.3	[195]
1×10^{-7} to 5×10^{-10} [kg O_2 m^{-3} s^{-1}]	0.1 to 1 wt%	-	Adopted	0.005 to 0.9	[206]
up to 292 [kg O_2 m^{-3} day^{-1}]	3.5%	Questa, USA [Mo]	Simulated	up to 0.08	[207]

* Conversion performed using the stoichiometry of Equation (11) and bulk waste rock density and porosity values provided in the corresponding references.

2.6.1. Static and Kinetic Testing

Static tests aim to quantify the acid-generating and acid-neutralizing capabilities of bulk mine waste materials [45]: paste-pH, acid-base accounting (ABA), and net acid generation (NAG) tests are common waste-rock classification methods. The detailed procedures, advantages, and shortcomings of static tests have been reviewed [45,46,132,208]. In brief, ABA is based on a determination of the total sulfur, sulfide, and carbon content of a waste-rock sample. The acid-producing potential (AP) is calculated by multiplying the sulfur or sulfide content by a stoichiometric factor [132], typically differing with sulfide reactivity (Section 2.5) [209,210]. Acid-neutralization potential (NP) is typically determined by titration with HCl (Sobek method [132,211,212]), although such NP determination has been repeatedly revised [213–215] and equally deploys correction factors [45,216–218]. Both AP and NP typically have units of kg $CaCO_3$ per ton waste rock and ultimately aggregate unresolved mineralogical heterogeneity into bulk waste-rock properties. From AP and NP, a net-neutralizing potential (NNP = AP–NP; kg $CaCO_3$/ton waste rock) or neutralizing potential ratio (NPR = NP/AP; unitless) can be calculated. Although different, site-specific cut-off values to assess acid-production risks are used, it is often assumed that NNP < −20 kg $CaCO_3$/ton or NPR < 1 indicate potential net acid-generating material, whereas material with NNP > 20 kg $CaCO_3$/ton or NPR > 3 is non-acid generating [132]. Considerable uncertainty remains associated with these NNP or NPR classification criteria and unexpected drainage acidification in high NPR materials, as well as non-acidic drainage from high-sulfide materials [219], have been reported. Being practical and cost-efficient, ABA analyses continue to be widely used and optimized, e.g., through (i) corrections for oxidation steps that dissolve organic material, (ii) corrections for samples with abundant forms of S other than sulfide [220], and (iii) improved determination of silicate NP [221,222].

Static testing is often complemented by kinetic testing to overcome the discrepancy between conditions for (short-term) laboratory testing on crushed, sieved samples and the long-term behavior of coarser waste rock observed in the field. Examples of kinetic tests include laboratory humidity cell tests with artificial wetting- and drying cycles [129,132], column tests [68], as well as field cell tests of a variety of sizes [128,177]. The benefits and limitations of humidity cell tests have been previously reviewed [176,223]. Kinetic tests can be used to predict long-term weathering dynamics, investigate specific geochemical processes, and test possible reclamation scenarios [175]. While long-term kinetic testing (i.e., months to years) is comparably expensive, kinetic tests are generally performed after static testwork has identified critical or uncertain samples. Field-based kinetic tests range from small (<1 m^3) to large-scale (thousands of m^3) experiments with waste-rock barrels, pads, or test piles, and although field tests can account for site-specific climatic conditions, they usually cannot be established until advanced stages of mine operation.

Several multi-scale field research programs with a combination of static and long-term kinetic tests have been conducted/completed recently, e.g., at the Aitik [195,200,224], Diavik [225–228], and Antamina [115,166,177,199,229,230] mines. These programs have investigated waste-rock weathering dynamics and drainage quality evolution in small-scale laboratory tests, meso-scale columns, and up to full-scale field experiments. A key advantage of such programs is the opportunity to investigate geochemical reactions and transport processes under quasi-controlled conditions across a range of scales, thereby providing a means to verify the reliability and accuracy of static tests for model parameterization (Section 4.2) and practical industrial-scale predictions. Additional multiscale investigations will be crucial to address remaining knowledge gaps, improve our conceptual understanding of waste rock weathering processes, and inform practical drainage quality prediction models.

2.6.2. Macroscale Geochemical Heterogeneity

Full-scale waste-rock piles, especially in complex geological settings [231], may exhibit a high degree of geochemical heterogeneity, arising from the mixing of different waste-rock materials and depositional practices (i.e., push- versus end-dumping [232]). On the one hand, such heterogeneities can complicate the development of predictive water quality models when the location and nature of more reactive waste materials is unknown. However, strategic placement of different materials (e.g., co-location or blending of acid-producing and acid-neutralizing materials) can also be an effective strategy to mitigate acid drainage risks [233]. Geochemical material heterogeneity can cause unpredictable weathering dynamics [115,234–236] and representative sampling is critical to characterize the spatial distribution of mineral reactivity that can significantly affect the overall drainage signature of composite systems. Relevant geological, lithological, and alteration units must be sampled relative to the amounts and particle size of each material [45,132]: inadequate sampling can contribute to substantial variability or incorrect assessment of waste-rock reactivity. Sampling and analysis of waste-rock properties or material heterogeneity must be fit-for-purpose to inform the deployed prediction model: generalization and extrapolation from static tests may be insufficient when done in lieu of appropriate kinetic testing, as illustrated by the wide range of weathering rates observed in the field (Table 3).

Thus, bulk waste-rock parameters can be used to aggregate variability in mineralogical composition of waste rock but static test results or bulk weathering indicators may not reflect in-situ sulfide oxidation rates and acid-generation in spatially heterogeneous systems, e.g., when retention by secondary minerals is insufficiently quantified. More accurate predictions of waste-rock drainage dynamics in full-scale systems may be achieved when static tests, detailed mineralogical analyses and targeted kinetic tests are effectively combined.

3. Physical Transport Processes

In addition to the geochemical weathering and attenuation processes discussed above, waste-rock drainage quality and quantities are controlled by the transport of gas, water, and heat through porous waste-rock piles: aqueous transport controls the ultimate export of solubilized solutes from the waste-rock pile, and gas transport can affect weathering rates through controlling oxygen supply required to sustain sulfide oxidation.

3.1. Aqueous Transport

Water flow within waste-rock piles is highly site-specific and varies with climate, (internal) pile structure, and hydrogeological waste-rock properties. At the macroscale (Figure 3), drainage of a waste-rock pile can be described by a water balance [16]:

$$P = E + R + G - R + \Delta S \tag{16}$$

in which P is precipitation, E is evapo(transpi)ration, R is runoff, G is groundwater exchange, R is reaction, and ΔS is change in internal storage (i.e., infiltration or drainage). Modern waste-rock piles are preferably placed onto relatively impermeable or lined surfaces to minimize interaction with groundwater if present at the site. Further, the highest oxidation rates listed in Table 3 indicate that water consumption by weathering reactions is maximally on the order of 0.01–0.1 kg H_2O m^{-3} year^{-1}, and thus only relevant to the water budget for tall piles stored at very dry conditions (<200 mm year^{-1}).

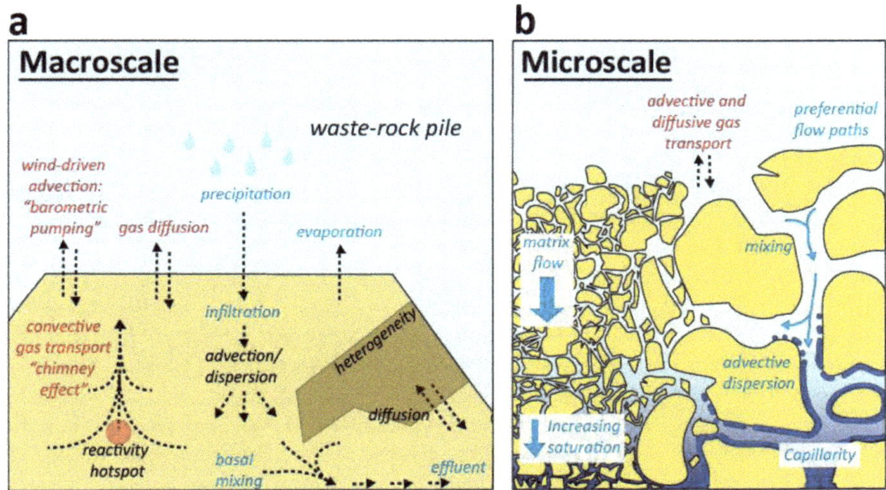

Figure 3. Schematic illustration of relevant aqueous (blue) and gaseous (red) mass transport processes in waste-rock piles, distinguishing between macroscale (**a**) and porescale or microscale (**b**) processes.

Evaporation has been investigated for engineered waste-rock covers [237,238] but less for bare waste-rock piles [239]. Near-surface water and energy balances have been used to estimate evaporation rates for waste-rock piles by combining rainfall gauging, soil-water characteristic curves and suction measurements, in-situ moisture sensors, scintillometry, and eddy covariance measurements or surface heat mapping [240]. Evapo(transpi)ration rates inferred from local meteorological measurements are poorly representative for the quite unique, coarse-textured nature of waste rock and empirical relationships to estimate evaporation (e.g., Penman-Monteith) must be calibrated for such surface properties [241]. Water and energy balancing at various mine sites have indicated that evaporation may be the dominant component (i.e., >70%) of the water balance of waste-rock piles [241,242]. A comparison of twelve waste-rock piles [240] indicated that effective evaporation was lower for bare waste-rock piles compared to covered or vegetated piles. However, evaporation rates from waste-rock piles strongly differ from variability in local climatic conditions (wind speed, insolation [243]), slope aspects, as well as surface waste-rock properties (particle size, roughness, albedo). Further study is required to resolve the controls on evaporative fluxes posed by thermal gradients [239,244] (e.g., quantifying surface heating using infrared cameras) versus those of vapor pressure versus diffusive gradients [245] (e.g., mapping of saturation and the evaporative 'drying front' [246]).

Surface runoff from waste-rock piles may be deliberately minimized to avoid erosion on the pile batters, in which case it is typically insignificant for the water balance. Yet, certain reclamation strategies rely on maximizing runoff from compacted (subsurface) cover layers to limit percolation into underlying reactive material [236,247]. Material compaction (e.g., in traffic surfaces) can facilitate surface runoff and ponding, especially under flashy precipitation patterns. Eventually, the non-runoff or evaporated fraction of precipitation will infiltrate the unsaturated waste-rock pile. Downward propagation of precipitation forms a wetting front that becomes drainage constituting the base seepage. Changes in internal water storage can be significant when drainage fronts successively migrate through unsaturated waste rock as a result of seasonal precipitation patterns or during so-called wetting-up phases in newly deposited piles [225,229,248].

At the porescale (Figure 3), the drainage flow regime from a waste-rock pile is the result of infiltration through a typically large range of grain and pore sizes. Particle sizes in waste rock can vary from greater than 1-m boulders to sub-millimeter clay-sized fractions (i.e., wide particle size distributions and porosity ranges) [16]. Such variability must be accounted for conceptual hydrological models aimed at capturing waste-rock flow dynamics and parameters such as hydraulic conductivity can be highly non-uniform across poorly-sorted heterogeneous waste-rock piles [229,249]. Infiltration is often (conceptually) separated into a slower matric flow component, in which water flows under capillary and gravity forces as described by Richards' equation, and a faster macropore (or non-capillary, preferential) flow component, which is more rapid and channelized [250]. Wetting fronts move according to kinematic velocity under dominantly matrix flow conditions but can travel hundreds of times faster than the average velocity as a result of preferential flow [251]. Preferential flow phenomena have been extensively studied and reviewed [252,253], including in mine waste rock [249,254–256].

Tracer tests with internal or externally applied conservative solutes can be used to identify and quantify preferential flow phenomena. Calibration of hydraulic parameters from tracer tests is often preferred over conventional field tests (e.g., infiltrometer, permeameter), which are challenging for large-scale systems and boulder-sized particles. Approaches to describe non-uniform flow in unsaturated porous media exist include advanced dual-domain (dual-porosity, dual-permeability) models that have meanwhile been applied to waste rock [229,230]. While preferential flow will be critical for the prediction of water quantity hydrographs and flushing dynamics, water quality may be dominantly controlled by matrix flow that facilitates longer contact times with finer-grainer particles with elevated reactive surface areas and thus solute mobilization [16]. Finally, freeze-and-thaw cycles may completely alter the hydrological regime in waste-rock piles and can thereby critically affect drainage dynamics [257,258]. The cryohydrogeology of waste-rock piles is strongly coupled to the thermal regime through phase transitions (Section 3.3) and has been described by model approaches that include the Clapeyron equation or Soil-Freezing Characteristic Curves [259]. Thus, waste-rock flow dynamics are highly site-specific and further research is required to optimize and parameterize hydrological models for full-scale waste-rock systems.

3.2. Gas Transport

Oxygen consumption from sulfide oxidation and CO_2 production from carbonate dissolution may induce poregas pressures and compositions that differ from atmospheric conditions, particularly in large waste-rock piles with reactive material [194,198,199,260]. In smaller-scale (laboratory) experiments with less-reactive waste rock, poregas variations are typically neglected yet rarely experimentally determined. Gradients in poregas composition and pressure trigger gas transfer through diffusion and advection, respectively (Figure 3), which affect internal poregas distributions and the generally outward transport of CO_2 versus inward replenishment of O_2. The contribution of dissolved oxygen transport in percolating water is usually negligible (low solubility) [16] and convection as a combined heat and gas transfer mechanism is discussed below.

Molecular diffusion typically dominates gas transport in low permeability waste-rock systems: the controls of oxygen diffusion on sulfide oxidation rates have been estimated from mass-balance calculations (Fick's laws) [192,194,261], and more advanced modeling of combined gas advection-diffusion with spatially-discretized diffusion parameters [262]. In addition to diffusion, pressure gradients can result in advective gas transport in porous media [263]. Estimates of advective gas transport using e.g., pneumatic heads and Darcy's law for gas, have shown that barometric fluctuations can be relevant for gas transport particularly through more permeable coarse rubble zones at the bases of waste-rock piles [194,243,264,265].

Both diffusive and advective gas transfer are strongly controlled by the properties of the waste rock, including its permeability, degree of porewater saturation, and spatial heterogeneity therein [194,266]. Various empirical and semi-empirical formulae to estimate diffusivity and permeability from particle size distributions or porosity ranges exist [267–269] but these often invoke tortuosity or constrictivity parameters that are poorly defined [266]. Typical effective permeability and gas-diffusivity ranges for waste rock have been reviewed [16] and vary orders of magnitude between and within sites [194]. As a result, gas transfer is usually only quantitatively constrained for waste-rock piles, even though oxygen supply may be rate-limiting in full-scale prediction models. Site-specific field determination of gas transport properties, especially in dual-porosity media, is particularly challenging for full-sized waste-rock systems across large areal extents. Further research is therefore required to assess the applicability and practicality of in-situ sensing or gas injection tracer tests for such parameterization, considering that a quantitative understanding of gas transport limitations can be used to better assess weathering rates and optimize drainage quality predictions.

3.3. Heat Transport

Exothermic sulfide oxidation (e.g., 1000–1500 kJ·mol^{-1} for pyrite [270]) can cause internal temperatures in waste-rock piles to rise to tens of degrees above average ambient temperatures, e.g., up to 65 °C at some sites [204]. Thermal profiles have been measured in a large number of waste-rock piles [204,226,271,272], often by means of instrumented boreholes [194]. Heat transfer from reactive 'hotspots' in waste-rock piles is dominated by convective heat transfer, the basis of the so-called 'chimney effect' [206,273] (Figure 3). Convection is the combination of conduction (heat diffusion) and advection (bulk fluid flow); radiative transfer and viscous dissipation are typically assumed negligible in waste rock. Natural convection arises from buoyancy forces in waste-rock piles generated by oxidation reactions that produce heat and alter the density of the poregas (e.g., through water vapor [206]). Temperature gradients caused by internal heating trigger upward convection against the downward diffusive transport of oxygen into a waste-rock pile. Mass transfer estimates and modeling studies have shown that convective heat and gas transfer can sustain high waste-rock weathering rates [204,274]. The onset and strength of convective transport is determined by the temperature gradient (local atmospheric temperature versus the spatial distribution of reactive waste rock [236,273]) and waste-rock permeability. As a result, parameterization of heat transfer properties is not straightforward [16] and effective heat transfer properties vary widely as a function of porewater saturation and porosity [204,236,266,272]. Further work is required to relate heat transfer mechanisms to waste-rock properties and pile construction methods, and thereby help quantify the contribution of convection to overall weathering rates and ultimately drainage quality.

3.4. Physical Heterogeneity

Physical heterogeneity within waste-rock piles induces variations in their hydrogeological properties (water, air, and heat transport). In-situ investigation of temperature, poregas composition, moisture distribution, infiltration, and solute transport can help assess these heterogeneities on practical scales [275–277], e.g., by mapping flow channeling or diversion along structural features such as intermittent traffic surfaces or tipping benches. Recent improvements in our ability to measure in-situ properties at increasing resolution have helped characterize heterogeneity for improved drainage quality predictions and numerical models (Section 4.2).

Soil moisture sensing techniques have been recently reviewed [278]. Point measurement techniques provide precise measurements of hydrogeological properties but are challenging to extrapolate to larger areas because of the waste-rock heterogeneities. Previous work on the hydrogeological properties of waste-rock piles have used point measurements of water content, for instance using time-domain reflectometry (TDR) sensors [16,279,280], frequency-domain (FDR) sensors [281], and soil water potential sensors [282,283], or by oven-drying discrete waste-rock samples taken at several depths in boreholes [162,163]. Remote sensing techniques offer larger scales of investigation, but often at the expense of lower resolutions that might be inadequate to discriminate the internal fine structure of waste-rock piles. Nevertheless, electrical resistivity techniques (ERT; [275,281]), ground penetrating radar (GPR; [284–286]), electromagnetic induction (EMI; [284,287]), fibre optic distributed temperature sensing (FO-DTS; [288]), or oxygen/hydrogen isotopic signatures [162] have all been applied successfully to investigate hydrogeological dynamics in waste-rock piles. In addition, basal lysimeters are often used to record internal mixing and drainage at the bottom of waste-rock piles [115,280,282]. In addition to direct sensing and measurement techniques, physical and geochemical heterogeneity in waste-rock piles may be reverse-engineered from truck movement and placement records [289]. Ideally, multiple of the aforementioned methods are combined to gain a more comprehensive understanding of the overall hydrogeological behavior of heterogeneous waste-rock piles.

3.5. Coupling Between Geochemical Reactions and Physical Transport

The overall waste-rock drainage from a composite, heterogeneous waste-rock pile often reflects the mixed aggregate signal of individual reaction-infiltration pathways, each resulting from a sequence of interacting processes. The drainage composition arising at each individual pathway results from exposure to a usually wide range of physicochemical properties and the transport time of the infiltrating fluid. Even with a perfect description of the spatial heterogeneity in relevant properties (Section 3.4), prediction of such a fluid composition constitutes a challenging task, due to the coupled nature of the physicochemical processes described above [16,290]. Figure 4 schematically illustrates the different physical transport processes and geochemical reactions controlling waste-rock drainage quality and quantity and how they interact.

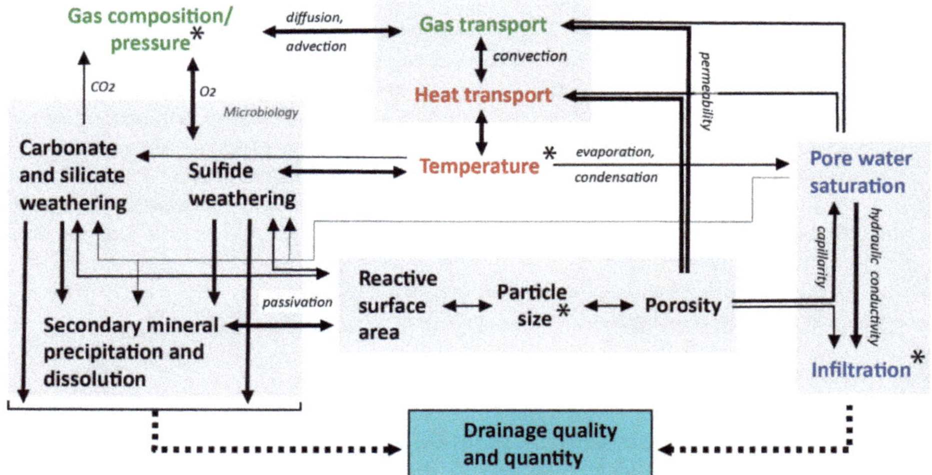

Figure 4. Schematic illustration of major couplings and feedbacks between geochemical reactions and physical transport processes. Solid arrow lines indicate interactions or a control-hierarchy between different processes, with the width of the arrow being (qualitatively) proportional to the magnitude or strength of the feedback. External controls include that of site meteorology on infiltration, temperature, and gas pressure and that of site geology and communition method on waste-rock particle sizes, as indicated by the asterisks.

A selection of these couplings has been previously investigated [16,205,266]. Many processes and associated feedbacks operate virtually simultaneously. For instance, sulfide oxidation produces acidity that induces carbonate and silicate weathering and secondary mineral precipitation, all of which affect pH and the subsequent progress of oxidation [172]. Sulfide oxidation may undergo orders-of-magnitude acceleration under acidifying conditions, due to a fractional-order rate dependence on proton activity and due to increased oxidation by ferric ion. At the same time, sulfide oxidation is an exothermic reaction: the reaction rates of sulfide oxidation may increase by several factors between every 10 °C increase in temperature, as per Arrhenius' relation [194]. Temperature also impacts geochemical reactions rates through minor changes in equilibrium and gas solubility constants, and fluid properties such as density and viscosity that can affect transport rates [266]. While some couplings thus enforce controls on magnitudes of percentages or fractions at most, others impose orders-of-magnitude influences. The task at hand is to assess the strength of the coupling, and many feedback mechanisms remain poorly quantified. One example is the cryo-hydrogeological feedback between latent heat, gas, and fluid flow behavior and reaction rates: ice is approximately 3.8 times more heat conductive than water at 0 °C and in cold climates, gas migration through partially frozen waste-rock piles can maintain oxidation reactions and control freeze–thaw dynamics [257,258].

The intricately coupled processes in waste rock can exert both positive and negative feedbacks towards poorer drainage quality (Figure 4). As detailed above, sulfide oxidation releases heat and acidity, both of which accelerate oxidation (Equation (1)) and induce a convective supply of oxygen that can accelerate oxidation ('chimney effect', Section 3.3). For such positive feedback loops, small parameter variability can yield significant reinforcing perturbations in long-term predictions. In contrast, drainage quality deterioration may also be lessened by negative feedbacks that decelerate the oxidation process. For instance, waste-rock weathering rates can be lowered by mineral surface passivation and pore clogging, a decreased microbial activity at elevated temperatures, or reduction of oxygen replenishment by mass transport limitations [194]. Finally, while certain feedbacks operate directly or on comparatively short time-scales (e.g., flow channeling and preferential flow), others can require years to become relevant (e.g., progressive dissolution of reactive grain sizes [291,292] or secondary mineral precipitation that affect reactive surface area [290,293] and porosity/permeability [266]).

Processes and their feedbacks can be accounted for in different layers of complexity. Firstly, they are dependent on the molecular-scale to mineral-scale physical chemistry, i.e., passivation, porosity, and grain size evolution. Secondly, large heterogeneous systems can display feedbacks that are controlled by the large-scale structural make-up of the waste-rock pile (e.g., local oxygen content and oxidation rates that vary with infiltrating precipitation fronts through compacted layers). The scale of the studied system and according prediction model thus will determine which processes and feedbacks need to be considered and parameterized.

4. Practical Waste-Rock Drainage Predictions

Prediction of drainage type, amount, and timing is required to reduce or prevent the potential detrimental environmental impacts of waste-rock weathering [132]. Such prediction is typically required already during mine planning/permitting stages, in which full-scale waste-rock storage facilities do not yet exist. Drainage quality and quantity predictions for long-term planning need to integrate a combination of geochemical and physical transport processes, while geochemistry and mineralogy determine the occurrence, abundance, and reactivity of potential solutes and hydrogeology determines their transport and mobility [51], and both are highly interconnected as discussed above.

The scope of investigation required to inform drainage predictions differs based on the cost-benefit context or engineering problem at hand. For instance, static test results and mass-balance estimates may suffice for shorter-term, smaller-scope problems ('black box' models), but long-term and full-size water management strategies for entire mine sites require comprehensive data sets and/or coupled reactive-transport models [13,294]. Experience is required to identify what data is needed for the practical problem at hand and how prediction model outcomes can be interpreted or not ("all models are wrong, some are useful" [295]). The development of mine waste-rock drainage models typically includes baseline scenarios and multiple (conceptual) prediction models that are iteratively updated as data becomes available during operation [296].

4.1. Scaling Phenomena

Direct extrapolation of waste-rock properties from small-scale and short-duration experiments into long-term prediction models for full-size systems can lead to erroneous drainage quantity and quality estimates because smaller-scale static tests of kinetic experiments do not reflect the complex or variable weathering conditions in the field and extent of material heterogeneity encountered in full-scale piles. In addition, many physicochemical processes have undetectable impacts below certain spatiotemporal dimensions [297]: e.g., convection does not exist in humidity cell tests. So-called scaling factors have been proposed to facilitate the use of laboratory parameters onto practice-relevant waste-rock systems [130,298], but these are often semi-empirical correction factors based on limited mechanistic evidence and remain poorly verified across different mine sites.

Upscaling phenomena arise when small-scale processes are extrapolated onto larger scales over orders-of-magnitude, without accounting for changes in continuity or heterogeneity in the studied system. Upscaling of drainage predictions must therefore address the difference between micro- versus macroscale parameters for kinetic or transport processes [299]: the task at hand is to identify which processes undergo relevant scale transitions that are not captured by model parameters (e.g., the onset of thermal convection or transitioning out of a uniform Darcian flow domain).

As discussed in Section 3.4, quantification of spatial heterogeneity in waste-rock piles is pertinent for successful drainage predictions. However, the degree of system heterogeneity may increase with its spatial dimensions [235], and further study (e.g., using variogram models) is required to assess for which parameters and to which extent increasing heterogeneity causes scaling phenomena. An opportunity to investigate scale transitions in waste-rock piles may be provided by dimensionless numbers: quantities relating the spatiotemporal scales of physicochemical processes. Examples include the Damköhler number (transport versus reaction timescales) and Rayleigh number (buoyancy-driven flow; convection), which has been used to assess waste-rock weathering dynamics in piles of various dimensions [206]. Future work on upscaling phenomena is required to improve drainage prediction models.

4.2. Reactive-Transport Models (RTMs)

Reactive transport modelling can be used to quantify the physicochemical processes and interactions between reactive fluids and engineered or geological porous media and has been applied for coupled geochemical and transport phenomena from the micrometer scale to the watershed scale [290,300]. In contrast to the geochemical (aqueous-phase) equilibrium models described above (Section 2.4.1), reactive-transport models (RTMs) are based on mass and energy conservation relations and process-based equations that describe transport (hydraulic advection and dispersion, diffusion, convection) as well as chemical reactions across solids, liquids, and gases [135,231,266,301]. As such, RTMs can account for time-discretized mass and heat transport as well as surface complexation, liquid-gas partitioning, mineral dissolution and precipitation reactions and various biogeochemical reactions [290,292]. Reactive-transport models have been widely deployed for the investigation of waste-rock weathering processes [135,203,296,302] and environmental impact assessment of (abandoned) mines [303], using codes like MIN3P [304], PHREEQC [305], Hytec [306], Geochemist's Workbench [136], Toughreact [307], Crunchflow, or PFlotran [308].

Due to a typically modular approach, RTMs allow for the incremental inclusion of processes and their interactions [290] and constitute powerful investigative tools. They offer the possibility to quantitatively assess the relevance of certain processes through sensitivity analyses and can be used to investigate the effects of potential engineering solutions to mitigate risks associated to poor drainage quality, such as co-disposal techniques [235] and fluid control layers or covers [135,231,309]. While the versatility of RTMs may be their prime advantage, it also constitutes a limitation: the choice and parametrization of the considered processes. An immediate drawback is that multiple descriptions or solutions of a problem can yield the same drainage composition, which is referred to as non-uniqueness [310]. Setting-up an RTM requires modelling choices, ranging from the selection of relevant processes (heat generation, phase-changes, permeability evolution), solid phases, and their reactive properties (kinetic reactions) to the physical description (flow behavior, physical heterogeneity). While the challenge of such selections increases with the scale of the studied problem, the description of individual subcomponents alone can be difficult (e.g., the initial waste-rock mineralogy can consist of multiple (amorphous) phases present at levels below analytical detection limits for which thermodynamic or kinetic data is unavailable). In addition, RTMs often rely on mathematical equations for physical or chemical processes that simplify potential feedbacks or aggregate unresolved heterogeneity (e.g., through adoption of bulk-averaged parameters) [297]. For instance, the dependence of modelled waste-rock hydraulics on measurable parameters (e.g., matric suction) is often modeled with Van Genuchten relations [250,311,312], which are dependent on particle size [313] and therefore prone to change over time.

With increasing computational power and availability of big data, it has become possible to represent large-scale pile dynamics with stochastic [234,235] or (fully-) coupled process-based RTMs [227,301]. Emerging applications for mine waste-rock systems include dual-continuum approaches [230]. In contrast, relatively few applications of reactive-transport models with fully-coupled thermo-hydro-chemical descriptions (e.g., thermally-driven gas migration) have been performed on large heterogeneous 3D waste-rock systems. A primary challenge in reactive-transport drainage prediction models is not only to identify the processes that are relevant in the studied waste-rock domain, but also to accurately parametrize them: more advanced models for large heterogeneous piles are inherently more difficult to experimentally parametrize. Mechanistic reactive-transport models can be parameterized practically through synoptic sampling (e.g., tracer-tests [51]), and they have the potential to inform practical optimization of the design and management or waste-rock piles, or the mitigation of poor drainage quality. While RTMs have proven crucial to assess process-relevance and waste-rock weathering dynamics under various conditions [231], their development and verification is also laborious. In order to justify and establish their use for practical predictive purposes and management decision-making, it is critical that the relative sensitivities of the model inputs and outputs are translated into actual design parameters.

5. Concluding Remarks

In this review, we discussed critical hydrogeochemical processes affecting the weathering of mine waste rock and its potential environmental impacts. Geochemical and mineralogical reactions, in addition to water, air, and heat transport, are all of key importance in determining waste-rock weathering rates and thus drainage quality as well as quantity. The research reviewed here has been instrumental in providing an understanding of the controls on these individual physicochemical processes, as well as their coupling. Studying the mobility of mine waste pollutants and the underlying mechanisms from the micro- to the macroscale remains important, as expanding mining operations around the world pose increasing potential environmental risks: forward-thinking in the design of long-term dumps that could pose potential leachate quality problems is critical. An improved and quantitative understanding of the factors controlling mine waste-rock drainage dynamics, paired with a growing ability to map in-situ heterogeneity in full-scale systems and the power to harness large-scale, high-resolution data in practical models, will allow engineers and practitioners to develop more robust prediction models and sustainable management decisions. Continuing research can facilitate optimized waste management and thereby prevent waste-rock related environmental deterioration in the future.

Author Contributions: Conceptualization, B.V.; Investigation, B.V., N.S., B.P., H.J.; Writing-Review & Editing, B.V., N.S., B.P., H.J. All authors have read and agreed to the published version of the manuscript.

Funding: This research received no external funding.

Conflicts of Interest: The authors declare no conflict of interest.

References

1. Hudson-Edwards, K.A.; Jamieson, H.E.; Lottermoser, B.G. Mine Wastes: Past, Present, Future. *Elements* **2011**, *7*, 375–380. [CrossRef]
2. Blowes, D.W.; Ptacek, C.J.; Jambor, J.L.; Weisener, C.G. The geochemistry of acid mine drainage. *Environ. Geochem.* **2003**, *9*, 149–204. [CrossRef]
3. Dino, G.A.; Cavallo, A.; Rossetti, P.; Garamvölgyi, E.; Sándor, R.; Coulon, F. Towards Sustainable Mining: Exploiting Raw Materials from Extractive Waste Facilities. *Sustainability* **2020**, *12*, 2383. [CrossRef]
4. Lottermoser, B.G. Recycling, Reuse and Rehabilitation of Mine Wastes. *Elements* **2011**, *7*, 405–410. [CrossRef]

5. Lèbre, É.; Corder, G.D.; Golev, A. Sustainable practices in the management of mining waste: A focus on the mineral resource. *Miner. Eng.* **2017**, *107*, 34–42. [CrossRef]

6. Bian, Z.; Miao, X.; Lei, S.; Chen, S.; Wang, W.; Struthers, S. The challenges of reusing mining and mineral-processing wastes. *Science* **2012**, *337*, 702–703. [CrossRef]

7. Byrne, P.; Hudson-Edwards, K.A.; Bird, G.; Macklin, M.G.; Brewer, P.A.; Williams, R.D.; Jamieson, H.E. Water quality impacts and river system recovery following the 2014 Mount Polley mine tailings dam spill, British Columbia, Canada. *Appl. Geochem.* **2018**, *91*, 64–74. [CrossRef]

8. Santamarina, J.C.; Torres-Cruz, L.A.; Bachus, R.C. Why coal ash and tailings dam disasters occur. *Science* **2019**, *364*, 526–528. [CrossRef]

9. Fischer, S.; Rosqvist, G.; Chalov, S.; Jarsjö, J. Disproportionate water quality impacts from the century-old nautanen copper mines, Northern Sweden. *Sustainability* **2020**, *12*, 1394. [CrossRef]

10. Demchak, J.; Skousen, J.; McDonald, L.M. Longevity of acid discharges from underground mines located above the regional water table. *J. Environ. Qual.* **2004**, *33*, 656–668. [CrossRef]

11. Lottermoser, B.G. *Mine Wastes: Characterization, Treatment and Environmental Impacts*, 3rd ed.; Springer: Berlin/Heidelberg, Germany, 2010; ISBN 3-642-12419-4. [CrossRef]

12. Mudd, G.M. The Environmental sustainability of mining in Australia: Key mega-trends and looming constraints. *Resour. Policy* **2010**, *35*, 98–115. [CrossRef]

13. Wolkersdorfer, C.; Nordstrom, D.K.; Beckie, R.D.; Cicerone, D.S.; Elliot, T.; Edraki, M.; Valente, T.; França, S.C.A.; Kumar, P.; Oyarzún, R.; et al. Guidance for the Integrated Use of Hydrological, Geochemical, and Isotopic Tools in Mining Operations. *Mine Water Environ.* **2020**, *39*, 204–228. [CrossRef]

14. Hudson-Edwards, A.K.; Dold, B. Mine waste characterization, management and remediation. *Minerals* **2015**, *5*, 82–85. [CrossRef]

15. Jamieson, H.E. Geochemistry and mineralogy of solid mine waste: Essential knowledge for predicting environmental impact. *Elements* **2011**, *7*, 381–386. [CrossRef]

16. Amos, R.T.; Blowes, D.W.; Bailey, B.L.; Sego, D.C.; Smith, L.; Ritchie, A.I.M. Waste-rock hydrogeology and geochemistry. *Appl. Geochem.* **2015**, *57*, 140–156. [CrossRef]

17. Lindsay, M.B.J.; Moncur, M.C.; Bain, J.G.; Jambor, J.L.; Ptacek, C.J.; Blowes, D.W. Geochemical and mineralogical aspects of sulfide mine tailings. *Appl. Geochem.* **2015**, *57*, 157–177. [CrossRef]

18. Ghorbani, Y.; Franzidis, J.-P.; Petersen, J. Heap leaching technology—current state, innovations, and future directions: A review. *Miner. Process. Extr. Metall. Rev.* **2016**, *37*, 73–119. [CrossRef]

19. Marsden, J.O.; Botz, M.M. Heap leach modeling—A review of approaches to metal production forecasting. *Miner. Metall. Process.* **2017**, *34*, 53–64. [CrossRef]

20. Pradhan, N.; Nathsarma, K.C.; Srinivasa Rao, K.; Sukla, L.B.; Mishra, B.K. Heap bioleaching of chalcopyrite: A review. *Miner. Eng.* **2008**, *21*, 355–365. [CrossRef]

21. Dold, B. Sustainability in metal mining: From exploration, over processing to mine waste management. *Rev. Environ. Sci. Bio/Technol.* **2008**, *7*, 275. [CrossRef]

22. Johnson, D.B.; Hallberg, K.B. Acid mine drainage remediation options: A review. *Sci. Total Environ.* **2005**, *338*, 3–14. [CrossRef] [PubMed]

23. Park, I.; Tabelin, C.B.; Jeon, S.; Li, X.; Seno, K.; Ito, M.; Hiroyoshi, N. A review of recent strategies for acid mine drainage prevention and mine tailings recycling. *Chemosphere* **2019**, *219*, 588–606. [CrossRef]

24. Evangelou, V.P.; Zhang, Y.L. A review: Pyrite oxidation mechanisms and acid mine drainage prevention. *Crit. Rev. Environ. Sci. Technol.* **1995**, *25*, 141–199. [CrossRef]

25. Cornelis, G.; Johnson, C.A.; Van Gerven, T.; Vandecasteele, C. Leaching mechanisms of oxyanionic metalloid and metal species in alkaline solid wastes: A review. *Appl. Geochem.* **2008**, *23*, 955–976. [CrossRef]

26. Chandra, A.P.; Gerson, A.R. The mechanisms of pyrite oxidation and leaching: A fundamental perspective. *Surf. Sci. Rep.* **2010**, *65*, 293–315. [CrossRef]

27. Li, Y.; Kawashima, N.; Li, J.; Chandra, A.P.; Gerson, A.R. A review of the structure, and fundamental mechanisms and kinetics of the leaching of chalcopyrite. *Adv. Colloid Interface Sci.* **2013**, *197–198*, 1–32. [CrossRef]

28. Parbhakar-Fox, A.; Lottermoser, B. Principles of Sulfide Oxidation and Acid Rock Drainage. In *Environmental Indicators in Metal Mining*; Springer: Berlin/Heidelberg, Germany, 2016; pp. 15–34. ISBN 978-3-319-42729-4.

29. Bosecker, K. Bioleaching: Metal solubilization by microorganisms. *FEMS Microbiol. Rev.* **1997**, *20*, 591–604. [CrossRef]

30. Nordstrom, D.K.; Southam, G. Geomicrobiology of sulfide mineral oxidation. In *Reviews in Mineralogy*; Banfield, J.F., Nealson, K.H., Eds.; Mineralogical Society of America: Chantilly, VA, USA, 1997; Volume 35, pp. 361–390.

31. Rawlings, D.E. Heavy metal mining using microbes. *Annu. Rev. Microbiol.* **2002**, *56*, 65–91. [CrossRef]

32. Baker, B.J.; Banfield, J.F. Microbial communities in acid mine drainage. *FEMS Microbiol. Ecol.* **2003**, *44*, 139–152. [CrossRef]

33. Johnson, D.B.; Hallberg, K.B. The microbiology of acidic mine waters. *Res. Microbiol.* **2003**, *154*, 466–473. [CrossRef]

34. Watling, H.R. The bioleaching of sulphide minerals with emphasis on copper sulphides—A review. *Hydrometallurgy* **2006**, *84*, 81–108. [CrossRef]

35. Rawlings, D.E.; Johnson, D.B. The microbiology of biomining: Development and optimization of mineral-oxidizing microbial consortia. *Microbiology* **2007**, *153*, 315–324. [CrossRef] [PubMed]

36. Johnson, D.B. Biomining-biotechnologies for extracting and recovering metals from ores and waste materials. *Curr. Opin. Biotechnol.* **2014**, *30*, 24–31. [CrossRef] [PubMed]

37. Huang, L.-N.; Kuang, J.-L.; Shu, W.-S. Microbial ecology and evolution in the acid mine drainage model system. *Trends Microbiol.* **2016**, *24*, 581–593. [CrossRef]

38. Kaksonen, A.H.; Boxall, N.J.; Gumulya, Y.; Khaleque, H.N.; Morris, C.; Bohu, T.; Cheng, K.Y.; Usher, K.M.; Lakaniemi, A.M. Recent progress in biohydrometallurgy and microbial characterisation. *Hydrometallurgy* **2018**, *180*, 7–25. [CrossRef]

39. Ledin, M.; Pedersen, K. The environmental impact of mine wastes—roles of microorganisms and their significance in treatment of mine wastes. *Earth-Sci. Rev.* **1996**, *41*, 67–108. [CrossRef]

40. Akcil, A.; Koldas, S. Acid Mine Drainage (AMD): Causes, treatment and case studies. *J. Clean. Prod.* **2006**, *14*, 1139–1145. [CrossRef]

41. Neculita, C.M.; Zagury, G.J.; Bussière, B. Passive treatment of acid mine drainage in bioreactors using sulfate-reducing bacteria: Critical review and research needs. *J. Environ. Qual.* **2007**, *36*, 1–16. [CrossRef]

42. RoyChowdhury, A.; Sarkar, D.; Datta, R. Remediation of acid mine drainage-impacted water. *Curr. Pollut. Rep.* **2015**, *1*, 131–141. [CrossRef]

43. Jamieson, H.E.; Walker, S.R.; Parsons, M.B. Mineralogical characterization of mine waste. *Appl. Geochem.* **2015**, *57*, 85–105. [CrossRef]

44. Piatak, N.M.; Parsons, M.B.; Seal, R.R. Characteristics and environmental aspects of slag: A review. *Appl. Geochem.* **2015**, *57*, 236–266. [CrossRef]

45. Parbhakar-Fox, A.; Lottermoser, B.G. A critical review of acid rock drainage prediction methods and practices. *Miner. Eng.* **2015**, *82*, 107–124. [CrossRef]

46. Dold, B. Acid rock drainage prediction: A critical review. *J. Geochem. Explor.* **2017**, *172*, 120–132. [CrossRef]

47. Kefeni, K.K.; Msagati, T.A.M.; Mamba, B.B. Acid mine drainage: Prevention, treatment options, and resource recovery: A review. *J. Clean. Prod.* **2017**, *151*, 475–493. [CrossRef]

48. Lottermoser, B.G. *Environmental Indicators in Metal. Mining*; Springer: Berlin/Heidelberg, Germany, 2017; ISBN 978-3-319-42729-4. [CrossRef]

49. Moodley, I.; Sheridan, C.M.; Kappelmeyer, U.; Akcil, A. Environmentally sustainable acid mine drainage remediation: Research developments with a focus on waste/by-products. *Miner. Eng.* **2018**, *126*, 207–220. [CrossRef]

50. Rajaram, V.; Dutta, S.; Parameswaran, K. *Sustainable Mining Practices: A Global Perspective*; A.A. Balkema: Leiden, The Netherlands, 2005; ISBN 9058096890.

51. Nordstrom, D.K. Hydrogeochemical processes governing the origin, transport and fate of major and trace elements from mine wastes and mineralized rock to surface waters. *Appl. Geochem.* **2011**, *26*, 1777–1791. [CrossRef]

52. Öhlander, B.; Chatwin, T.; Alakangas, L. Management of sulfide-bearing waste, a challenge for the mining industry. *Minerals* **2012**, *2*, 1–10. [CrossRef]

53. Simate, G.S.; Ndlovu, S. Acid mine drainage: Challenges and opportunities. *J. Environ. Chem. Eng.* **2014**, *2*. [CrossRef]

54. Nordstrom, D.K.; Blowes, D.W.; Ptacek, C.J. Hydrogeochemistry and microbiology of mine drainage: An update. *Appl. Geochem.* **2015**, *57*, 3–16. [CrossRef]

55. Batterham, R.J. The mine of the future—Even more sustainable. *Miner. Eng.* **2017**, *107*, 2–7. [CrossRef]

56. Aznar-Sánchez, J.A.; García-Gómez, J.J.; Velasco-Muñoz, J.F.; Carretero-Gómez, A. Mining waste and its sustainable management: Advances in worldwide research. *Minerals* **2018**, *8*, 284. [CrossRef]

57. Li, J.; Kawashima, N.; Fan, R.; Schumann, R.C.; Gerson, A.R.; Smart, R.S.C. Method for distinctive estimation of stored acidity forms in acid mine wastes. *Environ. Sci. Technol.* **2014**, *48*, 11445–11452. [CrossRef] [PubMed]

58. Vaughan, D.J. Minerals Sulphides. In *Reference Module in Earth Systems and Environmental Sciences*; Elsevier: Amsterdam, The Netherlands, 2013; ISBN 978-0-12-409548-9.

59. Elberling, B.; Schippers, A.; Sand, W. Bacterial and chemical oxidation of pyritic mine tailings at low temperatures. *J. Contam. Hydrol.* **2000**, *41*, 225–238. [CrossRef]

60. Singer, P.C.; Stumm, W. Acidic mine drainage: The rate-determining step. *Science* **1970**, *167*, 1121–1123. [CrossRef] [PubMed]

61. Dos Santos, E.C.; de Mendonça Silva, J.C.; Duarte, H.A. Pyrite oxidation mechanism by oxygen in aqueous medium. *J. Phys. Chem. C* **2016**, *120*, 2760–2768. [CrossRef]

62. Tabelin, C.B.; Suchol, V.; Mayumi, I.; Naoki, H.; Toshifumi, I. Pyrite oxidation in the presence of hematite and alumina: II. Effects on the cathodic and anodic half-cell reactions. *Sci. Total Environ.* **2017**, *581–582*, 126–135. [CrossRef]

63. Schippers, A.; Breuker, A.; Blazejak, A.; Bosecker, K.; Kock, D.; Wright, T.L. The biogeochemistry and microbiology of sulfidic mine waste and bioleaching dumps and heaps, and novel Fe(II)-oxidizing bacteria. *Hydrometallurgy* **2010**, *104*, 342–350. [CrossRef]

64. Luther, G.W.; Findlay, A.; MacDonald, D.; Owings, S.; Hanson, T.; Beinart, R.; Girguis, P. Thermodynamics and kinetics of sulfide oxidation by oxygen: A look at inorganically controlled reactions and biologically mediated processes in the environment. *Front. Microbiol.* **2011**, *2*, 62. [CrossRef]

65. Schrenk, M.O.; Edwards, K.J.; Goodman, R.M.; Hamers, R.J.; Banfield, J.F. Distribution of thiobacillus ferrooxidans and Leptospirillum ferrooxidans: Implications for generation of acid mine drainage. *Science* **1998**, *279*, 1519–1522. [CrossRef]

66. Korehi, H.; Blöthe, M.; Schippers, A. Microbial diversity at the moderate acidic stage in three different sulfidic mine tailings dumps generating acid mine drainage. *Res. Microbiol.* **2014**, *165*, 713–718. [CrossRef]

67. Hallberg, K.B. New perspectives in acid mine drainage microbiology. *Hydrometall.* **2010**, *104*, 448–453. [CrossRef]

68. Blackmore, S.; Vriens, B.; Sorensen, M.; Power, I.M.; Smith, L.; Hallam, S.J.; Mayer, U.K.; Beckie, R.D. Microbial and geochemical controls on waste rock weathering and drainage quality. *Sci. Total Environ.* **2018**, *640–641*, 1004–1014. [CrossRef] [PubMed]

69. Rimstidt, J.D.; Vaughan, D.J. Pyrite oxidation: A state-of-the-art assessment of the reaction mechanism. *Geochim. Cosmochim. Acta* **2003**, *67*, 873–880. [CrossRef]

70. Schippers, A.; Sand, W. Bacterial leaching of metal sulfides proceeds by two indirect mechanisms via thiosulfate or via polysulfides and sulfur. *Appl. Environ. Microbiol.* **1999**, *65*, 319–321. [CrossRef] [PubMed]

71. Borilova, S.; Mandl, M.; Zeman, J.; Kucera, J.; Pakostova, E.; Janiczek, O.; Tuovinen, O.H. Can sulfate be the first dominant aqueous sulfur species formed in the oxidation of pyrite by acidithiobacillus ferrooxidans? *Front. Microbiol.* **2018**, *9*, 3134. [CrossRef]

72. Johnson, A.C.; Romaniello, S.J.; Reinhard, C.T.; Gregory, D.D.; Garcia-Robledo, E.; Revsbech, N.P.; Canfield, D.E.; Lyons, T.W.; Anbar, A.D. Experimental determination of pyrite and molybdenite oxidation kinetics at nanomolar oxygen concentrations. *Geochim. Cosmochim. Acta* **2019**, *249*, 160–172. [CrossRef]

73. Kameia, G.; Ohmotob, H. The kinetics of reactions between pyrite and O2-bearing water revealed from in situ monitoring of DO, Eh and pH in a closed system. *Geochim. Cosmochim. Acta* **2000**, *64*, 2585–2601. [CrossRef]

74. Janzen, M.P.; Nicholson, R.V.; Scharer, J.M. Pyrrhotite reaction kinetics: Reaction rates for oxidation by oxygen, ferric iron, and for nonoxidative dissolution. *Geochim. Cosmochim. Acta* **2000**, *64*, 1511–1522. [CrossRef]

75. Nicholson, R.V.; Gillham, R.W.; Reardon, E.J. Pyrite oxidation in carbonate-buffered solution: 1. Experimental kinetics. *Geochim. Cosmochim. Acta* **1988**, *52*, 1077–1085. [CrossRef]

76. Williamson, M.A.; Rimstidt, J.D. The kinetics and electrochemical rate-determining step of aqueous pyrite oxidation. *Geochim. Cosmochim. Acta* **1994**, *58*, 5443–5454. [CrossRef]

77. Rimstidt, J.D.; Chermak, J.A.; Gagen, P.M. Rates of Reaction of Galena, Sphalerite, Chalcopyrite, and Arsenopyrite with Fe(III) in Acidic Solutions. In *Environmental Geochemistry of Sulfide Oxid*; American Chemical Society: Washington, DC, USA, 1993; Volume 550, pp. 1–2. ISBN 9780841227729.

78. Walker, F.P.; Schreiber, M.E.; Rimstidt, J.D. Kinetics of arsenopyrite oxidative dissolution by oxygen. *Geochim. Cosmochim. Acta* **2006**, *70*, 1668–1676. [CrossRef]

79. Heidel, C.; Tichomirowa, M.; Breitkopf, C. Sphalerite oxidation pathways detected by oxygen and sulfur isotope studies. *Appl. Geochem.* **2011**, *26*, 2247–2259. [CrossRef]

80. Olson, G.J.; Clark, T.R. Bioleaching of molybdenite. *Hydrometallurgy* **2008**, *93*, 10–15. [CrossRef]

81. Jerz, J.K.; Rimstidt, J.D. Pyrite oxidation in moist air Associate editor: M. A. McKibben. *Geochim. Cosmochim. Acta* **2004**, *68*, 701–714. [CrossRef]

82. Moses, C.O.; Herman, J.S. Pyrite oxidation at circumneutral pH. *Geochim. Cosmochim. Acta* **1991**, *55*, 471–482. [CrossRef]

83. Sun, H.; Chen, M.; Zou, L.; Shu, R.; Ruan, R. Study of the kinetics of pyrite oxidation under controlled redox potential. *Hydrometallurgy* **2015**, *155*, 13–19. [CrossRef]

84. Long, H.; Dixon, D.G. Pressure oxidation of pyrite in sulfuric acid media: A kinetic study. *Hydrometallurgy* **2004**, *73*, 335–349. [CrossRef]

85. Boon, M.; Heijnen, J.J. Chemical oxidation kinetics of pyrite in bioleaching processes. *Hydrometallurgy* **1998**, *48*, 27–41. [CrossRef]

86. McKibben, M.A.; Barnes, H.L. Oxidation of pyrite in low temperature acidic solutions: Rate laws and surface textures. *Geochim. Cosmochim. Acta* **1986**, *50*, 1509–1520. [CrossRef]

87. Lehner, S.; Savage, K. The effect of As, Co, and Ni impurities on pyrite oxidation kinetics: Batch and flow-through reactor experiments with synthetic pyrite. *Geochim. Cosmochim. Acta* **2008**, *72*, 1788–1800. [CrossRef]

88. Navarro, C.A.; von Bernath, D.; Jerez, C.A. Heavy Metal Resistance Strategies of Acidophilic Bacteria and Their Acquisition: Importance for Biomining and Bioremediation. *Biol. Res.* **2013**, *46*, 363–371. [CrossRef]

89. Chen, J.; He, F.; Zhang, X.; Sun, X.; Zheng, J.; Zheng, J. Heavy metal pollution decreases microbial abundance, diversity and activity within particle-size fractions of a paddy soil. *FEMS Microbiol. Ecol.* **2014**, *87*, 164–181. [CrossRef] [PubMed]

90. Hendry, M.J.; Biswas, A.; Essilfie-Dughan, J.; Chen, N.; Day, S.J.; Barbour, S.L. Reservoirs of Selenium in Coal Waste Rock: Elk Valley, British Columbia, Canada. *Environ. Sci. Technol.* **2015**, *49*, 8228–8236. [CrossRef] [PubMed]

91. Chon, H.-T.; Hwang, J.-H. Geochemical Characteristics of the Acid Mine Drainage in the Water System in the Vicinity of the Dogye Coal Mine in Korea. *Environ. Geochem. Health* **2000**, *22*, 155–172. [CrossRef]

92. Black, A.; Craw, D. Arsenic, copper and zinc occurrence at the Wangaloa coal mine, southeast Otago, New Zealand. *Int. J. Coal Geol.* **2001**, *45*, 181–193. [CrossRef]

93. Qureshi, A.; Maurice, C.; Öhlander, B. Potential of coal mine waste rock for generating acid mine drainage. *J. Geochem. Explor.* **2016**, *160*, 44–54. [CrossRef]

94. Equeenuddin, S.M.; Tripathy, S.; Sahoo, P.K.; Panigrahi, M.K. Hydrogeochemical characteristics of acid mine drainage and water pollution at Makum Coalfield, India. *J. Geochem. Explor.* **2010**, *105*, 75–82. [CrossRef]

95. Sahoo, P.K.; Tripathy, S.; Equeenuddin, S.M.; Panigrahi, M.K. Geochemical characteristics of coal mine discharge vis-à-vis behavior of rare earth elements at Jaintia Hills coalfield, northeastern India. *J. Geochem. Explor.* **2012**, *112*, 235–243. [CrossRef]

96. Szczepanska, J.; Twardowska, I. Distribution and environmental impact of coal-mining wastes in Upper Silesia, Poland. *Environ. Geol.* **1999**, *38*, 249–258. [CrossRef]

97. Banks, S.B.; Banks, D. Abandoned mines drainage: Impact assessment and mitigation of discharges from coal mines in the UK. *Eng. Geol.* **2001**, *60*, 31–37. [CrossRef]

98. Hughes, J.; Craw, D.; Peake, B.; Lindsay, P.; Weber, P. Environmental characterisation of coal mine waste rock in the field: An example from New Zealand. *Environ. Geol.* **2007**, *52*, 1501–1509. [CrossRef]

99. Calkins, W.H. The chemical forms of sulfur in coal: A review. *Fuel* **1994**, *73*, 475–484. [CrossRef]

100. Kazadi Mbamba, C.; Harrison, S.T.L.; Franzidis, J.-P.; Broadhurst, J.L. Mitigating acid rock drainage risks while recovering low-sulfur coal from ultrafine colliery wastes using froth flotation. *Miner. Eng.* **2012**, *29*, 13–21. [CrossRef]

101. Banerjee, D. Acid drainage potential from coal mine wastes: Environmental assessment through static and kinetic tests. *Int. J. Environ. Sci. Technol.* **2014**, *11*, 1365–1378. [CrossRef]

102. Dutta, M.; Saikia, J.; Taffarel, S.R.; Waanders, F.B.; de Medeiros, D.; Cutruneo, C.M.N.L.; Silva, L.F.O.; Saikia, B.K. Environmental assessment and nano-mineralogical characterization of coal, overburden and sediment from Indian coal mining acid drainage. *Geosci. Front.* **2017**, *8*, 1285–1297. [CrossRef]

103. Ghosh, W.; Dam, B. Biochemistry and molecular biology of lithotrophic sulfur oxidation by taxonomically and ecologically diverse bacteria and archaea. *FEMS Microbiol. Rev.* **2009**, *33*, 999–1043. [CrossRef]

104. Acharya, C.; Sukla, L.B.; Misra, V.N. Biodepyritisation of coal. *J. Chem. Technol. Biotechnol.* **2004**, *79*, 1–12. [CrossRef]

105. Schippers, A.; Rohwerder, T.; Sand, W. Intermediary sulfur compounds in pyrite oxidation: Implications for bioleaching and biodepyritization of coal. *Appl. Microbiol. Biotechnol.* **1999**, *52*, 104–110. [CrossRef]

106. Davalos, A.; Pecina, E.T.; Soria, M.; Carrillo, F.R. Kinetics of Coal Desulfurization in An Oxidative Acid Media. *Int. J. Coal Prep. Util.* **2009**, *29*, 152–172. [CrossRef]

107. Juszczak, A.; Domka, F.; Kozłowski, M.; Wachowska, H. Microbial desulfurization of coal with Thiobacillus ferrooxidans bacteria. *Fuel* **1995**, *74*, 725–728. [CrossRef]

108. Hoffmann, M.R.; Faust, B.C.; Panda, F.A.; Koo, H.H.; Tsuchiya, H.M. Kinetics of the Removal of Iron Pyrite from Coal by Microbial Catalysis. *Appl. Environ. Microbiol.* **1981**, *42*, 259–271. [CrossRef] [PubMed]

109. Liang, L.; McNabb, J.A.; Paulk, J.M.; Gu, B.; McCarthy, J.F. Kinetics of iron(II) oxygenation at low partial pressure of oxygen in the presence of natural organic matter. *Environ. Sci. Technol.* **1993**, *27*, 1864–1870. [CrossRef]

110. Kalin, M.; Cairns, J.; McCready, R. Ecological engineering methods for acid mine drainage treatment of coal wastes. *Resour. Conserv. Recycl.* **1991**, *5*, 265–275. [CrossRef]

111. Kaksonen, A.H.; Puhakka, J.A. Sulfate reduction based bioprocesses for the treatment of acid mine drainage and the recovery of metals. *Eng. Life Sci.* **2007**, *7*, 541–564. [CrossRef]

112. Lefticariu, L.; Walters, E.R.; Pugh, C.W.; Bender, K.S. Sulfate reducing bioreactor dependence on organic substrates for remediation of coal-generated acid mine drainage: Field experiments. *Appl. Geochem.* **2015**, *63*, 70–82. [CrossRef]

113. Küsel, K. Microbial cycling of iron and sulfur in acidic coal mining lake sediments. *Water Air Soil Pollut. Focus* **2003**, *3*, 67–90. [CrossRef]

114. Ayora, C.; Caraballo, M.A.; Macias, F.; Rötting, T.S.; Carrera, J.; Nieto, J.-M. Acid mine drainage in the Iberian Pyrite Belt: 2. Lessons learned from recent passive remediation experiences. *Environ. Sci. Pollut. Res.* **2013**, *20*, 7837–7853. [CrossRef]

115. Vriens, B.; Peterson, H.E.; Laurenzi, L.; Smith, L.; Aranda, C.; Mayer, K.U.; Beckie, R.D. Long-term monitoring of waste-rock weathering at Antamina, Peru. *Chemosphere* **2019**, *215*, 858–869. [CrossRef]

116. Maree, J.P.; de Beer, M.; Strydom, W.F.; Christie, A.D.M.; Waanders, F.B. Neutralizing Coal Mine Effluent with Limestone to Decrease Metals and Sulphate Concentrations. *Mine Water Environ.* **2004**, *23*, 81–86. [CrossRef]

117. Alakangas, L.; Andersson, E.; Mueller, S. Neutralization/prevention of acid rock drainage using mixtures of alkaline by-products and sulfidic mine wastes. *Environ. Sci. Pollut. Res.* **2013**, *20*, 7907–7916. [CrossRef]

118. Davies, H.; Weber, P.; Lindsay, P.; Craw, D.; Peake, B.; Pope, J. Geochemical changes during neutralisation of acid mine drainage in a dynamic mountain stream, New Zealand. *Appl. Geochem.* **2011**, *26*, 2121–2133. [CrossRef]

119. Chou, L.; Garrels, R.M.; Wollast, R. Comparative study of the kinetics and mechanisms of dissolution of carbonate minerals. *Chem. Geol.* **1989**, *78*, 269–282. [CrossRef]

120. Morse, J.W.; Arvidson, R.S. The dissolution kinetics of major sedimentary carbonate minerals. *Earth-Sci. Rev.* **2002**, *58*, 51–84. [CrossRef]

121. Kaufmann, G.; Dreybrodt, W. Calcite dissolution kinetics in the system $CaCO_3$–H_2O–CO_2 at high undersaturation. *Geochim. Cosmochim. Acta* **2007**, *71*, 1398–1410. [CrossRef]

122. Stumm, W.; Wollast, R. Coordination chemistry of weathering: Kinetics of the surface-controlled dissolution of oxide minerals. *Rev. Geophys.* **1990**, *28*, 53–69. [CrossRef]

123. Schwertmann, U. Solubility and dissolution of iron oxides. *Plant. Soil* **1991**, *130*, 1–25. [CrossRef]

124. White, A.F.; Brantley, S.L. Chemical weathering rates of silicate minerals; an overview. *Rev. Mineral. Geochem.* **1995**, *31*, 1–22.

125. Gruber, C.; Kutuzov, I.; Ganor, J. The combined effect of temperature and pH on albite dissolution rate under far-from-equilibrium conditions. *Geochim. Cosmochim. Acta* **2016**, *186*, 154–167. [CrossRef]

126. Rimstidt, J.D.; Dove, P.M. Mineral/solution reaction rates in a mixed flow reactor: Wollastonite hydrolysis. *Geochim. Cosmochim. Acta* **1986**, *50*, 2509–2516. [CrossRef]

127. Kirste, D.; Pearce, J.; Golding, S. Parameterizing Geochemical Models: Do Kinetics of Calcite Matter? *Procedia Earth Planet. Sci.* **2017**, *17*, 606–609. [CrossRef]

128. Vriens, B.; Skierszkan, E.K.; St-Arnault, M.; Salzsauler, K.A.; Aranda, C.; Mayer, K.U.; Beckie, R.D. Mobilization of metal(loid) oxyanions through circumneutral waste-rock drainage. *ACS Omega* **2019**, *4*, 10205–10215. [CrossRef]

129. Plante, B.; Benzaazoua, M.; Bussière, B. Predicting Geochemical Behaviour of Waste Rock with Low Acid Generating Potential Using Laboratory Kinetic Tests. *Mine Water Environ.* **2011**, *30*, 2–21. [CrossRef]

130. Plante, B.; Bussière, B.; Benzaazoua, M. Lab to field scale effects on contaminated neutral drainage prediction from the Tio mine waste rocks. *J. Geochem. Explor.* **2014**, *137*, 37–47. [CrossRef]

131. Nicholson, R.V.; Rinker, M.J. Metal leaching from sulphide mine waste under neutral pH conditions. In Proceedings of the 5th International Conference on Acid Rock Drainage (ICARD), Denver, CO, USA, 21–24 May 2000; pp. 951–958.

132. Price, W.A. Prediction Manual for Drainage Chemistry from Sulphidic Geologic Materials. MEND Report 1.20.1. Available online: https://www.fs.usda.gov/Internet/FSE_DOCUMENTS/stelprdb5336546.pdf (accessed on 1 July 2020).

133. Tabelin, C.B.; Silwamba, M.; Paglinawan, F.C.; Mondejar, A.J.S.; Duc, H.G.; Resabal, V.J.; Opiso, E.M.; Igarashi, T.; Tomiyama, S.; Ito, M.; et al. Solid-phase partitioning and release-retention mechanisms of copper, lead, zinc and arsenic in soils impacted by artisanal and small-scale gold mining (ASGM) activities. *Chemosphere* **2020**, *260*, 127574. [CrossRef]

134. Tamoto, S.; Tabelin, C.B.; Igarashi, T.; Ito, M.; Hirojohsi, N. Short and long term release mechanisms of arsenic, selenium and boron from a tunnel-excavated sedimentary rock under in situ conditions. *J. Cont. Hydr.* **2015**, *175–176*, 60–71. [CrossRef]

135. Demers, I.; Molson, J.; Bussière, B.; Laflamme, D. Numerical modeling of contaminated neutral drainage from a waste-rock field test cell. *Appl. Geochem.* **2013**, *33*, 346–356. [CrossRef]

136. Tabelin, C.B.; Sasaki, R.; Igarashi, T.; Park, I.; Tamoto, S.; Arima, T.; Ito, M.; Hiroyoshi, N. Simultaneous leaching of arsenite, arsenate, selenite and selenate, and their migration in tunnel-excavated sedimentary rocks: II. Kinetic and reactive transport modeling. *Chemosphere* **2017**, *188*, 444–454. [CrossRef]

137. Nordstrom, D.K.; Archer, D.G. *Arsenic Thermodynamic Data and Environmental Geochemistry—Arsenic in Ground Water: Geochemistry and Occurrence*; Welch, A.H., Stollenwerk, K.G., Eds.; Springer: Boston, MA, USA, 2003; pp. 1–25. ISBN 978-0-306-47956-4.

138. Kocourková, E.; Sracek, O.; Houzar, S.; Cempírek, J.; Losos, Z.; Filip, J.; Hršelová, P. Geochemical and mineralogical control on the mobility of arsenic in a waste rock pile at Dlouhá Ves, Czech Republic. *J. Geochem. Explor.* **2011**, *110*, 61–73. [CrossRef]

139. Wei, X.; Zhang, S.; Shimko, J.; Dengler, R.W., II. Mine drainage: Treatment technologies and rare earth elements. *Water Environ. Res.* **2019**, *91*, 1061–1068. [CrossRef]

140. González, V.; Vignati, D.; Leyval, C.; Giamberini, L. Environmental fate and ecotoxicity of lanthanides: Are they a uniform group beyond chemistry? *Environ. Int.* **2014**, *71*, 148–157. [CrossRef] [PubMed]

141. Romero-Freire, A.; Turlin, F.; André-Mayer, A.-S.; Pelletier, M.; Cayer, A.; Giamberini, L. Giamberini Biogeochemical Cycle of Lanthanides in a Light Rare Earth Element-Enriched Geological Area (Quebec, Canada). *Minerals* **2019**, *9*, 573. [CrossRef]

142. Ayora, C.; Macías, F.; Torres, E.; Lozano, A.; Carrero, S.; Nieto, J.-M.; Pérez-López, R.; Fernández-Martínez, A.; Castillo-Michel, H. Recovery of rare earth elements and yttrium from passive-remediation systems of acid mine drainage. *Environ. Sci. Technol.* **2016**, *50*, 8255–8262. [CrossRef]

143. Royer-Lavallée, A.; Neculita, C.M.; Coudert, L. Removal and potential recovery of rare earth elements from mine water. *J. Ind. Eng. Chem.* **2020**, *89*, 47–57. [CrossRef]

144. Edahbi, M.; Plante, B.; Benzaazoua, M. Environmental challenges and identification of the knowledge gaps associated with REE mine wastes management. *J. Clean. Prod.* **2019**, *212*, 1232–1241. [CrossRef]

145. Jamieson, H.; Laidlow, A.; Parsons, M. Characterization of U and REE Mobility Downstream of U Tailings near Bancroft, Ontario. In Proceedings of the International Conference on Acid Rock Drainage (ICARD), Santiago, Chile, 21–24 April 2015.

146. Edahbi, M.; Plante, B.; Benzaazoua, M.; Pelletier, M. Geochemistry of rare earth elements within waste rocks from the Montviel carbonatite deposit, Québec, Canada. *Environ. Sci. Pollut. Res.* **2018**, *25*, 10997–11010. [CrossRef]

147. Edahbi, M.; Plante, B.; Benzaazoua, M.; Kormos, L.; Pelletier, M. Rare earth elements (La, Ce, Pr, Nd, and Sm) from a carbonatite deposit: Mineralogical characterization and geochemical behavior. *Minerals* **2018**, *8*, 55. [CrossRef]

148. Edahbi, M.; Plante, B.; Benzaazoua, M.; Ward, M.; Pelletier, M. Mobility of rare earth elements in mine drainage: Influence of iron oxides, carbonates, and phosphates. *Chemosphere* **2018**, *199*, 647–654. [CrossRef]

149. Kauppila, P.; Räisänen, M. Mineralogical and geochemical alteration of Hitura sulphide mine tailings with emphasis on nickel mobility and retention. *J. Geochem. Explor.* **2008**, *97*, 1–20. [CrossRef]

150. Kauppila, P.; Räisänen, M.; Johnson, R. Geochemical Characterisation of Seepage and Drainage Water Quality from Two Sulphide Mine Tailings Impoundments: Acid Mine Drainage versus Neutral Mine Drainage. *Mine Water Environ.* **2008**, *28*, 30–49. [CrossRef]

151. Radkova, A.; Jamieson, H.; Campbell, K. Antimony mobility during the early stages of stibnite weathering in tailings at the Beaver Brook Sb deposit, Newfoundland. *Appl. Geochem.* **2020**, *115*, 104528. [CrossRef]

152. Lindsay, M.B.J.; Condon, P.D.; Jambor, J.L.; Lear, K.G.; Blowes, D.W.; Ptacek, C.J. Mineralogical, geochemical, and microbial investigation of a sulfide-rich tailings deposit characterized by neutral drainage. *Appl. Geochem.* **2009**, *24*, 2212–2221. [CrossRef]

153. Jamieson, H.E.; Bromstad, M.; Nordstrom, D.K. Extremely arsenic-rich, pH-neutral waters from the Giant mine, Canada. In Proceedings of the First International Conference on Mine Water Solutions in Extreme Environments, Lima, Peru, 15–17 April 2013; pp. 82–94.

154. Kwong, Y.T.J.; Percival, J.B.; Soprovich, E.A. *Arsenic Mobilization and Attenuation in Near-Neutral Drainage—Implications for Tailings and Waste Rock Management for Saskatchewan Uranium Mines*; Canadian Institute of Mining, Metallurgy and Petroleum: Montreal, QC, Canada, 2000; ISBN 1-894475-05-4.

155. Plante, B.; Benzaazoua, M.; Bussière, B.; Biesinger, M.C.; Pratt, A.R. Study of Ni sorption onto Tio mine waste rock surfaces. *Appl. Geochem.* **2010**, *25*, 1830–1844. [CrossRef]

156. Powell, K.J.; Brown, P.L.; Byrne, R.H.; Gajda, T.; Hefter, G.; Sjöberg, S.; Wanner, H. Chemical speciation of environmentally significant metals with inorganic ligands Part 2: The Cu^{2+} + OH^-, Cl^-, CO_3^{2-}, SO_4^{2-}, and PO_4^{3-} systems (IUPAC Technical Report). *Pure Appl. Chem.* **2007**, *79*, 895–950. [CrossRef]

157. Powell, K.J.; Brown, P.L.; Byrne, R.H.; Gajda, T.; Hefter, G.; Leuz, A.-K.; Sjöberg, S.; Wanner, H. Chemical speciation of environmentally significant metals with inorganic ligands. Part 5: The Zn^{2+} + OH^-, Cl^-, CO_3^{2-}, SO_4^{2-}, and PO_4^{3-} systems (IUPAC Technical Report). *Pure Appl. Chem.* **2013**, *85*, 2249–2311. [CrossRef]

158. Grafe, M.; Eick, M.J.; Grossl, P.R.; Saunders, A.M. Adsorption of Arsenate and Arsenite on Ferrihydrite in the Presence and Absence of Dissolved Organic Carbon. *J. Environ. Qual.* **2002**, *31*, 1115–1123. [CrossRef]

159. Zhou, Y.-F.; Haynes, R.J. Sorption of Heavy Metals by Inorganic and Organic Components of Solid Wastes: Significance to Use of Wastes as Low-Cost Adsorbents and Immobilizing Agents. *Crit. Rev. Environ. Sci. Technol.* **2010**, *40*, 909–977. [CrossRef]

160. Stumm, W.; Morgan, J.J. *Aquatic Chemistry—Chemical Equilibria and Rates in Natural Waters*, 3rd ed.; Wiley-Interscience: Hoboken, NJ, USA, 1995; ISBN 978-0471511854.

161. Skierszkan, E.K.; Stockwell, J.S.; Dockrey, J.W.; Weis, D.; Beckie, R.D.; Mayer, K.U. Molybdenum (Mo) stable isotopic variations as indicators of Mo attenuation in mine waste-rock drainage. *Appl. Geochem.* **2017**, *87*, 71–83. [CrossRef]

162. Bao, Z.; Blowes, D.W.; Ptacek, C.J.; Bain, J.; Holland, S.P.; Wilson, D.; Wilson, W.; MacKenzie, P. Faro waste rock project: Characterizing variably saturated flow behavior through full-scale waste-rock dumps in the continental subarctic region of northern Canada using field measurements and stable isotopes of water. *Water Resour. Res.* **2020**, *56*. [CrossRef]

163. St-Arnault, M.; Vriens, B.; Klein, B.; Blaskovich, R.; Aranda, C.; Mayer, K.U.; Beckie, R.D. Geochemical and mineralogical assessment of localized reactive zones through a full scale heterogeneous waste rock pile. *Miner. Eng.* **2020**, *145*, 106089. [CrossRef]

164. Al, T.A.; Martin, C.J.; Blowes, D.W. Carbonate-mineral/water interactions in sulfide-rich mine tailings. *Geochim. Cosmochim. Acta* **2000**, *64*, 3933–3948. [CrossRef]

165. Tabelin, C.B.; Corpuz, R.D.; Igarashi, T.; Villacorte-Tabelin, M.; Diaz Alorro, R.; Yoo, K.; Raval, S.; Ito, M.; Hiroyoshi, N. Acid mine drainage formation and arsenic mobility under strongly acidic conditions: Importance of soluble phases, iron oxyhydroxides/oxides and nature of oxidation layer on pyrite. *J. Hazard. Mat.* **2020**, *399*, 122844. [CrossRef]

166. Hirsche, D.T.; Blaskovich, R.; Mayer, K.U.; Beckie, R.D. A study of Zn and Mo attenuation by waste-rock mixing in neutral mine drainage using mixed-material field barrels and humidity cells. *Appl. Geochem.* **2017**, *84*, 114–125. [CrossRef]

167. Conlan, M.J.W.; Mayer, K.U.; Blaskovich, R.; Beckie, R.D. Solubility controls for molybdenum in neutral rock drainage. *Geochem. Explor. Environ. Anal.* **2012**, *12*, 21–32. [CrossRef]

168. Moncur, M.C.; Jambor, J.L.; Ptacek, C.J.; Blowes, D.W. Mine drainage from the weathering of sulfide minerals and magnetite. *Appl. Geochem.* **2009**, *24*, 2362–2373. [CrossRef]

169. Plumlee, G.S. The environmental geology of mineral deposis. In *The Environmental Geochemistry of Mineral Deposits. Part A: Processes, Techniques and Health Issues*; Plumlee, G.S., Longsdon, M.S., Eds.; Society of Economic Geologists: Littleton, CO, USA, 1999; pp. 71–116.

170. Weisener, C.G.; Weber, P.A. Preferential oxidation of pyrite as a function of morphology and relict texture. *N. Z. J. Geol. Geophys.* **2010**, *53*, 167–176. [CrossRef]

171. Dold, B. Dissolution kinetics of schwertmannite and ferrihydrite in oxidized mine samples and their detection by differential X-ray diffraction (DXRD). *Appl. Geochem.* **2003**, *18*, 1531–1540. [CrossRef]

172. Huminicki, D.M.C.; Rimstidt, J.D. Iron oxyhydroxide coating of pyrite for acid mine drainage control. *Appl. Geochem.* **2009**, *24*, 1626–1634. [CrossRef]

173. V. Nicholson, R.; Gillham, R.W.; Reardon, E.J. Pyrite oxidation in carbonate-buffered solution: 2. Rate control by oxide coatings. *Geochim. Cosmochim. Acta* **1990**, *54*, 395–402. [CrossRef]

174. Stott, M.B.; Watling, H.R.; Franzmann, P.D.; Sutton, D. The role of iron-hydroxy precipitates in the passivation of chalcopyrite during bioleaching. *Miner. Eng.* **2000**, *13*, 1117–1127. [CrossRef]

175. Roy, V.; Demers, I.; Plante, B.; Thériault, M. Kinetic Testing for Oxidation Acceleration and Passivation of Sulfides in Waste Rock Piles to Reduce Contaminated Neutral Drainage Generation Potential. *Mine Water Environ.* **2020**, *39*, 242–255. [CrossRef]

176. Maest, A.S.; Nordstrom, D.K. A geochemical examination of humidity cell tests. *Appl. Geochem.* **2017**, *81*, 109–131. [CrossRef]

177. St-Arnault, M.; Vriens, B.; Klein, B.; Mayer, K.U.; Beckie, R.D. Mineralogical controls on drainage quality during the weathering of waste rock. *Appl. Geochem.* **2019**, *108*, 104376. [CrossRef]

178. Fan, R.; Short, M.D.; Zeng, S.-J.; Qian, G.; Li, J.; Schumann, R.C.; Kawashima, N.; Smart, R.S.C.; Gerson, A.R. The Formation of Silicate-Stabilized Passivating Layers on Pyrite for Reduced Acid Rock Drainage. *Environ. Sci. Technol.* **2017**, *51*, 11317–11325. [CrossRef] [PubMed]

179. Zhou, Y.; Fan, R.; Short, M.D.; Li, J.; Schumann, R.C.; Xu, H.; Smart, R.S.C.; Gerson, A.R.; Qian, G. Formation of aluminum hydroxide-doped surface passivating layers on pyrite for acid rock drainage control. *Environ. Sci. Technol.* **2018**, *52*, 11786–11795. [CrossRef] [PubMed]

180. Li, X.; Hiroyoshi, N.; Tabelin, C.B.; Naruwa, K.; Harada, C.; Ito, M. Suppressive effects of ferric-catecholate complexes on pyrite oxidation. *Chemosphere* **2019**, *214*, 70–78. [CrossRef] [PubMed]

181. Park, I.; Tabelin, C.B.; Seno, K.; Jeon, S.; Ito, M.; Hiroyoshi, N. Simultaneous suppression of acid mine drainage formation and arsenic release by Carrier-microencapsulation using aluminum-catecholate complexes. *Chemosphere* **2018**, *205*, 414–425. [CrossRef] [PubMed]

182. Berry, V.K.; Murr, L.E.; Hiskey, J.B. Galvanic interaction between chalcopyrite and pyrite during bacterial leaching of low-grade waste. *Hydrometall.* **1978**, *3*, 309–326. [CrossRef]

183. Kwong, Y.T.J.; Swerhone, G.W.; Lawrence, J.R. Galvanic sulphide oxidation as a metal-leaching mechanism and its environmental implications. *Geochem. Explor. Environ. Anal.* **2003**, *3*, 337–343. [CrossRef]

184. Qian, G.; Fan, R.; Short, M.D.; Schumann, R.C.; Li, J.; St.C. Smart, R.; Gerson, A.R. The effects of galvanic interactions with pyrite on the generation of acid and metalliferous drainage. *Environ. Sci. Technol.* **2018**, *52*, 5349–5357. [CrossRef]

185. Chopard, A.; Plante, B.; Benzaazoua, M.; Bouzahzah, H.; Marion, P. Geochemical investigation of the galvanic effects during oxidation of pyrite and base-metals sulfides. *Chemosphere* **2017**, *166*, 281–291. [CrossRef]

186. Hudson-Edwards, K.A.; Schell, C.; Macklin, M.G. Mineralogy and geochemistry of alluvium contaminated by metal mining in the Rio Tinto area, southwest Spain. *Appl. Geochem.* **1999**, *14*, 1015–1030. [CrossRef]

187. Parbhakar-Fox, A.; Lottermoser, B.; Bradshaw, D. Evaluating waste rock mineralogy and microtexture during kinetic testing for improved acid rock drainage prediction. *Miner. Eng.* **2013**, *52*, 111–124. [CrossRef]

188. Parbhakar-Fox, A.K.; Edraki, M.; Walters, S.; Bradshaw, D. Development of a textural index for the prediction of acid rock drainage. *Miner. Eng.* **2011**, *24*, 1277–1287. [CrossRef]

189. Brough, C.P.; Warrender, R.; Bowell, R.J.; Barnes, A.; Parbhakar-Fox, A. The process mineralogy of mine wastes. *Miner. Eng.* **2013**, *52*, 125–135. [CrossRef]

190. Wang, H.; Dowd, P.A.; Xu, C. A reaction rate model for pyrite oxidation considering the influence of water content and temperature. *Miner. Eng.* **2019**, *134*, 345–355. [CrossRef]

191. Dold, B. Evolution of Acid Mine Drainage Formation in Sulphidic Mine Tailings. *Minerals* **2014**, *4*, 621–641. [CrossRef]

192. Sracek, O.; Gélinas, P.; Lefebvre, R.; Nicholson, R. V Comparison of methods for the estimation of pyrite oxidation rate in a waste rock pile at Mine Doyon site, Quebec, Canada. *J. Geochem. Explor.* **2006**, *91*, 99–109. [CrossRef]

193. Hollings, P.; Hendry, M.J.; Nicholson, R.V.; Kirkland, R.A. Quantification of oxygen consumption and sulphate release rates for waste rock piles using kinetic cells: Cluff lake uranium mine, northern Saskatchewan, Canada. *Appl. Geochem.* **2001**, *16*, 1215–1230. [CrossRef]

194. Vriens, B.; St.Arnault, M.; Laurenzi, L.; Smith, L.; Mayer, K.U.; Beckie, R.D. Localized Sulfide oxidation limited by oxygen availability in a full-scale waste-rock pile. *Vadose Zone J.* **2018**, *17*, 1–68. [CrossRef]

195. Eriksson, N.; Destouni, G. Combined effects of dissolution kinetics, secondary mineral precipitation, and preferential flow on copper leaching from mining waste rock. *Water Resour. Res.* **1997**, *33*, 471–483. [CrossRef]

196. Strömberg, B.; Banwart, S. Weathering kinetics of waste rock from the Aitik copper mine, Sweden: Scale dependent rate factors and pH controls in large column experiments. *J. Contam. Hydrol.* **1999**, *39*, 59–89. [CrossRef]

197. Lapakko, K. Comparison of Duluth Complex rock dissolution in the laboratory and field. In Proceedings of the Proceedings American Society of Mining and Reclamation, Pittsburgh, PA, USA, 24–29 April 1994; pp. 419–428.

198. Lorca, M.E.; Mayer, K.U.; Pedretti, D.; Smith, L.; Beckie, R.D. Spatial and temporal fluctuations of pore-gas composition in sulfidic mine waste rock. *Vadose Zone J.* **2016**, *15*. [CrossRef]

199. Vriens, B.; Smith, L.; Mayer, K.U.; Beckie, R.D. Poregas distributions in waste-rock piles affected by climate seasonality and physicochemical heterogeneity. *Appl. Geochem.* **2019**, *100*, 305–315. [CrossRef]

200. Linklater, C.M.; Sinclair, D.J.; Brown, P.L. Coupled chemistry and transport modelling of sulphidic waste rock dumps at the Aitik mine site, Sweden. *Appl. Geochem.* **2005**, *20*, 275–293. [CrossRef]

201. Strömberg, B.; Banwart, S. Kinetic modelling of geochemical processes at the Aitik mining waste rock site in northern Sweden. *Appl. Geochem.* **1994**, *9*, 583–595. [CrossRef]

202. Harries, J.R.; Ritchie, A.I.M. The use of temperature profiles to estimate the pyritic oxidation rate in a waste rock dump from an opencut mine. *Water. Air. Soil Pollut.* **1981**, *15*, 405–423. [CrossRef]

203. Molson, J.W.; Fala, O.; Aubertin, M.; Bussière, B. Numerical simulations of pyrite oxidation and acid mine drainage in unsaturated waste rock piles. *J. Contam. Hydrol.* **2005**, *78*, 343–371. [CrossRef]

204. Lefebvre, R.; Hockley, D.; Smolensky, J.; Geélinas, P. Multiphase transfer processes in waste rock piles producing acid mine drainage. 1: Conceptual model and system characterization. *J. Contam. Hydrol.* **2001**, *52*, 137–164. [CrossRef]

205. Lefebvre, R.; Hockley, D.; Smolensky, J.; Lamontagne, A. Multiphase transfer processes in waste rock piles producing acid mine drainage: 2. Applications of numerical simulation. *J. Contam. Hydrol.* **2001**, *52*, 165–186. [CrossRef]

206. Kuo, E.Y.; Ritchie, A.I.M. The impact of convection on the overall oxidation rate in sulfidic waste rock dumps. In *Proceedings Mining and the Environment II*; Goldsack, D., Belzile, N., Yerwood, P., Hall, G., Eds.; Laurentian University: Sudbury, ON, Canada, 1999; pp. 211–220.

207. Lefebvre, R.; Lamontagne, A.; Wels, C. Numerical simulations of acid drainage in the Sugar Shack South rock pile, Questa Mine, New Mexico, USA. In Proceedings of the Proceedings 2nd Joint IAH-CNC and CGS Groundwater Specialty Conference, 54th Canadian Geotechnical Conference, Calgary, AB, Canada, 16–19 September 2001.

208. Karlsson, T.; Räisänen, M.L.; Lehtonen, M.; Alakangas, L. Comparison of static and mineralogical ARD prediction methods in the Nordic environment. *Environ. Monit. Assess.* **2018**, *190*, 719. [CrossRef]

209. Chopard, A.; Benzaazoua, M.; Bouzahzah, H.; Plante, B.; Marion, P. A contribution to improve the calculation of the acid generating potential of mining wastes. *Chemosphere* **2017**, *175*, 97–107. [CrossRef] [PubMed]

210. Schumann, R.; Stewart, W.; Miller, S.; Kawashima, N.; Li, J.; Smart, R. Acid–base accounting assessment of mine wastes using the chromium reducible sulfur method. *Sci. Total Environ.* **2012**, *424*, 289–296. [CrossRef] [PubMed]

211. Weber, P.A.; Thomas, J.E.; Skinner, W.M.; Smart, R.S.C. A methodology to determine the acid-neutralization capacity of rock samples. *Can. Mineral.* **2005**, *43*, 1183–1192. [CrossRef]

212. Paktunc, A.D. Mineralogical constraints on the determination of neutralization potential and prediction of acid mine drainage. *Environ. Geol.* **1999**, *39*, 103–112. [CrossRef]

213. Skousen, J.; Renton, J.; Brown, H.; Evans, P.; Leavitt, B.; Brady, K.; Cohen, L.; Ziemkiewicz, P. Neutralization potential of overburden samples containing siderite. *J. Environ. Qual.* **1997**, *26*, 673–681. [CrossRef]

214. Weber, P.A.; Thomas, J.E.; Skinner, W.M.; Smart, R.S.C. Improved acid neutralisation capacity assessment of iron carbonates by titration and theoretical calculation. *Appl. Geochem.* **2004**, *19*, 687–694. [CrossRef]

215. Skousen, J.; Simmons, J.; McDonald, L.M.; Ziemkiewicz, P. Acid–base accounting to predict post-mining drainage quality on surface mines. *J. Environ. Qual.* **2002**, *31*, 2034–2044. [CrossRef] [PubMed]

216. Morin, K.A.; Hutt, N.M. On the Nonsense of Arguing the Superiority of an Analytical Method for Neutralization Potential. Minesite Drainage Assessment Group, case study# 32. Available online: www.mdag.com/case_studies/cs32.html (accessed on 1 July 2020).

217. Bouzahzah, H.; Benzaazoua, M.; Plante, B.; Bussiere, B. A quantitative approach for the estimation of the "fizz rating" parameter in the acid-base accounting tests: A new adaptations of the Sobek test. *J. Geochem. Explor.* **2015**, *153*, 53–65. [CrossRef]

218. Bouzahzah, H.; Benzaazoua, M.; Bussiere, B.; Plante, B. Prediction of acid mine drainage: Importance of mineralogy and the test protocols for static and kinetic tests. *Mine Water Environ.* **2014**, *33*, 54–65. [CrossRef]

219. Gerson, A.R.; Rolley, P.J.; Davis, C.; Feig, S.T.; Doyle, S.; Smart, R.S.C. Unexpected non-acid drainage from sulfidic rock waste. *Sci. Rep.* **2019**, *9*, 4357. [CrossRef]

220. Pope, J.; Weber, P.; Mackenzie, A.; Newman, N.; Rait, R. Correlation of acid base accounting characteristics with the Geology of commonly mined coal measures, West Coast and Southland, New Zealand. *N. Z. J. Geol. Geophys.* **2010**, *53*, 153–166. [CrossRef]

221. Miller, S.D.; Stewart, W.S.; Rusdinar, Y.; Schumann, R.E.; Ciccarelli, J.M.; Li, J.; Smart, R.S.C. Methods for estimation of long-term non-carbonate neutralisation of acid rock drainage. *Sci. Total Environ.* **2010**, *408*, 2129–2135. [CrossRef] [PubMed]

222. Jambor, J.L.; Dutrizac, J.E.; Raudsepp, M. Measured and computed neutralization potentials from static tests of diverse rock types. *Environ. Geol.* **2007**, *52*, 1173–1185. [CrossRef]

223. Sapsford, D.J.; Bowell, R.J.; Dey, M.; Williams, K.P. Humidity cell tests for the prediction of acid rock drainage. *Miner. Eng.* **2009**, *22*, 25–36. [CrossRef]

224. Strömberg, B.; Banwart, S.A. Experimental study of acidity-consuming processes in mining waste rock: Some influences of mineralogy and particle size. *Appl. Geochem.* **1999**, *14*, 1–16. [CrossRef]

225. Smith, L.J.D.; Bailey, B.L.; Blowes, D.W.; Jambor, J.L.; Smith, L.; Sego, D.C. The Diavik waste rock project: Initial geochemical response from a low sulfide waste rock pile. *Appl. Geochem.* **2013**, *36*, 210–221. [CrossRef]

226. Pham, N.H.; Sego, D.C.; Arenson, L.U.; Blowes, D.W.; Amos, R.T.; Smith, L. The Diavik Waste Rock Project: Measurement of the thermal regime of a waste-rock test pile in a permafrost environment. *Appl. Geochem.* **2013**, *36*, 234–245. [CrossRef]

227. Wilson, D.; Amos, R.; W. Blowes, D.; B. Langman, J.; Smith, L.; C. Sego, D. Diavik Waste Rock Project: Scale-up of a reactive transport model for temperature and sulfide-content dependent geochemical evolution of waste rock. *Appl. Geochem.* **2018**, *96*, 177–190. [CrossRef]

228. Langman, B.J.; Moore, L.M.; Ptacek, J.C.; Smith, L.; Sego, D.; Blowes, W.D. Diavik Waste Rock Project: Evolution of mineral weathering, element release, and acid generation and neutralization during a five-year humidity cell experiment. *Minerals* **2014**, *4*, 257–278. [CrossRef]

229. Blackmore, S.; Smith, L.; Ulrich Mayer, K.; Beckie, R.D. Comparison of unsaturated flow and solute transport through waste rock at two experimental scales using temporal moments and numerical modeling. *J. Contam. Hydrol.* **2014**, *171*, 49–65. [CrossRef]

230. Blackmore, S.; Pedretti, D.; Mayer, K.U.; Smith, L.; Beckie, R.D. Evaluation of single- and dual-porosity models for reproducing the release of external and internal tracers from heterogeneous waste-rock piles. *J. Contam. Hydrol.* **2018**, *214*, 65–74. [CrossRef] [PubMed]

231. Lahmira, B.; Lefebvre, R.; Aubertin, M.; Bussière, B. Effect of heterogeneity and anisotropy related to the construction method on transfer processes in waste rock piles. *J. Contam. Hydrol.* **2016**, *184*, 35–49. [CrossRef] [PubMed]

232. Pearce, S.; Birkham, T.; O'Kane, M.; Dobchuk, D. *Linking Waste Rock Dump Construction and Design with AMD Risk: A Quantitative Approach*; British Columbia Mine Reclamation Symposium, The University of British Columbia: Vancouver, BC, Canada, 2016. [CrossRef]

233. Parbhakar-Fox, A.; Fox, N.; Jackson, L.; Cornelius, R. Forecasting geoenvironmental risks: Integrated applications of mineralogical and chemical data. *Minerals* **2018**, *8*, 541. [CrossRef]

234. Pedretti, D.; Mayer, K.U.; Beckie, R.D. Stochastic multicomponent reactive transport analysis of low quality drainage release from waste rock piles: Controls of the spatial distribution of acid generating and neutralizing minerals. *J. Contam. Hydrol.* **2017**, *201*, 30–38. [CrossRef] [PubMed]

235. Pedretti, D.; Mayer, K.U.; Beckie, R.D. Controls of uncertainty in acid rock drainage predictions from waste rock piles examined through Monte-Carlo multicomponent reactive transport. *Stoch. Environ. Res. Risk Assess.* **2020**, *34*, 219–233. [CrossRef]

236. Lahmira, B.; Lefebvre, R.; Aubertin, M.; Bussière, B. Effect of material variability and compacted layers on transfer processes in heterogeneous waste rock piles. *J. Contam. Hydrol.* **2017**, *204*, 66–78. [CrossRef] [PubMed]

237. Swanson, D.A.; Barbour, S.L.; Wilson, G.W.; O'Kane, M. Soil-atmosphere modelling of an engineered soil cover for acid generating mine waste in a humid, alpine climate. *Can. Geotech. J.* **2003**, *40*, 276–292. [CrossRef]

238. Nicholls, E.M.; Drewitt, G.B.; Fraser, S.; Carey, S.K. The influence of vegetation cover on evapotranspiration atop waste rock piles, Elk Valley, British Columbia. *Hydrol. Process.* **2019**, *33*, 2594–2606. [CrossRef]

239. Carey, S.K.; Barbour, S.L.; Hendry, M.J. Evaporation from a waste-rock surface, Key Lake, Saskatchewan. *Can. Geotech. J.* **2005**, *42*, 1189–1199. [CrossRef]

240. Birkham, T.; O'Kane, M.; Goodbrand, A.; Barbour, S.L.; Carey, S.K.; Straker, J.; Baker, T.; Klein, R. Near-surface water balances of waste rock dumps. In Proceedings of the British Columbia Mine Reclamation Symposium, Prince George, BC, Canada, 22–25 September 2014.

241. Peterson, H.E. Unsaturated Hydrology, Evaporation, and Geochemistry of Neutral and Acid Rock Drainage in Highly Heterogeneous Mine Waste Rock at the Antamina Mine, Peru. Ph.D. Thesis, The University of British Columbia, Vancouver, BC, Canada, 2014.

242. Rohde, T.K.; Defferrard, P.L.; Lord, M. Store and release cover water balance for the south waste rock dump at Century mine. In Proceedings of the 11th International Conference on Mine Closure 2016, Perth, Australia, 15–17 March 2016; Fourie, A.B., Fourie, A.B., Tibbett, M., Tibbett, M., Eds.; Australian Centre for Geomechanics PP—Perth: Perth, Australia, 2016; pp. 47–59.

243. Chi, X.; Amos, R.T.; Stastna, M.; Blowes, D.W.; Sego, D.C.; Smith, L. The Diavik Waste Rock Project: Implications of wind-induced gas transport. *Appl. Geochem.* **2013**, *36*, 246–255. [CrossRef]

244. Sjoberg, D.B.; Lee, B.S.; Jian, Z. Prediction of water vapor movement through waste rock. *J. Geotech. Geoenvironmental Eng.* **2004**, *130*, 293–302. [CrossRef]

245. Haghighi, E.; Or, D. Evaporation from porous surfaces into turbulent airflows: Coupling eddy characteristics with pore scale vapor diffusion. *Water Resour. Res.* **2013**, *49*, 8432–8442. [CrossRef]

246. Wilson, G.W.; Fredlund, D.G.; Barbour, S.L. The effect of soil suction on evaporative fluxes from soil surfaces. *Can. Geotech. J.* **1997**, *34*, 145–155. [CrossRef]

247. Ramasamy, M.; Power, C. Evolution of acid mine drainage from a coal waste rock pile reclaimed with a simple soil cover. *Hydrology* **2019**, *6*, 83. [CrossRef]

248. Neuner, M.; Smith, L.; Blowes, D.W.; Sego, D.C.; Smith, L.J.D.; Fretz, N.; Gupton, M. The Diavik waste rock project: Water flow through mine waste rock in a permafrost terrain. *Appl. Geochem.* **2013**, *36*, 222–233. [CrossRef]

249. Trinchero, P.; Beckie, R.; Sanchez-Vila, X.; Nichol, C. Assessing preferential flow through an unsaturated waste rock pile using spectral analysis. *Water Resour. Res.* **2011**, *47*, W07532. [CrossRef]

250. Šimůnek, J.; Jarvis, N.J.; van Genuchten, M.T.; Gärdenäs, A. Review and comparison of models for describing non-equilibrium and preferential flow and transport in the vadose zone. *J. Hydrol.* **2003**, *272*, 14–35. [CrossRef]

251. Nimmo, J.R. Simple predictions of maximum transport rate in unsaturated soil and rock. *Water Resour. Res.* **2007**, *43*. [CrossRef]

252. Gerke, H.H. Preferential flow descriptions for structured soils. *J. Plant. Nutr. Soil Sci.* **2006**, *169*, 382–400. [CrossRef]

253. Jarvis, N.J. A review of non-equilibrium water flow and solute transport in soil macropores: Principles, controlling factors and consequences for water quality. *Eur. J. Soil Sci.* **2007**, *58*, 523–546. [CrossRef]

254. Shahhosseini, M.; Doulati Ardejani, F.; Amini, M.; Ebrahimi, L. The spatial assessment of acid mine drainage potential within a low-grade ore dump: The role of preferential flow paths. *Environ. Earth Sci.* **2019**, *79*, 28. [CrossRef]

255. Nichol, C.; Smith, L.; Beckie, R. Field-scale experiments of unsaturated flow and solute transport in a heterogeneous porous medium. *Water Resour. Res.* **2005**, *41*, 1–11. [CrossRef]

256. Eriksson, N.; Gupta, A.; Destouni, G. Comparative analysis of laboratory and field tracer tests for investigating preferential flow and transport in mining waste rock. *J. Hydrol.* **1997**, *194*, 143–163. [CrossRef]

257. Sinclair, S.A.; Pham, N.; Amos, R.T.; Sego, D.C.; Smith, L.; Blowes, D.W. Influence of freeze–thaw dynamics on internal geochemical evolution of low sulfide waste rock. *Appl. Geochem.* **2015**, *61*, 160–174. [CrossRef]

258. Langman, J.B.; Blowes, D.W.; Amos, R.T.; Atherton, C.; Wilson, D.; Smith, L.; Sego, D.C.; Sinclair, S.A. Influence of a tundra freeze-thaw cycle on sulfide oxidation and metal leaching in a low sulfur, granitic waste rock. *Appl. Geochem.* **2017**, *76*, 9–21. [CrossRef]

259. Kurylyk, B.L.; Watanabe, K. The mathematical representation of freezing and thawing processes in variably-saturated, non-deformable soils. *Adv. Water Resour.* **2013**, *60*, 160–177. [CrossRef]

260. Birkham, T.K.; Hendry, M.J.; Wassenaar, L.I.; Mendoza, C.A.; Lee, E.S. Characterizing Geochemical Reactions in Unsaturated Mine Waste-Rock Piles Using Gaseous O_2, CO_2, $^{12}CO_2$, and $^{13}CO_2$. *Environ. Sci. Technol.* **2003**, *37*, 496–501. [CrossRef]

261. Elberling, B.; Nicholson, R.V.; Scharer, J.M. A combined kinetic and diffusion model for pyrite oxidation in tailings: A change in controls with time. *J. Hydrol.* **1994**, *157*, 47–60. [CrossRef]

262. Pantelis, G.; Ritchie, A.I.M. Rate-limiting factors in dump leaching of pyritic ores. *Appl. Math. Model.* **1992**, *16*, 553–560. [CrossRef]

263. Massmann, J.; Farrier, D. Effects of atmospheric pressures on gas transport in the vadose zone. *Water Resour. Res.* **1992**, *28*, 777–791. [CrossRef]

264. Amos, R.T.; Blowes, D.W.; Smith, L.; Sego, D.C. Measurement of Wind-Induced Pressure Gradients in a Waste Rock Pile. *Vadose Zone J.* **2009**, *8*, 953–962. [CrossRef]

265. Lahmira, B.; Lefebvre, R.; Hockley, D.; Phillip, M. Atmospheric Controls on Gas Flow Directions in a Waste Rock Dump. *Vadose Zone J.* **2014**, *13*. [CrossRef]

266. Pantelis, G.; Ritchie, A.I.M.; Stepanyants, Y.A. A conceptual model for the description of oxidation and transport processes in sulphidic waste rock dumps. *Appl. Math. Model.* **2002**, *26*, 751–770. [CrossRef]

267. Collin, M.; Rasmuson, A. A comparison of gas diffusivity models for unsaturated porous media. *Soil Sci. Soc. Am. J.* **1988**, *52*, 1559–1565. [CrossRef]

268. Millington, R.J.; Quirk, J.P. Permeability of porous solids. *Trans. Faraday Soc.* **1961**, *57*, 1200–1207. [CrossRef]

269. Wang, T.; Huang, Y.; Chen, X.; Chen, X. Using grain-size distribution methods for estimation of air permeability. *Groundwater* **2016**, *54*, 131–142. [CrossRef]

270. Rohwerder, T.; Schippers, A.; Sand, W. Determination of reaction energy values for biological pyrite oxidation by calorimetry. *Thermochim. Acta* **1998**, *309*, 79–85. [CrossRef]

271. Sracek, O.; Choquette, M.; Gélinas, P.; Lefebvre, R.; Nicholson, R. V Geochemical characterization of acid mine drainage from a waste rock pile, Mine Doyon, Québec, Canada. *J. Contam. Hydrol.* **2004**, *69*, 45–71. [CrossRef]

272. Tan, Y.; Ritchie, A.I.M. In situ determination of thermal conductivity of waste rock dump material. *Water. Air. Soil Pollut.* **1997**, *98*, 345–359. [CrossRef]

273. Ning, L.; Zhang, Y. Onset of thermally induced gas convection in mine wastes. *Int. J. Heat Mass Transf.* **1997**, *40*, 2621–2636. [CrossRef]

274. Gou, W.; Parizek, R.R.; Rose, A.W. The role of thermal convection in resupplying O_2 to strip coal-mine spoil. *Soil Sci.* **1994**, *158*, 47–55. [CrossRef]

275. Anterrieu, O.; Chouteau, M.; Aubertin, M. Geophysical characterization of the large-scale internal structure of a waste rock pile from a hard rock mine. *Bull. Eng. Geol. Environ.* **2010**, *69*, 533–548. [CrossRef]

276. Dobriyal, P.; Qureshi, A.; Badola, R.; Hussain, S.A. A review of the methods available for estimating soil moisture and its implications for water resource management. *J. Hydrol.* **2012**, *458*, 110–117. [CrossRef]

277. Appels, W.M.; Ireson, A.M.; Barbour, S.L. Impact of bimodal textural heterogeneity and connectivity on flow and transport through unsaturated mine waste rock. *Adv. Water Resour.* **2018**, *112*, 254–265. [CrossRef]

278. Babaeian, E.; Sadeghi, M.; Jones, S.B.; Montzka, C.; Vereecken, H.; Tuller, M. Ground, proximal, and satellite remote sensing of soil moisture. *Rev. Geophys.* **2019**, *57*, 530–616. [CrossRef]

279. Nichol, C.; Smith, L.; Beckie, R. Time domain reflectrometry measurements of water content in coarse waste rock. *Can. Geotech. J.* **2003**, *40*, 137–148. [CrossRef]

280. Smith, L.J.D.; Moncur, M.C.; Neuner, M.; Gupton, M.; Blowes, D.W.; Smith, L.; Sego, D.C. The Diavik waste rock project: Design, construction, and instrumentation of field-scale experimental waste-rock piles. *Appl. Geochem.* **2013**, *36*, 187–199. [CrossRef]

281. Dimech, A.; Chouteau, M.; Aubertin, M.; Bussière, B.; Martin, V.; Plante, B. Three-dimensional time-lapse geoelectrical monitoring of water infiltration in an experimental mine waste rock pile. *Vadose Zone J.* **2019**, *18*. [CrossRef]

282. Dubuc, J.; Pabst, T.; Aubertin, M. An assessment of the hydrogeological response of the flow control layer installed on the experimental waste rock pile at the lac Tio mine. In Proceedings of the GeoOttawa 2017—70th Canadian Geotechnical Conference, Ottawa, ON, Canada, 1–4 October 2017.

283. Keller, J.; Busker, L.; Milczarek, M.; Rice, R.; Williamson, M. Monitoring of the geochemical evolution of waste rock facilities at Newmont's Phoenix Mine. In Proceedings of the VI International Seminar on Mine Closure, Lake Louise, AB, Canada, 18–21 September 2011.

284. Poisson, J.; Chouteau, M.; Aubertin, M.; Campos, D. Geophysical experiments to image the shallow internal structure and the moisture distribution of a mine waste rock pile. *J. Appl. Geophys.* **2009**, *67*, 179–192. [CrossRef]

285. Van Dam, R.; Gutierrez, L.; Mclemore, V.; Wilson, G.; Hendrickx, J.; Walker, B. Near Surface Geophysics for the Structural Analysis of a Mine Rock Pile, Northern New Mexico. *J. Am. Soc. Min. Reclam.* **2005**, *2005*, 1178–1201. [CrossRef]

286. Liu, X.; Chen, J.; Cui, X.; Liu, Q.; Cao, X.; Chen, X. Measurement of soil water content using ground-penetrating radar: A review of current methods. *Int. J. Digit. Earth* **2019**, *12*, 95–118. [CrossRef]

287. Power, C.; Tsourlos, P.; Ramasamy, M.; Nivorlis, A.; Mkandawire, M. Combined DC resistivity and induced polarization (DC-IP) for mapping the internal composition of a mine waste rock pile in Nova Scotia, Canada. *J. Appl. Geophys.* **2018**, *150*, 40–51. [CrossRef]

288. Wu, R.; Martin, V.; McKenzie, J.; Broda, S.; Bussière, B.; Aubertin, M.; Kurylyk, B.L. Laboratory-scale assessment of a capillary barrier using fibre optic distributed temperature sensing (FO-DTS). *Can. Geotech. J.* **2020**, *57*, 115–126. [CrossRef]

289. Li, Y.; Topal, E.; Ramazan, S. Optimising the long-term mine waste management and truck schedule in a large-scale open pit mine. *Min. Technol.* **2016**, *125*, 35–46. [CrossRef]

290. Seigneur, N.; Ulrich Mayer, K.; Steefel, C.I. Reactive transport in evolving porous media. In *Reviews in Mineralogy and Geochemistry*; Mineralogical Society: Twickenham, UK, 2019; pp. 197–238.

291. Beckingham, L.E.; Mitnick, E.H.; Steefel, C.I.; Zhang, S.; Voltolini, M.; Swift, A.M.; Yang, L.; Cole, D.R.; Sheets, J.M.; Ajo-Franklin, J.B.; et al. Evaluation of mineral reactive surface area estimates for prediction of reactivity of a multi-mineral sediment. *Geochim. Cosmochim. Acta* **2016**, *188*, 310–329. [CrossRef]

292. Steefel, C.I.; Lichtner, P.C. Multicomponent reactive transport in discrete fractures: I. Controls on reaction front geometry. *J. Hydrol.* **1998**, *209*, 186–199. [CrossRef]

293. Wunderly, D.M.; Blowes, D.W.; Frind, O.E.; Ptacek, C. Sulfide mineral oxidation and subsequent reactive transport of oxidation products in mine tailings impoundments: A numerical model. *Water Resour. Res.* **1996**, *32*, 3173–3187. [CrossRef]

294. Maest, A.S.; Kuipers, J.R.; Travers, C.L.; Atkins, D.A. Predicting Water Quality at Hardrock Mines; 90 pages, Kuipers & Associates and Buka Environmental. Available online: http://pebblescience.org/Pebble-Mine/acid-drainage-pdfs/PredictionsReportFinal.pdf (accessed on 1 July 2020).

295. Box, G.E.P. Science and statistics. *J. Am. Stat. Assoc.* **1976**, *71*, 791–799. [CrossRef]
296. Muniruzzaman, M.; Kauppila, P.M.; Karlsson, T. *Water Quality Prediction of Mining Waste Facilities Based on Predictive Models*; GTK Open File Research Report 16/2018; GTK: Espoo, Finland, 2018; Available online: http://tupa.gtk.fi/raportti/arkisto/16_2018.pdf (accessed on 1 August 2020).
297. Beckie, R. Analysis of Scale Effects in Large-Scale Solute-Transport Models. In *Scale Dependence and Scale Invariance in Hydrology*; Sposito, G., Ed.; Cambridge University Press: Cambridge, UK, 1998; pp. 314–334. ISBN 9780521571258.
298. Malmström, M.E.; Destouni, G.; Banwart, S.A.; Strömberg, B.H.E. Resolving the scale-dependence of mineral weathering rates. *Environ. Sci. Technol.* **2000**, *34*, 1375–1378. [CrossRef]
299. Morin, K.A.; Hutt, N.M. *Environmental Geochemistry of Minesite Drainage: Practical Theory and Case Studies, Digital Edition*; MDAG Publishing: Surrey, BC, Canada, 2001.
300. Molins, S.; Knabner, P. Multiscale approaches in reactive transport modeling. *Rev. Mineral. Geochem.* **2019**, *85*, 27–48. [CrossRef]
301. Molson, J.; Aubertin, M.; Bussière, B. Reactive transport modelling of acid mine drainage within discretely fractured porous media: Plume evolution from a surface source zone. *Environ. Model. Softw.* **2012**, *38*, 259–270. [CrossRef]
302. Pabst, T.; Molson, J.; Aubertin, M.; Bussière, B. Reactive transport modelling of the hydro-geochemical behaviour of partially oxidized acid-generating mine tailings with a monolayer cover. *Appl. Geochem.* **2017**, *78*, 219–233. [CrossRef]
303. Tomiyama, S.; Igarashi, T.; Tabelin, C.B.; Tangviroon, P.; Hiroyuki, I. Modeling of the groundwater flow system in excavated areas of an abandoned mine. *J. Cont. Hydr.* **2020**, *230*, 103617. [CrossRef]
304. Mayer, K.U.; Frind, E.O.; Blowes, D.W. Multicomponent reactive transport modeling in variably saturated porous media using a generalized formulation for kinetically controlled reactions. *Water Resour. Res.* **2002**, *38*, 13–21. [CrossRef]
305. Parkhurst, D.L.; Appelo, C.A.J. User's guide to PHREEQC (Version 2): A computer program for speciation, batch-reaction, one-dimensional transport, and inverse geochemical calculations. USGS Water Resources Investigations Report 99-4259. Available online: https://pubs.er.usgs.gov/publication/wri994259 (accessed on 1 July 2020).
306. van der Lee, J.; De Windt, L.; Lagneau, V.; Goblet, P. Module-oriented modeling of reactive transport with HYTEC. *Comput. Geosci.* **2003**, *29*, 265–275. [CrossRef]
307. Xu, T.; Sonnenthal, E.; Spycher, N.; Pruess, K. *TOUGHREACT User's Guide: A Simulation Program for Non-Isothermal Multiphase Reactive Geochemical Transport in Variable Saturated Geologic Media*; University of California: Berkeley, CA, USA, 2004. [CrossRef]
308. Steefel, C.I.; Appelo, C.A.J.; Arora, B.; Jacques, D.; Kalbacher, T.; Kolditz, O.; Lagneau, V.; Lichtner, P.C.; Mayer, K.U.; Meeussen, J.C.L.; et al. Reactive transport codes for subsurface environmental simulation. *Comput. Geosci.* **2015**, *19*, 445–478. [CrossRef]
309. Raymond, K.; Seigneur, N.; Su, D.; Plante, B.; Poaty, B.; Bussiere, B.; Mayer, K. Numerical modeling of a laboratory-scale waste rock pile featuring an engineered cover system. *Minerals* **2020**, *10*, 652. [CrossRef]
310. Li, L.; Maher, K.; Navarre-Sitchler, A.; Druhan, J.; Meile, C.; Lawrence, C.; Moore, J.; Perdrial, J.; Sullivan, P.; Thompson, A.; et al. Expanding the role of reactive transport models in critical zone processes. *Earth-Sci. Rev.* **2017**, *165*, 280–301. [CrossRef]
311. Wösten, J.H.M.; van Genuchten, M.T. Using texture and other soil properties to predict the unsaturated soil hydraulic functions. *Soil Sci. Soc. Am. J.* **1988**, *52*, 1762–1770. [CrossRef]
312. Arya, L.M.; Leij, F.J.; Shouse, P.J.; van Genuchten, M.T. Relationship between the Hydraulic Conductivity Function and the Particle-Size Distribution. *Soil Sci. Soc. Am. J.* **1999**, *63*, 1063–1070. [CrossRef]
313. Mishra, S.; Parker, J.C.; Singhal, N. Estimation of soil hydraulic properties and their uncertainty from particle size distribution data. *J. Hydrol.* **1989**, *108*, 1–18. [CrossRef]

 *minerals*

Article

Effective Treatment of Acid Mine Drainage with Microbial Fuel Cells: An Emphasis on Typical Energy Substrates

Chenbing Ai [1,2,3], Zhang Yan [2,4,5], Shanshan Hou [2,4], Xiaoya Zheng [2,4], Zichao Zeng [2], Charles Amanze [2,4], Zhimin Dai [6,7], Liyuan Chai [1,3], Guanzhou Qiu [2,4] and Weimin Zeng [2,4,*]

[1] School of Metallurgy and Environment, Central South University, Changsha 410083, Hunan, China; chenbingai@csu.edu.cn (C.A.); chailiyuan@csu.edu.cn (L.C.)
[2] School of Minerals Processing and Bioengineering, Central South University, Changsha 410083, Hunan, China; 201910107647@mail.scut.edu.cn (Z.Y.); 185611026@csu.edu.cn (S.H.); zhengxiaoya@csu.edu.cn (X.Z.); zzc1073790321@csu.edu.cn (Z.Z.); charles.amanze@csu.edu.cn (C.A.); qgz@csu.edu.cn (G.Q.)
[3] Chinese National Engineering Research Center for Control and Treatment of Heavy Metal Pollution, Central South University, Changsha 410083, Hunan, China
[4] Key Laboratory of Biometallurgy of Ministry of Education, Central South University, Changsha 410083, China
[5] College of Environmental Science and Engineering, Fujian Key Laboratory of Pollution Control & Resource Reuse, Fujian Normal University, Fuzhou 350007, Fujian Province, China
[6] Central South Water Science and Technology Co., LTD, Changsha 410083, China; daizhimin0414@163.com
[7] National City Supply Water Quality Monitoring Network Changsha Monitoring Station, Changsha 410083, China
*   Correspondence: zengweimin1024@126.com; Tel.: +86-0731-88877472

Received: 10 April 2020; Accepted: 10 May 2020; Published: 15 May 2020

Abstract: Acid mine drainage (AMD), characterized by a high concentration of heavy metals, poses a threat to the ecosystem and human health. Bioelectrochemical system (BES) is a promising technology for the simultaneous treatment of organic wastewater and recovery of metal ions from AMD. Different kinds of organic wastewater usually contain different predominant organic chemicals. However, the effect of different energy substrates on AMD treatment and microbial communities of BES remains largely unknown. Here, results showed that different energy substrates (such as glucose, acetate, ethanol, or lactate) affected the startup, maximum voltage output, power density, coulombic efficiency, and microbial communities of the microbial fuel cell (MFC). Compared with the maximum voltage output (55 mV) obtained by glucose-fed-MFC, much higher maximum voltage output (187 to 212 mV) was achieved by MFCs fed individually with other energy substrates. Acetate-fed-MFC showed the highest power density (195.07 mW/m^2), followed by lactate (98.63 mW/m^2), ethanol (52.02 mW/m^2), and glucose (3.23 mW/m^2). Microbial community analysis indicated that the microbial communities of anodic electroactive biofilms changed with different energy substrates. The *unclassified_f_Enterobacteriaceae* (87.48%) was predominant in glucose-fed-MFC, while *Geobacter* species only accounted for 0.63%. The genera of *Methanobrevibacter* (23.70%), *Burkholderia-Paraburkholderia* (23.47%), and *Geobacter* (11.90%) were the major genera enriched in the ethanol-fed-MFC. *Geobacter* was most predominant in MFC enriched by lactate (45.28%) or acetate (49.72%). Results showed that the abundance of exoelectrogens *Geobacter* species correlated to electricity-generation capacities of electroactive biofilms. Electroactive biofilms enriched with acetate, lactate, or ethanol effectively recovered all Cu^{2+} ion (349 mg/L) of simulated AMD in a cathodic chamber within 53 h by reduction as Cu^0 on the cathode. However, only 34.65% of the total Cu^{2+} ion was removed in glucose-fed-MFC by precipitation with anions and cations rather than Cu^0 on the cathode.

Keywords: acid mine drainage; copper recovery; microbial fuel cell; electricity generation; microbial community

---

1. Introduction

Acid mine drainage (AMD) is one typical pollutant of water in many countries that have historic or current mining activities. Sulfide minerals present in mining wastes (e.g., open pits, mining waste rock, and tailings) are inevitably oxidized to form AMD when exposed to water, air, and chemolithotrophic acidophiles [1–3]. AMD is characterized by a high acidity and high concentration of toxic heavy metals/metalloids [2]. Thus, if it is not managed properly, AMD can undoubtedly cause considerable water and soil contamination, massive biodiversity loss in the aquatic ecosystem, and severe health impacts on nearby communities [4]. In order to achieve the long-term environmental sustainability regarding mining activities, effective and efficient technologies that can tackle the remediation of AMD are highly required.

Alkaline neutralizing chemicals, such as limestone and slaked lime, are conventionally adopted to treat AMD by decreasing the extreme acidity and precipitating the dissolved various poisonous metals/metalloids as hydroxides [5]. Despite effective remediation, the large volumes of sludges resulted from precipitation containing heavy metals/metalloids, which are categorized as hazardous materials and need further safe disposal. Other active and passive remediation technologies, such as bioremediation, phytoremediation, electrodialysis, wetlands, and adsorption, are also commonly used to treat AMD [4]. However, all those technologies have the drawbacks of either low remediation efficiency or high cost. Besides, some of those technologies generally produce new wastes (e.g., sludge, brines, and spent media), which require further treatment.

In fact, the high concentration of dissolved metals in AMD can be recovered by the bioelectrochemical system (BES) as valuable products to offset the cost of treatment. Therefore, the bioelectrochemical system is a promising technology for the treatment of AMD. The bioelectrochemical system is a special biological treatment process of sewage wastewater, which mainly utilizes the catalytic activity of electroactive microorganisms [6]. Under anaerobic conditions, electroactive microorganisms degrade organic pollutants and transmit electrons through external circuits to generate electricity [7]. As a new form of biomass energy utilization and pollutant removal, bioelectrochemical systems have received extensive attention due to their non-polluting characteristics [8,9]. Compared with a single strain, the electrogenic microbial consortium has many more advantages, such as higher electricity generation efficiency, a wider range of organic substrate, and higher coulombic efficiency [10,11]. Therefore, the enrichment and acclimation of electrogenic microbial consortium from environmental samples is a conventional and effective way to increase the power density of bioelectrochemical systems. Previous studies have shown that different energy substrates used to enrich electrogenic consortium can modulate the microbial community of electroactive biofilms [12,13]. However, studies focusing on the effects of typical energy substrates on the capacities of AMD treatment and microbial communities of BES are not available.

The purposes of this study were to compare the impacts of four typical energy substrates on the performance, microbial communities, and capacities of AMD treatment of enriched electroactive biofilms. Here, single-chamber microbial fuel cells were inoculated with anaerobic sludge and fed with glucose, acetate, ethanol, or lactate, respectively, as energy substrates to enrich electroactive biofilms. The performance of enriched electroactive biofilms was evaluated after the maximum voltage output was reached. The microbial communities of enriched electroactive biofilms were analyzed by high throughput sequence technology. The AMD treatment capacities of enriched electroactive biofilms were evaluated in dual-chamber microbial fuel cells. In addition, the mechanism for copper removal on the surface of the cathode was explored. These results indicated that the effects of organic chemical (that is usually contained in organic wastewater) on the enrichment of electroactive biofilm should be first evaluated in order to obtain an efficiently simultaneous treatment of organic wastewater and AMD.

2. Materials and Methods

2.1. The Configuration of Microbial Fuel Cell (MFC) Reactors

A single-chamber MFC reactor was adopted to enrich electroactive biofilms (Figure 1A). The cube-shaped single-chamber MFC reactor with a cylindrical chamber (3 cm diameter × 4 cm length) was made of Perspex. Each MFC reactor (with a working volume 28 mL) consisted of a carbon brush (1.5 cm in radius × 3 cm in length) as anode and a carbon cloth with disk shape (projected surface area of 7.07 cm^2) as a cathode. To save cost, the expensive Pt catalyst usually used to coat cathode was not adopted in this study [14]. The anode and cathode were connected by an external resistance of 1000 Ω by titanium wire. In order to remove contaminants on the surface, both carbon brush and carbon cloth were soaked overnight in acetone, followed by washing with distilled water and baked in a muffle furnace at 450 °C for 30 min. The dual-chamber MFC reactor was adopted to treat the simulated acid mine drainage (Figure 1B). The dual-chamber MFC reactor consisted of an anode chamber (28 mL) and a cathode chamber (15 mL). The anodes enriched in these single-chamber MFCs were then used in the double-chamber MFC. The two chambers were separated by an anion exchange membrane (Hangzhou Grion Environmental Technology, Co., Ltd, Hangzhou, China). The cathode and the electroactive anode of the dual-chamber MFC reactor were connected by an external resistance of 10 Ω. The cathode of the dual-chamber MFC reactor was made of carbon cloth with a rectangle shape (2.5 cm in length × 0.9 cm in width) immersed in simulated AMD.

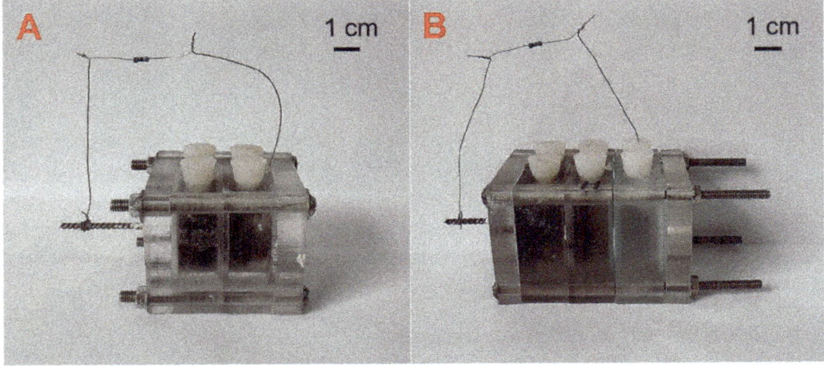

Figure 1. The single-chamber (**A**) and double-chamber (**B**) microbial fuel cell (MFC) reactors.

2.2. Startup and Operation of MFC

The single-chamber MFCs were inoculated with anaerobic sludge obtained from a municipal wastewater treatment plant. Abiotic single-chamber MFCs without the inoculum of anaerobic sludge were set up. Duplicate single-chamber MFC reactors were set up for each energy substrate. The medium used to enrich electroactive biofilms contained 20 mM energy substrate (glucose, acetate, ethanol, or lactate, respectively), trace element solution (100 μL/L), and Wolfe's vitamins (0.5 mL/L) in 50 mM phosphate buffer (4.56 g/L, Na_2HPO_4; 2.45 g/L, NaH_2PO_4; 0.31 g/L, NH_4Cl; 0.13 g/L, KCl; 0.02 g/L, $CaCl_2$), as modified from previous study [15]. The trace element solution contained the following chemicals per liter: 3.00 g $MgSO_4$, 0.25 g $FeSO_4 \cdot 7H_2O$, 0.15 g $ZnCl_2$, 0.60 g $MnSO_4 \cdot H_2O$, 0.01 g H_3BO_3, 0.01 g $CuSO_4 \cdot 2H_2O$, 0.03 g $NiCl_2 \cdot 6H_2O$, 0.03 g Na_2MoO_4, 0.20 g $CoCl_2$, 0.03 g $Na_2WO_4 \cdot 2H_2O$, and 0.15g $KAl(SO_4)_2 \cdot 12H_2O$. All the chemicals used in this study were analytic pure (Sinopharm Chemical Reagent Co., Ltd, Shanghai, China). These MFCs were operated in a fed-batch mode in a temperature-controlled incubator (30 °C). The medium was replaced once the output voltage of MFC declined below 20 mV. The medium used to maintain the growth of enriched bioelectroactive biofilms in the anode chamber of these dual-chamber MFCs was identical with that used for the single-chamber MFCs. The cathode chamber was fed with the simulated AMD that was diluted from the leachate of chalcopyrite bioleaching with acid water (pH 1.80) [16]. The simulated AMD mainly contained 348.87 mg/L Cu^{2+}, 45.06 mg/L Fe^{3+}, and 7.03 mg/L Fe^{2+} with a pH value of 1.80. The Cu^{2+} and Fe^{3+} could be served as terminal electron acceptors. Abiotic double-chamber MFCs without the enriched electroactive biofilm were set up. Duplicate double-chamber MFC reactors containing the electroactive biofilms enriched with each of these different energy substrates were set up to treat the simulated AMD.

2.3. Analysis and Calculations

The voltage across the 1000 Ω external resistance of single-chamber MFCs was recorded every 50 s by the data acquisition unit (ADAM-4017 Analog Input Model, Advantech Co., Ltd, Shenzhen, China) connected to the computer. The power density and polarization curve of single-chamber MFCs were analyzed and calculated, as described in a previous study [17]. The power density was normalized to the geometrical surface area of the anode. Coulombic efficiency of single-chamber MFCs was calculated according to a previous study [18]. Electrochemical impedance spectroscopy (EIS) was applied to determine the internal resistance of these single-chamber MFCs enriched with different energy substrates using a potentiostat (Gamry reference 600+ workstation, Philadelphia, Pennsylvania, USA). The EIS measurements were conducted using a three-electrode configuration, with a saturated Ag/AgCl reference electrode and the anode serving as the working electrode and the cathode as the counter electrode. For each experimental condition, the EIS measurement was conducted in the frequency range from 1000 kHz to 0.01 Hz with an AC amplitude of 5 mV and analyzed by the software of Zview. The concentration of Fe^{2+} and Fe^{3+} in the cathode chamber was determined using the phenanthroline method [19]. The concentration of Cu^{2+} was quantified with bis-cyclohexanone oxalyldihydrazone (BCO) [20]. The pH value of catholyte was measured with a pH-meter (SJ-4A, Leichi, Shanghai, China).

Scanning electron micrograph (SEM, JSM-6490LV, JEOL, Tokyo, Japan) was adopted to observe the enriched electroactive biofilms and the structure of cathode surfaces. The energy dispersive X-ray spectrometry (EDXS; Elect super, EDAX AMETEK, Kleve, Germany) equipped for SEM was used to examine the morphologies and compositions of the deposits on cathode electrodes after the treatment of AMD. The products deposited on the cathode electrode were determined by the X-ray powder diffraction (XRD) (D8 Advance, Bruker Corporation, Karlsruhe, Germany), in which data were recorded in the 2θ range of 10 to 80 degree with a step of 0.02 degree.

2.4. Genomic DNA Extraction and MiSeq Sequencing of Bioelectroactive Biofilms

The electroactive biofilms enriched with different energy substrates in MFCs with stable output voltages were sampled to extract the total genomic DNA by the DNeasy PowerSoil DNA Isolation Kit (QIAGEN, Chatsworth, CA, USA). Illumina adapter sequence, together with the universal primer pair 515FmodF (5'-GTGYCAGCMGCCGCGGTAA-3') and 806RmodR (5'-GGACTACNVGGGTWTCTAAT-3'), were used to amplify the V4 region of the bacterial and archaeal 16S rDNA genes. PCR amplification was performed on Applied Biosystems GeneAmp® 9700 thermal cycler (ABI Inc., Foster City, CA, USA). PCR system (25 μL) consisted of 1 μL of template DNA, 1 μL (10 nM) of each primer, 9.5 μL of DNase-free deionized water, and 12.5 μL of 2× Taq PCR Master Mix (TransGen, Beijing, China). Triplicate amplifications for each genomic DNA sample were amplified and blended to minimize potential biases of amplification, which were separated by agarose gel electrophoresis (2%, w/v) and recovered using AxyPrep DNA gel extraction kit (Axygen Scientific Inc., Union City, CA, USA). The concentration of the recovered PCR products was measured using QuantiFluor™-ST Fluorometer (Promega Corporation, Madison, WI, USA). Sequencing libraries were prepared and sequenced by the Illumina MiSeq platform with the sequencing strategy PE250 (Shanghai Majorbio Bio-pharm Technology Co., Ltd, Shanghai, China).

The raw data of 16S rRNA gene sequences from MiSeq sequencing was in FASTQ format. The Illumina adapter and other specific sequences were trimmed before the following process. Then, the pair-end reads with at least 10 bp overlap, and lower than 5% mismatches were merged using the Fast Length Adjustment of SHort reads (FLASH) software [21]. The sequences shorter than 240 bp, chimeric sequences, and low-quality sequences were filtered, trimmed, and removed [22]. Operational taxonomic units (OTUs) were obtained based on the threshold of 97% similarity by using UPARSE [23]. The taxonomy of OTU representative sequences was phylogenetically assigned to taxonomic classifications by the Ribosomal Database Project (RDP) classifier at the threshold of 70% for confidence based on the Bayesian algorithm [24]. Community richness, Ace and Shannon indices, and Chao1 richness estimates were obtained by MOTHUR analysis [25].

3. Results and Discussion

3.1. Effect of Different Energy Substrates on Single Chamber MFC Performance

Different energy substrates (i.e., glucose, acetate, ethanol, or lactate) affected the startup, maximum voltage output, power density, and coulombic efficiency of single-chamber microbial fuel cells (Figure 2). The output voltage of the MFCs enriched with lactate as an energy substrate began to be detectable only 40 h after the inoculation with anaerobic sludge (Figure 2A). However, in order to generate a detectable output voltage, 120, 210, or 220 h was required, respectively, for the MFCs enriched with ethanol, acetate, or glucose. Compared with the maximum voltage output (55 mV) obtained by glucose-fed-MFC, much higher maximum voltage output (187 to 212mV) was achieved by MFCs fed individually with the other three energy substrates. Around 400 h after the initial inoculation, the output voltage of each MFC reached the maximum. Thereafter, the output voltage could rapidly increase to the maximum value immediately after the removal of planktonic microorganisms by replenishing with growth medium containing each energy substrate (Figure 2A). This rapid recovery of maximum output voltage indicated that the current was mainly generated by the sessile microorganisms on the surface of the anode. Acetate-fed-MFC showed the highest power density (195.07 mW/m^2), followed by lactate (98.63 mW/m^2), ethanol (52.02 mW/m^2), and glucose (3.23 mW/m^2) (Figure 2B). As indicated by the polarization test, the output voltage of acetate-fed-MFC was much higher than those of other MFCs at different external resistance (Figure 2C). On the contrary, the output voltage of glucose-fed-MFC was the lowest (Figure 2C). Coulombic efficiencies of these MFCs were dependent on the energy substrates. The MFCs enriched with lactate had the highest coulombic efficiency (33.34%), followed by the MFCs enriched with ethanol (14.30%), acetate (12.53%), and glucose (1.98%). The lowest coulombic efficiency obtained by the glucose-fed-MFCs was consistent with the previous studies because the glucose is a fermentable substrate that can be utilized by diverse microorganisms besides the exoelectrogens enriched in the electroactive biofilms under the anaerobic condition [12,26].

The ohmic resistance and charge transfer resistances of these MFCs were obtained by electrochemical impedance spectroscopy (EIS) (Figure 3). As described in the previous study, the impedance at the high-frequency limit is the ohmic resistance, and the diameter of the semicircle is the charge transfer resistance [27]. The ohmic resistance of MFC containing the bioelectroactive biofilms enriched with glucose was 19.55 Ω. However, the MFCs containing the bioelectroactive biofilms enriched with acetate (2.81 Ω), lactate (3.39 Ω), and ethanol (5.31 Ω) had much lower ohmic resistance (Figure 3). The charge transfer resistances of the MFCs containing different bioelectroactive biofilms were also dependent on the energy substrate used for enrichment. The charge transfer resistances of the MFCs containing different bioelectroactive biofilms enriched with glucose, acetate, ethanol, and lactate were 33.67 Ω, 7.39 Ω, 15.00 Ω, and 15.38 Ω, respectively. Discrepancy regarding startup, maximum voltage output, power density, coulombic efficiency, and charge transfer resistances of single-chamber MFCs enriched with different energy substrate implied that the electroactive biofilms enriched on the surface of anode were different in terms of the microbial community.

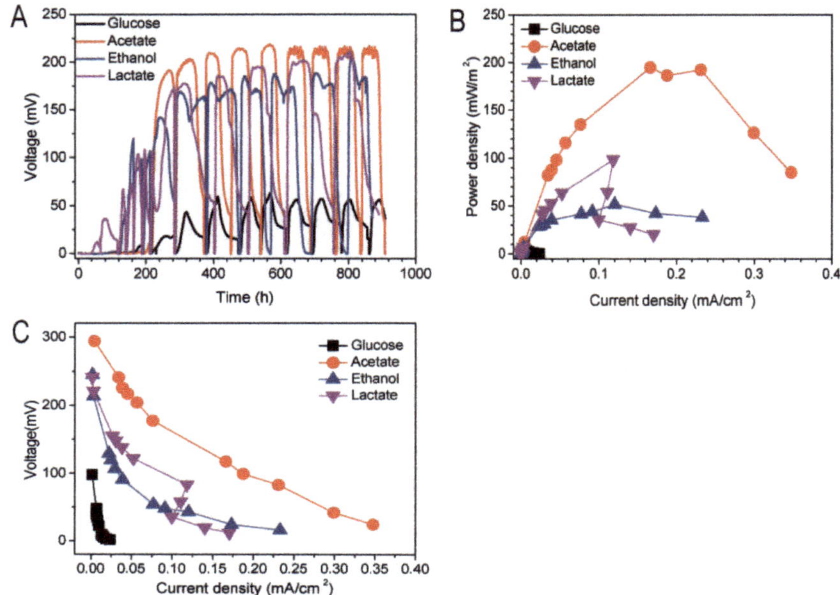

Figure 2. The output voltage (**A**), power density (**B**), and polarization curve (**C**) of MFCs containing bioelectroactive biofilms enriched with different energy substrates.

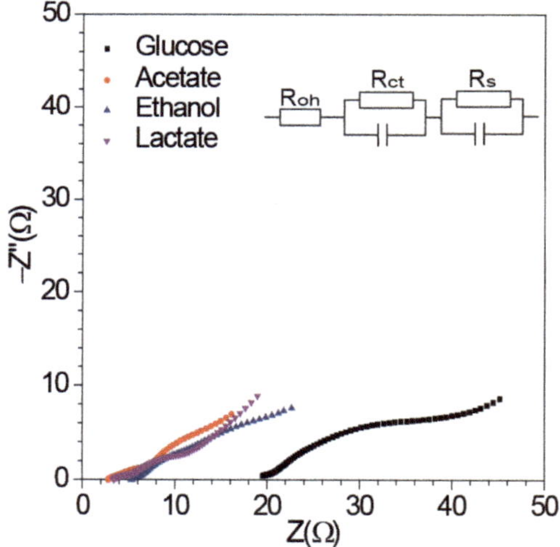

Figure 3. Electrochemical impedance spectroscopy (EIS) analysis of the MFCs containing the bioelectroactive biofilms enriched with different energy substrates.

3.2. Microbial Community of Anodic Bioelectroactive Biofilms

In contrast to the abiotic control, electroactive biofilms were enriched on the surface of the anode of MFCs when they reached the maximum output voltage, as revealed by SEM analysis (Figure 4). The existence of electroactive biofilms on the surface of anode demonstrated the importance of electroactive biofilms for the generation of electricity. These electroactive biofilms consisted of microorganisms with different cell morphologies. This indicated the diversity of electroactive biofilm regarding the microbial community.

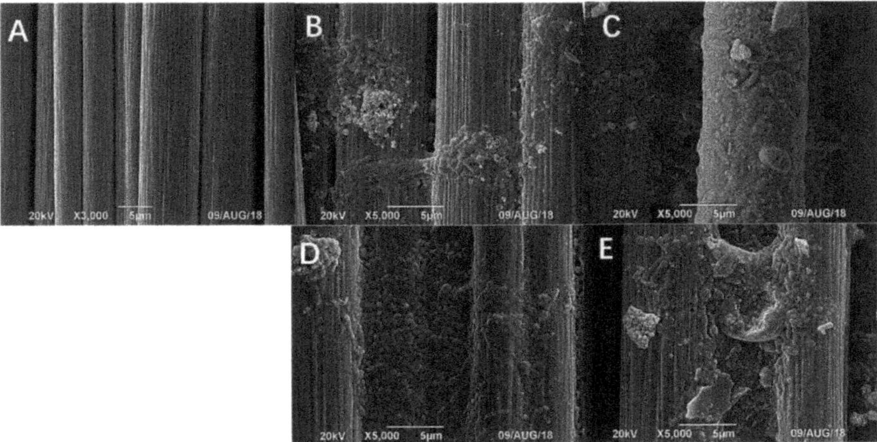

Figure 4. SEM of the bioelectroactive biofilms enriched with different energy substrates. (**A**: abiotic control; **B**: glucose; **C**: acetate; **D**: ethanol; **E**: lactate).

In order to investigate the microbial community of electroactive biofilms enriched by different energy substrates (i.e., glucose, acetate, ethanol, or lactate), approximately 32,430 to 97,701 high-quality sequencing reads were obtained from each sample (Table 1). A total number of 710 OTU was detected in the inoculated anaerobic sludge (Figure 5). During the enrichment of the bioelectroactive biofilms process, there was a succession of microorganisms at the OTU level. After MFCs reached the stable maximum output voltage, there were 590,482,286 and 205 OTUs in the bioelectroactive biofilms enriched by acetate, lactate, ethanol, and glucose, respectively (Figure 5). Both the microbial abundance and microbial diversity of these electroactive biofilms enriched by different energy substrates were less than that of the inoculated anaerobic sludge, as indicated by the Shannon index and Simpson index listed in Table 1. These data indicated that the microbial abundance and microbial diversity of these electroactive biofilms were dependent on the energy substrate.

Table 1. The α-diversity of enriched bioelectroactive biofilms.

Sample	Reads	Sobs	Shannon	Simpson	Ace	Chao	Coverage
Inoculum	40818	710	5.2069	0.0138	732.31	735.02	0.9988
Glucose	97701	205	0.7931	0.7633	363.79	347.38	0.9993
Acetate	45985	590	2.9848	0.2534	670.25	672.18	0.9975
Ethanol	43875	286	2.7412	0.1233	428.09	371.00	0.9981
Lactate	32430	482	2.9694	0.2130	589.26	586.79	0.9963

The most dominant phyla were *Proteobacteria*, *Bacteroidetes*, and *Saccharibacteria* in the inoculated anaerobic sludge (Figure 6). Both *Proteobacteria* and *Bacteroidetes* remained as the dominant phyla in these enriched electroactive biofilms. The proportion of *Proteobacteria* increased significantly, while the proportion of *Bacteroidetes* decreased remarkably in these electroactive biofilms (Figure 6A). *Firmicutes* was enriched as one of the dominant phyla in these electroactive biofilms. It was worth mentioning that *Euryarchaeota* was enriched in these electroactive biofilms, especially in the electroactive biofilms fed with ethanol as an energy substrate.

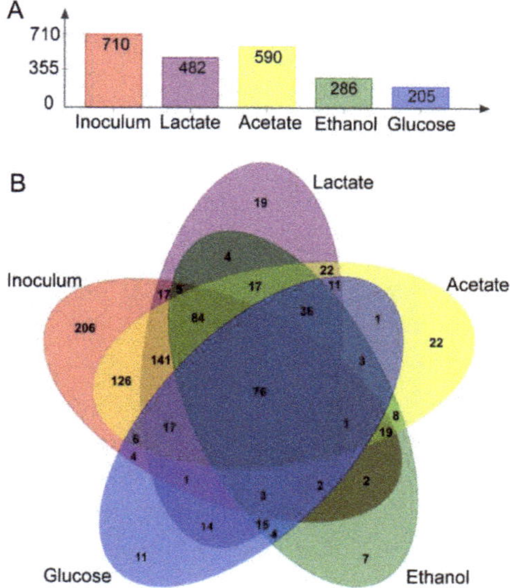

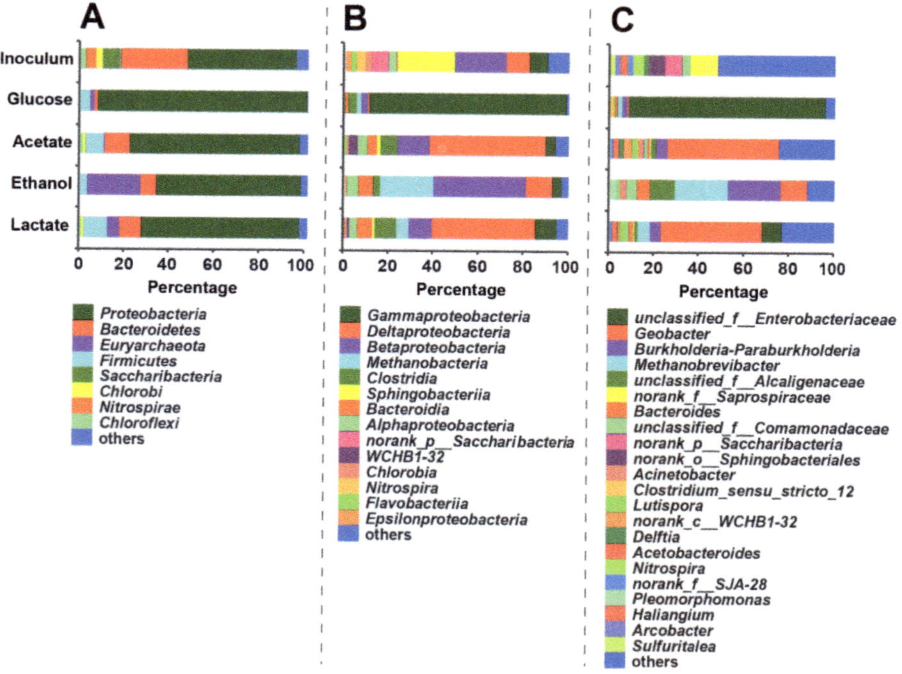

Figure 5. The number (**A**) and Venn diagram analysis (**B**) of the operational taxonomic unit (OTU) of these bioelectroactive biofilms enriched with different energy substrates.

Figure 6. Microbial community of the bioelectroactive biofilms enriched with different energy substrates at phylum (**A**), class (**B**), and genus (**C**) levels.

The major classes in the electroactive biofilms were different from that of the anaerobic sludge (Figure 6B). *Gammaproteobacteria*, *Deltaproteobacteria*, and *Betaproteobacteria* were the three major classes within the inoculated anaerobic sludge and the electroactive biofilms enriched, respectively, with acetate, ethanol, or lactate. However, only *Gammaproteobacteria* constituted as the major class of the electroactive biofilms enriched with glucose (87.70%). *Sphingobacteriia* (25.48%) and *norank_p_Saccharibacteria* (7.66%) were two major classes that existed in the inoculated anaerobic sludge, both of which were shifted as minor constituents in these electroactive biofilms.

The major genera in anodic electroactive biofilms were modulated by energy substrates (Figure 6C). The *unclassified_f_Enterobacteriaceae* (87.48%) was predominant in the glucose-fed-MFC, while *Geobacter* species only accounted for 0.63%. The genera of *Methanobrevibacter* (23.70%), *Burkholderia-Paraburkholderia* (23.47%), and *Geobacter* (11.90%) were the major genera enriched in the ethanol-fed-MFC. *Geobacter* was most predominant in the MFC enriched by lactate (45.28%) or acetate (49.72%), which corroborated with a previous study [28]. Results showed that the abundance of classic exoelectrogens *Geobacter* species correlated to the electricity-generation capacities of electroactive biofilms. It is worth mentioning that the *Euryarchaeota* was enriched in these electroactive biofilms, especially in the electroactive biofilms fed with ethanol as an energy substrate (Table 2). Recent studies have shown that quorum sensing (QS) plays an important role in shaping the dynamics of microbial community structure and enhancing the electron transfer process in the anodic electroactive biofilms of MFCs [29,30].

Table 2. The ratio of archaea species in the bioelectroactive biofilms enriched with different energy substrates.

Archaeal Genus	Percentage in Anode Biofilm (%)				
	Inoculum	Glucose	Acetate	Ethanol	Lactate
Methanobacterium	0	0.01	0	0.04	0.15
Methanobrevibacter	0	1.82	0	23.70	5.23
Methanosaeta	0	0	0.02	0	0.05
Methanosarcina	0	0	0.19	0	0
Methanomassiliicoccus	0	0	0.25	0	0.02

The expression of functional genes in either single strain or microbial consortium has been altered by various physicochemical parameters [16,31,32]. Therefore, it is necessary to identify and compare the important genes involved in the electron transfer for electricity generation of these electrochemical biofilms in MFCs by comparative metagenomic and transcriptomic analyses in the future. The extracellular polymeric substances (EPS) are important for the functional roles of single strain and consortium [30,33,34]. The EPS of electroactive biofilm contains proteins, glycoproteins, extracellular DNA, glycolipids, and humic substances [30]. Previous studies have shown that cytochrome proteins, pili, and nanowire in EPS are directly involved in electron transfer [30,35]. Characterization of the compositions and redox properties of the EPS of these enriched electrochemical biofilms will provide novel insights into the functional role of EPS in mediating electron transfer.

3.3. Contribution of Electroactive Biofilms on Anolyte's Chemical Oxygen Demand Removal and Catholyte's Copper Recovery

Different ratio of chemical oxygen demand (COD) was depleted in the anodic chamber for the electroactive biofilms enriched by glucose (51.32%), acetate (82.00%), ethanol (72.49%), or lactate (35.95%), respectively, in 53 h after replenishing with fresh growth medium for copper recovery in dual-chamber MFCs (Figure 7A). A high concentration of COD (1909 mg/L) was removed in the anolyte of MFC fed with glucose as an energy substrate. Considering the lowest electricity production in each batch, most of the COD removed in the anolyte of glucose-fed-MFC was ascribed to the anaerobic growth by non-electrogenic microorganisms. It is worth mentioning that the number of planktonic microorganisms in MFC fed with glucose was much higher than those in the MFCs fed with other energy substrates.

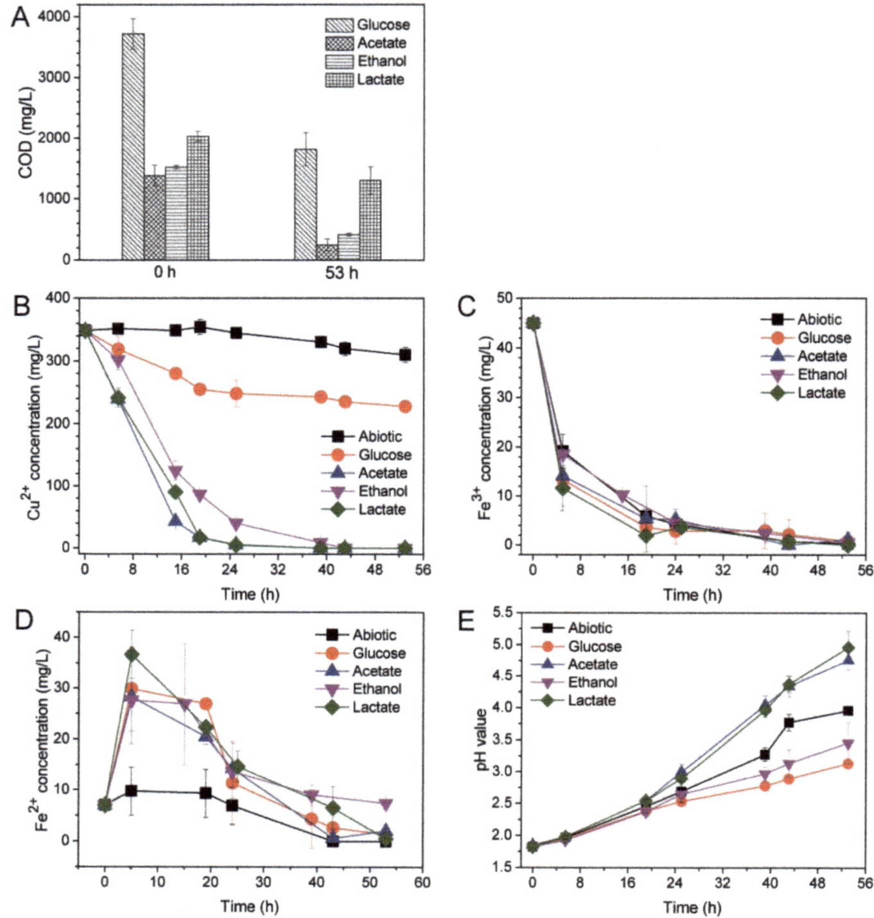

Figure 7. The changes in chemical oxygen demand (COD) in anode chamber (**A**), and cupric ion (**B**), ferric iron (**C**), ferrous iron (**D**), and pH value (**E**) of cathode chamber of the MFCs containing electroactive biofilms enriched with different energy substrates.

The dual-chamber MFCs containing the electroactive biofilms enriched with acetate, ethanol, or lactate, respectively, could effectively recover copper from the acid mine drainage (Figure 7B). The copper in the catholyte of these MFCs decreased significantly after the initiation of the treatment of AMD. At the 39th h, no detectable copper ion was found in catholyte of MFCs containing the electroactive biofilms enriched with acetate or lactate. At the 43rd h, the copper ion in the catholyte of MFC containing the electroactive biofilms fed by ethanol was also completely recovered. However, the dual-chamber MFC containing electroactive biofilms enriched with glucose was deficient in the recovery of copper (Figure 7B). Only part of the copper ion (34.65%) was removed at the 53rd h, with a high concentration of Cu^{2+} (228.00 mg/L) remaining in the catholyte. The high concentration of Cu^{2+} (310 mg/L) remained in the catholyte of abiotic control at the end of this experiment. Iron ions in the stimulated AMD were mainly Fe^{3+} (Figure 7C,D). The decrease of Fe^{3+} concentration in the catholyte of MFCs containing electroactive biofilms was partially ascribed to the bioelectrochemical reduction at the cathode to Fe^{2+} (Figure 7D). The decrease of iron ions in the catholyte of abiotic control probably resulted from the elevated pH value (Figure 7E). The pH values in catholyte of all these MFCs with electroactive biofilms were increased during the treatment of AMD. The increase in pH value was likely ascribed to the diffusion of anions from the anolyte across the anion exchange member and reacted with the protons in the catholyte. Therefore, the decrease of iron ions in the catholyte of MFCs with electroactive biofilms was also affected by the increased pH values.

3.4. Morphologies of Electrode and XRD Analysis

The color of cathodes of dual-chamber MFCs containing the electroactive biofilms enriched with acetate, ethanol, or lactate, respectively, turned from black to brown after 53 h of treatment of AMD (Figure 8). This phenomenon indicated the bioelectrochemical reduction of copper on the surface of the cathode. However, the color of the cathode of abiotic control and MFCs containing electroactive biofilms fed with glucose remained as black (Figure 8).

Figure 8. The cathodes of the MFCs after the treatment of simulated AMD. (**1:** abiotic; **2:** glucose; **3:** acetate; **4:** ethanol; **5:** lactate).

In order to better understand the copper recovery mechanism, the cathodes of dual-chamber MFCs after the treatment of AMD for 53 h were analyzed with SEM and XRD. The SEM micrographs of cathode surfaces of these MFCs containing the electroactive biofilms enriched with acetate, ethanol, or lactate were similar in terms of structure and morphology, which were different from that of the cathodes of abiotic control and the glucose-fed-MFCs (Figure 9). No deposit was observed on the cathodic surface of abiotic MFCs, which was further confirmed by the EDS analysis (Figure 9A). There were many thin segregates on the surface of cathodes of glucose-fed MFCs. Further, EDS analysis of the composition of these segregates clearly showed the characteristic peaks of Cu signals at 0.98, 8.06, and 8.87 KeVs, which confirmed the formation of Cu products (Figure 9B). Besides the Cu, many other elements (i.e., P, S, Cl, Na and Ca) were detected as compared with the surface of cathodes of abiotic control MFCs. This indicated that part of the cupric ion was precipitated with other anions and cations on the surface of the cathode, which was not observed in previous studies. The EDS analysis showed that the deposits on the cathodic surface of MFCs containing the electroactive biofilms enriched with acetate, ethanol, or lactate mainly contained the element of Cu (Figure 9C–E).

The XRD patterns of the cathodic surface of MFCs containing the electroactive biofilms enriched with acetate, ethanol, or lactate clearly demonstrated the metal copper (Cu^0) with characteristic peaks at 43.3, 50.4, and 74.1 degrees in 2-Theta (Figure 10). However, these characteristic peaks for metal copper (Cu^0) were absent for the cathode from the abiotic control MFCs and the MFCs fed with glucose. This further indicated that no copper was deposited on the cathodic surface of these MFCs. The decrease of copper in the catholyte of the abiotic control MFCs and the MFCs fed with glucose was probably ascribed to the precipitation with other anions or cations.

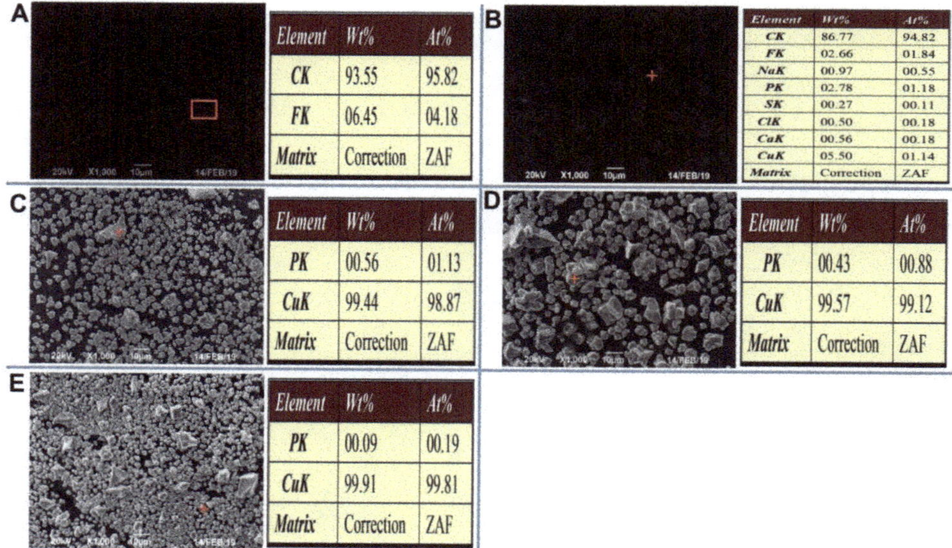

Figure 9. SEM pictures and EDS analysis of the cathodes of MFCs after the treatment of simulated AMD. (**A**: abiotic; **B**: glucose; **C**: acetate; **D**: ethanol; **E**: lactate).

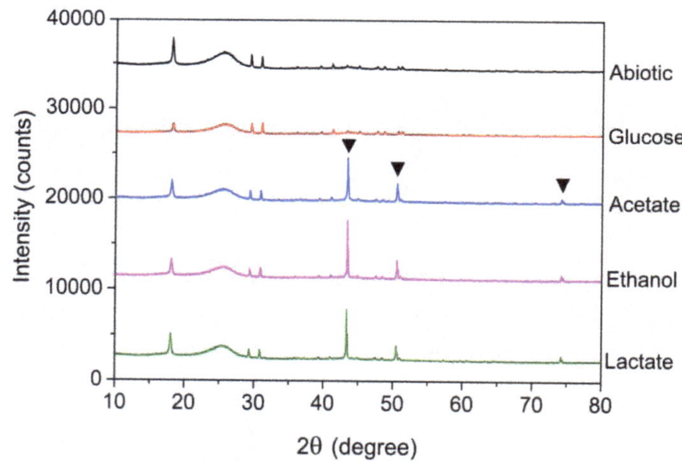

Figure 10. XRD analysis of the cathodes of MFCs after the treatment of simulated AMD (▼indicates characteristic peaks of Cu^0).

3.5. Comparison of this Study with Previous Studies

The organic wastewater (individually simulated by four typical pure chemicals) and simulated AMD were simultaneously treated in dual-chamber MFCs in this study. The effect of different energy substrates on anodic electroactive biofilms enrichment, bioelectrochemical activity, microbial communities, and AMD treatment was compared. For the scale-up of the BESs to treat the real industrial AMD in mining sites, these pure organic chemicals should be replaced by the real organic wastewater available near the pollution site in order to greatly reduce the costs. Different sources of real organic wastewater usually contain different predominant organic chemicals (such as these typical chemicals used in this study). Therefore, it is necessary to evaluate the effects of different energy substrates on anodic electroactive biofilms enrichment (both bioelectrochemical activity and microbial communities) and AMD treatment.

It is a fact that various organics (either pure chemical or real organic wastewater) have been studied as an energy substrate for MFCs [36]. However, only some studies have focused on the comparison of the electrochemical performance of MFCs enriched with different organic substrates (Table 3). On comparing our study with these studies, we found out the following differences: (1). The organic substrates used were generally different in other studies; (2). The microbial community structure of the anodic electroactive biofilms enriched with different energy substrates was studied with high throughput sequencing technique in this study. However, two other studies analyzed the anodic electroactive biofilms with traditional culture-dependent technique or denaturing gradient gel electrophoresis (DGGE) [12,37]. In addition, simultaneous treatment of different organic wastewater and simulated AMD was analyzed in this study.

MFCs have been adopted to recover some heavy metals from wastewater [38,39]. In order to be consistent with the target heavy metal in our study, the studies that focused on treatment with Cu^{2+} were selected and compared (Table 3). On comparing our study with these studies, we found out the following differences: (1). Only a single energy substrate was used in these studies; (2). The microbial community structures of the anodic electroactive biofilm in these studies were scarcely studied. There is a study that analyzed the microbial community structure of the anodic electroactive biofilm under the stress of different concentrations of Cu^{2+} in municipal wastewater in single-chamber MFC [14]. However, the simulated AMD was treated in the cathode chamber, while the electroactive biofilm was in the anode chamber in this study (two chambers were separated by an anion exchange membrane). Therefore, the stress of the electroactive biofilm should be negligible. Collectively, results obtained in this study are insightful for the enrichment of electroactive biofilms for AMD treatment.

Table 3. Comparison of this study with other related studies.

Energy Substrates	Research Focuses		Reference
	Microbial Community of Anodic Biofilm	Wastewater (Containing Cu^{2+})	
acetate, lactate, ethanol, glucose	Yes, high throughput sequencing technique.	Yes	This study
acetate, butyrate, propionate, glucose	Yes, using the culture-dependent technique.	No	[12]
acetate, ethanol, glucose	No	No	[40]
acetate, butyrate, glucose	Yes, using denaturing gradient gel electrophoresis.	No	[37]
glucose, methanol, propyl alcohol	Yes, high throughput sequencing technique.	No	[13]
acetate, glucose, starch, dextran, butyrate	No	No	[41]
glucose, lactose, cheese	No	No	[42]
glucose, acetate, xylose	No	No	[43]
acetate	No	Yes	[44]
acetate	No	Yes	[45]
glucose	No	Yes	[46]
acetate	No	Yes	[47]
acetate	No	Yes	[48]
acetate	No	Yes	[17]
acetate	No	Yes	[49]
acetate	No	Yes	[50]
acetate	No	Yes	[51]
acetate	Yes, high throughput sequencing technique; anodic biofilm was directly exposed to Cu^{2+}.	Yes	[14]

4. Conclusions

This study showed that different energy substrates affected the startup, maximum voltage output, power density, coulombic efficiency, ohmic resistance, and the charge transfer resistance of MFC. The microbial community structures of these electroactive biofilms were modulated by energy substrates during the enrichment. The abundance of classic exoelectrogens *Geobacter* species correlated with the electricity-generation capacities of different electroactive biofilms. *Geobacter* species constituted as the predominant components of the electroactive biofilms enriched with acetate, ethanol, or lactate, which existed as minor species in glucose-fed electroactive biofilms (0.63%). The MFCs containing the glucose-fed electroactive biofilms were deficient in the extraction of copper from AMD. On the contrary, the MFCs containing the electroactive biofilms enriched with acetate, ethanol, or lactate recovered almost all the Cu^{2+} from the AMD by electrochemical reduction as metal copper (Cu^0) on the surface of the cathode. These results indicated that the effects of organic chemical (that is usually contained in organic wastewater) on the enrichment of electroactive biofilm should be first evaluated in order to obtain an efficient simultaneous treatment of organic wastewater and AMD. Further research works are needed to assess the technical feasibility of the bioelectrochemical system to treat AMD, such as scale-up the reactor and run in continuous mode.

Author Contributions: Data curation, Z.Z.; Formal analysis, C.A. (Chenbing Ai); Investigation, C.A. (Chenbing Ai), Z.Y., S.H., X.Z., and C.A. (Charles Amanze); Methodology, Z.Y.; Project administration, G.Q.; Resources, Z.D.; Supervision, W.Z.; Writing—original draft, C.A. (Chenbing Ai); Writing—review and editing, L.C., G.Q., and W.Z. All authors have read and agreed to the published version of the manuscript.

Funding: This work was supported by the postdoctoral research funding of Central South University (Grant No. 207154), the National Natural Science Foundation of China (Grant No. 31470230, 51320105006, 51604308), the Youth Talent Foundation of Hunan Province of China (No.2017RS3003), Natural Science Foundation of Hunan Province of China (No.2018JJ2486), Key Research and Development Projects in Hunan Province (2018WK2012).

Conflicts of Interest: The authors declare no conflict of interest.

References

1. Ai, C.; Liang, Y.; Miao, B.; Chen, M.; Zeng, W.; Qiu, G. Identification and analysis of a novel gene cluster involves in Fe^{2+} oxidation in *Acidithiobacillus ferrooxidans* ATCC 23270, a typical biomining acidophile. *Curr. Microbiol.* **2018**, *75*, 818–826. [CrossRef] [PubMed]
2. Park, I.; Tabelin, C.B.; Jeon, S.; Li, X.; Seno, K.; Ito, M.; Hiroyoshi, N. A review of recent strategies for acid mine drainage prevention and mine tailings recycling. *Chemosphere* **2019**, *219*, 588–606. [CrossRef] [PubMed]
3. Igarashi, T.; Herrera, P.S.; Uchiyama, H.; Miyamae, H.; Iyatomi, N.; Hashimoto, K.; Tabelin, C.B. The two-step neutralization ferrite-formation process for sustainable acid mine drainage treatment: Removal of copper, zinc and arsenic, and the influence of coexisting ions on ferritization. *Sci. Total Environ.* **2020**, *715*, 136877. [CrossRef] [PubMed]
4. Simate, G.S.; Ndlovu, S. Acid mine drainage: Challenges and opportunities. *J. Environ. Chem. Eng.* **2014**, *2*, 1785–1803. [CrossRef]
5. Tolonen, E.T.; Sarpola, A.; Hu, T.; Ramo, J.; Lassi, U. Acid mine drainage treatment using by-products from quicklime manufacturing as neutralization chemicals. *Chemosphere* **2014**, *117*, 419–424. [CrossRef] [PubMed]
6. Logan Bruce, E.; Rabaey, K. Conversion of wastes into bioelectricity and chemicals by using microbial electrochemical technologies. *Science* **2012**, *337*, 686–690. [CrossRef]
7. Ai, C.; Hou, S.; Yan, Z.; Zheng, X.; Amanze, C.; Chai, L.; Qiu, G.; Zeng, W. Recovery of metals from acid mine drainage by bioelectrochemical system inoculated with a novel exoelectrogen, *Pseudomonas* sp. E8. *Microorganisms* **2020**, *8*, 41. [CrossRef]
8. Santoro, C.; Arbizzani, C.; Erable, B.; Ieropoulos, I. Microbial fuel cells: From fundamentals to applications. A review. *J. Power Sources* **2017**, *356*, 225–244. [CrossRef]
9. Wang, H.; Ren, Z.J. A comprehensive review of microbial electrochemical systems as a platform technology. *Biotechnol. Adv.* **2013**, *31*, 1796–1807. [CrossRef]
10. Logan, B.E.; Regan, J.M. Electricity-producing bacterial communities in microbial fuel cells. *Trends Microbiol.* **2006**, *14*, 512–518. [CrossRef]
11. Kiely, P.D.; Regan, J.M.; Logan, B.E. The electric picnic: Synergistic requirements for exoelectrogenic microbial communities. *Curr. Opin. Biotech.* **2011**, *22*, 378–385. [CrossRef] [PubMed]
12. Chae, K.J.; Choi, M.J.; Lee, J.W.; Kim, K.Y.; Kim, I.S. Effect of different substrates on the performance, bacterial diversity, and bacterial viability in microbial fuel cells. *Bioresour. Technol.* **2009**, *100*, 3518–3525. [CrossRef] [PubMed]

13. Zhang, S.H.; Qiu, C.H.; Fang, C.F.; Ge, Q.L.; Hui, Y.X.; Han, B.; Pang, S. Characterization of bacterial communities in anode microbial fuel cells fed with glucose, propyl alcohol and methanol. *Appl. Biochem. Micro.* **2017**, *53*, 250–257. [CrossRef]

14. Wu, Y.; Zhao, X.; Jin, M.; Li, Y.; Li, S.; Kong, F.; Nan, J.; Wang, A. Copper removal and microbial community analysis in single-chamber microbial fuel cell. *Bioresour. Technol.* **2018**, *253*, 372–377. [CrossRef]

15. Shehab, N.; Li, D.; Amy, G.L.; Logan, B.E.; Saikaly, P.E. Characterization of bacterial and archaeal communities in air-cathode microbial fuel cells, open circuit and sealed-off reactors. *Appl. Microbiol. Biot.* **2013**, *97*, 9885–9895. [CrossRef]

16. Ai, C.; Yan, Z.; Chai, H.; Gu, T.; Wang, J.; Chai, L.; Qiu, G.; Zeng, W. Increased chalcopyrite bioleaching capabilities of extremely thermoacidophilic *Metallosphaera sedula* inocula by mixotrophic propagation. *J. Ind. Microbiol. Biotechnol.* **2019**, *46*, 1113–1127. [CrossRef]

17. Cheng, S.A.; Wang, B.S.; Wang, Y.H. Increasing efficiencies of microbial fuel cells for collaborative treatment of copper and organic wastewater by designing reactor and selecting operating parameters. *Bioresour. Technol.* **2013**, *147*, 332–337. [CrossRef]

18. Logan B., E.; Hamelers, B.; Rozendal, R.; Schröder, U.; Keller, J.; Freguia, S.; Aelterman, P.; Verstraete, W.; Rabaey, K. Microbial fuel cells methodology and technology. *Environ. Sci. Technol.* **2006**, *40*, 5181–5192. [CrossRef]

19. Herrera, L.; Ruiz, P.; Aguillon, J.C.; Fehrmann, A. A new spectrophotometric method for the determination of ferrous iron in the presence of ferric iron. *J. Chem. Technol. Biotechnol.* **1989**, *44*, 171–181. [CrossRef]

20. Wetlesen, C.U. Rapid spectrophotometric determination of copper in iron, steel and ferrous alloys. *Anal. Chim. Acta* **1957**, *16*, 268–270. [CrossRef]

21. Magoc, T.; Salzberg, S.L. FLASH: Fast length adjustment of short reads to improve genome assemblies. *Bioinformatics* **2011**, *27*, 2957–2963. [CrossRef] [PubMed]

22. Edgar, R.C.; Haas, B.J.; Clemente, J.C.; Quince, C.; Knight, R. UCHIME improves sensitivity and speed of chimera detection. *Bioinformatics* **2011**, *27*, 2194–2200. [CrossRef]

23. Edgar, R.C. UPARSE: Highly accurate OTU sequences from microbial amplicon reads. *Nat. Methods* **2013**, *10*, 996–998. [CrossRef] [PubMed]

24. Wang, Q.; Garrity, G.M.; Tiedje, J.M.; Cole, J.R. Naive Bayesian classifier for rapid assignment of rRNA sequences into the new bacterial taxonomy. *Appl. Environ. Microb.* **2007**, *73*, 5261–5267. [CrossRef]

25. Schloss, P.D.; Gevers, D.; Westcott, S.L. Reducing the effects of PCR amplification and sequencing artifacts on 16S rRNA-based studies. *PLoS ONE* **2011**, *6*, e27310. [CrossRef] [PubMed]

26. Christwardana, M.; Frattini, D.; Accardo, G.; Yoon, S.P.; Kwon, Y. Early-stage performance evaluation of flowing microbial fuel cells using chemically treated carbon felt and yeast biocatalyst. *Appl. Energy* **2018**, *222*, 369–382. [CrossRef]

27. He, Z.; Mansfeld, F. Exploring the use of electrochemical impedance spectroscopy (EIS) in microbial fuel cell studies. *Energy Environ. Sci.* **2009**, *2*, 215–219. [CrossRef]

28. Paitier, A.; Godain, A.; Lyon, D.; Haddour, N.; Vogel, T.M.; Monier, J.M. Microbial fuel cell anodic microbial population dynamics during MFC start-up. *Biosens. Bioelectron.* **2017**, *92*, 357–363. [CrossRef]

29. Cai, W.; Zhang, Z.; Ren, G.; Shen, Q.; Hou, Y.; Ma, A.; Deng, Y.; Wang, A.; Liu, W. Quorum sensing alters the microbial community of electrode-respiring bacteria and hydrogen scavengers toward improving hydrogen yield in microbial electrolysis cells. *Appl. Energy* **2016**, *183*, 1133–1141. [CrossRef]

30. Christwardana, M.; Frattini, D.; Duarte, K.D.Z.; Accardo, G.; Kwon, Y. Carbon felt molecular modification and biofilm augmentation via quorum sensing approach in yeast-based microbial fuel cells. *Appl. Energy* **2019**, *238*, 239–248. [CrossRef]

31. Ai, C.; McCarthy, S.; Eckrich, V.; Rudrappa, D.; Qiu, G.; Blum, P. Increased acid resistance of the archaeon, *Metallosphaera sedula* by adaptive laboratory evolution. *J. Ind. Microbiol. Biotechnol.* **2016**, *43*, 1455–1465. [CrossRef] [PubMed]

32. Peng, T.; Zhou, D.; Liu, Y.; Yu, R.; Qiu, G.; Zeng, W. Effects of pH value on the expression of key iron/sulfur oxidation genes during bioleaching of chalcopyrite on thermophilic condition. *Ann. Microbiol.* **2019**, *69*, 627–635. [CrossRef]

33. Ai, C.; Yan, Z.; Zhou, H.; Hou, S.; Chai, L.; Qiu, G.; Zeng, W. Metagenomic Insights into the Effects of seasonal temperature variation on the activities of activated sludge. *Microorganisms* **2019**, *7*, 713. [CrossRef] [PubMed]

34. Stöckl, M.; Teubner, N.C.; Holtmann, D.; Mangold, K.M.; Sand, W. Extracellular polymeric substances from *Geobacter sulfurreducens* biofilms in microbial fuel cells. *Acs Appl. Mater. Interfaces* **2019**, *11*, 8961–8968. [CrossRef]

35. Reguera, G.; McCarthy, K.D.; Mehta, T.; Nicoll, J.S.; Tuominen, M.T.; Lovley, D.R. Extracellular electron transfer via microbial nanowires. *Nature* **2005**, *435*, 1098–1101. [CrossRef]

36. Pant, D.; Van Bogaert, G.; Diels, L.; Vanbroekhoven, K. A review of the substrates used in microbial fuel cells (MFCs) for sustainable energy production. *Bioresour. Technol.* **2010**, *101*, 1533–1543. [CrossRef]

37. Zhang, Y.; Min, B.; Huang, L.; Angelidaki, I. Electricity generation and microbial community response to substrate changes in microbial fuel cell. *Bioresour. Technol.* **2011**, *102*, 1166–1173. [CrossRef]

38. Nancharaiah, Y.V.; Venkata Mohan, S.; Lens, P.N. Metals removal and recovery in bioelectrochemical systems: A review. *Bioresour. Technol.* **2015**, *195*, 102–114. [CrossRef]

39. Mathuriya, A.S.; Yakhmi, J.V. Microbial fuel cells to recover heavy metals. *Environ. Chem. Lett.* **2014**, *12*, 483–494. [CrossRef]

40. Sharma, Y.; Li, B. The variation of power generation with organic substrates in single-chamber microbial fuel cells (SCMFCs). *Bioresour. Technol.* **2010**, *101*, 1844–1850. [CrossRef]

41. Min, B.; Logan, B.E. Continuous electricity generation from domestic wastewater and organic substrates in a flat plate microbial fuel cell. *Environ. Sci. Technol.* **2004**, *38*, 5809–5814. [CrossRef] [PubMed]

42. Antonopoulou, G.; Stamatelatou, K.; Bebelis, S.; Lyberatos, G. Electricity generation from synthetic substrates and cheese whey using a two chamber microbial fuel cell. *Biochem. Eng. J.* **2010**, *50*, 10–15. [CrossRef]

43. Thygesen, A.; Poulsen, F.W.; Min, B.; Angelidaki, I.; Thomsen, A.B. The effect of different substrates and humic acid on power generation. *Bioresour. Technol.* **2009**, *100*, 1186–1191. [CrossRef]

44. Heijne, A.T.; Liu, F.; van der Weijden, R.; Weijma, J.; Buisman, C.J.N.; Hamelers, H.V.M. Copper recovery combined with electricity production in a microbial fuel cell. *Environ. Sci. Technol.* **2010**, *44*, 4376–4381. [CrossRef] [PubMed]

45. Tao, H.C.; Zhang, L.J.; Gao, Z.Y.; Wu, W.M. Copper reduction in a pilot-scale membrane-free bioelectrochemical reactor. *Bioresour. Technol.* **2011**, *102*, 10334–10339. [CrossRef] [PubMed]

46. Tao, H.C.; Liang, M.; Li, W.; Zhang, L.J.; Ni, J.R.; Wu, W.M. Removal of copper from aqueous solution by electrodeposition in cathode chamber of microbial fuel cell. *J. Hazard. Mater.* **2011**, *189*, 186–192. [CrossRef] [PubMed]

47. Wu, D.; Huang, L.; Quan, X.; Li Puma, G. Electricity generation and bivalent copper reduction as a function of operation time and cathode electrode material in microbial fuel cells. *J. Power Sources* **2016**, *307*, 705–714. [CrossRef]

48. Zhang, L.; Tao, H.; Wei, X.; Lei, T.; Li, J.; Wang, A.; Wu, W. Bioelectrochemical recovery of ammonia–copper(II) complexes from wastewater using a dual chamber microbial fuel cell. *Chemosphere* **2012**, *89*, 1177–1182. [CrossRef]

49. Rodenas Motos, P.; Ter Heijne, A.; van der Weijden, R.; Saakes, M.; Buisman, C.J.; Sleutels, T.H. High rate copper and energy recovery in microbial fuel cells. *Front. Microbiol.* **2015**, *6*, 527. [CrossRef]

50. Zhang, Y.; Yu, L.; Wu, D.; Huang, L.; Zhou, P.; Quan, X.; Chen, G. Dependency of simultaneous Cr(VI), Cu(II) and Cd(II) reduction on the cathodes of microbial electrolysis cells self-driven by microbial fuel cells. *J. Power Sources* **2015**, *273*, 1103–1113. [CrossRef]

51. Luo, H.; Liu, G.; Zhang, R.; Bai, Y.; Fu, S.; Hou, Y. Heavy metal recovery combined with H_2 production from artificial acid mine drainage using the microbial electrolysis cell. *J. Hazard. Mater.* **2014**, *270*, 153–159. [CrossRef] [PubMed]

minerals

Article

Performance Evaluation of Fe-Al Bimetallic Particles for the Removal of Potentially Toxic Elements from Combined Acid Mine Drainage-Effluents from Refractory Gold Ore Processing

Elham Aghaei [1], Zexiang Wang [1], Bogale Tadesse [1], Carlito Baltazar Tabelin [2], Zakaria Quadir [3] and Richard Diaz Alorro [1,*]

1   Western Australian School of Mines: Minerals, Energy and Chemical Engineering, Faculty of Science and Engineering, Curtin University, Kalgoorlie, WA 6403, Australia; elham.aghaei@postgrad.curtin.edu.au (E.A.); zexiang.wang@postgrad.curtin.edu.au (Z.W.); bogale.tadesse@curtin.edu.au (B.T.)
2   School of Minerals and Energy Resources Engineering, Faculty of Engineering, University of New South Wales, Sydney, NSW 2052, Australia; c.tabelin@unsw.edu.au
3   John de Laeter Centre and School of Civil and Mechanical Engineering, Faculty of Science and Engineering, Curtin University, Bentley, WA 6845, Australia; zakaria.quadir@curtin.edu.au
*   Correspondence: Richard.Alorro@curtin.edu.au; Tel.: +61-890886187

Citation: Aghaei, E.; Wang, Z.; Tadesse, B.; Tabelin, C.B.; Quadir, Z.; Alorro, R.D. Performance Evaluation of Fe-Al Bimetallic Particles for the Removal of Potentially Toxic Elements from Combined Acid Mine Drainage-Effluents from Refractory Gold Ore Processing. *Minerals* 2021, 11, 590. https://doi.org/10.3390/min11060590

Academic Editor: Juan Antelo

Received: 19 April 2021
Accepted: 27 May 2021
Published: 31 May 2021

Publisher's Note: MDPI stays neutral with regard to jurisdictional claims in published maps and institutional affiliations.

Abstract: Acid mine drainage (AMD) is a serious environmental issue associated with mining due to its acidic pH and potentially toxic elements (PTE) content. This study investigated the performance of the Fe-Al bimetallic particles for the treatment of combined AMD-gold processing effluents. Batch experiments were conducted in order to eliminate potentially toxic elements (including Hg, As, Cu, Pb, Ni, Zn, and Mn) from a simulated waste solution at various bimetal dosages (5, 10, and 20 g/L) and time intervals (0 to 90 min). The findings show that metal ions with greater electrode potentials than Fe and Al have higher affinities for electrons released from the bimetal. Therefore, a high removal (>95%) was obtained for Hg, As, Cu, and Pb using 20 g/L bimetal in 90 min. Higher uptakes of Hg, As, Cu, and Pb than Ni, Zn, and Mn also suggest that electrochemical reduction and adsorption by Fe-Al (oxy) hydroxides as the primary and secondary removal mechanisms, respectively. The total Al^{3+} dissolution in the experiments with a higher bimetal content (10 and 20 g/L) were insignificant, while a high release of Fe ions was recorded for various bimetal dosages. Although the secondary Fe pollution can be considered as a drawback of using the Fe-Al bimetal, this issue can be tackled by a simple neutralization and Fe precipitation process. A rapid increase in the solution pH (initial pH 2 to >5 in 90 min) was also observed, which means that bimetallic particles can act as a neutralizing agent in AMD treatment system and promote the precipitation of the dissolved metals. The presence of chloride ions in the system may cause akaganeite formation, which has shown a high removal capacity for PTE. Moreover, nitrate ions may affect the process by competing for the released electrons from the bimetal owing to their higher electrode potential than the metals. Finally, the Fe-Al bimetallic material showed promising results for AMD remediation by electrochemical reduction of PTE content, as well as acid-neutralization/metal precipitation.

Keywords: acid mine drainage; gold processing effluents; Fe-Al bimetallic particles; electrochemical reduction

1. Introduction

Acid mine drainage (AMD) refers to acidic runoff rich in high concentrations of metal ions, such as iron (Fe), manganese (Mn), zinc (Zn), copper (Cu), lead (Pb), nickel (Ni), arsenic (As), cadmium (Cd), aluminum (Al), and mercury (Hg) [1–3]. AMD is associated with mining and mineral processing activities and comes from the natural oxidation of sulfide-bearing minerals (such as pyrite) exposed to water, oxygen, and microbes [4,5]. AMD is considered one of the most prevalent causes of environmental pollution which stems from its high acidity (pH < 3) and toxic metal content [6]. Tailings waste from processing of refractory gold ores is one of the major areas of concern as it contains sulfide species and is very likely to produce AMD over time, especially in dry climates and high evaporation rates [7]. Therefore, parts of tailings with sulfide minerals content exposed to air will start to oxidize during summer to form AMD.

To tackle the issue of AMD, many attempts have been made to limit the generation and release of AMD by protecting sulfide minerals from air, water, and bacteria and minimizing their interactions [5,8–13]. However, due to practical constraints involved in the prevention strategy [14], the next available option is AMD treatment by either active or passive methods [2]. The most common active methods include neutralization using caustic soda (sodium hydroxide), calcium hydroxide ($Ca(OH)_2$) or limestone ($CaCO_3$), as well as adsorption, ion exchange, and crystallization [6,15]. However, the interest in improving the efficiency of AMD remediation techniques motivated researchers to develop passive methods, which involves biological and chemical treatment of AMD using wetlands [16], permeable reactive barriers [17,18], compost reactors, and bioreactors, and cost-effective materials such as recycled concrete aggregates [14], sulfur-reducing bacteria (SRB) [19], and fly ash [20]. Studies with AMD have focused on neutralizing the acidity and heavy metal removal. However, in the case of combined AMD-waste effluents resulting from refractory gold processing with Cl^- and NO_3 content, it has not been considered anywhere before.

When exploring the most appropriate treatment techniques, it is crucial to consider the use of non-toxic, cost-effective, and high-performance materials with the lowermost potential of hazardous wastes/bi-products generation. Accordingly, zero-valent iron (ZVI) has been considered a promising element for removal of heavy metals and PTE from the aquatic environment [21] and the most common reactive material used in permeable reactive barriers (PRBs) for remediating AMD and contaminated groundwater [22]. Depending on the environmental conditions (pH, redox, and oxic-anoxic conditions), type and concentration of dissolved constituents; ZVI can remove heavy metals and PTE from solutions effectively through adsorption, surface complexation, reductive precipitation, and co-precipitation [3,22]. However, one major drawback with this kind of application is the decreasing reactivity and performance of ZVI in the long-term due to iron corrosion and surface passivation by an iron oxy-hydroxide film [21,22]. Recently, iron-based bimetallic materials have been developed aimed at improving the reactivity and efficiency of ZVI in removing PTE. In this regard, due to the synergistic effect of Fe and Al, the Fe-Al bimetal has shown remarkably improved reductive ability for the contaminants [23]. The potential difference between Fe and Al ($E^0(Al^{3+}/Al^0) = -1.667$ V and $E^0(Fe^{2+}/Fe^0) = -0.44$ V) promotes better electron transfer within the bimetallic system and slows the passivation of the Fe surface, resulting in a higher reducing capacity for target contaminants [24].

A number of studies have examined the performance of Fe-Al bimetallic particles for their ability to remove heavy metals, including Cr(VI) [25], As(III) [26], U(VI) [27] from waste solutions. Their findings demonstrated the high capacity, selectivity, and rapid removal rate of target metal ions by the bimetal, predominantly through electrochemical reduction. Moreover, in a study by Han et al. (2016) [28], a higher removal efficiency for aqueous heavy metal ions (Cr(VI), Cd(II), Ni(II), Cu(II), and Zn(II)) was achieved by acid-washed ZVAl/ZVI mixture in PRBs compared to acid-washed ZVAl or ZVI alone. One significant finding to emerge from this previous study was that the Fe-Al bimetal formation during the reaction has been identified as a major contributing factor to the high removal efficiency. Despite previous studies describing Fe-Al bimetal as a potential technique in wastewater remediation, the direct application of this bimetal in AMD treatment has not been reported to date. According to standard electrode potential of Fe, Al and metals found in AMD such as Pb, Cu, Hg, and Zn, it is clear that the Fe-Al bimetallic material is an effective medium for treating AMD. Compared to common passive treatment methods, which suffer from long processing time [29], Fe-Al bimetallic particles are fast and effective for metal removal. Moreover, both Al and Fe are among the most abundant elements on the earth, and the amount of required bimetal for AMD remediation, and waste generated, is very small. Therefore, this paper evaluates the performance of the Fe-Al bimetal for acid-neutralization and removal of potentially toxic elements from simulated AMD combined with gold processing effluents by considering the influencing parameters including bimetal dosage and reaction time.

2. Materials and Methods

2.1. Preparation of Combined AMD-Waste Effluent from Refractory Gold Processing

The combined AMD prepared in this study represents the combination of AMD and effluents resulting from the processing of refractory gold ores containing sulfide minerals in the Goldfields region of Western Australia. Parts of tailings from these processing plants, exposed to atmospheric conditions, are very liable to generate AMD over time. The gold processing tailings dam contains Cl^-, and NO_3^- ions because of using hydrochloric acid (HCl) or nitric acid (HNO_3) in the acid washing stage [30] and lead nitrate in cyanidation [31,32]. Moreover, scaling up the mining and processing operations has risen the demand for groundwater sources. The available source of process water in Australia, especially in arid regions, is hypersaline groundwater with high Cl^- content [33,34]. With regard to what mentioned above, the combined AMD solution was prepared using 1000 mg/L single-element standard solutions of Mn, Pb, As, Ni, Cu, Zn, and Hg in 2% nitric acid (Sigma-Aldrich), as well as calcium chloride ($CaCl_2$), sodium chloride (NaCl) and ferrous sulfate ($FeSO_4$). The initial pH of the prepared solution was adjusted to 2 using sodium hydroxide (NaOH). The synthetic AMD was with initial metal concentrations shown in Table 1.

Table 1. Initial solute concentrations (mg/L) in the synthetic AMD-gold processing effluents.

Hg	Al	Ca	Mn	Na	Fe	As	Pb	SO_4^{2-}	Ni	Cu	Zn
32.5	1.4	273.6	36.3	10,896.0	388.4	9.3	91.4	652.0	18.2	53.4	31.9

2.2. Synthesis of Fe-Al Bimetallic Particles

All the reagents used in this study were of analytical grade. The Fe-Al bimetallic particles were synthesized using ZVAl powder (D90 = 86.5 μm) obtained from Barnes (NSW, Australia), and ferric chloride ($FeCl_3·6H_2O$, >99% purity) purchased from Chemsupply (SA, Australia). Fe-Al bimetals were prepared by optimizing the procedure used by Chen et al. (2008) [24] and Fu et al. (2015) [25], which are based on the electrochemical reduction and deposition of iron on the ZVAl surface. The first step was to remove the unreactive layer of aluminum oxide from ZVAl using acid washing, in which 20 mL of 1 M hydrochloric acid (HCl) was added to flasks containing 3 gr ZVAl in a shaking incubator and agitated for 15 min at 40 °C and 110 rpm. This treatment was followed by the cementation step by adding 30 mL Fe^{3+} solutions with a certain concentration (to give 0.5 g Fe to 1 gr Al) to the flasks and agitating for 30 min. Then, the Fe/Al particles were recovered and rinsed with deionized water, and dried in a vacuum desiccator. The residual Fe and Al concentrations in the solution was measured (data not shown) to calculate the total Fe and Al content of recovered bimetallic particles in each preparation batch as 1 g Fe/2.1 g Al (±0.05 for 3 samples) (Equations (1) and (2)).

$$Fe_T = Fe_0 - Fe_r \tag{1}$$

$$Al_T = Al_0 - Al_r \tag{2}$$

where

Fe_T and Al_T are the total Fe and Al content of the bimetal,
Fe_0 and Al_0 are applied Fe and Al content, and
Fe_r and Al_r are residual Fe and Al ions concentrations.

2.3. Analytical Techniques

The concentration of dissolved ions were analyzed using inductively coupled plasma optical emission spectroscopy (ICP-OES) and mass spectroscopy (ICP-MS). X-ray powder diffraction (XRD) of bimetallic particles was performed using an Olympus diffractometer (Olympus Scientific Solutions Americas, USA) with Co-Kα radiation source in the range between 5 and 55° (2θ). To characterize the size distribution of ZVAl powder and bimetallic particles, Mastersizer Malvern 3000 was used (Malvern Instruments Ltd., Malvern, UK). The structure and elemental mapping of the bimetal were determined using a Tescan Clara field emission scanning electron microscope (SEM) equipped with an energy dispersive spectrometer (EDS) manufactured by the Oxford Instrument, Oxfordshire, UK).

2.4. Experimental Procedure

To investigate the combined AMD treatment using Fe-Al bimetallic particles, batch experiments were conducted in an incubator shaker at 110 rpm and 25 °C with varying time intervals from 10 to 90 min. In each batch, a specific amount of Fe-Al bimetals (5, 10, and 20 g/L) was added to Erlenmeyer flasks containing 25 mL of the prepared waste solution. No acid or alkali was subsequently added to control the pH during the reaction. All experiments were conducted in duplicate, and average values were presented. After a specified time, the solution content of each flask was recovered by filtration and analyzed for heavy metal concentrations. The percentage of heavy metal removal (% R) was calculated using Equation (3):

$$\%R = \frac{C_0 - C}{C_0} \times 100 \tag{3}$$

where C_0: Initial heavy metal ion concentration, mg/L, and C: Residual heavy metal ion concentration, mg/L.

Stabilities of pollutants were modelled by the Geochemist's Workbench® [35] with the THERMODDEM database [36] based on measured solute activities in the experiments.

3. Results and Discussion

3.1. Characterization of the Fe-Al Bimetallic Particles

The particle size distribution of the ZVAl and the synthesized Fe-Al bimetallic material is illustrated in Figure 1. The graph shows that there was an increase in the particle size of the Fe-Al bimetal compared to ZVAl. Ninety percent of the ZVAl distribution has a smaller particle size of 86.5 μm (D_{90}) while this value increased to 134 μm after acid washing and loading with Fe. Moreover, the diffraction peaks at 45.0° for both Al and Fe and 52.5° for Fe in the XRD pattern (Figure 2) confirmed the presence of both Al and Fe in the bimetal structure.

In addition, the core-shell structure of the bimetal has been detected in the SEM mapping (Figure 3). From the EDS spectra shown in Figure 3, it can be seen that Al is mostly found in the core while Fe is the dominant element on the surface of Al.

3.2. pH Monitoring

The pH plays a vital role in the AMD treatment as increasing in pH can lead to the dissolved metal and hydroxides precipitation [37]. Figure 4 shows the experimental data for the solution pH at a different time and bimetal dosage. As illustrated in the graph, a clear trend of increasing pH with time from 0 to 30 min for all bimetal dosage at initial pH 2 was observed. However, from 30 to 90 min, a slight change in the pH was recorded. In addition, for the combined AMD treated with the greater bimetal dosages, the higher pH values were obtained. The pH of the solution containing 20 g/L of bimetal reached more than 5.5 after 30 min, although it exhibited a slight decrease from 60 to 90 min. The Eh of the solutions was 0.5 V at initial pH of 2 just before adding the bimetal and decreased to minimum value of around 0.21 V in 90 min for all bimetal dosages.

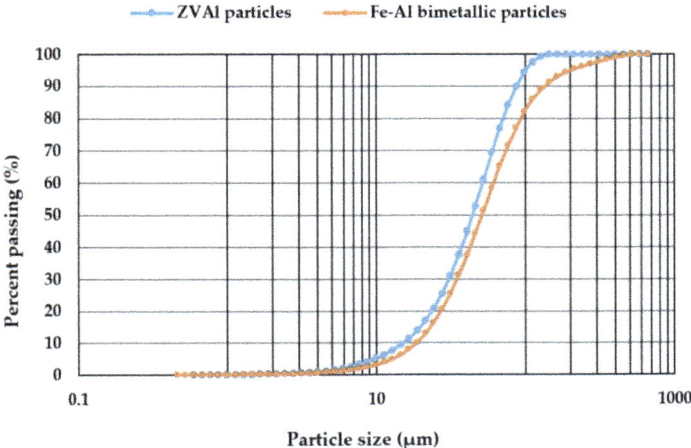

Figure 1. Cumulative particle size distribution of ZVAl and Fe-Al bimetallic material.

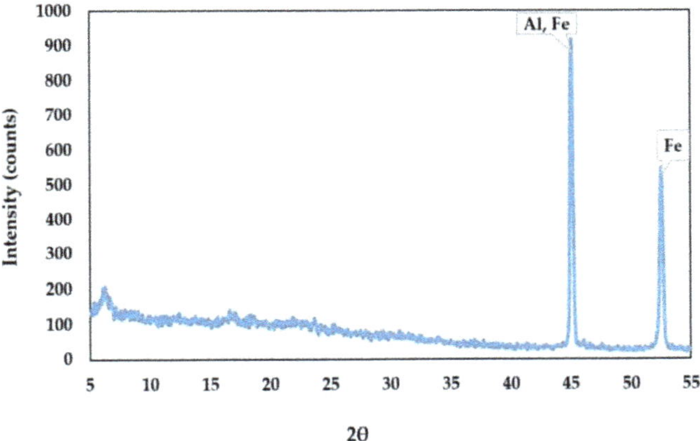

Figure 2. XRD pattern for the synthesized Fe-Al bimetal.

In the acidic aqueous system containing Fe-Al bimetallic particles and dissolved oxygen, the oxidation of ZVAl to Al^{3+} ($E^0(Al^{3+}/Al^0) = -1.667$ V) and ZVI to Fe^{2+} ($E^0(Fe^{2+}/Fe^0) = -0.44$ V) (Equations (4) and (5)) was accompanied by oxygen reduction in the presence of protons (H^+) and the generation of hydrogen peroxide (H_2O_2) ($E^0(O_2/H_2O_2)= +0.695$ V) [38] (Equations (6) and (7)) [38,39]. Hydrogen peroxide subsequently accelerated the ZVAl corrosion to Al^{3+} (Equation (8)) [40,41] and triggered a Fenton reaction, where Fe^{2+} and H_2O_2 reacted to form Fe^{3+}, hydroxyl radicals ($OH^{\cdot}$), and hydroxyl ions (OH^-) (Equation (9)) [38]. In addition, more OH^- released into the solution, where H_2O in the solution picked up electrons. The evolution of H_2 gas resulted from H_2O/H^+ reduction in the solution was also evident in the experiments (Equations (10) and (11)). To sum up, increasing the solution pH is obviously related to the release of OH^- ions into the solution via several reactions in the solution, as discussed above.

$$Al^0 \rightarrow Al^{3+} + 3e^- \tag{4}$$

$$Fe^0 \rightarrow Fe^{2+} + 2e^- \tag{5}$$

$$O_2 + H^+ + e^- \rightarrow HO_2 \tag{6}$$

$$2HO_2^{\cdot} \rightarrow H_2O_2 + O_2 \tag{7}$$

$$Al^0 + 3H_2O_2 \rightarrow Al^{3+} + 3OH^{\cdot} + OH^- \tag{8}$$

$$Fe^{2+} + H_2O_2 \rightarrow Fe^{3+} + OH^{\cdot} + OH^- \tag{9}$$

$$2H_2O + 2e^- \rightarrow H_2 + 2OH^- \tag{10}$$

$$2H^+ + 2e^- \rightarrow H_2 \tag{11}$$

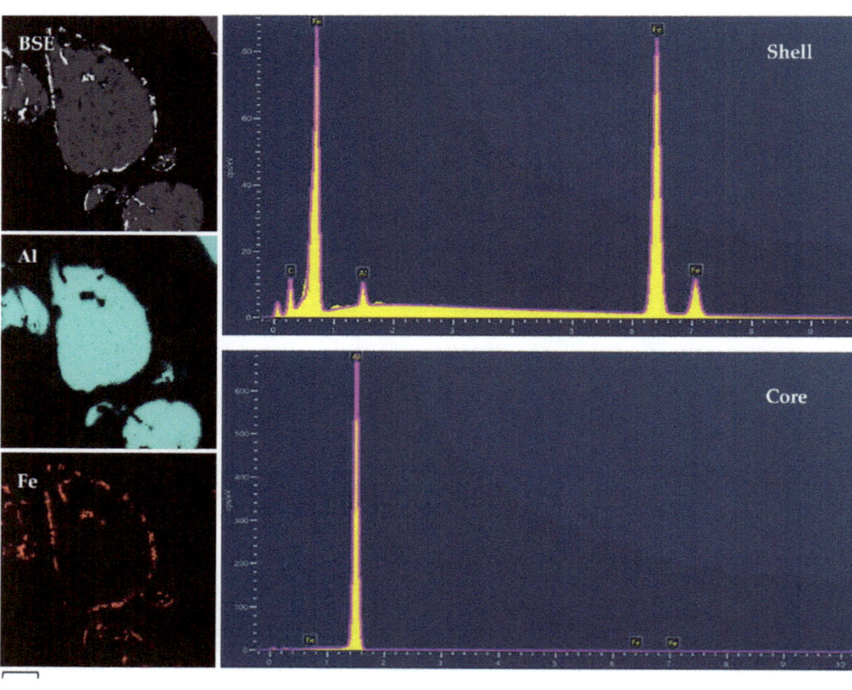

Figure 3. SEM image and EDS spectra of Fe-Al bimetallic material with a core-shell structure.

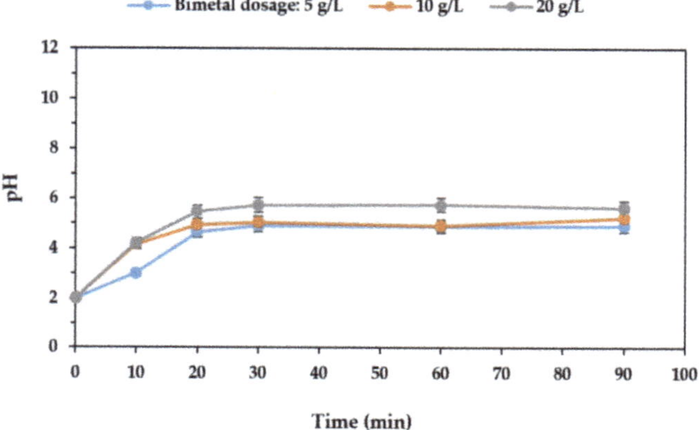

Figure 4. Variation in the pH of the combined AMD solution treated by the Fe-Al bimetal at a different time and bimetal dosage.

3.3. Metal Removal by the Fe-Al Bimetallic Material

The percent removal of various metals from the synthetic combined AMD treated by the Fe-Al bimetallic material for 90 min was compared and illustrated in Figure 5. What stands out in this figure is the higher uptake of Hg, As, Cu and Pb at all bimetal levels compared with Zn, Ni, and Mn. Experiments with 20 g/L of the bimetal resulted in significant removal of Hg (99.74%), As (99.80%), Cu (98.20%), and Pb (95.50%), while it dropped to 69.50% removal for Zn, 22.34% for Ni, and <5% for Mn. As previously stated, the higher standard redox potential of the aqueous contaminants than the Al and Fe is the underlying cause of a greater removal rate. Table 2 displays the standard reduction potential of various aqueous species in the experiments at 25 °C. The data are arranged in the increasing order of E^0, which means an increase in the tendency of species to get reduced. Therefore, under competitive conditions in the process, the loss of Hg, As, Cu, and Pb was higher than Ni, Zn, and Mn, as they have a greater attraction for electrons released from the bimetal.

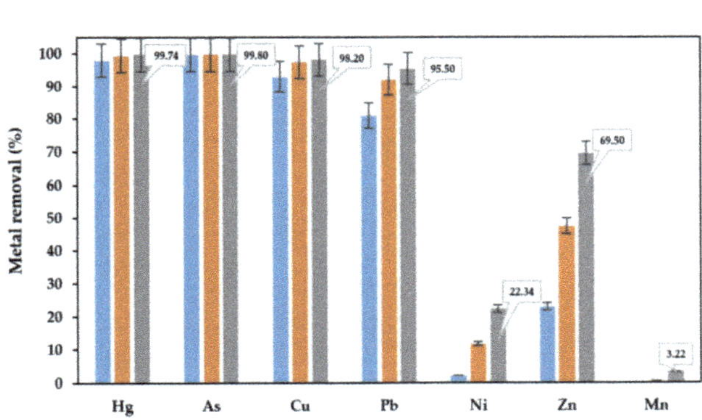

Figure 5. A comparison of the removal of various PTE from the synthetic combined AMD by Fe-Al bimetallic particles after 90 min and initial pH 2.

Table 2. Standard reduction potential of different species in aqueous solution at 25 °C [24,42].

Aqueous Species	Reduction Half Reactions	E^0 (V)
Aluminum (Al)	$Al^{3+} + 3e^- \rightarrow Al_{(S)}$	−1.68
Manganese (Mn)	$Mn^{2+} + 2e^- \rightarrow Mn_{(S)}$	−1.18
Zinc (Zn)	$Zn^{2+} + 2e^- \rightarrow Zn_{(S)}$	−0.76
Iron (Fe(II))	$Fe^{2+} + 2e^- \rightarrow Fe_{(S)}$	−0.44
Nickel (Ni)	$Ni^{2+} + 2e^- \rightarrow Ni_{(S)}$	−0.28
Lead (Pb)	$Pb^{2+} + 2e^- \rightarrow Pb_{(S)}$	−0.13
Copper (Cu(I))	$Cu^{2+} + e^- \rightarrow Cu^+$	+0.15
Arsenic (As(III))	$H_3AsO_3 + 3H^+ + 3e^- \rightarrow As + 3H_2O$	+0.24
Copper (Cu(II))	$Cu^{2+} + 2e^- \rightarrow Cu_{(S)}$	+0.34
Arsenic (As(V))	$H_3AsO_4 + 2H^+ + 2e^- \rightarrow HAsO_2 + 4H_2O$	+0.56
Iron (Fe(III))	$Fe^{3+} + e^- \rightarrow Fe^{2+}$	+0.77
Mercury (Hg)	$Hg^{2+} + 2e^- \rightarrow Hg_{(l)}$	+0.86

The variation of residual metal concentrations over time at different bimetal dosages (5, 10, and 20 g/L) and initial pH of 2 are shown in Figure 6. The initial Hg concentration in the solution (Figure 6a) dropped significantly in 10 min at all bimetal dosages, although the residual Hg(II) was slightly higher in the experiments with 5g/L bimetal (2.14 mg/L) compared to 10 and 20 g/L (<0.2 mg/L). In addition, ZVAl and ZVI on the bimetal surface, Fe^{2+} ions (E° (Fe^{3+}/Fe^{2+}) = +0.77) have also been considered as a reducing agent for Hg(II) elimination from the solution [43]. Moreover, the increase in the solution pH to >4.5 in 20 min at various bimetal dosages, resulted in the precipitation of Fe (oxy)hydroxides on the bimetal surface (Figure 7a), which can sequester Hg(II) from the solution [43].

A similar trend to Hg was observed for the residual As and Cu concentrations (Figure 6b,c) within 90 min of AMD treatment using the bimetal. Despite that within 20 min of the process using 5 g/L bimetal As uptake was lower than that of 10, and 20 g/L, the removal rate was almost the same from 20 to 90 min. In addition, the initial Cu concentration of 53.44 mg/L went down to 3.5, 1.4, and 1 mg/L at 90 min for 5, 10, and 20 g/L bimetal, respectively. The higher bimetal concentrations performed more effectively for the Pb(II) reduction, so that the best result was obtained by using 20 g/L of the bimetal at 60 min (96% removal) (Figure 6d).

Prior studies [26,44] have reported the possible mechanisms for As removal by the Fe-Al bimetal as follows: (1) the adsorption of part of free As(III) by Fe-Al oxides on the bimetal surface at the initial stages of the process; (2) the oxidation of the majority of As(III) to As(V) by reactive oxygen species generated in the system, and the subsequent adsorption of As(V) by the Fe-Al (oxy)hydroxides on the bimetal surface; (3) the reduction of the adsorbed As(V) to As(III), and then to As(0) by Fe and Al (either directly or through the galvanic cell effect) in the anoxic inner layer of the bimetallic particles. Therefore, the observed discrepancy in the As uptake within 20 min of the process using different bimetal dosages corresponds most directly to the solution pH and the formation of the Fe-Al oxy-hydroxides on the bimetal (Figure 7a,b). For the AMD treated with 10 and 20 g/L bimetal, the pH values reached more than 4 in 10 min, while the same pH was recorded after 20 min for the experiment with 5 g/L bimetal (Figure 4). Considering that under acidic and circumneutral pH, the solubility, mobility, and toxicity of the As(III) is higher than As(V) species [45,46], the Fe-Al bimetal seems to be an effective material for the As remediation from the contaminated water.

According to the redox potential (Table 2), Cu(II) and Pb(II) can be easily reduced to Cu^0 and Pb^0 by both Fe and Al. Moreover, the reduction of Cu(II) to Cu(I) is also thermodynamically favored, resulting in the formation of insoluble Cu_2O [42]. However, Igarashi, et al. (2020) [47] have shown that Cu precipitation at pH < 6 is unfavorable and its reduction is mostly attributed to co-precipitation with Fe and Al oxy-hydroxides. The metal contaminants with more negative redox potential than Fe, including Ni, Zn, and Mn, are hard to be reduced by Fe. The reduction by ZVAl and adsorption may be the predominant removal mechanism as the bimetal surface became more negatively charged with increasing OH^- concentration, which enhanced the attraction between heavy metal ions and ZVI or ZVAl [28]. The higher removal of Zn, compared to Ni (Figure 6e,f), may be attributed to the adsorption by precipitated iron oxides on the bimetallic particles, which has been considered as a major Zn(II) removal mechanism by ZVI [48]. Moreover, due to the competitive effects, the uptake of Ni, Zn, and Mn by the bimetal may be hindered by Hg, As, Cu, and Pb ions. In addition, as can be seen in Figure 7c–e, the pH increase in the system was not sufficient to drive Mn, Zn, and Ni precipitation as they exist as aqueous species under the experimental conditions. As can be seen in Figure 6e–g, concentrations of Zn, Ni, and Mn decreased over the first 10 min of the reaction before increasing again from 10 to 30 min. The re-dissolution of these metal ions can be explained by the re-oxidation of deposited metals with the accumulated Fe(III) ((Fe^{3+}/Fe^{2+}) = +0.77). In addition, the concentration of heavy metal ions such as Zn, Ni, and Cu do not seem to be affected by the high concentration of Cl^- anion in the pH range of the experiments (pH of 2-4) as it causes an increase in the metal's solubility [49–51]. However, in the case of Hg, calomel (Hg_2Cl_2) precipitation may contribute to its reduction from the solution [52,53].

Regarding the presence of nitrate ions in the system, it should be noted that nitrate has a higher electron affinity than metal ions in the process meaning it is more likely to gain electrons released from the bimetal in the competitive system. In the Fe-Al bimetallic system, both Al and Fe are able to reduce nitrate ion to nitrite (NO_2^-) (E° = 0.965 V), then ammonia (NH_4^+) (E° = 0.897 V) or nitrogen gas [54,55]. Further research should be undertaken to investigate the precise effect of NO_3^- and Cl^- ions on heavy metal removal from AMD and on the bimetal performance.

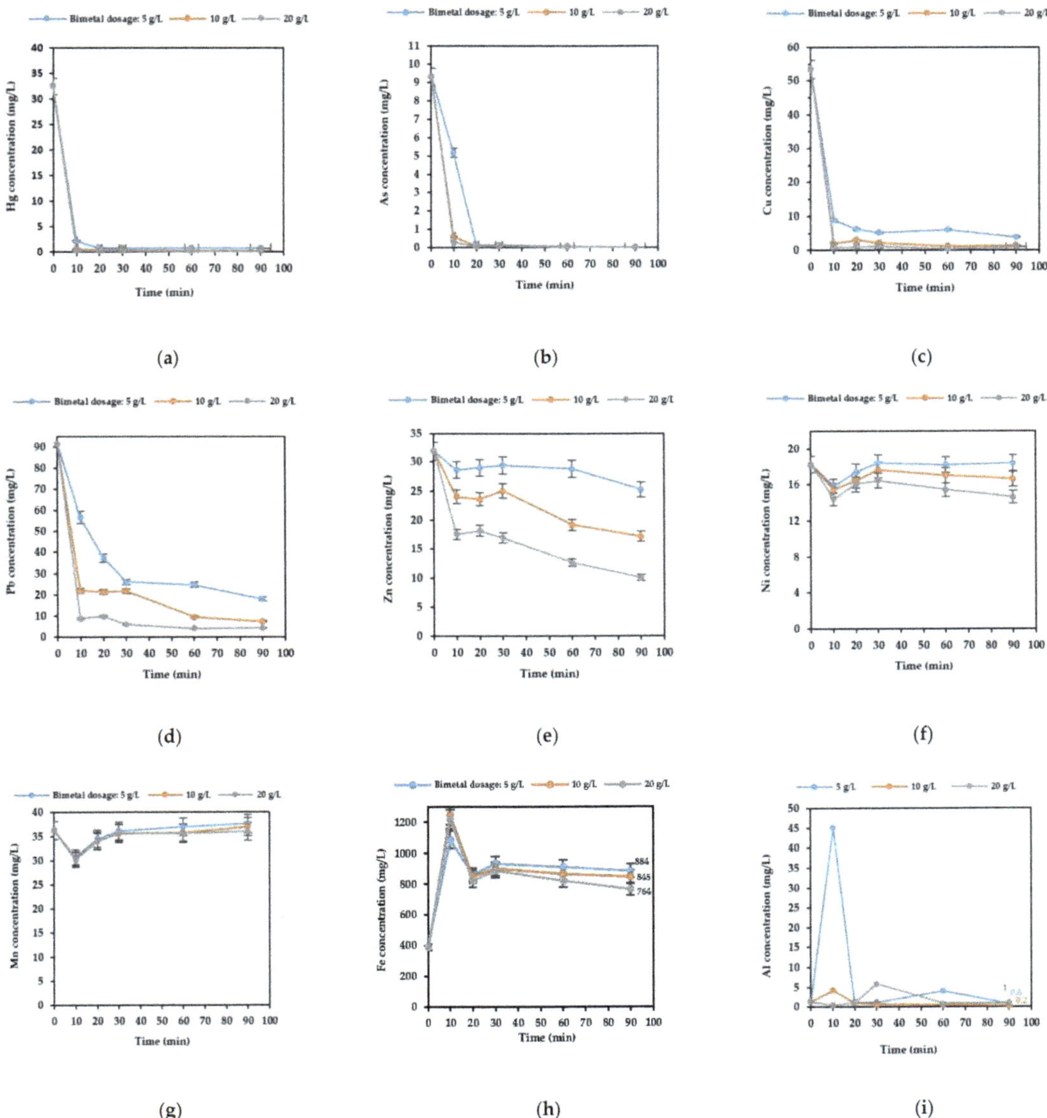

Figure 6. Variation of residual metal concentrations (**a**) As, (**b**) Hg, (**c**) Cu, (**d**) Pb, (**e**) Zn, (**f**) Ni, (**g**) Mn, (**h**) Fe, and (**i**) Al over time at different bimetal dosage (5, 10, and 20 g/L) and initial pH of 2.

The Fe corrosion (Equation (5)) led to a rise in its concentration in the solution within 10 min of the process (Figure 6h). However, it decreased after 10 min, which may be attributed to the precipitation of Fe ions by increasing the pH (Figure 7a). The total Fe ions concentration after 90 min of the process using 5, 10, and 20 g/L of the bimetal are 884, 845, and 764 mg/L, respectively. The pH rise driven by increasing the bimetal dosage can be the main reason for the lower dissolved Fe ions concentrations. As can be seen in Figure 7a (the dashed rectangle) in the experimental Eh range (0.5–0.21 V) and pH > 4,

Fe may precipitate as Schwertmannite or Magnetite. Moreover, depending on the Fe^{2+}, Fe^{3+}, and Cl^- concentrations, pH, and temperature of the reaction mixture, chloride ions may incorporate into the iron (oxy)hydroxide structure to form akaganetite [56,57], which has shown desirable sorption properties for PTE, such as As and Zn [58,59]. So, further research is needed to better understand the possibility of akaganetite formation under the experimental conditions of this study.

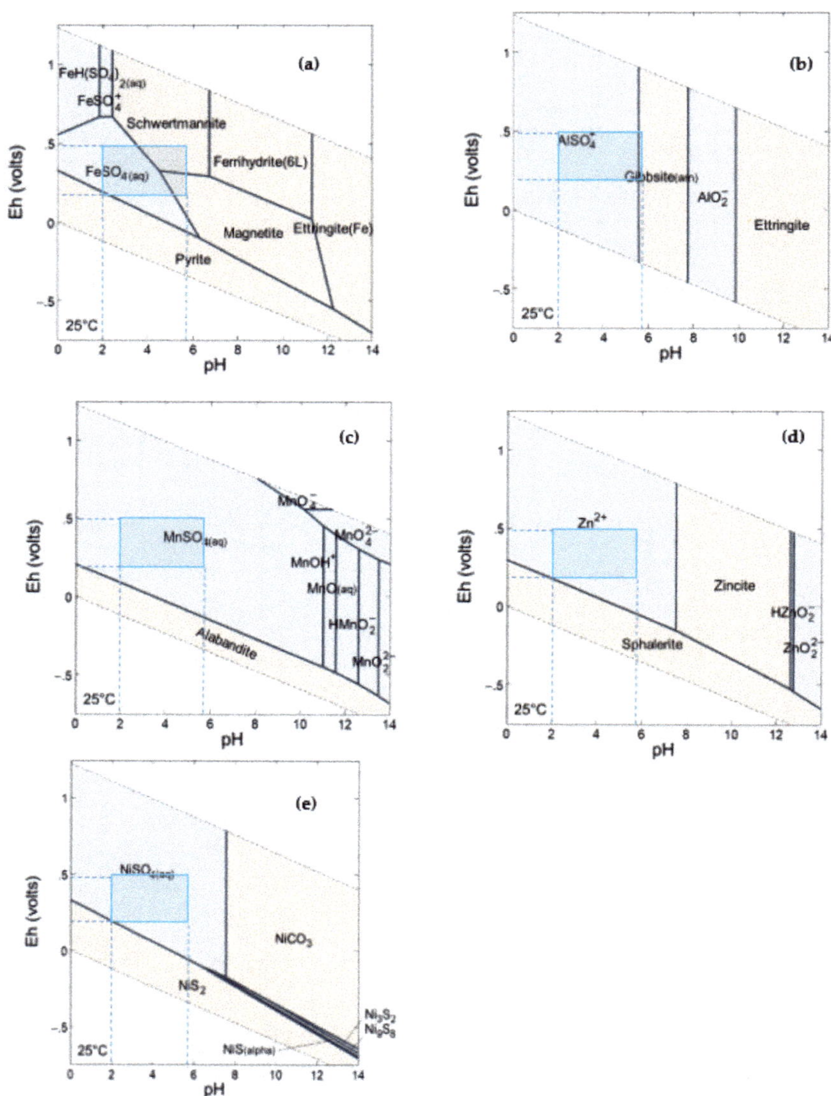

Figure 7. Eh-pH predominance diagram of (**a**) Fe^{3+} (activity = $10^{-2.65}$), (**b**) Al^{3+} (activity = $10^{-4.53}$), (**c**) Mn^{2+} (activity = $10^{-3.99}$), (**d**) Zn^{2+} (activity = $10^{-3.9}$), and (**e**) of Ni^{2+} (activity = $10^{-3.7}$) at 25 °C, 1.013 bars and activities of SO_4^{2-}, Na^+ and Ca^{2+} equal to $10^{-1.6}$, $10^{-0.57}$ and $10^{-2.4}$, respectively. Carbonate was modelled in the system by equilibrating it with the average CO_2 in air (Fugacity = $10^{-3.5}$). The dashed rectangle refers to experimental conditions in this study.

Considering that the maximum recommended level of Fe in drinking water by the World Health Organization (WHO) is 2 mg/L [25], the release of Fe ions from the bimetal after the reaction with AMD is significant. However, Fe is less harmful compared to toxic heavy metals content in AMD and can be removed from the solution by a secondary neutralization process and converted to usable iron oxides as a raw material in pigments, ceramics, etc. [37]. The Al^{3+} concentrations in the final solutions are negligible, except for the experiment with 5 g/L of the bimetal within 10 min of the process in which the total dissolved AL^{3+} was 45 mg/L (Figure 6i). It is attributed to the solution pH, which is less than 3, and Al^{3+} is the predominant species (Figure 7b). In addition, the Al^{3+} concentration in the experiment using 10 g/L of the bimetal and the reaction time of 90 min meet the established limit in drinking water by WHO (0.2 mg/L) [25].

4. Conclusions

In this investigation, the synthesized Fe-Al bimetallic material has demonstrated high efficiency for a rapid removal of potentially toxic elements from the combined AMD-waste solutions resulting from refractory gold production. Owing to a greater tendency for electrons released from the bimetal, higher removal rate was obtained for Hg, As, Cu, and Pb than the Ni, Zn, and Mn. Experiments with 20 g/L of the bimetal resulted in significant removal of Hg (99.74%), As (99.80%), Cu (98.20%), and Pb (95.50%) in 90 min, while it dropped to 69.50% removal for Zn, 22.34% for Ni, and <5% for Mn. Therefore, the electrochemical reduction of PTE by the bimetal seems to be the major contributing mechanism. The findings of this study also indicate that the higher bimetal dosages result in the greater heavy metal uptake from the solution. Moreover, the corrosion of Fe and Al in the bimetallic system and consequently the release of Fe(III), Al(III), and OH^- ions into the solution led to the formation of Fe-Al (oxy)hydroxides which could sequester PTE, such as Hg and Zn, via adsorption. In addition, with respect to electrode potential of metal species, Fe(III) ions engaged in the re-oxidation of deposited Zn, and more significantly Mn and Ni, led to an increase in their concentrations after 10 min. The increase in the initial pH of 2 to more than 5 in 90 min using Fe-Al bimetallic particles is promising in AMD remediation as it can reduce the amount of alkaline reagents. Nearly no Al ions were detected in the solutions at higher bimetal concentrations. Although the Fe release from the bimetal was high, it can be precipitated and converted to a valuable by-product such as iron pigments. However, more research on this topic needs to be undertaken to identify the influencing parameters, characterize and analyze the surface chemistry of bimetallic particles after reaction with PTE and measure the elemental composition, and chemical and electronic state of the elements on the bimetal. Chloride ions in the studied system may affect the process by akaganeite formation and changing the stability of PTE. Moreover, the higher electrode potential of nitrate compared to Fe, Al, and other metals in the process, could mean it has a higher tendency to gain electrons and get reduced. However, further studies regarding the precise effect of Cl^- and NO_3^- on PTE removal and bimetal performance is strongly recommended. The reversibility of the process and reusability of the bimetal also warrant additional investigation.

Author Contributions: Conceptualization, R.D.A., C.B.T., E.A.; methodology, E.A. and R.D.A.; validation and analysis, E.A., Z.W., R.D.A. and Z.Q.; writing—original draft preparation, E.A. and Z.W.; writing—review and editing, R.D.A., B.T., C.B.T. and E.A.; supervision, R.D.A. and B.T.; All authors have read and agreed to the published version of the manuscript.

Funding: This research received no external funding.

Data Availability Statement: Data sharing is not applicable to this article.

Acknowledgments: Curtin University's Strategic Scholarship is gratefully acknowledged for the PhD scholarship granted to Elham Aghaei.

Conflicts of Interest: The authors declare no conflict of interest.

References

1. Park, I.; Tabelin, C.B.; Jeon, S.; Li, X.; Seno, K.; Ito, M.; Hiroyoshi, N. A review of recent strategies for acid mine drainage prevention and mine tailings recycling. *Chemosphere* **2019**, *219*, 588–606. [CrossRef] [PubMed]
2. Moodley, I.; Sheridan, C.M.; Kappelmeyer, U.; Akcil, A. Environmentally sustainable acid mine drainage remediation: Research developments with a focus on waste/by-products. *Miner. Eng.* **2018**, *126*, 207–220. [CrossRef]
3. Wilkin, R.T.; McNeil, M.S. Laboratory evaluation of zero-valent iron to treat water impacted by acid mine drainage. *Chemosphere* **2003**, *53*, 715–725. [CrossRef]
4. Wills, B.A.; Finch, J.A. Chapter 16—Tailings Disposal. In *Wills' Mineral Processing Technology (Eighth Edition)*; Wills, B.A., Finch, J.A., Eds.; Butterworth-Heinemann: Boston, FL, USA, 2016; pp. 439–448. [CrossRef]
5. Kefeni, K.K.; Msagati, T.A.M.; Mamba, B.B. Acid mine drainage: Prevention, treatment options, and resource recovery: A review. *J. Clean. Prod.* **2017**, *151*, 475–493. [CrossRef]
6. Naidu, G.; Ryu, S.; Thiruvenkatachari, R.; Choi, Y.; Jeong, S.; Vigneswaran, S. A critical review on remediation, reuse, and resource recovery from acid mine drainage. *Environ. Pollut.* **2019**, *247*, 1110–1124. [CrossRef] [PubMed]
7. Dold, B. Evolution of Acid Mine Drainage Formation in Sulphidic Mine Tailings. *Minerals* **2014**, *4*, 621–641. [CrossRef]
8. Diao, Z.; Shi, T.; Wang, S.; Huang, X.; Zhang, T.; Tang, Y.; Zhang, X.; Qiu, R. Silane-based coatings on the pyrite for remediation of acid mine drainage. *Water Res.* **2013**, *47*, 4391–4402. [CrossRef] [PubMed]
9. Shu, X.; Dang, Z.; Zhang, Q.; Yi, X.; Lu, G.; Guo, C.; Yang, C. Passivation of metal-sulfide tailings by covalent coating. *Miner. Eng.* **2013**, *42*, 36–42. [CrossRef]
10. Alakangas, L.; Andersson, E.; Mueller, S. Neutralization/prevention of acid rock drainage using mixtures of alkaline by-products and sulfidic mine wastes. *Environ. Sci. Pollut. Res.* **2013**, *20*, 7907–7916. [CrossRef]
11. Nason, P.; Johnson, R.H.; Neuschütz, C.; Alakangas, L.; Öhlander, B. Alternative waste residue materials for passive in situ prevention of sulfide-mine tailings oxidation: A field evaluation. *J. Hazard. Mater.* **2014**, *267*, 245–254. [CrossRef]
12. Jin, S.; Fallgren, P.H.; Morris, J.M.; Cooper, J.S. Source Treatment of Acid Mine Drainage at a Backfilled Coal Mine Using Remote Sensing and Biogeochemistry. *Water Air Soil Pollut.* **2008**, *188*, 205–212. [CrossRef]
13. Li, X.; Hiroyoshi, N.; Tabelin, C.B.; Naruwa, K.; Harada, C.; Ito, M. Suppressive effects of ferric-catecholate complexes on pyrite oxidation. *Chemosphere* **2019**, *214*, 70–78. [CrossRef]
14. Jones, S.N.; Cetin, B. Evaluation of waste materials for acid mine drainage remediation. *Fuel* **2017**, *188*, 294–309. [CrossRef]
15. Bortnikova, S.; Gaskova, O.; Yurkevich, N.; Saeva, O.; Abrosimova, N. Chemical Treatment of Highly Toxic Acid Mine Drainage at A Gold Mining Site in Southwestern Siberia, Russia. *Minerals* **2020**, *10*, 867. [CrossRef]
16. Pat-Espadas, A.M.; Loredo Portales, R.; Amabilis-Sosa, L.E.; Gómez, G.; Vidal, G. Review of Constructed Wetlands for Acid Mine Drainage Treatment. *Water* **2018**, *10*, 1685. [CrossRef]
17. Fytas, K. Use of permeable reactive barriers to treat acid mine effluents. *Int. J. Min. Reclam. Environ.* **2010**, *24*, 206–215. [CrossRef]
18. Gibert, O.; Rötting, T.; Cortina, J.L.; de Pablo, J.; Ayora, C.; Carrera, J.; Bolzicco, J. In-situ remediation of acid mine drainage using a permeable reactive barrier in Aznalcóllar (Sw Spain). *J. Hazard. Mater.* **2011**, *191*, 287–295. [CrossRef] [PubMed]
19. Kaksonen, A.H.; Puhakka, J.A. Sulfate Reduction Based Bioprocesses for the Treatment of Acid Mine Drainage and the Recovery of Metals. *Eng. Life Sci.* **2007**, *7*, 541–564. [CrossRef]
20. Gitari, M.W.; Petrik, L.F.; Etchebers, O.; Key, D.L.; Iwuoha, E.; Okujeni, C. Treatment of acid mine drainage with fly ash: Removal of major contaminants and trace elements. *J. Environ. Sci. Health Part A Toxic/Hazard. Subst. Environ. Eng.* **2006**, *41*, 1729–1747. [CrossRef]
21. Wu, Y.; Guan, C.-Y.; Griswold, N.; Hou, L.-Y.; Fang, X.; Hu, A.; Hu, Z.-Q.; Yu, C.-P. Zero-valent iron-based technologies for removal of heavy metal(loid)s and organic pollutants from the aquatic environment: Recent advances and perspectives. *J. Clean. Prod.* **2020**, *277*, 123478. [CrossRef]
22. Obiri-Nyarko, F.; Grajales-Mesa, S.J.; Malina, G. An overview of permeable reactive barriers for in situ sustainable groundwater remediation. *Chemosphere* **2014**, *111*, 243–259. [CrossRef] [PubMed]
23. Nidheesh, P.V.; Khatri, J.; Anantha Singh, T.S.; Gandhimathi, R.; Ramesh, S.T. Review of zero-valent aluminium based water and wastewater treatment methods. *Chemosphere* **2018**, *200*, 621–631. [CrossRef] [PubMed]
24. Chen, L.-H.; Huang, C.-C.; Lien, H.-L. Bimetallic iron–aluminum particles for dechlorination of carbon tetrachloride. *Chemosphere* **2008**, *73*, 692–697. [CrossRef] [PubMed]
25. Fu, F.; Cheng, Z.; Dionysiou, D.D.; Tang, B. Fe/Al bimetallic particles for the fast and highly efficient removal of Cr(VI) over a wide pH range: Performance and mechanism. *J. Hazard. Mater.* **2015**, *298*, 261–269. [CrossRef]
26. Cheng, Z.; Fu, F.; Dionysiou, D.D.; Tang, B. Adsorption, oxidation, and reduction behavior of arsenic in the removal of aqueous As(III) by mesoporous Fe/Al bimetallic particles. *Water Res.* **2016**, *96*, 22–31. [CrossRef]
27. Xiang, S.; Cheng, W.; Nie, X.; Ding, C.; Yi, F.; Asiri, A.M.; Marwani, H.M. Zero-valent iron-aluminum for the fast and effective U(VI) removal. *J. Taiwan Inst. Chem. Eng.* **2018**, *85*, 186–192. [CrossRef]
28. Han, W.; Fu, F.; Cheng, Z.; Tang, B.; Wu, S. Studies on the optimum conditions using acid-washed zero-valent iron/aluminum mixtures in permeable reactive barriers for the removal of different heavy metal ions from wastewater. *J. Hazard. Mater.* **2016**, *302*, 437–446. [CrossRef]
29. Iakovleva, E.; Mäkilä, E.; Salonen, J.; Sitarz, M.; Wang, S.; Sillanpää, M. Acid mine drainage (AMD) treatment: Neutralization and toxic elements removal with unmodified and modified limestone. *Ecol. Eng.* **2015**, *81*, 30–40. [CrossRef]

30. Marsden, J.O.; House, C.I. *Chemistry of Gold Extraction*, 2nd ed.; SME: Littleton, CO, USA, 2009.
31. Deschenes, G.; Lastra, R.; Brown, J.R.; Jin, S.; May, O.; Ghali, E. Effect of lead nitrate on cyanidation of gold ores: Progress on the study of the mechanisms. *Miner. Eng.* **2000**, *13*, 1263–1279. [CrossRef]
32. Deschênes, G.; McMullen, J.; Ellis, S.; Fulton, M.; Atkin, A. Investigation on the cyanide leaching optimization for the treatment of KCGM gold flotation concentrate—phase 1. *Miner. Eng.* **2005**, *18*, 832–838. [CrossRef]
33. Ali, R.; Turner, J. A Study of the Suitability of Saline Surface Water for Recharging the Hypersaline Palaeochannel Aquifers of the Eastern Goldfields of Western Australia. *Mine Water Environ.* **2004**, *23*, 110–118. [CrossRef]
34. Muir, D.M. *Gold Processing with Saline Water*; The Australasian Institute of Mining and Metallurgy: Carlton, Australia, 1994.
35. Bethke, C.M.; Yeakel, S. *The Geochemist's Workbench—A User's Guide to GSS, Rxn, Act2, Tact, Spec8, React, Gtplot, X1t, X2t, and Xtplot*; Aqueous Solutions LLC: Urbana, IL, USA, 2011.
36. Blanc, P.; Lassin, A.; Piantone, P.; Azaroual, M.; Jacquemet, N.; Fabbri, A.; Gaucher, E.C. Thermoddem: A geochemical database focused on low temperature water/rock interactions and waste materials. *Appl. Geochem.* **2012**, *27*, 2107–2116. [CrossRef]
37. Simate, G.S.; Ndlovu, S. Acid mine drainage: Challenges and opportunities. *J. Environ. Chem. Eng.* **2014**, *2*, 1785–1803. [CrossRef]
38. Lien, H.-L.; Yu, C.-H.; Kamali, S.; Sahu, R.S. Bimetallic Fe/Al system: An all-in-one solid-phase Fenton reagent for generation of hydroxyl radicals under oxic conditions. *Sci. Total Environ.* **2019**, *673*, 480–488. [CrossRef] [PubMed]
39. Wu, S.; Yang, S.; Liu, S.; Zhang, Y.; Ren, T.; Zhang, Y. Enhanced reactivity of zero-valent aluminum with ball milling for phenol oxidative degradation. *J. Colloid Interface Sci.* **2020**, *560*, 260–272. [CrossRef]
40. Bokare, A.D.; Choi, W. Zero-valent aluminum for oxidative degradation of aqueous organic pollutants. *Environ. Sci. Technol.* **2009**, *43*, 7130–7135. [CrossRef]
41. Yang, S.; Zheng, D.; Ren, T.; Zhang, Y.; Xin, J. Zero-valent aluminum for reductive removal of aqueous pollutants over a wide pH range: Performance and mechanism especially at near-neutral pH. *Water Res.* **2017**, *123*, 704–714. [CrossRef] [PubMed]
42. O'Carroll, D.; Sleep, B.; Krol, M.; Boparai, H.; Kocur, C. Nanoscale zero valent iron and bimetallic particles for contaminated site remediation. *Adv. Water Resour.* **2013**, *51*, 104–122. [CrossRef]
43. Vernon, J.D.; Bonzongo, J.-C.J. Volatilization and sorption of dissolved mercury by metallic iron of different particle sizes: Implications for treatment of mercury contaminated water effluents. *J. Hazard. Mater.* **2014**, *276*, 408–414. [CrossRef] [PubMed]
44. Meng, C.; Mao, Q.; Luo, L.; Zhang, J.; Wei, J.; Yang, Y.; Tan, M.; Peng, Q.; Tang, L.; Zhou, Y. Performance and mechanism of As(III) removal from water using Fe-Al bimetallic material. *Sep. Purif. Technol.* **2018**, *191*, 314–321. [CrossRef]
45. Liu, F.; Yang, W.; Li, W.; Zhao, G.-C. Simultaneous Oxidation and Sequestration of Arsenic(III) from Aqueous Solution by Copper Aluminate with Peroxymonosulfate: A Fast and Efficient Heterogeneous Process. *ACS Omega* **2021**, *6*, 1477–1487. [CrossRef] [PubMed]
46. Tabelin, C.B.; Igarashi, T.; Villacorte-Tabelin, M.; Park, I.; Opiso, E.M.; Ito, M.; Hiroyoshi, N. Arsenic, selenium, boron, lead, cadmium, copper, and zinc in naturally contaminated rocks: A review of their sources, modes of enrichment, mechanisms of release, and mitigation strategies. *Sci. Total Environ.* **2018**, *645*, 1522–1553. [CrossRef]
47. Igarashi, T.; Herrera, P.S.; Uchiyama, H.; Miyamae, H.; Iyatomi, N.; Hashimoto, K.; Tabelin, C.B. The two-step neutralization ferrite-formation process for sustainable acid mine drainage treatment: Removal of copper, zinc and arsenic, and the influence of coexisting ions on ferritization. *Sci. Total Environ.* **2020**, *715*, 136877. [CrossRef]
48. Rangsivek, R.; Jekel, M.R. Removal of dissolved metals by zero-valent iron (ZVI): Kinetics, equilibria, processes and implications for stormwater runoff treatment. *Water Res.* **2005**, *39*, 4153–4163. [CrossRef]
49. Beverskog, B.; Puigdomenech, I. *Pourbaix Diagrams for the System Copper-Chlorine at 5–100 °C*; Swedish Nuclear Power Inspectorate: Stockholm, Sweden, 1998; p. 56.
50. He, D.; Zeng, L.; Zhang, G.; Guan, W.; Cao, Z.; Li, Q.; Wu, S. Extraction behavior and mechanism of nickel in chloride solution using a cleaner extractant. *J. Clean. Prod.* **2020**, *242*, 118517. [CrossRef]
51. Stec, M.; Jagustyn, B.; Słowik, K.; Ściążko, M.; Iluk, T. Influence of High Chloride Concentration on pH Control in Hydroxide Precipitation of Heavy Metals. *J. Sustain. Metall.* **2020**, *6*, 239–249. [CrossRef]
52. Grassi, S.; Netti, R. Sea water intrusion and mercury pollution of some coastal aquifers in the province of Grosseto (Southern Tuscany—Italy). *J. Hydrol.* **2000**, *237*, 198–211. [CrossRef]
53. Spyropoulou, A.; Lazarou, Y.G.; Laspidou, C. Mercury Speciation in the Water Distribution System of Skiathos Island, Greece. *Proceedings* **2018**, *2*, 668. [CrossRef]
54. Liu, Y.; Wang, J. Reduction of nitrate by zero valent iron (ZVI)-based materials: A review. *Sci. Total Environ.* **2019**, *671*, 388–403. [CrossRef]
55. Esfahani, A.R.; Datta, T. Nitrate removal from water using zero-valent aluminium. *Water Environ. J.* **2020**, *34*, 25–36. [CrossRef]
56. Scheck, J.; Lemke, T.; Gebauer, D. The Role of Chloride Ions during the Formation of Akaganéite Revisited. *Minerals* **2015**, *5*, 778–787. [CrossRef]
57. Rémazeilles, C.; Refait, P. On the formation of β-FeOOH (akaganéite) in chloride-containing environments. *Corros. Sci.* **2007**, *49*, 844–857. [CrossRef]
58. Zhao, J.; Lin, W.; Chang, Q.; Li, W.; Lai, Y. Adsorptive characteristics of akaganeite and its environmental applications: A review. *Environ. Technol. Rev.* **2012**, *1*, 114–126. [CrossRef]
59. Deliyanni, E.A.; Bakoyannakis, D.N.; Zouboulis, A.I.; Peleka, E. Removal of Arsenic and Cadmium by Akaganeite Fixed-Beds. *Sep. Sci. Technol.* **2003**, *38*, 3967–3981. [CrossRef]

 *minerals*

Article

Effects of Ferrous Sulfate Addition on the Selective Flotation of Scheelite over Calcite and Fluorite

Moon Young Jung [1], Jay Hyun Park [2],* and Kyoungkeun Yoo [3],* 

[1] Department of Bio Environmental Engineering, Semyung University, 65 Semyung-ro, Chungcheongbuk-do 27136, Korea; myjung@semyung.ac.kr
[2] Institute of Mine Reclamation Technology, Mine Reclamation Corporation(MIRECO), 2 Segye-ro, Gangwon-do 26464, Korea
[3] Department of Energy & Resources Engineering, Korea Maritime and Ocean University (KMOU), 727 Taejong-ro, Busan 49112, Korea
* Correspondence: jayhp@mireco.or.kr (J.H.P.); kyoo@kmou.ac.kr (K.Y.)

Received: 15 September 2020; Accepted: 29 September 2020; Published: 30 September 2020

Abstract: The addition of ferrous sulfate as a depressant for Ca-bearing minerals such as calcite and fluorite during scheelite flotation was investigated to recover scheelite from tungsten mine tailings, using Hallimond-tube flotation tests, zeta-potential measurement and Fourier-transform infrared (FT-IR) analyses. The flotation tests indicate that the selectivity of scheelite recovery was the largest over calcite and fluorite under the following conditions: 0.5 g sample, 50 g/ton AF65, 1×10^{-3} M sodium oleate, 1200 g/ton SF2 (sodium silicate and ferrous sulfate) depressant with the 8:2 ratio of sodium silicate and ferrous sulfate, 50 mL/min air injection rate, 5 min flotation time, and pH 8. The selectivity of scheelite flotation increased when the amount of SF2 depressant addition increased to 1200 g/ton, but it decreased by adding 1400 g/ton SF2, which would result from the precipitation of iron components. In the zeta potential results, the zeta potentials of scheelite with the collector show similar results regardless of the addition of SF2, while the change of zeta potentials of calcite and fluorite by adding NaO collector diminished when SF2 was added. In FT-IR analyses, the spectrum of NaO in scheelite results was observed regardless of the addition of SF2, while the spectra of NaO in calcite and fluorite results disappeared when SF2 was added. These results suggest that the addition of SF2 prevents the adsorption of NaO on the surface of calcite and fluorite. Therefore, the addition of SF2 could enhance the selectivity of scheelite flotation over calcite and fluorite.

Keywords: scheelite; calcite; fluorite; depressant; ferrous sulfate; selective flotation

1. Introduction

The number of abandoned metal mines in South Korea has been found to be 2089, and it was reported that 1268 of the abandoned metal mines have detrimentally affected the environments around the abandoned mines [1]. Sangdong mine is a representative scheelite mine in South Korea, and 12 million tons of tailings was generated during a 40-year operation after it had been abandoned [2,3]. In South Korea, Mine Reclamation Corporation (MIRECO), a state-owned company, has developed technologies to recover valuable materials from tailings [4], which could reduce potential risk of tailings. Although some research was performed to use the tailing of Sandong mine as constructive materials [2], technology for the recovery of valuable materials such as scheelite should be required.

Scheelite has been found to be a representative tungsten ore [5]. Tungsten is being widely used in various industries such as the high-temperature technology, chemical, lighting, X-ray technology, and machine manufacturing industries, owing to its low vapor pressure, high melting point, excellent electricity and thermal conductivity, high density, high abrasion resistance, and excellent

X-ray performance [6,7]. Therefore, the recovery of scheelite from tailings is important not only in obtaining tungsten, but also in reducing the volume of tailings. Gravity separation such as jig process has been reported to recover scheelite [8], but, generally, the flotation process has been selected for beneficiation of scheelite [9].

It is known that scheelite can be selectively floated by using fatty acid and a fatty acid derivative as a collector [10,11]. However, since scheelite ($CaWO_4$) contains calcium components, it is closely related to other Ca-bearing minerals such as calcite ($CaCO_3$) and fluorite (CaF_2). Scheelite has similar physicochemical properties, such as the existence of the same cations (Ca^{2+}) in the mineral, solubility, hardness, specific gravity, and point of zero charge [12,13], which causes similar chemisorption behavior of the fatty acid and a fatty acid derivative as a collector on the Ca-bearing minerals [9,14–16]. Generally, it has been found that sodium silicate is the most widely used depressant for selective flotation of scheelite [15], and assistant depressant such as ferrous sulfate has been investigated [15,16].

Although the mechanism of the depression of calcite [15] and fluorite [16] during scheelite flotation, using a mixture of sodium silicate and ferrous sulfate as a depressant, was reported by investigating the minerals with X-ray photoelectron spectroscopy (XPS), more experiments will be required to understand the depression mechanism of sodium silicate and ferrous sulfate. Therefore, in the present study, the optimum flotation condition was obtained from Hallimond-tubes flotation tests, and then the effect of ferrous sulfate addition with sodium silicate was investigated with zeta potential measurement and FT-IR analyses using scheelite, calcite and fluorite conditioned at the optimum condition.

2. Materials and Methods

2.1. Materials

The samples that were used in this study were the concentrates of scheelite ($CaWO_4$), calcite ($CaCO_3$), and fluorite (CaF_2), which were obtained from domestic mines such as Sandong mine, Imha mine, and Sinpo mine, respectively. Figure 1 shows the X-ray diffraction (XRD) analysis results for scheelite, calcite, and fluorite. The analysis results show that all the three samples are pure, with almost absolutely no impurities.

All reagents used in this study are of reagent grade except AF65 (commercial grade). Sodium oleate (Samchun Chemical Co., Ltd., Yeosu, Korea), which is known to be used often in the flotation research on scheelite and Ca-bearing minerals [15,16], and AF65 (Cytec Industries Inc., Woodland Park, NJ, USA) were used as a collector and a frother, respectively. A mixture of ferrous sulfate heptahydrate ($FeSO_4{\cdot}7H_2O$, Samchun Chemical Co., Ltd., Yeosu, Korea), or oxalic acid ($C_2H_2O_4{\cdot}2H_2O$, Samchun Chemical Co., Ltd., Yeosu, Korea) with sodium silicate (Na_2SiO_3, Samchun Chemical Co., Ltd., Yeosu, Korea) was used as a depressant. Table 1 shows the summary of the depressant used in this study.

Table 1. Depressants used in this study.

Depressant	Abbreviation	Chemical Composition
Sodium silicate		Na_2SiO_3
Sodium silicate + ferrous sulfate	SF2	$Na_2SiO_3 + FeSO_4$
Sodium silicate + oxalic acid	SO	$Na_2SiO_3 + C_2H_2O_4$

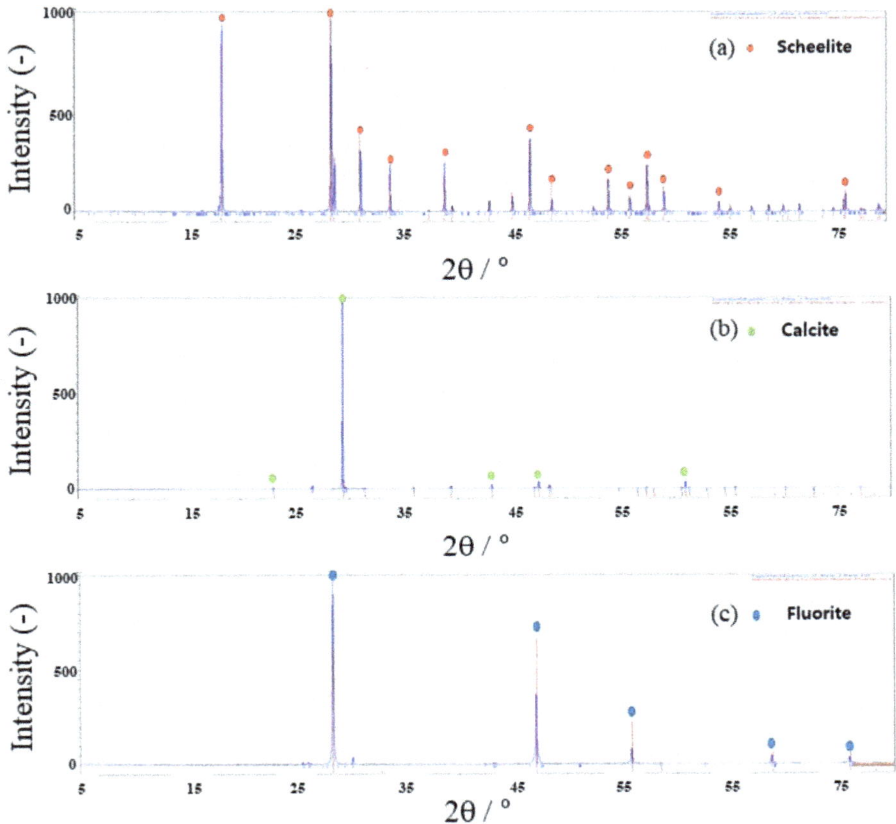

Figure 1. X-ray diffraction (XRD) analysis results for Ca-bearing minerals: (**a**) scheelite (CaWO$_4$), (**b**) calcite (CaCO$_3$), and (**c**) fluorite (CaF$_2$).

2.2. Flotation Test Procedures

Flotation tests on singular mineral (scheelite, calcite, and fluorite) were using Hallimond tubes with 300 mL of working volume under the following conditions: 0.5 g mineral sample (75 μm–150 μm), 50 g/ton AF65 as a frother, 1×10^{-3} M sodium oleate as a collector, 50 mL/min air injection rate, 5-min conditioning for each reagent addition, and 5 min flotation time. The pulp was stirred with a magnetic bar during the conditioning and flotation tests. In a typical run, the pH of 300 mL distilled water with the sample was adjusted at 8 with HCl and NaOH, and then depressant, collector, and frother were added in order. After the flotation tests, the products were collected by filtering with a filter paper and then dried at 105 °C. The floatability (%) was calculated as follows:

$$\text{Floatability (\%)} = M_{\text{final}} \text{ (g)} / M_{\text{initial}} \text{ (g)} \times 100 \tag{1}$$

where M_{final} and M_{initial} indicate mass of final and initial added mineral, respectively.

2.3. Zeta Potential and Fourier-Transform Infrared (FTIR) Analyses

The zeta potentials of the mineral particles were measured with Beckman Coulter's Delsa Nano C. The mineral samples were finely ground with an agate mortar, and less than 5 μm particles collected using a filter (pore size: 5 μm) were used. The zeta potential was measured using a 1.0×10^{-3} M

NaCl solution as the supporting electrolyte. The pH value was adjusted with HCl and NaOH and was measured after the Zeta potential tests.

The adsorption of the collector and depressants on the surfaces of Ca-bearing minerals was examined using Jasco's FTIR spectrometer. After the 0.5 g mineral sample with less than 200 mesh was added to 300 mL distilled water, the flotation reagents (collector and depressant) were added to the suspension, which was conditioned for 5 min. Then, the solid-liquid separated mineral samples were dried in vacuum before FTIR analyses.

3. Results and Discussion

Bo et al. reported that sodium silicate forms hydrophilic colloidal silicic acid, which is selectively adsorbed on the surface of calcite and fluorite, and that, in particular, the formation of hydrophilic colloids became smoother in their study when the mixing ratio of sodium silicate and oxalic acid was 3:1 [17]. Based on this finding, the depressant that contains the sodium silicate and oxalic acid in a ratio of 7.5:2.5, 5:5, 2.5:7.5, and 0:10 was used to investigate the effect of the depressants on the floatability of scheelite, calcite and fluorite. Figure 2 shows the floatability of the minerals increased, but the selectivity decreased with decreasing the ratio of sodium silicate.

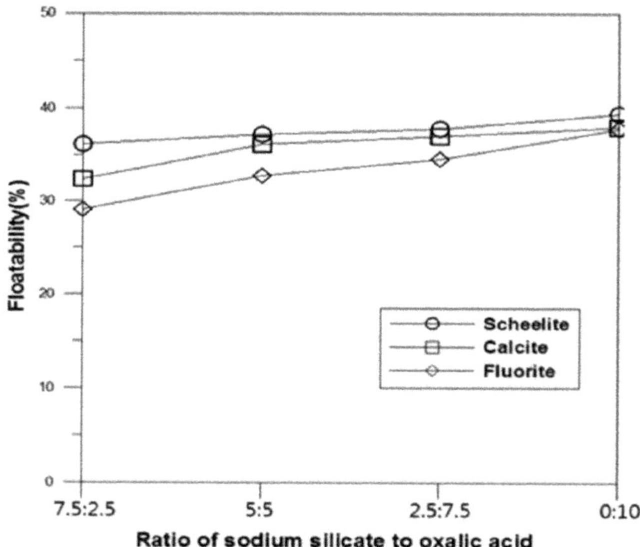

Figure 2. Floatability of scheelite, calcite, and fluorite in the function of the mixing ratio of sodium silicate and oxalic acid from 7.5:2.5 to 0:10.

The difference in floatability was the highest at the 7.5:2.5 sodium silicate-oxalic acid mixing ratio, and the result shows a good agreement with the conventional study [17], where the 3:1 ratio was suggested as an optimum condition. At this ratio of 7.5:2.5, however, the floatabilities of scheelite, calcite and fluorite were 36%, 32% and 29%, respectively. Thus, satisfactory selection efficiency could not be expected when using the depressant of sodium silicate and oxalic acid.

The combination of sodium silicate and ferrous sulfate (SF2) was proposed as a depressant for the selective flotation of scheelite [15,16]. The depressant SF2 was prepared in a ratio of 6:4 to 9:1 of 5% sodium silicate and 1% ferrous sulfate. The flotation tests with SF2 were performed under the following conditions: 0.5 g sample, 50 g/ton AF65, 1×10^{-3} M sodium oleate, 200 g/ton SF2 depressant, 50 mL/min air injection rate, 5 min flotation time, and pH 8, and the results are shown in Figure 3.

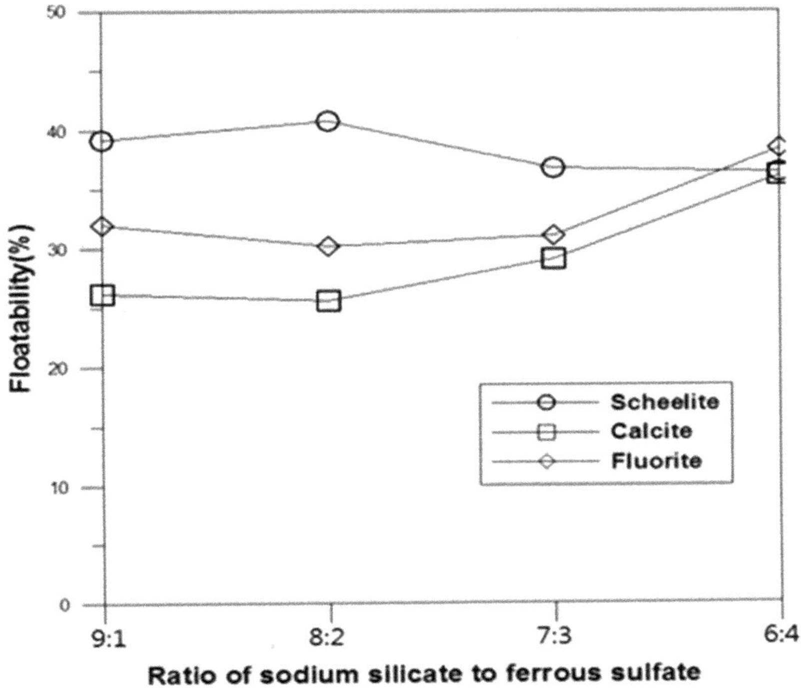

Figure 3. Floatability of scheelite, calcite, and fluorite in the function of the mixing ratio of sodium silicate and ferrous sulfate from 6:4 to 9:1.

The floatability of scheelite decreased but those of calcite and fluorite increased with increasing the ratio of ferrous sulfate. The precipitation was observed in the solution, and the amount of precipitate increased with the ratio of ferrous sulfate. Iron component was detected in the precipitate, and iron precipitated due to the oxidation of ferrous to ferric ions by aeration because ferric ion could precipitate over pH 3.5 [18]. At a ratio of 8:2 for sodium silicate and ferrous sulfate, the floatabilities of scheelite, calcite and fluorite were 40.8%, 25.6% and 30.2%, respectively, and the selectivity was the largest at this ratio. The selectivity decreased with increasing the ratio of ferrous sulfate because the effect of ferrous sulfate addition diminished due to the precipitation.

The effects of SF2 depressant dosage on the floatability of scheelite, calcite, and fluorite was investigated under the following conditions: 0.5 g sample, 50 g/ton AF65, 1×10^{-3} M sodium oleate, 0–1400 g/ton SF2 depressant with the 8:2 ratio of sodium silicate and ferrous sulfate, 50 mL/min air injection rate, 5 min flotation time, and pH 8, and the results were shown in Figure 4. With increasing SF2 dosage, the floatability of scheelite decreased gradually while those of calcite and fluorite did sharply. That is, when the SF2 dosage increased from 200 to 1200 g/ton, the floatability of scheelite decreased by 7% (from 40.8 to 33.6%) but the floatability of calcite decreased sharply by 21% (from 25.6 to 4.6%) and that of fluorite decreased sharply by approximately 20% (from 30.2 to 10.6%). When the SF2 dosage was 1400 g/ton, the floatability values of calcite and fluorite were around 5%, suggesting an excellent suppressing effect, but that of scheelite was also severely suppressed to around 20%. Since the selectivity is the largest at 1200 g/ton of SF2 dosage, the SF2 dosage was fixed at 1200 g/ton in the following tests.

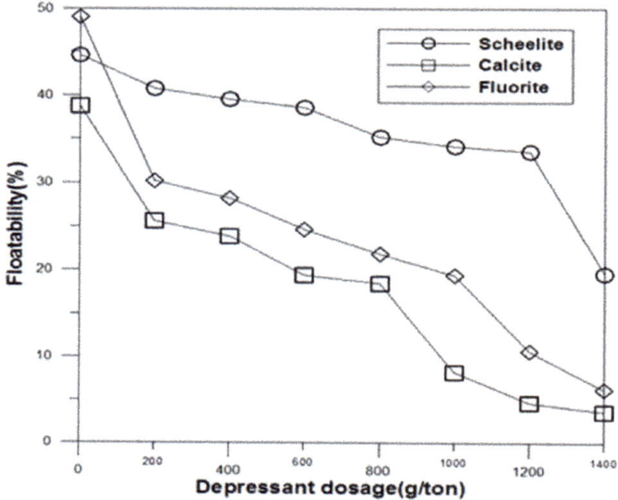

Figure 4. Floatability of scheelite, calcite, and fluorite in the function of SF2 dosage.

The generally known order of addition of flotation reagents is to add a depressant before a collector. The Ca-bearing minerals were recovered through bulk flotation using the collector sodium oleate, and then selective flotation was performed for scheelite from the bulk flotation product. Therefore, the effect of the order of flotation-reagent addition on the floatability was investigated under the following conditions: 0.5 g sample, 50 g/ton AF65, 1×10^{-3} M sodium oleate, 1200 g/ton SF2 depressant with the 8:2 ratio of sodium silicate and ferrous sulfate, 50 mL/min air injection rate, 5 min flotation time, and pH 8. The results are shown in Figure 5.

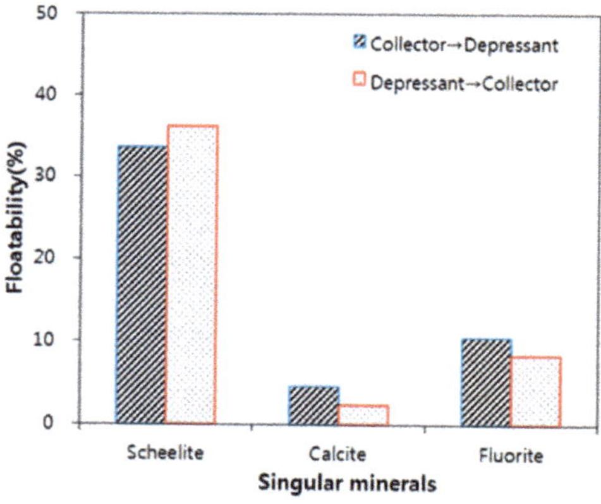

Figure 5. Floatability of scheelite, calcite, and fluorite in the function of the order of flotation-reagent addition.

As shown in Figure 5, the floatability of scheelite was higher than those of calcite and fluorite. The floatability of calcite and fluorite reduced when the depressant was added before the collector, while that of scheelite was enhanced by adding the depressant before the collector, where the selectivity became larger. Therefore, the effect of collector, which was added in the bulk flotation, should be considered to obtain the better selectivity of scheelite. Based on the results, the optimum flotation conditions with high selection efficiency for scheelite are 0.5 g sample, 50 g/ton AF65, 1×10^{-3} M sodium oleate, 1200 g/ton SF2 depressant with the 8:2 ratio of sodium silicate and ferrous sulfate,

50 mL/min air injection rate, 5 min flotation time, and pH 8, and the addition order of flotation reagents is to add the depressant before the collector.

Figures 6–8 show the zeta potential measurement results for scheelite, calcite, and fluorite with or without 1×10^{-3} M sodium oleate (collector) and 1200 g/ton SF2 (depressant). Each measurement was performed after 5 min conditioning when the collector or the depressant was added, and 1×10^{-3} M NaCl was added as a supporting solution. In the zeta potentials of scheelite, calcite, and fluorite without the collector and depressant, the measurement results showed that scheelite did not have a point of zero charge (PZC), but that the points of zero charge for calcite and fluorite are 8.6 and 4.7, respectively. Figure 6 indicates the Zeta potential results of scheelite with pH, which shows a good agreement with conventional studies [14,19]. The zeta potentials of scheelite without the collector and depressant are negatively charged over the entire pH range (2–11), and the zeta potentials are lower negatively when the flotation reagents are added. Since the zeta potentials of scheelite with the collector or with the collector and depressant show similar behaviors over the entire pH range, the addition of depressant does not have a significant effect on the adsorption of the collector on the surface of scheelite.

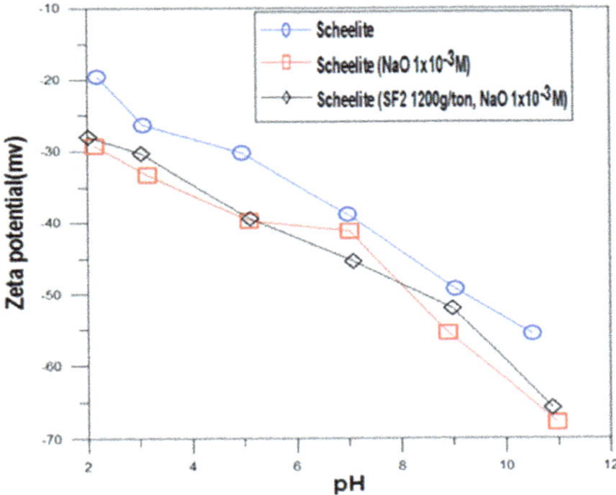

Figure 6. Zeta potentials of scheelite with and without the collector and depressant.

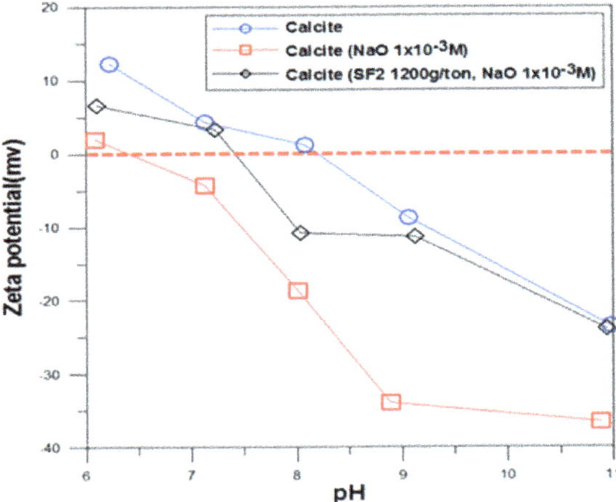

Figure 7. Zeta potentials of calcite with and without the collector and depressant.

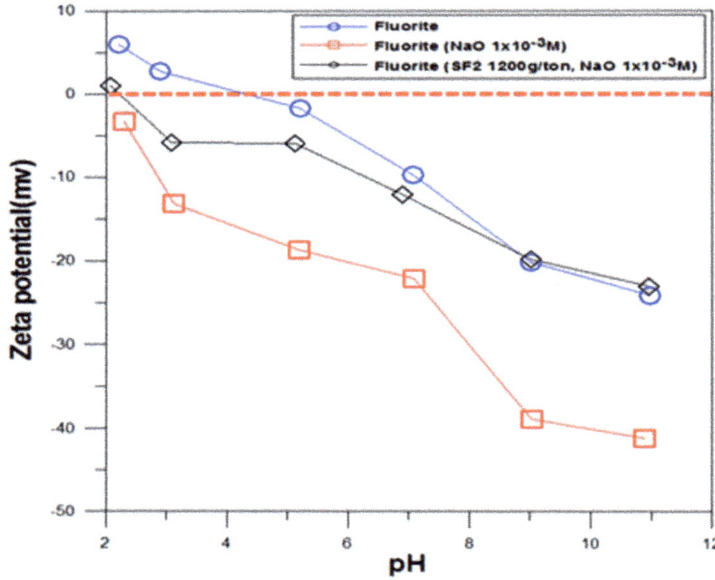

Figure 8. Zeta potentials of fluorite with and without the collector and depressant.

As shown in Figures 7 and 8, the zeta potentials of the calcite and fluorite only with the collector are the lowest at each pH compared with the zeta potentials of the minerals with or without the collector and depressant. The zeta potentials of calcite and fluorite with or without the collector and depressant show similar behaviors with increasing pH, which indicates that the addition of the SF2 depressant interfered the adsorption of the collector on the surface of minerals. Therefore, the addition of SF2 could enhance the selectivity of scheelite flotation.

FTIR analysis was performed to examine the adsorption of the flotation reagents on the surface of the minerals conditioned at the optimum flotation conditions, and the results were shown in Figures 9–11 for scheelite, calcite, and fluorite, respectively. Figure 9 shows the FTIR analysis results for scheelite, where the spectrum of sodium oleate forms the bands at 1559.1 and 2922.6 cm^{-1}. The band at 2922.6 cm^{-1} is known to be formed by the stretching vibrations of CH_2 and CH_3 [17], and the band at 1559.1 cm^{-1} is caused by the carbonyl mode (C=O), which mainly indicates a physically adsorbed species [7,20,21]. The spectrum of scheelite without flotation reagents (NaO and SF2) shows the bands in the 650–1000 cm^{-1} range, and, when the scheelite sample was conditioned in the 1×10^{-3} M sodium oleate solution, a band was observed at 2871.3 cm^{-1}, which is similar to the 2922.6 cm^{-1} of sodium oleate. In the experiment where the SF2 depressant was added before sodium oleate, a band was formed at 2900.7 cm^{-1}. These results suggest that the chemisorption of sodium oleate on the scheelite is maintained regardless of the addition of SF2. However, in the cases of calcite and fluorite, as shown in Figures 10 and 11, respectively, the spectra of calcite and fluorite with the collector (NaO) show the band at 2931.5 and 2910.4 cm^{-1}, which are due to the stretching vibration of sodium oleate, indicating good adsorption, but they disappeared in the minerals with the collector after the addition of the depressant (SF2). These results indicate that the addition of SF2 depressant prevents the adsorption of the NaO collector, which shows a good agreement with the experiment results of zeta potential measurement. Therefore, the addition of SF2 depressant could enhance the selectivity of scheelite over Ca-bearing minerals such as calcite and fluorite.

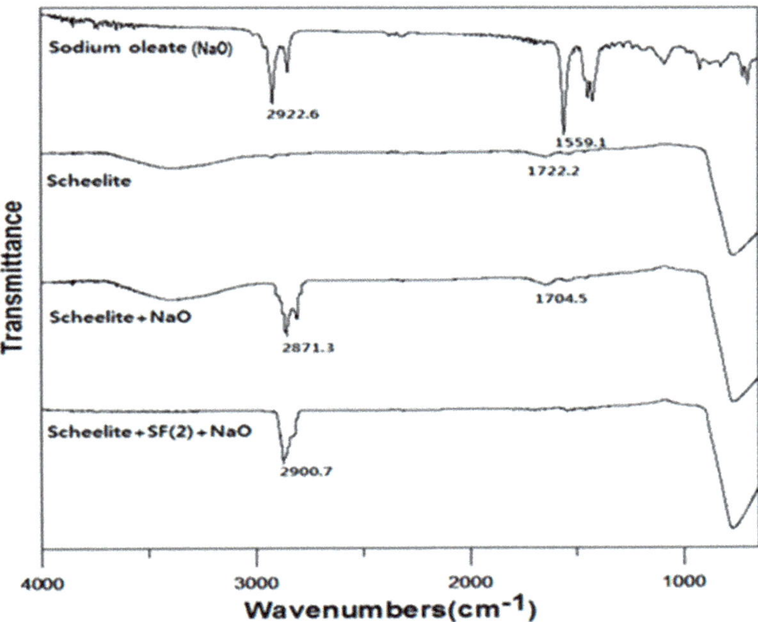

Figure 9. Fourier-Transform Infrared (FTIR) analysis result of scheelite at the optimum flotation conditions.

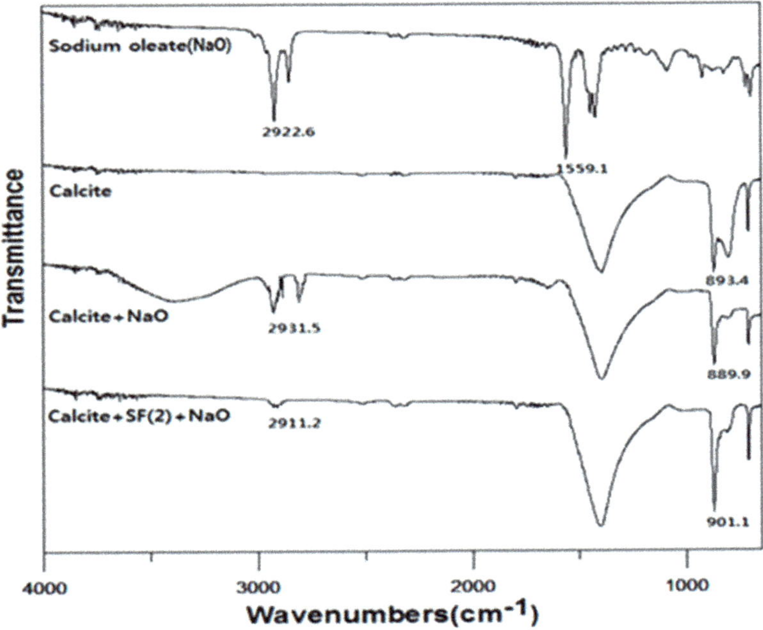

Figure 10. FTIR analysis result of calcite at the optimum flotation conditions.

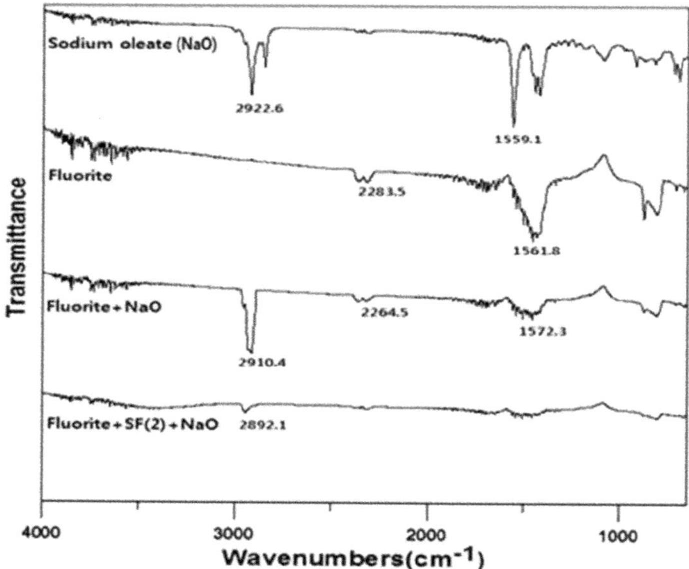

Figure 11. FTIR analysis result of fluorite at the optimum flotation conditions.

4. Conclusions

The addition of ferrous sulfate as a depressant for Ca-bearing minerals such as calcite and fluorite during scheelite flotation was investigated using Hallimond-tube flotation tests, zeta-potential measurement and FT-IR analyses.

The flotation tests indicate that the selectivity of scheelite recovery was the largest over calcite and fluorite under the following conditions: 0.5 g sample, 50 g/ton AF65, 1×10^{-3} M sodium oleate, 1200 g/ton SF2 (sodium silicate and ferrous sulfate) depressant with the 8:2 ratio of sodium silicate and ferrous sulfate, 50 mL/min air injection rate, 5 min flotation time, and pH 8. The selectivity of scheelite flotation increased when the amount of SF2 depressant addition increased to 1200 g/ton, but it decreased by adding 1400 g/ton SF2, which would result from the precipitation of iron components.

In the zeta potential and Fourier-transform infrared (FTIR) analyses results of scheelite, calcite, and fluorite conditioned under the optimum conditions, the adsorption of NaO (collector) on the surface of scheelite was observed regardless of SF2 depressant addition, while the zeta potentials of calcite and fluorite and FT-IR spectra with NaO/SF2 show similar behavior with calcite and fluorite without any floatation reagents. These results indicate that the addition of SF2 prevents the adsorption of NaO (collector) on the surface of calcite and fluorite, which are regarded as tailing during scheelite flotation, so the selectivity of scheelite flotation could be enhanced by adding the depressant consisting of sodium silicate and ferrous sulfate.

Author Contributions: Methodology, M.Y.J. and J.H.P.; writing-original draft preparation, M.Y.J. and K.Y.; project administration and funding acquisition, M.Y.J. and J.H.P.; data curation, M.Y.J. and K.Y.; writing-review and editing, providing ideas, M.Y.J., J.H.P. and K.Y. All authors have read and agreed to the published version of the manuscript.

Funding: This study was carried out as a research project for the "Establishment of WO₃ flotation concentration conditions in tungsten tailings" of Mineral Reclamation Corporation in 2015. The authors would like to express their sincere gratitude for the support given by such corporation.

Conflicts of Interest: The authors declare no conflict of interest.

References

1. Nguyen, T.T.; Yoo, K.; Jha, M.K.; Park, J.; Choi, U.; Choe, H.; Lee, J.C. Removal of Heavy Metals from Tailing in Citrate Solution with Ferric Chloride. *Mater. Trans.* **2018**, *59*, 1665–1668. [CrossRef]
2. Chu, Y.S.; Seo, S.G.; Choi, S.B.; Kim, G.M.; Hong, S.H. Properties of Cement Mortar as Particle Size and Hydrothermal Synthesis Temperature Using Scheelite Tailing. *J. Korean Inst. Resour. Recycl.* **2019**, *28*, 46–53. [CrossRef]
3. Kim, M.S.; Kang, H.C. A Study on Mineralogical Characterizations of Sangdong Mine Tailings. *J. Korean Soc. Miner. Energy Resour. Eng.* **2014**, *51*, 829–834. [CrossRef]
4. Yang, I.; Ji, W.; Park, J. Strategic investigation of development of mine reclamation technology based on third-stage road map. *J. Korean Soc. Miner. Energy Resour. Eng.* **2018**, *55*, 538–545. [CrossRef]
5. Ahn, H.H.; Lee, M.S. Hydrometallurgical Processes for the Recovery of Tungsten from Ores and Secondary Resources. *J. Korean Inst. Resour. Recycl.* **2018**, *27*, 3–10. [CrossRef]
6. Ilhan, S.; Kalpakli, A.O.; Kahruman, C.; Yusufoglu, I. The investigation of dissolution behavior of gangue materials during the dissolution of scheelite concentrate in oxalic acid solution. *Hydrometallurgy* **2013**, *136*, 15–26. [CrossRef]
7. Rao, K.H.; Forssberg, K.S.E. Mechanism of oleate interaction on salt-type minerals Part III. Adsorption, zeta potential and diffuse reflectance FT-IR studies of scheelite in the presence of sodium oleate. *Colloids Surf.* **1991**, *54*, 161–187. [CrossRef]
8. Baek, S.H.; Jeon, H.S. Application of Jig Separation for Pre-Concentration of Low-Grade Scheelite Ore. *Mater. Trans.* **2018**, *59*, 494–498. [CrossRef]
9. Kupka, N.; Rudolph, M. Froth flotation of scheelite–A review. *Int. J. Min. Sci. Techno.* **2018**, *28*, 373–384. [CrossRef]
10. Hu, Y.; Gao, Z.; Sun, W.; Liu, X. Anisotropic surface energies and adsorption behaviors of scheelite crystal. *Colloids Surf. A.* **2012**, *415*, 439–448. [CrossRef]
11. Shepeta, E.D.; Samatova, L.A.; Kondratev, S.A. Kinetics of calcium minerals flotation from Scheelite-carbonate ores. *J. Min. Sci.* **2012**, *48*, 746–753. [CrossRef]
12. Hu, Y.H.; Yang, F.; Sun, W. The flotation separation of scheelite from calcite using a quaternary ammonium salt as collector. *Miner. Eng.* **2011**, *24*, 82–84. [CrossRef]
13. Ozcan, O.; Bulutcu, A.N. Electrokinetic, infrared and flotation studies of scheelite and calcite with oxine, alkyl oxine, oleoyl sarcosine and quebracho. *Int. J. Miner. Process.* **1993**, *39*, 275–290. [CrossRef]
14. Gao, Z.; Bai, D.; Sun, W.; Cao, X.; Hu, Y. Selective flotation of scheelite from calcite and fluorite using a collector mixture. *Miner. Eng.* **2015**, *72*, 23–26. [CrossRef]
15. Deng, R.; Yang, X.; Hu, Y.; Ku, J.; Zuo, W.; Ma, Y. Effect of Fe (II) as assistant depressant on flotation separation of scheelite from calcite. *Miner. Eng.* **2018**, *118*, 133–140. [CrossRef]
16. Hu, Y.; Huang, Y.; Deng, R.; Ma, L.; Yin, W. Improvement Effect of $FeSO_4 \cdot 7H_2O$ on Flotation Separation of Scheelite from Fluorite. *ACS Omega.* **2019**, *4*, 11364–11371. [CrossRef]
17. Bo, F.; Xianping, L.; Jinqing, W.; Pengcheng, W. The flotation separation of scheelite from calcite using acidified sodium silicate as depressant. *Miner. Eng.* **2015**, *80*, 45–49. [CrossRef]
18. Na, H.; Eom, Y.; Hong, S.; Yoo, K. The Effects of Iron Powder Agglomeration on the Copper Removal Efficiency during Cementation Process for Treating Mine Drainages. *J. Korean Inst. Resour. Recycl.* **2019**, *28*, 74–79. [CrossRef]
19. Hu, Y.; Xu, Z. Interactions of amphoteric amino phosphoric acids with calcium-containing minerals and selective flotation. *Int. J. Miner. Process.* **2003**, *72*, 87–94. [CrossRef]
20. Rao, K.H.; Forssberg, K.S.E. Mechanism of fatty acid adsorption in salt-type mineral flotation. *Miner. Eng.* **1991**, *4*, 879–890.
21. Roonasi, P.; Yang, X.; Holmgren, A. Competition between sodium oleate and sodium silicate for a silicate/oleate modified magnetite surface studied by in situ ATR-FTIR spectroscopy. *J. Colloid Interf. Sci.* **2010**, *343*, 546–552. [CrossRef] [PubMed]

Article

Optimization of the Mix Formulation of Geopolymer Using Nickel-Laterite Mine Waste and Coal Fly Ash

Alberto Longos, Jr. [1,*], April Anne Tigue [1], Ithan Jessemar Dollente [1], Roy Alvin Malenab [1], Ivyleen Bernardo-Arugay [2], Hirofumi Hinode [3], Winarto Kurniawan [3] and Michael Angelo Promentilla [1,4,*]

[1] Department of Chemical Engineering, De La Salle University, Manila 1004, Philippines; april_tigue@dlsu.edu.ph (A.A.T.); ithan_dollente@dlsu.edu.ph (I.J.D.); roy.malenab@dlsu.edu.ph (R.A.M.)

[2] Department of Materials and Resources Engineering and Technology, MSU-Iligan Institute of Technology, Iligan 9200, Philippines; ivyleen.arugay@g.msuiit.edu.ph

[3] Department of Transdisciplinary Science and Engineering, Tokyo Institute of Technology, 2-12-1 Ookayama, Meguro-ku, Tokyo 152-8550, Japan; hinode2020040@gmail.com (H.H.); kurniawan.w.ab@m.titech.ac.jp (W.K.)

[4] Center for Engineering and Sustainable Development Research, De La Salle University, Manila 1004, Philippines

* Correspondence: alberto_longos@dlsu.edu.ph (A.L.J.); michael.promentilla@dlsu.edu.ph (M.A.P.); Tel.: +63-2-8524-4611 (M.A.P.)

Received: 13 November 2020; Accepted: 17 December 2020; Published: 21 December 2020

Abstract: Geopolymer cement has been popularly studied nowadays compared to ordinary Portland cement because it demonstrated superior environmental advantages due to its lower carbon emissions and waste material utilization. This paper focuses on the formulation of geopolymer cement from nickel–laterite mine waste (NMW) and coal fly ash (CFA) as geopolymer precursors, and sodium hydroxide (SH), and sodium silicate (SS) as alkali activators. Different mix formulations of raw materials are prepared to produce a geopolymer based on an I-optimal design and obtained different compressive strengths. A mixed formulation of 50% NMW and 50% CFA, SH-to-SS ratio of 0.5, and an activator-to-precursor ratio of 0.429 yielded the highest 28 d unconfined compressive strength (UCS) of 22.10 ± 5.40 MPa. Furthermore, using an optimized formulation of 50.12% NMW, SH-to-SS ratio of 0.516, and an activator-to-precursor ratio of 0.428, a UCS value of 36.30 ± 3.60 MPa was obtained. The result implies that the synthesized geopolymer material can be potentially used for concrete structures and pavers, pedestrian pavers, light traffic pavers, and plain concrete.

Keywords: geopolymer; laterite; alkali-activated; alumino-silicates; I-optimal; response surface methodology; optimization; mine waste

1. Introduction

The rapid increase in construction activity has been observed to meet the ever-increasing infrastructure demands [1]. In most construction activities, cement-based concrete is an essential and widely used material. The use of cement-based concrete, like ordinary Portland cement (OPC), is globally accepted due to ease of operation, excellent mechanical properties, and low-cost production compared to other construction materials [2]. However, OPC has drawbacks as it releases approximately one ton of CO_2, a greenhouse gas, to produce one ton of OPC [3]. It also has high energy consumption during production, and it consumes a significant amount of natural resources [2,3]. Due to increasing awareness of these issues, a viable alternative for the conventional Portland cement is currently being reviewed and studied by many researchers and scientists. Geopolymer cement is one of the emerging greener alternatives for the construction industry. It results from the chemical reaction between aluminosilicate waste materials and alkaline activators resulting in the inorganic polymer [3]. It is comprised of repeating units of silico-oxide (Si-O-Si), silico-aluminate (Si-O-Al-O-), ferro-silico-aluminate (-Fe-O-Si-O-Al-O-), or alumino-phosphate (-Al-O-P-O-), created through a process of geopolymerization [4].

Aluminosilicate sources, also called geopolymer precursors, can be sourced out from waste such as fly ash, blast furnace slag, silica fume, and rice husk or a combination of these precursors, which are rich from silicon (Si), aluminum (Al), or iron (Fe) in an amorphous form [5]. Mine waste has also emerged as a potential geopolymer precursor because it contains Si, Al, and Fe. Valorization of such waste would also reduce the environmental burden. For example, thermal and mechanical activations pretreatment were done to Ni-laterite mine waste from the Philippines to enhance its property as a geopolymer precursor [6]. Likewise, gold mine tailings in the Philippines are used to produce geopolymer bricks with a compressive strength of 5.5 MPa [7].

As properties of raw materials for geopolymer precursors could vary from one place to another, it is necessary to perform mix formulation studies to evaluate the potential application of such construction material. For example, using fly ash and granulated blast furnace as precursors, the optimal rational mix design resulted in an improved compressive strength comparable to OPC ranging from 32 to 66 MPa [1]. A statistical mix design of the experiment was also used to optimize the geopolymer properties from the ternary blend of red mud waste, rice husk ash, and diatomaceous earth [8]. Other studies showed that different mixes and combinations of fly ash-mine tailings (MT) mix [9], laterite–calcite, and laterite–slag mix [10], could increase the compressive strength. However, a binary blend of coal fly ash (CFA) and nickel–laterite mine waste (NMW) sourced out from the Philippines as geopolymer precursors have not been explored yet [9]. Thus, this study extends the work described in Longos et al. [6] and apply the statistical design of experiment [8] to determine the optimal mix formulation of coal fly ash (CFA) and nickel–laterite mine waste (NMW) with sodium hydroxide (SH)-sodium silicate (SS) as alkali activators.

2. Materials and Methods

2.1. Raw Material Preparation

Raw NMW was collected from a siltation pond of a nickel–laterite mining company, while CFA was obtained from a coal power plant located in Mindanao, Philippines. Raw materials were oven-dried at 105 °C for 24 h. Dried NMW showed clay-like characteristics, and the clumping of this clayey material facilitated the need for pre-grinding. The dried NMW was reduced in size using a Raijin portable attrition mill pulverizer with a power of 1500 Watts, blade diameter of 150 mm, and rotary speed of 1400 rpm. On the other hand, dried CFA already exhibited the needed fineness and would not need further grinding. Both raw material samples were then screened using a Tyler mesh sieve passing 50 mesh (0.297 mm). Analytical grade sodium silicate (water glass solution with 34.13% SiO_2, 14.65% Na_2O, 51.22% H_2O) with a silica modulus of 2.33, and sodium hydroxide flakes with 98% purity (manufactured by Formosa Plastic Corporation, Kaohsiung, Taiwan) were used in the study as the alkali activator components.

2.2. Raw Material Characterization Procedure

A particle size distribution (PSD) analysis of both raw materials was performed using a Tyler standard sieve series (Thomas Scientific, Swedesboro, NJ, USA.) ranging from mesh 4 to mesh 200 (4.75 mm to 0.075 mm) in a vibrating screen.

The chemical compositions of raw NMW and CFA were performed with X-ray fluorescence spectroscopy using Horiba Scientific XGT-7200 X-ray analytical microscope (Horiba Ltd., Kyoto, Japan) with an X-ray beam generation of 50 kV voltage and 35 A current.

The mineralogical analysis was also performed for both raw materials using a Multiflex Rigaku automated powder x-ray diffractometer (XRD) (Rigaku Corporation, Tokyo, Japan) ($\lambda_{Cu\,K\alpha}$ = 1.54 Å, Voltage = 40 kV, Current = 30.0 mA) with a measuring angle of 5–60°.

Scanning electron microscope (SEM) captured the morphological images and properties of raw materials using a FESEM Dual Beam Helios Nanolab 600i (FEI, Hillsboro, OR, USA.) with a voltage of 2.0 kV and beam current of 43 pA equipped with energy-dispersive X-ray spectroscopy (EDS) with a voltage of 15.0 kV and a beam current of 0.69 nA.

2.3. Toxicity Characteristic Leaching Procedure (TCLP)

Toxicity characteristic leaching procedure using US EPA Method 1311 was performed for both raw materials. This procedure is to determine the heavy metal leachability property, whether these materials are hazardous or not. The parameters used were liquid to solid ratio of 20:1 and an agitation speed of 30 rpm for 12 h. Leachate was then analyzed using an Agilent Technology inductively coupled plasma–mass spectrometry (ICP-MS) and Agilent 5100 inductively coupled plasma-optical emission spectroscopy (ICP-OES), both are done by a third-party laboratory.

2.4. Thermal Activation of Nickel-Laterite Mine Waste (NMW)

Pretreatment of NMW by thermal activation was performed first before experimental runs were conducted [7]. NMW samples were heat treated in the laboratory furnace at a ramping rate of 10 °C per minute to attain a temperature of 700 °C at a holding time of 2 h. The samples were left inside the furnace to be cooled down to room temperature after soaking at 700 °C.

2.5. Experimental Procedures and Runs

The design of the experiment was based on an I-optimal design, which is a mixture experiment intended to predict the responses for all possible formulations of the mixture and to identify optimal proportions for each of the ingredients minimizing the average variance of prediction. Table 1 shows the factors used in the mixture design like the activator-to-precursor ratio of 0.429 to 1.0 [8,11], NMW-CFA content (% NMW) of 50% to 100% [9,12] and SH-to-SS ratio of 1:2 to 2:1 [10,13].

Table 1. Parameters of each factor and level for geopolymer synthesis.

Factors	Low Level	Mid Level	High Level
1. Activator-to-precursor ratio	0.429	0.667	1.0
2. NMW-CFA content, as % NMW	50%	75%	100%
3. SH-to-SS ratio	1:2	1:1	2:1

The performance of the different factors was evaluated independently using runs randomly ordered by Design-Expert 11 (Design-Expert® software, version 11). A total of 18 runs were generated, with three as replicate points. The 18 experimental runs are shown in Table 2 with different combinations of factor levels.

2.6. Geopolymer Synthesis

For geopolymer preparation, run number 15 is the basis of the amounts of raw materials used. A 500 g of precursor (50% NMW + CFA) was prepared and set aside first for mixing later. With an activator-to-precursor of 0.438, the alkali activator was prepared first by mixing 71 g of 12 M sodium hydroxide (SH) with 143 g sodium silicate solution (SS). Then, 250 g of CFA was mixed with the prepared alkali activator. Manual mixing was done for at least 5 min until the consistency of the CFA-activator mixture was homogenized. Another 250 g NMW was then added to the mixture, and the second stage of manual mixing was done for at least 5 min until the consistency of the mixture was homogenized. During mixing, it must be noted that the mixture hardens immediately.

After stabilization, the geopolymer was placed in a square mold made of polyethylene material with a dimension of 50 mm × 50 mm × 50 mm. The prepared geopolymer can make 3 square molds. The molded sample was set for at least 24 h before it was demolded. The demolded sample was then placed in a polyethylene Ziploc. Next, the air was removed manually from the Ziploc before sealing. The sealed geopolymer samples were then placed in an oven at 80 °C for 24 h. Lastly, the samples were cured for 28 d at ambient temperature before further test and analysis.

Table 2. Experimental runs in standard order.

Std Order	Run Order	Factor 1: Activator-to-Precursor Ratio	Factor 2: NMW-CFA Content as % NMW	Factor 3: SH-to-SS Ratio
1	15	0.4286	50%	1:2
2	5	1.0000	50%	1:2
3	6	0.4286	75%	1:2
4	10	0.6667	100%	1:2
5	16	1.0000	100%	1:2
6	9	0.6667	50%	1:1
7	1	0.6667	50%	1:1
8	2	0.6667	75%	1:1
9	11	0.6667	75%	1:1
10	8	0.6667	75%	1:1
11	4	1.0000	75%	1:1
Std Order	Run Order	Factor 1: Activator-to-Precursor Ratio	Factor 2: NMW-CFA Content as % NMW	Factor 3: SH-to-SS Ratio
12	12	0.4286	100%	1:1
13	3	0.4286	50%	2:1
14	7	1.0000	50%	2:1
15	18	0.6667	75%	2:1
16	17	1.0000	75%	2:1
17	14	0.4286	100%	2:1
18	13	1.0000	100%	2:1

2.7. Unconfined Compressive Strength (UCS)

Unconfined compressive strength (UCS) was the response variable to evaluate the engineering property of the geopolymer specimens. It was performed following ASTM C109/C109M. This test method covers the determination of the compressive strength of hydraulic cement mortars, using 2-inch (50-mm) cube specimens to determine compliance with specifications.

3. Results and Discussions

3.1. Raw Material Characterization

3.1.1. X-ray Fluorescence Spectroscopy (Chemical Composition)

The elemental analysis of NMW and CFA is shown in Table 3, which is primarily composed of oxides of iron (Fe_2O_3), silicon (SiO_2), calcium (CaO), aluminum (Al_2O_3), magnesium (MgO), and nickel (NiO). Trace elements in both samples include oxides of manganese (MnO), titanium (TiO_2), potassium (K_2O), and silver (Ag_2O). NMW contains an oxide of chromium (Cr_2O_3), which is not present in CFA but contains oxides of strontium (SrO) and sulfur (SO_3). The composition of both raw materials showed that it could be a geopolymer precursor because of the high presence of silica and alumina.

Table 3. Chemical Composition of raw nickel–laterite mine waste (NMW) and coal fly ash (CFA).

Mass %	SiO_2	Al_2O_3	Fe_2O_3	CaO	MgO	NiO	Cr_2O_3	MnO	TiO_2	K_2O	Ag_2O	SrO	SO_3	LOI
NMW	20.54	2.79	47.68	5.46	4.23	1.94	0.85	0.38	0.25	0.35	0.04	-	-	15.50
CFA	26.12	8.01	22.70	29.35	1.98	0.03	-	0.23	0.97	0.89	0.12	0.30	5.31	4.0

- No detection in the analysis.

3.1.2. Particle Size Distribution

The particle size distributions of NMW and CFA are shown in Figure 1. The particle size of NMW is between 0.075 to about 2.36 mm with D_{50} (median diameter) of about 0.25 mm. At the same time, the particle size of CFA is between 0.0013 mm to about 4.75 mm with D_{50} (median diameter) of about 0.425 mm.

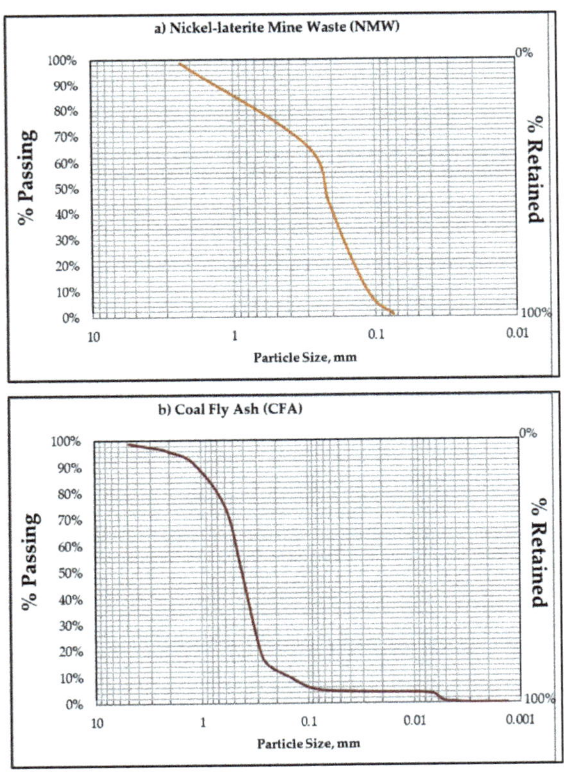

Figure 1. Particle size distribution (PSD) of (**a**) nickel–laterite mine waste and (**b**) coal fly ash.

3.1.3. Leachability of Metals Based on Toxicity Characteristics Leaching Procedure (TCLP)

Numerous studies have shown that under certain conditions, coal fly ash releases traces of heavy metals to the environment [14–16]. With this condition and the possibility of NMW leaching out heavy metals, the TCLP method was employed in this study for geopolymer precursors. This test determines the mobility of organic and inorganic analytes present in the liquid, solid, and multiphase waste [17]. Furthermore, it simulates the conditions that may be present in a landfill where water may pass through the landfilled waste and travel into the groundwater carrying the soluble materials with it. Table 4 shows the leachability of NMW and CFA. Indication suggests that the metallic components in both samples are well below the TCLP limits and considered non-hazardous according to the United States Environmental Protection Agency limits [18] and Philippines' DENR Administrative Order 2013-22 Standards for the Management of Hazardous Wastes [19].

However, it should be noted that initial leachability results warrant further study. For example, one study considers the limitations of such a test and emphasized that the test mimics leachate in landfills with lower pH and higher organic acid content than most of the municipal solid waste (MSW) [20]. Other factors need to be considered, as demonstrated by previous studies [21–23]. The mobility of trace elements and heavy metals in the environment depends mainly on the properties of solution and solids, such as pH, redox potential, chemical composition, surface properties, and mineral contents.

Table 4. Toxicity characteristic leaching procedure (TCLP) Analysis of raw materials samples in mg/L.

Raw Material	Ag	As	Ba	Cd	Cr	Hg	Ni	Pb	Se
CFA	0.00051	0.069	2.544	0.00042	0.035	0.00085	0.214	0.0027	0.0228
NMW	0.00045	0.00005	0.108	0.00037	0.19	0.0001	2.929	0.00335	0.001
TCLP limit [a]	5.0	5.0	100.0	1.0	5.0	0.2	-	5.0	1.0
Class A [b]	-	0.01	0.7	0.003	0.01	0.001	0.02	0.01	0.01
Class C [b]	-	0.02	3	0.005	0.01	0.002	0.2	0.05	0.02

[a] US Environmental Protection Agency [18] and Philippines' DAO 2013-22 Standards for the Management of Hazardous Wastes [19].; [b] Philippines DAO 2016-08 Water Quality Guidelines and General Effluent Standards [24]; - No regulatory limit.

Moreover, the TCLP result is compared with the Philippine regulatory standard for Class A and Class C water body quality (Table 4) based on DENR Administrative Order 2016-08 on Water Quality Guidelines and General Effluent Standards. These classifications and parameters are intended to maintain and preserve the quality of all water bodies based on their intended beneficial usage. Class A is intended for drinking water supply sources after conventional treatment, while Class C is intended for agriculture and irrigation [23,24]. While not conclusive, TCLP results suggest that both raw materials could potentially leach high quantities if beneficially used in an unconsolidated matter such as arsenic in CFA for both water classes. However, because TCLP simulates conditions within the environment (i.e., landfill) that are different from beneficial use scenarios, the values presented in Table 5 are higher than those obtained using non-buffered extractants such as distilled water [25]. The added benefit of comparing these values is that there is conservatism with the comparison.

Nevertheless, TCLP is still widely used as a test to determine if individual material is hazardous since the test function is a conservative predictor of leaching. In the Philippines, the same limit and test are also used to classify hazardous wastes, as stated in the DENR Administrative Order 2013-22 on Hazardous Waste Management Procedural Manual [19].

On the other hand, the leachability of the NMW-CFA geopolymer is also worthy to note as the alkalinity of the leachates of the precursors (NMW and CFA) or the geopolymer itself can influence the leachability behavior of inorganic pollutants [26]. With this, some of the heavy metals initially present may be more mobile due to higher pH values and warrant further investigation on the geopolymer [23]. For example, a study by Tigue et al. using a modified percolation test set-up has shown that the arsenic leached out in high concentration compared with other heavy metals present in the geopolymer sample [27]. On the other hand, some studies have also revealed that geopolymer technology is an effective technique in the immobilization of heavy metals. For instance, Ahmari et al. confirmed that heavy metals were found to be effectively immobilized in mine tailing geopolymer bricks [28]. Thus, this will be considered in future work, especially that the mobility of metals present in the material varies depending on the condition.

3.1.4. X-ray Diffractometer (XRD)

The XRD patterns of CFA and NMW are shown in Figure 2. It was detected that CFA contains minerals such as endellite, julgoldite, quartz, magnetite, troilite, and maghemite. The most substantial peak intensities in this pattern were quartz, which is a typical XRD pattern for coal fly ash similar to other studies [1,9]. On the other hand, NMW had various minerals identified; however, minerals that were dominantly detected include silhydrite, montmorillonite, kaolinite, santafeite, dickite, sodalite, szymanskiite, nontronite, moganite, tridymite, quartz, ferrosaponite, chegemite, maghemite, and goethite. Most minerals identified contain Fe, namely santafeite, nontronite, ferrosaponite, maghemite, and goethite, which can be matched with the XRF analysis with the highest composition at 47.68% Fe_2O_3. Table 5 summarized the minerals with their corresponding chemical formula. Most of the minerals for both samples contain Si and Al or aluminosilicate materials, which is a good indication for a geopolymer precursor. Moreover, broad peaks between 26–27°, 35–37° 2θ for CFA, and between 19–23°, 33–37° 2θ for raw NMW may correspond to the amorphous content of the material. It is believed that the amorphous content of the material has played a significant role in geopolymerization due to its reactive nature. This inference can be correlated with the study reported in Jaarsveld et al., wherein materials having high amorphous content were found to yield a geopolymer having a better mechanical property in binders [29].

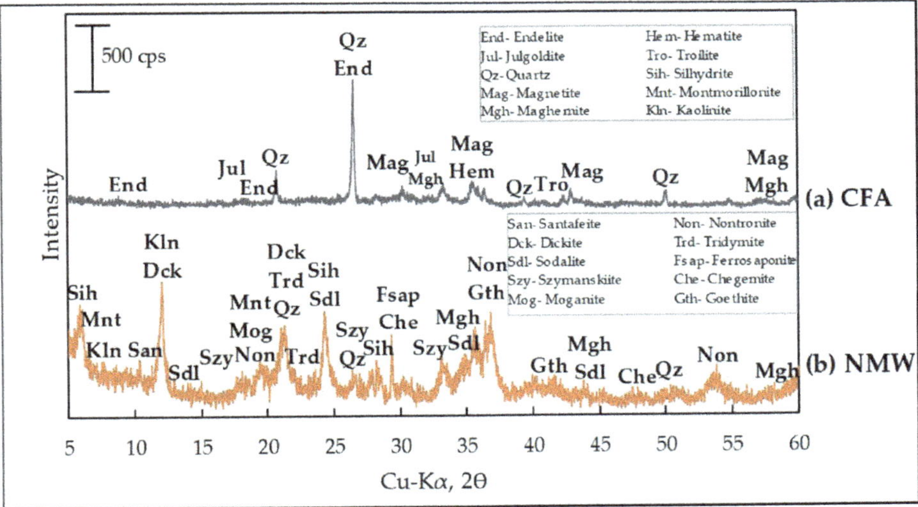

Figure 2. Mineralogical pattern (XRD) of raw materials (**a**) CFA and (**b**) NMW.

Table 5. List of detected minerals for NMW and CFA.

Mineral ID	Abbreviation [c]	Chemical Formula
CFA		
Endellite	End	$Al_2Si_2O_5(OH)_4 \cdot 2(H_2O)$
Julgoldite	Jul	$Ca_2Fe^{3+}(Fe^{3+},Al)_2(SiO_4)(Si_2O_7)(O,OH)_2 \cdot (H_2O)$
Quartz	Qz	SiO_2
Magnetite	Mag	Fe_3O_4
Maghemite	Mgh	$\gamma\text{-}Fe_2O_3$
Hematite	Hem	Fe_2O_3
Troilite	Tro	FeS

Table 5. *Cont.*

Mineral ID	Abbreviation [c]	Chemical Formula
NMW		
Silhydrite	Sih	$3SiO_2 \cdot (H_2O)$
Montmorillonite	Mnt	$(Na,Ca)0,3(Al,Mg)_2Si_4O_{10}(OH)_2 \cdot n(H_2O)$
Kaolinite	Kln	$Al_2Si_2O_5(OH)_4$
Santafeite	San	$(Mn,Fe,Al,Mg)2(Mn^{4+},Mn^{2+})_2(Ca,Sr,Na)_3(VO_4,AsO_4)_4(OH)_3 \cdot 2(H_2O)$
Dickite	Dck	$Al_2Si_2O_5(OH)_4$
Sodalite	Sdl	$Na_8Al_6Si_6O_{24}Cl_2$
Szymanskiite	Szy	$Hg^+16(Ni,Mg)_6(H_3O)_8(CO_3)_{12} \cdot 3(H_2O)$
Nontronite	Non	$Na0.3Fe^{3+}2(Si,Al)4O_{10}(OH)_2 \cdot n(H_2O)$
Moganite	Mog	SiO_2
Ferrosaponite	Fsap	$Ca0.3(Fe^{2+},Mg,Fe^{3+})_3(Si,Al)4O_{10}(OH)_2 \cdot 4(H_2O)$
Tridymite	Trd	SiO_2
Chegemite	Che	$Ca_7(SiO_4)_3(OH)_2$
Mineral ID	**Abbreviation [c]**	**Chemical Formula**
Maghemite	Mgh	$\gamma\text{-}Fe_2O_3$
Goethite	Gth	$Fe^{3+}O(OH)$

[c] Nomenclature of minerals is based on the International Mineralogical Association (IMA). Minerals not found in the table are abbreviated based on the format of Kretz as cited by Whitney et al. [30].

3.1.5. Scanning Electron Microscope

SEM images of raw materials were also captured, as shown in Figure 3. The structure of the raw NMW is platy and loose with sheets, which is favorable for water storage [31]. On the other hand, CFA images show that most of the particles are spherical (cenosphere) or are occurring as microspheres and are looser than the NMW particles. These microspheres increase the specific surface area of the fly ash [32]. Thus, there is a high probability that the total surface area of CFA is higher than a coarser platy structure, which may make the CFA more reactive than NMW.

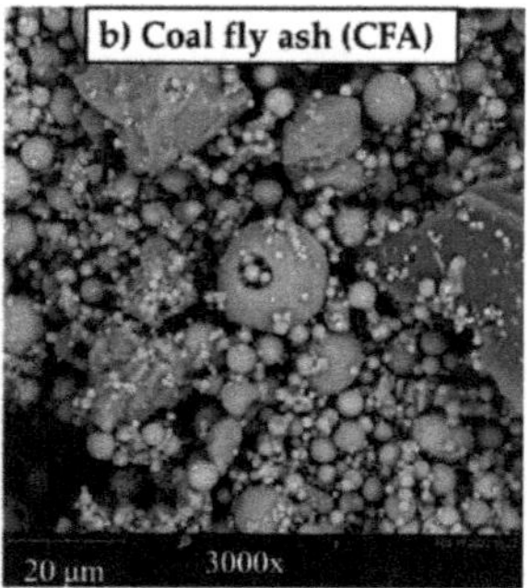

Figure 3. SEM images at magnification 3000× for (**a**) raw nickel–laterite mine waste and (**b**) coal fly ash.

3.2. Compressive Strength

The unconfined compressive strengths of the synthesized geopolymer were observed to range from 1.93 MPa up to 22.14 MPa after 28 d (Figure 4). The experimental formulation mix number 15 with an activator-to-precursor ratio of 0.429, 50% NMW, and with SH-to-SS ratio of 1:2 yielded the highest value making it the best mix sample among other runs. The sample with an activator-to-precursor ratio of 1, 100% NMW with 2 parts of sodium hydroxide and 1-part sodium silicate (SH-to-SS ratio of 2:1) resulted in deflocculation of NMW; hence it did not harden. This result is due to the following:

(1) the precursor is made from 100% NMW, which is not that reactive as CFA; (2) the high ratio of SH-to-SS and low reactive Si and Al content of NMW resulted in excess amounts of NaOH and water. Additional water generally improves the workability of the geopolymer paste made from a precursor with high water holding capacity such as NMW. However, the excess water content can cause a dilution effect, which affects the geopolymerization and, consequently, the number of active components that can be mobilized [33]. It was also observed that the higher viscosity of concentrated NaOH solution hindered the evaporation of excess water. This result means that the geopolymeric paste needs more time or higher curing temperature to gain better or full strength [34,35].

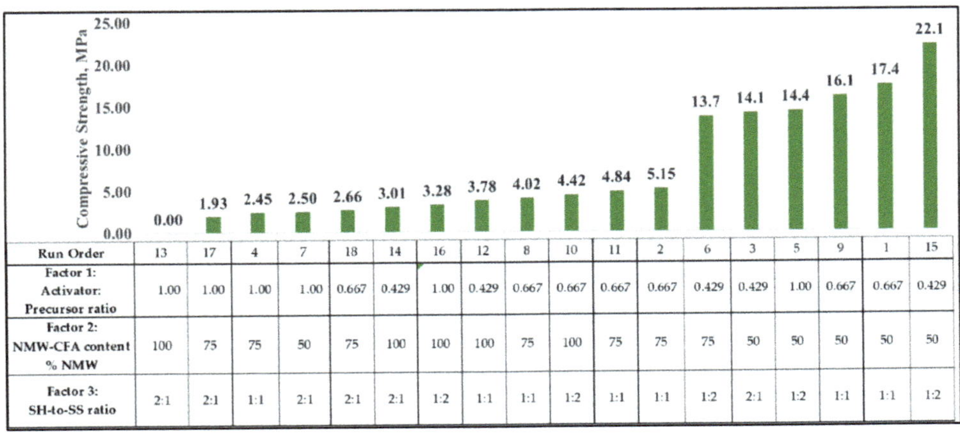

Run Order	13	17	4	7	18	14	16	12	8	10	11	2	6	3	5	9	1	15
Factor 1: Activator: Precursor ratio	1.00	1.00	1.00	1.00	0.667	0.429	1.00	0.429	0.667	0.667	0.667	0.667	0.429	0.429	1.00	0.667	0.667	0.429
Factor 2: NMW-CFA content % NMW	100	75	75	50	75	100	100	100	75	100	75	75	75	50	50	50	50	50
Factor 3: SH-to-SS ratio	2:1	2:1	1:1	2:1	2:1	2:1	1:2	1:1	1:1	1:2	1:1	1:1	1:2	2:1	1:2	1:1	1:1	1:2

Figure 4. Comparison of unconfined compressive strength at different mix proportions.

3.3. Model Statistics

The summary of model statistics suggested that the analysis model is quadratic with an adjusted R-squared of 92.69% and a predicted R-squared of 80.97%, as shown in Table 6.

Table 6. Summary of model statistics for UCS.

Source	Std. Dev.	R^2	Adjusted R^2	Predicted R^2	Remarks
Linear	3.10	0.823	0.785	0.704	
2FI	2.64	0.899	0.844	0.748	
Quadratic	**1.81**	**0.966**	**0.927**	**0.801**	**Suggested**
Cubic	0.716	0.998	0.989		Aliased

The analysis of variance (ANOVA) for the quadratic model, shown in Table 7, indicates that the model is significant. However, there are quadratic model terms with insignificant *p*-values, which the model must be reduced, as shown in Table 8.

Table 7. ANOVA for the quadratic model.ANOVA for the quadratic model.

Source	Sum of Squares	df	Mean Square	F-Value	p-Value	Remarks
Model	735.05	9	81.67	24.94	<0.0001	significant
A—activator-to-precursor ratio	99.69	1	99.69	30.44	0.0006	significant
B—NMW-CFA content	320.70	1	320.70	97.93	<0.0001	significant
C—SH-to-SS ratio	107.22	1	107.22	32.74	0.0004	significant
AB	24.69	1	24.69	7.54	0.0252	significant
AC	0.02	1	0.02	0.01	0.9451	not significant
BC	27.62	1	27.62	8.43	0.0198	significant
A^2	0.45	1	0.45	0.14	0.7200	not significant
B^2	25.45	1	25.45	7.77	0.0236	significant
C^2	4.22	1	4.22	1.29	0.2890	not significant
Residual	26.20	8	3.27			
Lack of fit	24.66	5	4.93	9.61	0.0458	significant
Pure error	1.54	3	0.51			
Cor total	761.25	17				

Table 8. ANOVA for the reduced quadratic model.

Source	Sum of Squares	df	Mean Square	F-Value	p-Value	Remarks
Model	727.58	6	121.26	39.61	<0.0001	significant
A—activator-to-precursor ratio	95.31	1	95.31	31.13	0.0002	significant
B—NMW-CFA content (% NMW)	315.15	1	315.15	102.94	<0.0001	significant
C—SH-to-SS ratio	100.18	1	100.18	32.72	0.0001	significant
AB	29.33	1	29.33	9.58	0.0102	significant
BC	28.54	1	28.54	9.32	0.0110	significant
B^2	44.58	1	44.58	14.56	0.0029	significant
Residual	33.68	11	3.06			
Lack of fit	32.14	8	4.02	7.83	0.0591	not significant
Pure error	1.54	3	0.5133			
Cor total	761.25					

3.4. Factors Affecting Compressive Strength

All three factors significantly affect the compressive strength of the synthesized geopolymer based on ANOVA (Table 8). Representative plots of the relationship of individual factors with their compressive strength are shown in Figure 5. All factors namely activator-to-precursor ratio (A), percentage NMW (B) and SH-to-SS ratio (C) show the same effect to the compressive strength which increases as individual factor decreases.

Factors affecting the compressive strength of geopolymers are related to the mechanisms involved in the alkaline activation of various aluminosilicate precursors. Garcia-Lodeiro et al. reviewed various geopolymerization models that can be summarized as follows: initially, the contact between the solid aluminosilicate source and the alkaline solution causes the glassy or amorphous precursor components to dissolve, releasing aluminates and silicates, probably as monomers. These monomers interact to form dimers, trimers, tetramers, and so on. When the solution reaches saturation, an aluminosilicate gel called N-A-S-H undergoes precipitation and restructuring. This step determines the composition, structure, and physical properties of the resulting geopolymer [36].

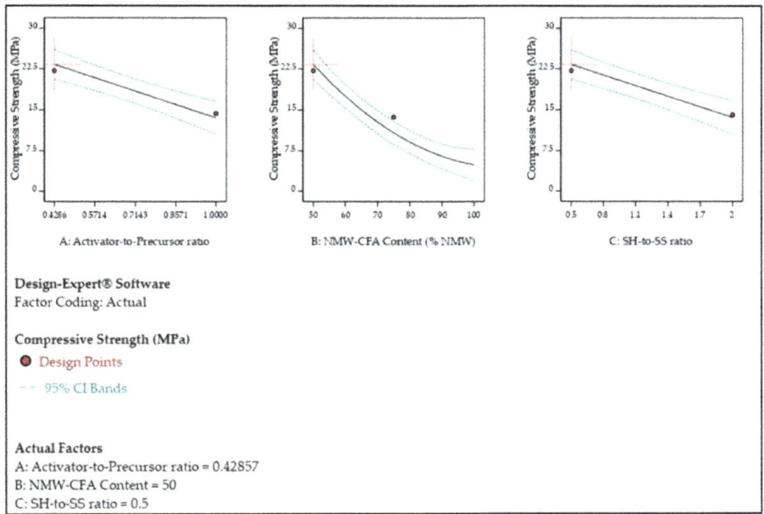

Figure 5. One-factor plot.

Based on the above mechanism, the amount and reactivity of precursors and alkaline activating solutions are crucial in geopolymer synthesis and properties. This result may have been because more precursor is present in the system to participate in the geopolymerization process. In the study of Pacheco-Torgal et al. [37], it suggested that aside from the composition of the precursor materials, the relative amount of the precursor and composition of the alkaline activators also affect the strength and other properties of geopolymers. This observation may be due to the increase in coal fly ash percentage, which is more reactive than the NMW. The decrease of SH-to-SS ratio means more SS in the solution, which means more SiO_3 content in the system can participate in the geopolymerization reaction.

Figure 6 shows the interaction graph of AB and BC. The interaction of factors A (activator-to-precursor ratio) and B (% NMW) shows that the compressive strength increases as both factors decrease (Figure 6a). The interaction AB shows that the trend of compressive strength is affected when %NMW (B) is changed from 50% to 100%. At 100% NMW, the change in compressive strength is less obvious when the activator-to-precursor ratio (A) is changed. Decreasing the % NMW (B) corresponds to the increase of CFA content in the system, in which CFA is more reactive for geopolymerization. Moreover, when the activator-to-precursor ratio (A) decreases, a higher precursor amount is present in the system. When the amount of precursor is increased, the geopolymerization reaction is boosted because of the high reactivity of CFA, increasing the compressive strength of the product. Similarly, Figure 6b also shows the interaction of factors B (% NMW) and C (SH-to-SS ratio) that compressive strength also increases as both factors B and C decreases. The interaction BC shows that when SH-to-SS ratio (C) is increased from 0.5 to 2, the trend of the compressive strength decreased slowly when % NMW (B) is increased from 50% to 100%. The possible explanation for this is that when % NMW is decreased, more CFA is present in the system. This interaction can be further explained by the mineralogy of the precursors. The XRD of the NMW shows that it has less active aluminosilicate components than CFA. Thus, increasing the amount of CFA in the precursor mix improves the reactivity and influence the compressive strength of the final geopolymer mixture. Moreover, when SH-to-SS is decreased, it means a high value of SS (Na_2SiO_3) is present. The SH may have easily dissolved the reactive aluminosilicates in both the NMW and CFA, so less hydroxide is needed before the geopolymerization process. Thus, SS is considerably needed as a binding source and promoter of hardening [38].

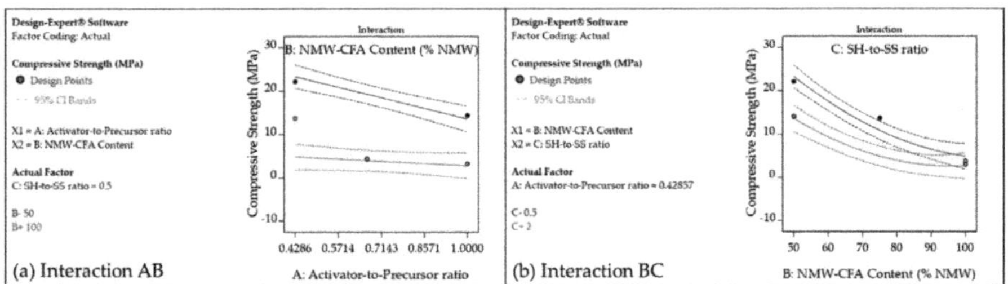

Figure 6. Interaction graph.

3.5. Regression Model

The regression result further explains the result of ANOVA. The resulting linear regression model based on these observed values is as follows:

$$Y = 86.337 - 30.408A - 1.310B - 11.408C + 0.268AB + 0.0996BC + 0.0052B^2 \tag{1}$$

where:

$Y = Compressive\ strength,\ MPa$

$A = Activator-to-precursor\ ratio$

$B = NMW-CFA\ content,\ \%\ NMW$

$C = SH-to-SS\ ratio$

From Equation (1), it can be observed that factor B (% NMW) can significantly affect the compressive strength of the geopolymer as it is present in 4 out of 7 coefficients in the equation. Based on the regression model, the compressive strength decreases linearly with B (% NMW) while increases with the following second-order terms: AB, BC, and B^2. However, as shown in Figure 5 (One-factor plot), the observed inverse relationship of compressive strength and B (% NMW) is more prevalent compared to the direct relationship of the 2nd degree involving B (% NMW) to the compressive strength. This model is expected since these are primary raw materials containing SiO_2 and Al_2O_3, which are necessary components to undergo geopolymerization. The decrease in % NMW corresponds to the increase of CFA content in the mixture. The raw material characterization of CFA suggests that it could be more reactive than NMW, and thus, the modification of the percentage from 100% to 50% NMW enhances the compressive strength. Moreover, CFA has higher SiO_2 and Al_2O_3 content. Thus, when the amount of CFA is increased, more aluminosilicate can participate in the reaction [5]. Furthermore, the increase of SiO_2 and Al_2O_3, with the additional presence of Fe^{3+} in the mixture, may also influence the compressive strength of the product by replacing or substituting Al^{3+} by Fe^{3+} in the octahedral sites of an aluminosilicate structure [39,40]. This model indicates that the proper selection of precursors needs to be determined using characterization methods. Lastly, the I-optimal design of the experiment can help us fine-tune the mix design of the geopolymer for the sampling and selection of a new precursor source [38].

3.6. Response Surface Methodology (Optimization)

The numerical optimization tool of the Design-Expert software was used to find the optimal point on the response surface to maximize the unconfined compressive strength of the synthesized geopolymer. The selected values were followed in the region where maximum strength can be seen, which from Figure 7, it can be observed that the maximum strength is approaching the minimum values of all the factors. With desirability of 1.0, the calculated optimized mix formulation is obtained with an activator-to-precursor ratio of 0.438, percent NMW of 50.1%, and an SH-to-SS ratio of 0.520. The predicted value is calculated at 22.9 MPa with a predicted R^2 equivalent of 0.890.

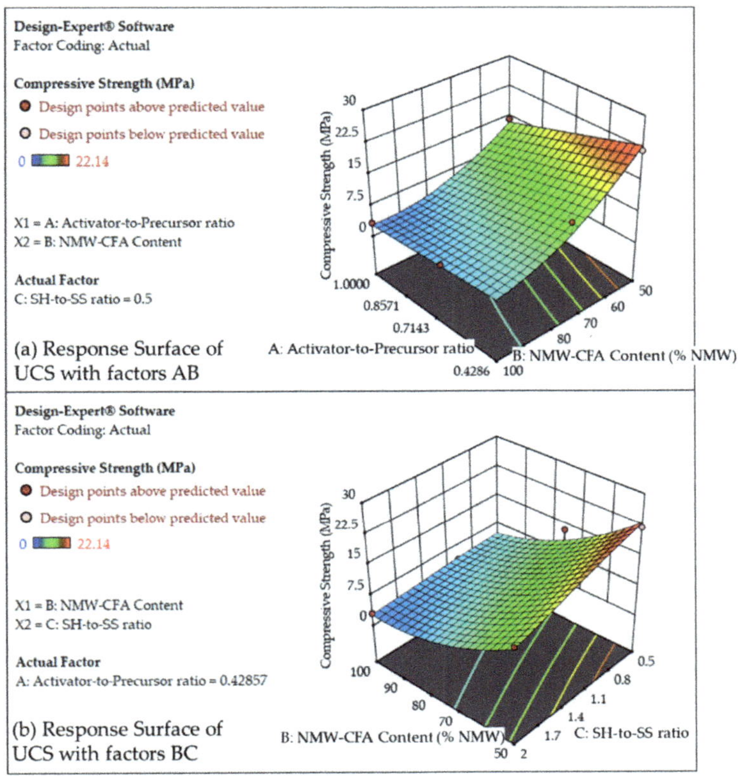

(a) Response Surface of UCS with factors AB

(b) Response Surface of UCS with factors BC

Figure 7. Response surface of unconfined compressive strength.

3.7. Confirmatory Run

Using the calculated optimized mix formulation obtained with an activator-to-precursor ratio of 0.438, percent NMW of 50.1%, and an SH-to-SS ratio of 0.520, a confirmatory run of the synthesized geopolymer was performed. The unconfined compressive strength of the confirmatory run was obtained to be 36.3 MPa with a deviation of −58.0% (Table 9).

Table 9. Predicted and observed values of unconfined compressive strength (UCS) of the confirmatory run.

A—Activator-to-Precursor Ratio	B: NMW-CFA Content (% NMW)	C: SH-to-SS Ratio	Predicted UCS MPa	Observed UCS MPa	% Deviation
0.438	50.1	0.520	22.9	36.3	−58.0%

The deviation of −58.0% could be attributed to the noise that was not controlled and measured during the experiment. Nonetheless, the result is in reasonable agreement with the predicted R^2 equivalent to 0.8902. The deviation may be attributed to the uncontrolled external factors, such as the type and strength of manual mixing of raw materials, the person who performed the mixing, and the UCS equipment used for the optimized sample analyzed by a third party.

3.8. Morphological Properties of Synthesized Geopolymer

Figure 8 shows the different images of synthesized geopolymer with different NMW-CFA content. Synthesized geopolymer using 100% NMW has several voids and more unreacted NMW, resulting in a lower compressive strength than a matrix with a compact structure. The morphology of geopolymer with 75% NMW has cemented surfaces but with fewer voids and fewer unreacted NMW, which can be a basis of a higher compressive strength than the previous geopolymer. Geopolymer with 50% NMW has the most apparent or widest distribution of cemented surfaces among the geopolymers, which signifies a higher compressive strength. Although, there are some spherical shapes seen, which is unreacted coal fly ash. On the other hand, the optimized sample has a larger cemented surface area, which explains its highest compressive strength for all the samples.

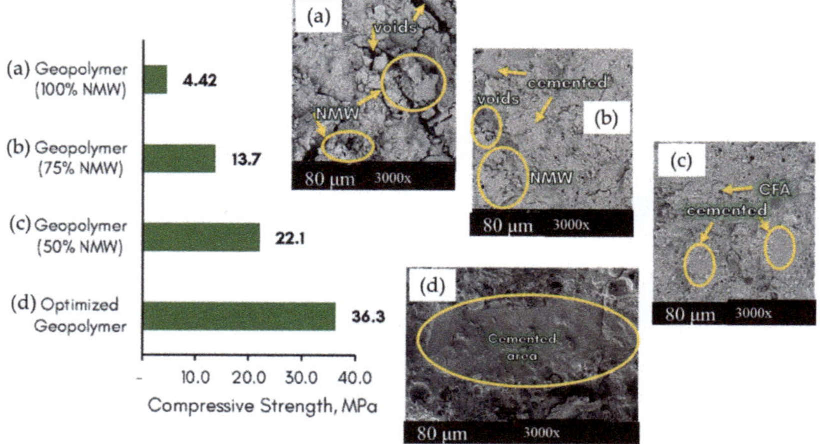

Figure 8. SEM images of synthesized geopolymer at different precursor mix (1000×) with corresponding UCS.

3.9. Potential Engineering Application

Table 10 shows the unconfined compressive strength of the synthesized geopolymer (optimized, 50% NMW, 75% NMW, and 100% NMW) and compared it to several types of concretes. Based on the standard unconfined compressive strengths, the synthesized geopolymers can have a potential application for concrete structures and concrete pavers, pedestrian and light traffic pavers, and plain concretes.

Table 10. Comparison of unconfined compressive strength from the standard materials.

Material	Mixture	Application	UCS (MPa)	Source
Class A Concrete	OPC-sand mixture	Concrete structures and concrete pavements	20.7	
Class C Concrete	OPC-sand mixture	Pedestrian and light traffic paver	20.7	DPWH and ASTM Standards [41,42]
Class B Concrete	OPC-sand mixture	Plain concrete for structure (curbs, gutter, sidewalks)	16.5	
Class F Concrete	OPC-sand mixture	Plain concrete for leveling	11.8	
Geopolymer	Optimized sample	Pavers, bricks	36.3 ± 3.6	This study, 2020
	50% NMW; 50% CFA	Pavers	22.1 ± 5.4	
	75% NMW; 25% CFA	Coring bricks	13.7 ± 2.9	
	100% NMW	Clay bricks	4.42 ± 0.3	

4. Conclusions

This paper presents an experimental study to produce an optimized geopolymer material that yields the highest value of unconfined compressive strength from the mixture of nickel–laterite mine waste (NMW), coal fly ash (CFA), and an alkali activator with components of sodium hydroxide (SH) and sodium silicate (SS). The optimum formulation mix was found to have an activator-to-precursor ratio of 0.428% NMW of 50.1%, and SH-to-SS ratio of 0.520, which produces a geopolymer with an average 28-day compressive strength of 36.3 MPa. This value is comparable to ordinary Portland cement for concrete structures and pavers, pedestrian pavers, light traffic pavers, and plain concrete.

SEM/EDX also showed that the optimum formulation has a cemented surface, resulting in a high unconfined compressive strength. The sample with low compressive strength was observed to have large voids in the microstructure, explaining its lower unconfined compressive strength.

Future work includes exploring the leachability of the NMW-CFA mixture to evaluate further the leachability of different mixtures, which may affect its pH during the test. Moreover, the leachability behavior of the NMW-CFA geopolymer product warrants further investigation as the nature of the material becomes more basic. Exploring the effect of the iron content of the NMW in the synthesis of the geopolymer, curing conditions, and other formulation mixes of the synthesized geopolymer should also consider in future studies. Other engineering properties can also be explored like water absorption, flexural strength, shrinkage, heat resistance, and porosity for other engineering applications like filtrations, panel boards, bricks, tiles, and other ceramic and building applications.

Author Contributions: Conceptualization, M.A.P.; methodology, A.L.J., A.A.T., and I.J.D.; validation, R.A.M., I.B.-A., and M.A.P.; formal analysis, A.L.J., and A.A.T.; investigation, M.A.P.; resources, M.A.P.; data curation, A.L.J., and A.A.T.; writing—original draft preparation, A.L.J.; writing—review and editing, M.A.P., I.B.-A., R.A.M., A.A.T., and I.J.D.; visualization, A.L.J., and I.J.D.; supervision, M.A.P., H.H., W.K., I.B.-A.; and R.A.M.; project administration, A.A.T.; funding acquisition, M.A.P. All authors have read and agreed to the published version of the manuscript.

Funding: This research was funded by the Department of Science and Technology-Philippine Council for Industry, Energy and Emerging Technology Research and Development (Project No. 07132) under the implementing agency of the Center for Engineering and Sustainable Development Research, De La Salle University-Manila and the APC was funded by Department of Science and Technology-Engineering Research and Development for Technology.

Acknowledgments: Authors acknowledge the following organizations: Department of Transdisciplinary Science and Engineering of Tokyo Institute of Technology; iNano Research Facility, De La Salle University; Advanced Device and Materials Testing Laboratory; Office of the Vice-Chancellor for Research and Innovation, De La Salle University; Geopolymers and Advanced Materials Engineering Research for Sustainability Laboratory (G.A.M.E.R.S. Lab), De La Salle University; Ceramics Engineering, MSU-Iligan Institute of Technology; Agata Mining Ventures, Inc. (AMVI); and STEAG Power, Inc.

References

1. Reddy, M.S.; Dinakar, P.; Rao, B.H. Mix design development of fly ash and ground granulated blast furnace slag based geopolymer concrete. *J. Build. Eng.* **2018**, *20*, 712–722. [CrossRef]
2. Patankar, S.V.; Ghugal Yuwaraj, M.; Jamkar, S.S. Mix Design of Fly Ash Based Geopolymer Concrete. In *Advances in Structural Engineering*; Springer: New Delhi, India, 2015; pp. 1619–1634. [CrossRef]
3. Li, N.; Shi, C.; Zhang, Z.; Wang, H.; Liu, Y. A review on mixture design methods for geopolymer concrete. *Compos. Part B Eng.* **2019**, *178*, 107490. [CrossRef]
4. Tchakouté, H.K.; Rüscher, C.H.; Hinsch, M.; Djobo, J.N.Y.; Kamseu, E.; Leonelli, C. Utilization of sodium waterglass from sugar cane bagasse ash as a new alternative hardener for producing metakaolin-based geopolymer cement. *Chemie der Erde* **2017**, *77*, 257–266. [CrossRef]
5. Ferdous, M.W.; Kayali, O.; Khennane, A. A Detailed Procedure of Mix Design for Fly Ash Based Geopolymer Concrete. In Proceedings of the Fourth Asia-Pacific Conference on FRP in Structures (APFIS 2013), Melbourne, Australia, 11–13 December 2013; pp. 11–13.

6. Longos, A.; Tigue, A.A.; Malenab, R.A.; Dollente, I.J.; Promentilla, M.A. Mechanical and thermal activation of nickel-laterite mine waste as a precursor for geopolymer synthesis. *Results Eng.* **2020**, *7*, 100148. [CrossRef]

7. Aseniero, J.P.J.; Opiso, E.M.; Banda, M.H.T.; Tabelin, C.B. Potential utilization of artisanal gold-mine tailings as geopolymeric source material: Preliminary investigation. *SN Appl. Sci.* **2019**, *1*, 35. [CrossRef]

8. Promentilla, M.A.B.; Thang, N.H.; Kien, P.T.; Hinode, H.; Bacani, F.T.; Gallardo, S.M. Optimizing Ternary-blended Geopolymers with Multi-response Surface Analysis. *Waste Biomass Valoriz.* **2016**, *7*, 929–939. [CrossRef]

9. Zhang, L.; Ahmari, S.; Zhang, J. Synthesis and characterization of fly ash modified mine tailings-based geopolymers. *Constr. Build. Mater.* **2011**, *25*, 3773–3781. [CrossRef]

10. Phetchuay, C.; Horpibulsuk, S.; Arulrajah, A.; Suksiripattanapong, C.; Udomchai, A. Strength development in soft marine clay stabilized by fly ash and calcium carbide residue based geopolymer. *Appl. Clay Sci.* **2016**, *127–128*, 134–142. [CrossRef]

11. Singh, B.; Ishwarya, G.; Gupta, M.; Bhattacharyya, S.K. Geopolymer concrete: A review of some recent developments. *Constr. Build. Mater.* **2015**, *85*, 78–90. [CrossRef]

12. Pavithra, P.; Srinivasula Reddy, M.; Dinakar, P.; Hanumantha Rao, B.; Satpathy, B.K.; Mohanty, A.N. A mix design procedure for geopolymer concrete with fly ash. *J. Clean. Prod.* **2016**, *133*, 117–125. [CrossRef]

13. Yankwa Djobo, J.N.; Elimbi, A.; Kouamo Tchakouté, H.; Kumar, S. Mechanical properties and durability of volcanic ash based geopolymer mortars. *Constr. Build. Mater.* **2016**, *124*, 606–614. [CrossRef]

14. Ibrahim, L.A.A. Chemical characterization and mobility of metal species in fly ash–water system. *Water Sci.* **2015**, *29*, 109–122. [CrossRef]

15. Akar, G.; Polat, M.; Galecki, G.; Ipekoglu, U. Leaching behavior of selected trace elements in coal fly ash samples from Yenikoy coal-fired power plants. *Fuel Process. Technol.* **2012**, *104*, 50–56. [CrossRef]

16. Koukouzas, N.; Ketikidis, C.; Itskos, G. Heavy metal characterization of CFB-derived coal fly ash. *Fuel Process. Technol.* **2011**, *92*, 441–446. [CrossRef]

17. US Environmental Agency (US EPA). Toxicity Characteristic Leaching Procedure Method 1311. Available online: https://www.epa.gov/sites/production/files/2015-12/documents/1311.pdf (accessed on 10 December 2020).

18. US EPA. Characteristics Introduction and Regulatory Definitions. Available online: https://www.epa.gov/sites/production/files/2015-10/documents/chap7_0.pdf (accessed on 21 February 2020).

19. DENR Department of Environment and Natural Resources (DENR). Administrative Order No. 2013-22 Revised Procedures and Standards for the Management of Hazardous Wastes (Revising DAO 2004-36). Available online: https://emb.gov.ph/wp-content/uploads/2018/06/dao-2013-22.pdf (accessed on 11 December 2020).

20. Intrakamhaeng, V.; Clavier, K.A.; Roessler, J.G.; Townsend, T.G. Limitations of the toxicity characteristic leaching procedure for providing a conservative estimate of landfilled municipal solid waste incineration ash leaching. *J. Air Waste Manag. Assoc.* **2019**, *69*, 623–632. [CrossRef]

21. Liu, Y.; Clavier, K.A.; Spreadbury, C.; Townsend, T.G. Limitations of the TCLP fluid determination step for hazardous waste characterization of US municipal waste incineration ash. *Waste Manag.* **2019**, *87*, 590–596. [CrossRef]

22. Tabelin, C.B.; Igarashi, T.; Villacorte-Tabelin, M.; Park, I.; Opiso, E.M.; Ito, M.; Hiroyoshi, N. Arsenic, selenium, boron, lead, cadmium, copper, and zinc in naturally contaminated rocks: A review of their sources, modes of enrichment, mechanisms of release, and mitigation strategies. *Sci. Total Environ.* **2018**, *645*, 1522–1553. [CrossRef]

23. Tabelin, C.; Silwamba, M.; Paglinawan, F.C.; Jane, A.; Mondejar, S.; Gia, H.; Joy, V.; Opiso, E.M. Solid-phase partitioning and release-retention mechanisms of copper, lead, zinc and arsenic in soils impacted by artisanal and small-scale gold mining (ASGM) activities. *Chemosphere* **2020**, *260*, 127574. [CrossRef]

24. Department of Environment and Natural Resources (DENR). Administrative Order No. 2016-08 on Water Quality Guidelines and General Effluent Standards of 2016. Available online: http://water.emb.gov.ph/wp-content/uploads/2016/06/DAO-2016-08-WQG-and-GES.pdf (accessed on 10 December 2020).

25. Alves, B.S.Q.; Dungan, R.S.; Carnin, R.L.P.; Galvez, R.; De Carvalho Pinto, C.R.S. Metals in waste foundry sands and an evaluation of their leaching and transport to groundwater. *Water Air. Soil Pollut.* **2014**, *225*. [CrossRef]

26. Özkök, E.; Davis, A.P.; Aydilek, A.H. Leaching of As, Cr, and Cu from High-Carbon Fly Ash–Soil Mixtures. *J. Environ. Eng.* **2013**, *139*, 1397–1408. [CrossRef]

27. Tigue, A.A.S.; Malenab, R.A.J.; Dungca, J.R.; Yu, D.E.C.; Promentilla, M.A.B. Chemical stability and leaching behavior of one-part geopolymer from soil and coal fly ash mixtures. *Minerals* **2018**, *8*, 411. [CrossRef]

28. Ahmari, S.; Zhang, L. Durability and leaching behavior of mine tailings-based geopolymer bricks. *Constr. Build. Mater.* **2013**, *44*, 743–750. [CrossRef]

29. Van Jaarsveld, J.G.S.; Van Deventer, J.S.J.; Lukey, G.C. The characterization of source materials in fly ash-based geopolymers. *Fuel Energy Abstr.* **2004**, *45*, 23. [CrossRef]

30. Whitney, D.L.; Evans, B.W. Abbreviations for Names of Rock-Forming Minerals Abbreviations for names of rock-forming minerals. *Am. Mineral.* **2015**, *95*, 185–187. [CrossRef]

31. Li, B.; Wang, H.; Wei, Y. The reduction of nickel from low-grade nickel laterite ore using a solid-state deoxidisation method. *Miner. Eng.* **2011**, *24*, 1556–1562. [CrossRef]

32. Liu, H.; Sun, Q.; Wang, B.; Wang, P.; Zou, J. Morphology and composition of microspheres in fly ash from the luohuang power plant, Chongqing, Southwestern China. *Minerals* **2016**, *6*, 30. [CrossRef]

33. Cherki El Idrissi, A.; Roziere, E.; Loukili, A.; Darson, S. Design of geopolymer grouts: The effects of water content and mineral precursor. *Eur. J. Environ. Civ. Eng.* **2018**, *22*, 628–649. [CrossRef]

34. Xu, H.; Van Deventer, J.S.J. The effect of alkali metals on the formation of geopolymeric gels from alkali-feldspars. *Colloids Surf. A Physicochem. Eng. Asp.* **2003**, *216*, 27–44. [CrossRef]

35. Livi, C.; Repette, W. Effect of NaOH concentration and curing regime on geopolymer. *Rev. IBRACON Estruturas e Mater.* **2017**, *10*, 1174–1181. [CrossRef]

36. Garcia-Lodeiro, I.; Palomo, A.; Fernández-Jiménez, A. *An Overview of the Chemistry of Alkali-Activated Cement-Based Binders*; Woodhead Publishing Limited: Cambridge, UK, 2015; ISBN 9781782422884.

37. Pacheco-Torgal, F.; Castro-Gomes, J.; Jalali, S. Alkali-activated binders: A review. Part 2. About materials and binders manufacture. *Constr. Build. Mater.* **2008**, *22*, 1315–1322. [CrossRef]

38. Solouki, A.; Viscomi, G.; Lamperti, R.; Tataranni, P. Quarry waste as precursors in geopolymers for civil engineering applications: A decade in review. *Materials* **2020**, *13*, 3146. [CrossRef] [PubMed]

39. Gomes, K.C.; Lima, G.S.T.; Torres, S.M.; De Barros, S.; Vasconcelos, I.F.; Barbosa, N.P. Iron distribution in geopolymer with ferromagnetic rich precursor. *Mater. Sci. Forum* **2010**, *643*, 131–138. [CrossRef]

40. Obonyo, E.A.; Kamseu, E.; Lemougna, P.N.; Tchamba, A.B.; Melo, U.C.; Leonelli, C. A sustainable approach for the geopolymerization of natural iron-rich aluminosilicate materials. *Sustainability* **2014**, *6*, 5535–5553. [CrossRef]

41. Association of Structural Engineers of the Philippines. *National Structural Code of the Philippines 2010: Buildings, Towers and other Vertical Structures*, 6th ed.; Association of Structural Engineers of the Philippines: Quezon City, Philippines, 2010; ISBN 2094-5477.

42. Japan International Cooperation Agency. The Urgent Development Study on the Project on Rehabilitation and Recovery from Typhoon Yolanda in the Philippines, Implemented by DOF, DPWH, DILG, Philippines. Available online: https://openjicareport.jica.go.jp/pdf/12283420_03.pdf (accessed on 21 February 2020).

 *minerals*

Article

Characterization of Slag Reprocessing Tailings-Based Geopolymers in Marine Environment

Jie Wu, Jing Li, Feng Rao * and Wanzhong Yin *

School of Zijin Mining, Fuzhou University, Fuzhou 350108, China; abbyduola@163.com (J.W.); lijing610082858@163.com (J.L.)
* Correspondence: fengrao@fzu.edu.cn (F.R.); yinwanzhong@163.com (W.Y.)

Received: 30 July 2020; Accepted: 10 September 2020; Published: 22 September 2020

Abstract: In this study, copper slag reprocessing tailings (CSRT) were synthesized into geopolymers with 40%, 50% and 60% metakaolin. The evolution of compressive strength and microstructures of CSRT-based geopolymers in a marine environment was investigated. Except for compressive strength measurement, the characterizations of X-ray diffraction (XRD), Fourier-transform infrared spectroscopy (FTIR), nuclear magnetic resonance (NMR) and scanning electron microscopy (SEM) were included. It was found that marine conditions changed the Si/Al ratio in the sodium-aluminosilicate-hydrate (N-A-S-H) gel backbone, promoted the geopolymerization process, led to more $Q^4(3Al)$, $Q^4(2Al)$ and $Q^4(1Al)$ gel formation and a higher compressive strength of the geopolymers. This provided a basis for the preparation of CSRT-based geopolymers into marine concrete.

Keywords: copper slag reprocessing tailings; geopolymer; microstructure; NMR

---

1. Introduction

Massive quantities of copper mine tailings are discharged from the beneficiation of the copper mine, which causes some environmental problems such as contamination of surrounding air, water bodies and soils, when disposed of in tailings storage facilities without rehabilitation [1,2]. Therefore, it is necessary to develop methods to reprocess the mine tailings to relieve their adverse impacts, whilst, at the same time, recycle resources and increase economic benefit. In the past decade, researchers used alkaline activation or the geopolymerization process for the consolidation of tailings [3,4]. Accordingly, the tailings are employed as raw materials in the alkaline activation process, in which the aluminosilicates in tailings are attacked by alkalis to decompose monomers and then to regroup and harden. For example, Ahmari et al. used sodium hydroxide as an alkaline activator to prepare copper mine tailings-based geopolymers, which were suitable for application of road base material at a room temperature [5]. Cristelo et al. applied fly ash as additive, and sodium hydroxide and sodium silicate as alkaline activators to prepare high-sulfur copper mine tailings-based geopolymer, which had a maximum compressive strength of 23.5 MPa [6]. Generally, mine tailings contain a substantial amount of crystal minerals and exhibit inert in geopolymeric reaction, so that the calcination of the tailings or addition of calcined materials (e.g., metakaolin and slag) is usually necessary; the reasons are: (1) the calcined materials (e.g., metakaolin and slag) have a large amount of silicon, aluminum and calcium, which are vital elements required for geopolymerization, (2) the calcined raw materials usually have a fast dissolution and gelation rate that makes geopolymers show high early compressive strength. [7]. The calcination induces dehydroxylation of the precursor materials and enhances their amorphous phase content, so that the dissolution of silicate and aluminate monomers is favorable in the formation of sodium-aluminosilicate-hydrate (N-A-S-H) gel [8].

Recently, geopolymer has become an emerging marine material, because of its particular microstructure. Some researchers considered the compact tetrahedral aluminosilicate structure might protect the geopolymer concrete from attack by corrosive ions in harsh marine conditions [9,10]. For instance, a metakaolin-based geopolymer was synthesized with sodium hydroxide and sodium silicate mixture as alkaline activator and exposed to seawater for 28 days, whose compressive strength exceeded 20 MPa [11]. Therefore, in this work, copper slag reprocessing tailings (CSRT)-based geopolymers were prepared and characterized in marine environment, in order to provide guidance for the preparation of tailings into marine concrete. Copper tailings have a significant amount of silicon, aluminum and calcium, which are crucial elements needed for geopolymerization. It was hypothesized that (1) the slag flotation tailings would have higher reactivity than other copper mine tailings in the geopolymerization process; (2) the formed geopolymers would have higher resistance against corrosion in the marine environment than ordinary Portland cement (OPC) concrete. After preparation, the CSRT-based geopolymers cured in air, artificial seawater and heat–cool cycles in seawater were analyzed.

2. Experimental

2.1. Materials

Copper mine tailings were obtained from the copper slag reprocessing tailings of Zijin Shan, Longyan, Fujian province, China. Kaolinite was purchased in Wuhan, Hubei province, China, which was used to prepare metakaolin through calcination at 800 °C for 6 h. Figure 1 gives the size distribution of the CSRT and metakaolin measured by a laser diffraction analyzer (LS-CWM, Omec, Zhuhai, China), in which the d_{50} and d_{85} are 16.81 and 36.71 µm for CSRT, and 2.38 and 3.52 µm for metakaolin, respectively. Table 1 shows the chemical composition of the CSRT and metakaolin analyzed by an X-ray fluorescence instrument (XRF, Axios advanced, PANalytical B.V., Almelo, The Netherlands). The CSRT mainly contained SiO_2 (28.19%) and Fe_2O_3 (57.83%), while SiO_2 (52.67%) and Al_2O_3 (41.37%) were the main components of the metakaolin. Figure 2 shows the main components of the CSRT and metakaolin characterized by X-ray diffraction (XRD, D8, Brucker, Karlsruhe, Germany). The CSRT mainly contained magnetite and fayalite, while metakaolin showed an amorphous peak at 2θ of 15–30° and crystal phase in quartz. Sodium silicate (ACS reagent grade), employed as an alkali activator, was bought from Aladdin Chemical Reagent (Shanghai, China) in the geopolymerization process. Deionized water was used throughout the whole process.

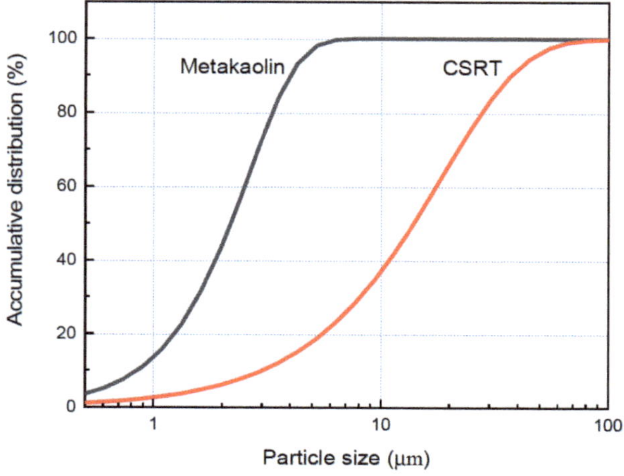

Figure 1. Size distribution of the tailings and metakaolin samples.

Table 1. Chemical composition of the, copper slag reprocessing tailings (CSRT) and metakaolin.

Components (%)	SiO$_2$	Al$_2$O$_3$	Fe$_2$O$_3$	MgO	CaO	TiO$_2$	Na$_2$O	K$_2$O
CSRT	28.19	4.24	57.83	1.101	1.906	0.303	0.525	0.81
Metakaolin	52.67	41.37	1.299	0.32	0.02	0.163	0.092	2.08

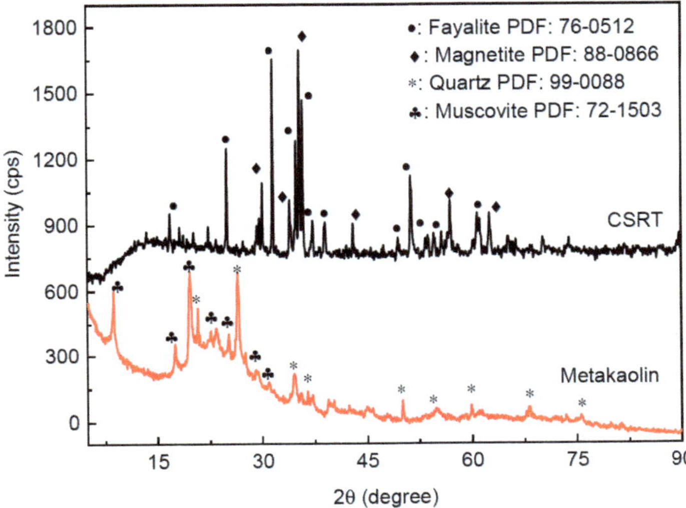

Figure 2. XRD patterns of the CSRT and metakaolin.

2.2. Methods

For the synthesis of the CSRT-based geopolymer, the activating solution was prepared by sodium silicate solution (6.17 M), and then mixed with the raw materials of CSRT and metakaolin for 10 min, which were of 40%, 50% and 60% metakaolin and a total solid of 192 g. Subsequently, the mixture was put into cubic steel molds (30 mm × 30 mm × 30 mm), which were vibrated to liberate air bubbles on a vibration table for 5 min. Next, the molds were sealed with a plastic sealing bag and cured at 60 °C for 12 h and left at room temperature for 7 days to complete the hydrate process of geopolymers to get the initial CSRT-based geopolymer. After that, the initial geopolymers were exposed or cured under three types of environment for 30 days: (1) in air; (2) in artificial seawater; (3) in a heat–cool seawater cycle. Geopolymers were exposed to artificial seawater at room temperature for 12 h and 12 h under the cycle in a freezer (−18 °C). Artificial seawater with 10-fold concentration was made with 292.5 g/L NaCl, 7.45 g/L KCl, 36 g/L MgSO$_4$. The artificial seawater was renewed every seven days throughout the whole curing process [12].

The compressive strength of the CSRT-based geopolymers was analyzed by a YAW-300 compression and flexure machine from Jinan Tianchen manufacture (Jinan, China). Three geopolymer samples were tested and the mean value was given in each measurement. The morphology and microstructure of the geopolymers were characterized using a scanning electron microscope (SEM, Quanta 250, FEI, Hillsboro, OR, USA) and XRD diffractometer. The geopolymer was ground to less than 75 µm to prepare specimens for XRD measurements, in which Cu-Kα1 radiation and a scanning rate of 0.1°/s from 5° to 90° of 2θ were used. The geopolymers were also analyzed for the microstructure by employing a Fourier-transform infrared spectroscopy (FTIR, Nicolet, Thermo Fisher Scientific, Waltham, MA, USA), of which transmittance spectra were obtained over a wavenumber of 400–4000 cm^{-1}; the resolution was 2 cm^{-1}. The interactions of silicate in the geopolymers were measured by a ^{29}Si nuclear magnetic resonance (NMR, AVANCE III, Bruker, Zurich, Switzerland) instrument. Solid-state ^{29}Si NMR spectra were collected at 99.35 MHz with a pulse width of 4 μs (i.e., 90 degrees for quantitative analysis) and a relaxation delay of 3 s. The spectra were referenced to tetramethylsilane (TMS).

3. Results and Discussion

Figure 3 shows the compressive strength of CSRT-based geopolymers prepared with different proportions of metakaolin and different exposures. The initial geopolymers, cured for 7 days, had a compressive strength of 2.3 MPa, 7.8 MPa and 15.3 MPa when the proportion of metakaolin was 40%, 50% and 60%, respectively. The low value in compressive strength suggested that the CSRT had low activity in the geopolymeric reaction. Moreover, metakaolin was obviously advantageous to increase the compressive strength of the geopolymer. Compared to the initial geopolymers, the compressive strength of the geopolymers that were exposed to air for 30 days had no obvious improvement or lessening of strength. While for the geopolymers exposed to seawater for 30 days, the compressive strength increased significantly to 10.7 MPa, 15.7 MPa and 32.6 MPa with 40%, 50% and 60% of metakaolin addition, respectively. For exposure to heat–cool cycles of seawater, the increase in compressive strength of geopolymers was less than those in seawater, which were 3.4 MPa, 10.2 MPa and 23 MPa with 40%, 50% and 60% of metakaolin addition, respectively. These results show that the compressive strength of the geopolymers was improved when exposed to seawater, which is in agreement with previous studies. Sotya et al. reported that the compressive strength of geopolymers was not affected by seawater immersion [11]. Jin et al. found that cations in seawater (e.g., Mg^{2+} and K^+) could diffuse into the aluminosilicate network structure of geopolymers to balance the negative charge, making the gel structure stable [13].

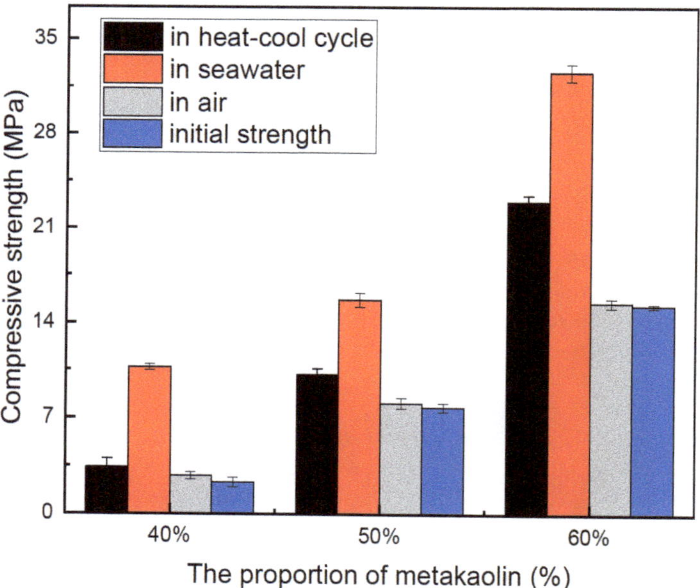

Figure 3. Compressive strength of the geopolymers prepared with different proportions of metakaolin and different exposures.

Figure 4 gives the XRD spectra of the geopolymers prepared with 60% metakaolin and which underwent different exposures for 30 days. Compared to the XRD patterns of raw materials (Figure 2), the crystal phases of quartz, magnetite, fayalite and muscovite remained after the geopolymeric reaction. While for exposure in seawater, the solution penetrated pores and precipitated halite (NaCl) was observed. The range of broad peak at 2θ of 15–30° in metakaolin transformed into 2θ of 22–35° in the geopolymers with different exposures, which indicates the formation of N-A-S-H gel [14,15].

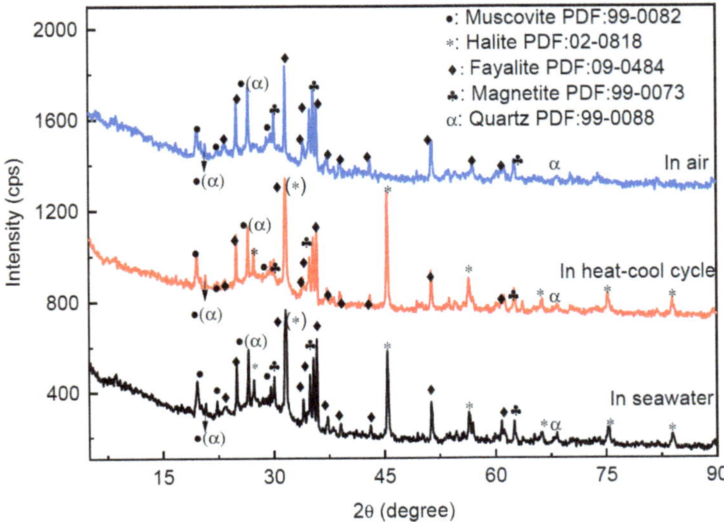

Figure 4. XRD spectra of the geopolymers prepared with 60% metakaolin.

Figure 5 shows the FTIR spectra of geopolymers prepared with 60% metakaolin and different exposures. The absorption peaks at 3440 cm^{-1} and 1644 cm^{-1} are due to stretching vibrations of OH$^-$, and H-OH bonds of free water, respectively, corresponding to adsorbed H$_2$O in geopolymers [16]. The absorption peak around 1453 cm^{-1} is assigned to asymmetric stretching of O-C-O bonds in CO$_3$$^{2-}$ groups, due to carbonation in the curing and exposing process of geopolymers [17]. The peak at 699 cm^{-1} is the Si-O bonds in quartz, suggesting the quartz particles are insoluble in the geopolymeric reaction [18]. The absorption peaks at 560 cm^{-1} and 460 cm^{-1} represent the zeolite framework in the structure of the geopolymer [19,20], which indicates zeolite frameworks are formed during the process. The band at about 1000 cm^{-1} is the Si-O-T bonds (T represents the tetrahedral Al or Si) in the geopolymer gel [21]. It is sensitive to the Si/Al ratio in the geopolymer backbone, which could shift toward the lower wavenumber at a low Si/Al ratio [22]. The wavenumber transformation indicates the exposure in the heat–cool cycle and seawater, which changed the evolution of the N-A-S-H gel. In addition, the higher intensity for the geopolymers exposed to seawater than in the heat–cool cycle and in air, suggests the formation of higher proportions of N-A-S-H gel.

NMR spectroscopy is an excellent analytical method for characterizing the short-range ordering and molecular structure of silicates [23,24]. It employs Gaussian peak deconvolution to overcome the lack of spectra resolution and separate and quantify Q^n(mAl) species ($0 \leq m \leq n \leq 4$, m, n = integer). The resonances at −74 and −79 ppm are assigned to (Q^0) and (Q^1), respectively, due to the presence of silicate monomer and dimer [25]. The resonance at −104 ppm represents $Q^3(R)$ when H in OH is substituted by alkali metal ion (Na$^+$ or K$^+$) in Q^3 [2], and the resonance at −114 ppm corresponds to the cristobalite in the geopolymers [26]. In geopolymer N-A-S-H gel, the $Q^4(4Al)$, $Q^4(3Al)$, $Q^4(2Al)$, $Q^4(1Al)$, $Q^4(0Al)$ resonate at around −84, −89, −93, −99 and −108 ppm, respectively [19].

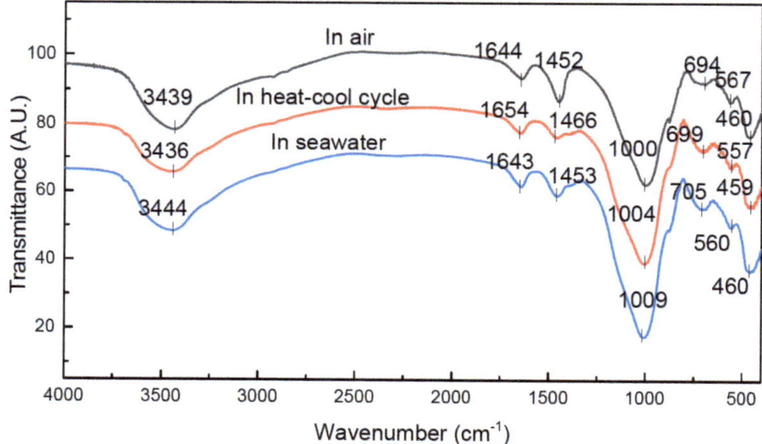

Figure 5. FTIR spectra of the geopolymers prepared with 60% metakaolin and different exposures.

Figure 6 gives the ^{29}Si NMR spectra and deconvolution of the raw metakaolin and CSRT-based geopolymers prepared with 60% metakaolin and exposed differently for 30 days. Metakaolin predominantly peaked at around −105 ppm. After geopolymerization, the ^{29}Si NMR lines shifted to less shielded values due to the geopolymeric reaction taking place to form N-A-S-H gel. For the geopolymers exposed to air, the heat–cool cycle and seawater, the Si sites in silicate monomers (Q^0) are 1.39%, 1.26% and 1.23%, those in silicate dimer (Q^1) are 8.63%, 6.65% and 6.10%, and in $Q^3(R)$ are 7.69%, 1.32% and 3.52%, but those in N-A-S-H gel ($Q^4(mAl)$, $0 \leq m \leq 4$) are 79.21%, 90.77%, and 87.78%, respectively. The Si sites in Q^0 and Q^1 slightly decrease and those in N-A-S-H gel increase when the geopolymers are exposed to seawater, indicating that the Q^0 and Q^1 transformed into N-A-S-H gel. It has been reported that the compressive strength relies on the amount of $Q^4(3Al)$, $Q^4(2Al)$ and $Q^4(1Al)$ [27]. The percentages of $Q^4(3Al)$, $Q^4(2Al)$ and $Q^4(1Al)$ are 51.08%, 63.78% and 64.78% for the exposures in air, heat–cool cycle and seawater, respectively, which explained the variation in compressive strength of geopolymers. These results suggest that exposure in seawater promotes the geopolymerization process.

Figure 7 shows the SEM morphology of the geopolymers that underwent different exposures for 30 days. The geopolymers exposed to both the heat–cool cycle and seawater showed homogenous and compact structures, indicating geopolymers with 60% metakaolin can be consolidated well. However, cracks were exhibited in geopolymers exposed to air. This result further confirmed that more N-A-S-H gel was formed in the geopolymers that were exposed to both the heat–cool cycle and seawater, thereby bridging the cracks, and thus increasing their compressive strength.

In this study, the geopolymers exposed to seawater showed a high formation of N-A-S-H gel, thereby exhibited a higher compressive strength than the geopolymers exposed to air. When geopolymers were exposed to the marine environment, alkaline ions (e.g., Mg^{2+}, Na^+) in seawater functioned as balancing cations to the negative charge of the tetra-coordinated aluminum, thereby promoting the geopolymerization process [28]. Therefore, for the geopolymers exposed to seawater, more Si-O-T gel was formed (Figure 5) and the Q^0 and Q^1 species were transformed into N-A-S-H gel (Figure 6).

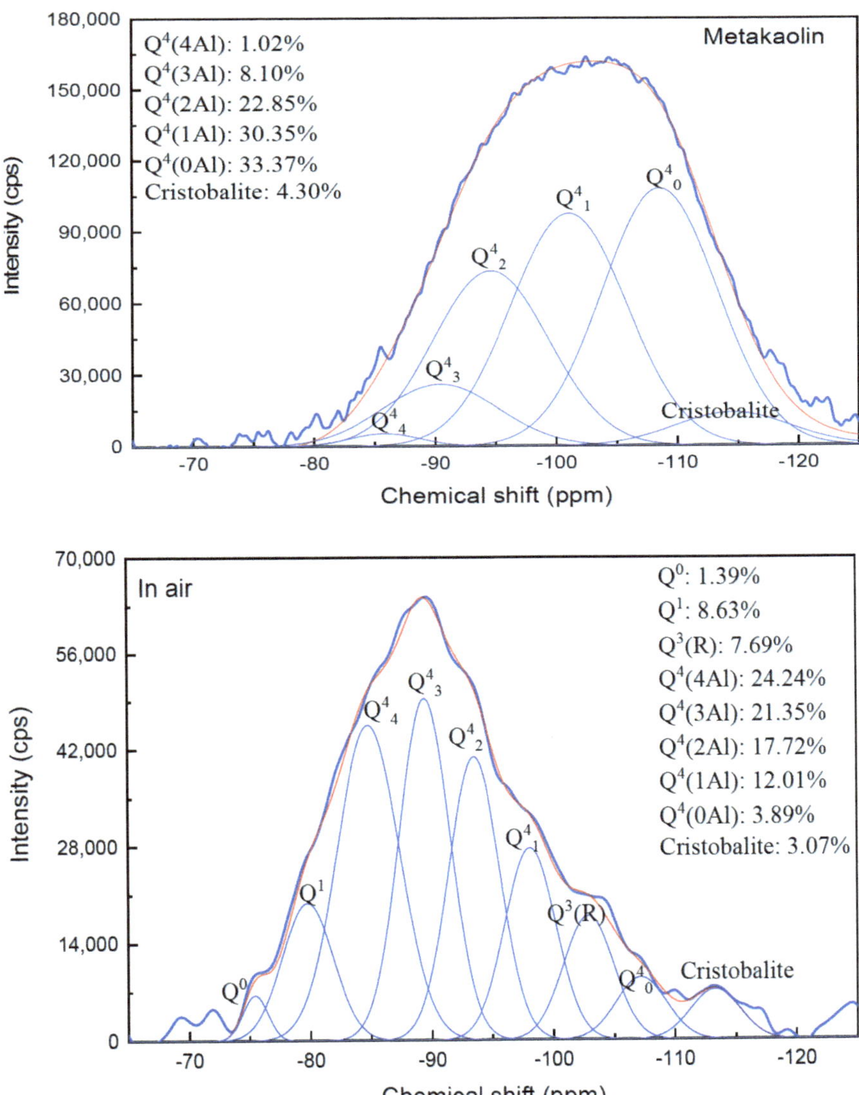

Figure 6. *Cont.*

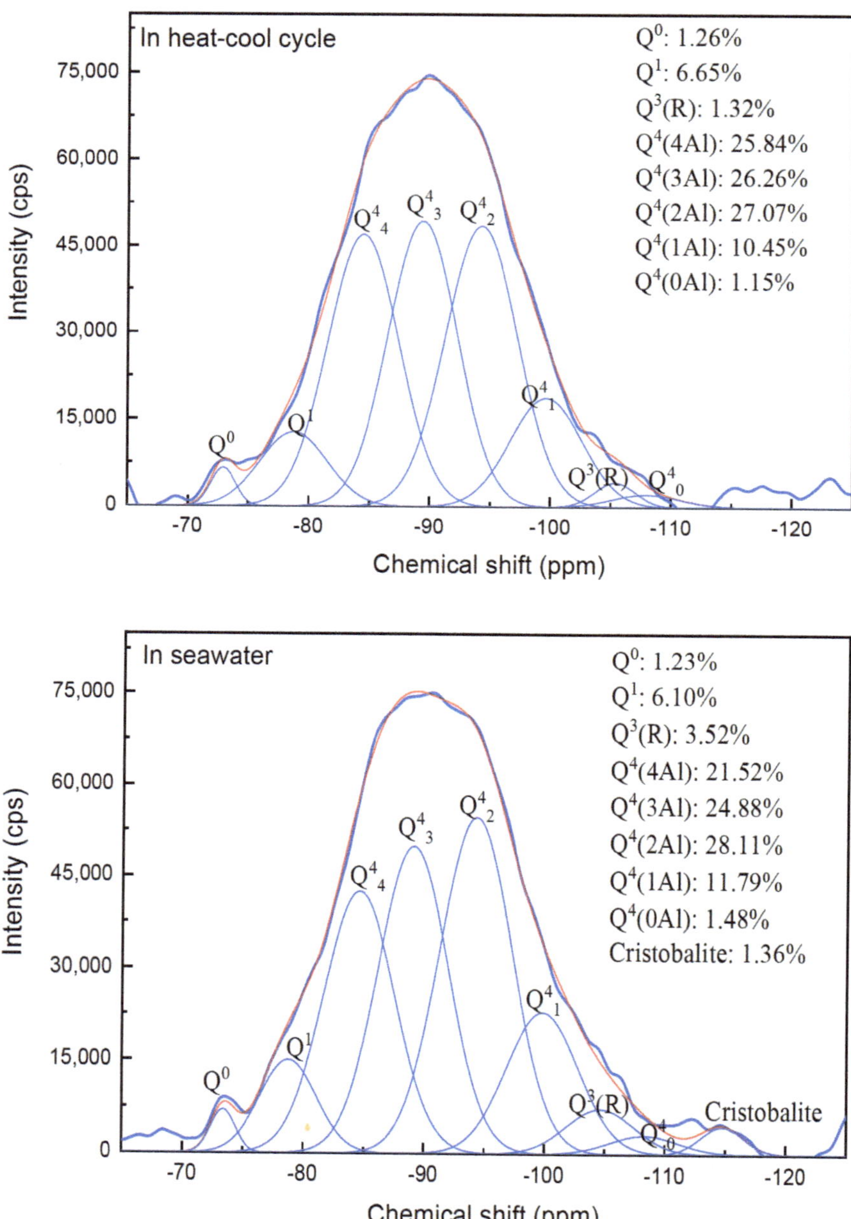

Figure 6. ^{29}Si NMR spectra and deconvolution of the geopolymers prepared with 60% metakaolin.

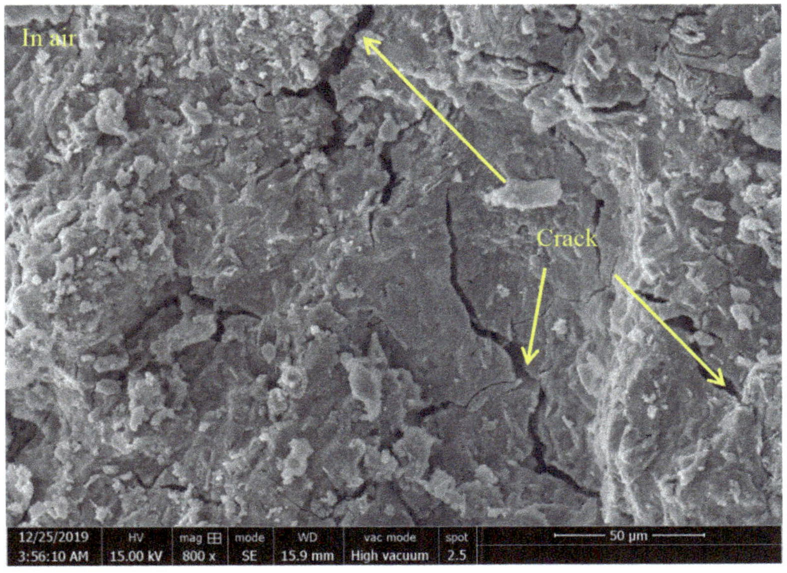

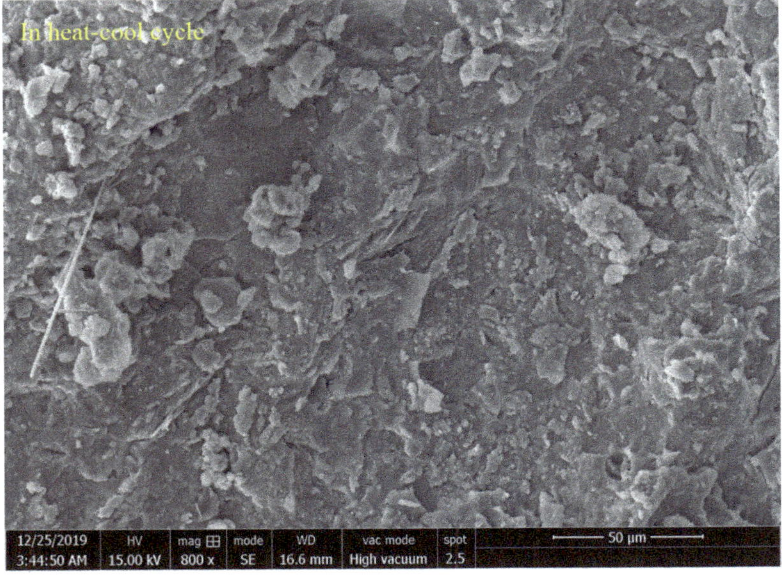

Figure 7. *Cont.*

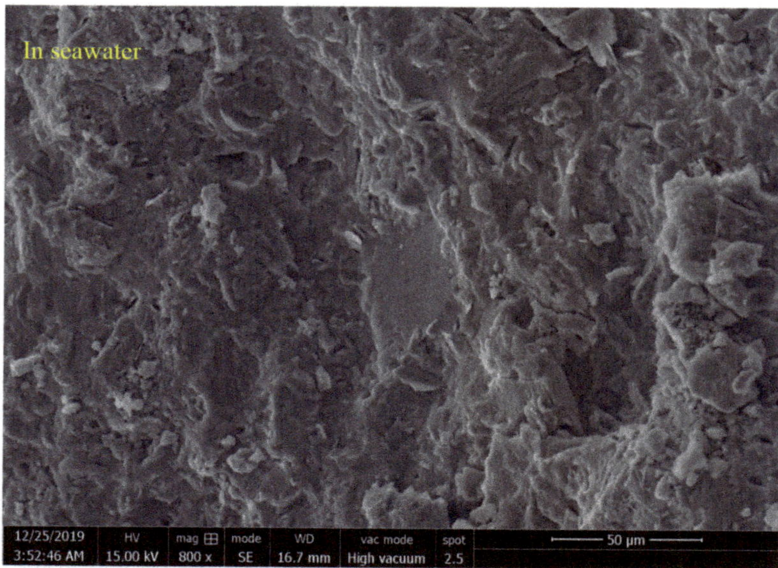

Figure 7. SEM images of the geopolymers prepared with 60% metakaolin.

4. Conclusions

Mine tailings after the flotation of copper slag are synthesized into geopolymers with the addition of metakaolin (40–60%), which enhances their compressive strength. For the geopolymers exposed to seawater, the Si/Al changed in the N-A-S-H gel backbone, more $Q^4(3Al)$, $Q^4(2Al)$ and $Q^4(1Al)$ gel and fewer cracks formed, resulting in a higher compressive strength being observed for these geopolymers than for those exposed to the heat–cool cycle of seawater and in air. This is attributed to the fact that alkaline ions in seawater balance the negative charge of the aluminum tetrahedrons in geopolymers, which promote the formation of N-A-S-H gel. As understanding the evolution of geopolymer in marine environments is important, this study is of significance and would benefit the development of marine concrete incorporating tailings. The optimization of compressive strength and long-term durability of the CRST-based geopolymer and tailoring of CRST-based geopolymer marine concrete should be studied in the future.

Author Contributions: J.W.: figures, study design, data collection, data analysis, and writing. J.L.: literature search, figures. F.R.: study design, data analysis, data interpretation, and writing. W.Y.: study design, data analysis, and data interpretation. All authors have read and agreed to the published version of the manuscript.

Funding: This study was financially supported by the National Natural Science Foundation of China under the project No. 51974093 and the Fuzhou University Qishan Scholar Talent Foundation under the Grant No. GXRC-18042, for which the authors are grateful.

References

1. Park, I.; Tabelin, C.B.; Jeon, S.; Li, X.; Seno, K.; Ito, M.; Hiroyoshi, N. A review of recent strategies for acid mine drainage prevention and mine tailings recycling. *Chemosphere* **2018**, *219*, 588–606. [CrossRef] [PubMed]
2. Tian, X.; Xu, W.; Song, S.; Rao, F.; Xia, L. Effects of curing temperature on the compressive strength and microstructure of copper tailing-based geopolymers. *Chemosphere* **2020**, *253*, 126754. [CrossRef] [PubMed]
3. Yang, F. *Geopolymerization of Copper Mine Tailings*; The University of Arizona: Tucson, AZ, USA, 2012.

4.	Ahmari, S.; Zhang, L.; Zhang, J. Effects of activator type/concentration and curing temperature on alkali-activated binder based on copper mine tailings. *J. Mater. Sci.* **2012**, *47*, 5933–5945. [CrossRef]

5.	Ahmari, S.; Chen, R.; Zhang, L. Utilization of mine tailings as road base material. In *American Society of Civil Engineers GeoCongress*; ASCE: Oakland, CA, USA, 2012; pp. 3654–3661.

6.	Cristelo, N.; Coelho, J.; Oliveira, M.; Consoli, N.C.; Palomo, Á.; Fernández-Jiménez, A. Recycling and application of mine tailings in alkali-activated cements and mortars-strength development and environmental assessment. *Appl. Sci.* **2020**, *10*, 2084. [CrossRef]

7.	Marescotti, P.; Carbone, C.; Capitani, L.D.; Grieco, G.; Lucchetti, G.; Servida, D. Mineralogical and geochemical characterisation of open-air tailing and waste-rock dumps from the Libiola Fe-Cu sulphide mine (Eastern Liguria, Italy). *Environ. Geol.* **2008**, *53*, 1613–1626. [CrossRef]

8.	Moukannaa, S.; Nazari, A.; Bagheri, A.; Loutou, M.; Sanjayan, J.G.; Hakkou, R. Alkaline fused phosphate mine tailings for geopolymer mortar synthesis: Thermal stability, mechanical and microstructural properties. *J. Non-Cryst. Solids.* **2019**, *511*, 76–85. [CrossRef]

9.	Kwan, W.H.; Cheah, C.B.; Ramli, M.; Chang, K.Y. Alkali-resistant glass fiber reinforced high strength concrete in simulated aggressive environment. *Mater. Constr.* **2018**, *63*, 62–71. [CrossRef]

10.	Albitar, M.; Mohamed Ali, M.; Visintin, P.; Drechsler, M. Durability evaluation of geopolymer and conventional concretes. *Constr. Build. Mater.* **2017**, *136*, 374–385. [CrossRef]

11.	Sotya, A.; Marta, N.D.; Wibowo, A.H.; Niken, S. Durability of geopolymer concrete upon seawater exposure. *Adv. Sci. Technol.* **2010**, *69*, 92–96.

12.	Li, X.; Rao, F.; Song, S.; Ma, Q. Deterioration in the microstructure of metakaolin-based geopolymers in marine environment. *J. Mater. Res. Technol.* **2019**, *8*, 2747–2752. [CrossRef]

13.	Jin, M.; Chen, Y.; Dong, H.; Chen, L.; Jin, Z. Seawater erosion resistance of geopolymer solidification MSWI fly ash. *Acta Mieraloliga Sin.* **2014**, *34*, 159–165. (In Chinese)

14.	Bernal, S.A.; Provis, J.L.; Walkley, B.; Nicolas, R.S.; Gehman, J.D.; Brice, D.G.; Kilcullen, A.R.; Duxson, P.; Deventer, J.S.J. Van Gel nanostructure in alkali-activated binders based on slag and fly ash, and effects of accelerated carbonation. *Cem. Concr. Res.* **2013**, *53*, 127–144. [CrossRef]

15.	Lee, N.K.; Koh, K.T.; An, G.H.; Ryu., G.S. Influence of binder composition on the gel structure in alkali activated fly ash/slag pastes exposed to elevated temperatures. *Ceram. Int.* **2017**, *43*, 2471–2480. [CrossRef]

16.	Bernal, S.A.; Provis, J.L.; Rose, V.; de Gutierrez, R.M. Evolution of binder structure in sodium silicate-activated slag-metakaolin blends. *Cem. Concr. Compos.* **2011**, *33*, 46–54. [CrossRef]

17.	Yang, T.; Xiao, Y.A.O.; Zuhua, Z.; Huajun, Z.H.U. Effects of NaOH solution concentration and reaction time on metakaolin geopolymerization. *J. Nanjing Univ. Technol.* **2013**, *35*, 21–25.

18.	Wan, Q.; Rao, F.; Song, S.; Cholico-González, D.F.; Ortiz, N.L. Combination formation in the reinforcement of metakaolin geopolymers with quartz sand. *Cem. Concr. Compos.* **2017**, *80*, 115–122. [CrossRef]

19.	Engelhardt, G.; Michel, D. *High-Resolution Solid-State NMR of Silicates and Zeolites*; Wily: New York, NY, USA, 1987.

20.	Mostafa, N.Y.; El-Hemaly, S.A.S.; Al-Wakeel, E.I.; El-Korashy, S.A.; Brown, P.W. Characterization and evaluation of the pozzolanic activity of Egyptian industrial by-products: I: Silica fume and dealuminated kaolin. *Cem. Concr. Res.* **2001**, *31*, 467–474. [CrossRef]

21.	Palomo, A.; Blanco-Varela, M.T.; Granizo, M.L.; Puertas, F.; Grutzeck, M.W. Chemical Stability of Cementitious Materials Based on Metakaolin. *Cem. Concr. Res.* **1999**, *29*, 997–1004. [CrossRef]

22.	Zheng, L.; Wang, W.; Shi, Y. The effects of alkaline dosage and Si/Al ratio on the immobilization of heavy metals in municipal solid waste incineration fly ash-based geopolymer. *Chemosphere* **2010**, *79*, 665–671. [CrossRef]

23.	Tian, X.; Rao, F.; Morales-Estrella, R.; Song, S. Effects of Aluminum Dosage on Gel Formation and Heavy Metal Immobilization in Alkali-Activated Municipal Solid Waste Incineration Fly Ash. *Energy Fuels* **2020**, *34*, 4727–4733. [CrossRef]

24.	Klinowski, J. Nuclear magnetic resonance studies of zeolites. *Prog. Nucl. Magn. Reson. Spectrosc.* **1984**, *16*, 237–309. [CrossRef]

25.	Singh, P.S.; Bastow, T.; Trigg, M. Structural studies of geopolymers by ^{29}Si and ^{27}Al MAS-NMR. *J. Mater. Sci.* **2005**, *40*, 3951–3961. [CrossRef]

26.	Fernández-Jiménez, A.; Palomo, A.; Sobrados, I.; Sanz, J. The role played by the reactive alumina content in the alkaline activation of fly ashes. *Microporous Mesoporous Mater.* **2006**, *91*, 111–119. [CrossRef]

27. Wan, Q.; Rao, F.; Song, S.; García, R.E.; Estrella, R.M.; Patiño, C.L.; Zhang, Y. Geopolymerization reaction, microstructure and simulation of metakaolin-based geopolymers at extended Si/Al ratios. *Cem. Concr. Compos.* **2017**, *79*, 45–52. [CrossRef]

28. Davidovits, J. Geopolymer chemistry and properties. In Proceedings of the 1st European Conference on Soft Mineralogy, Compiègne, France, 1–3 June 1988; Volume 1, pp. 25–48.

Article

The Use of Mining Waste Materials for the Treatment of Acid and Alkaline Mine Wastewater

Jacek Retka [1],*, Grzegorz Rzepa [2], Tomasz Bajda [2] and Lukasz Drewniak [1]

1   Institute of Microbiology, Faculty of Biology, University of Warsaw, Miecznikowa 1, 02-096 Warsaw, Poland; ldrewniak@biol.uw.edu.pl

2   Faculty of Geology, Geophysics and Environmental Protection, Department of Mineralogy, Petrography and Geochemistry, AGH University of Science and Technology, al. Mickiewicza 30, 30-059 Krakow, Poland; rzepa@agh.edu.pl (G.R.); bajda@agh.edu.pl (T.B.)

*   Correspondence: jretka@biol.uw.edu.pl

Received: 17 October 2020; Accepted: 24 November 2020; Published: 27 November 2020

Abstract: The mining of metal ores generates both liquid and solid wastes, which are increasingly important to manage. In this paper, an attempt was made to use waste rocks produced in the mining of zinc and lead to neutralizing acid mine drainage and alkaline flotation wastewater. Waste rock is a quartz-feldspar rock of hydrothermal origin. It is composed of, besides quartz and potassium feldspar (orthoclase), phyllosilicates (chlorite and mica), and sulfides (chiefly pyrite). To determine its physicochemical parameters and their variability, acid mine water and flotation wastewater were monitored for 12 months. Acid mine drainage (AMD) is characterized by a low pH (~3), high zinc concentration (~750 mg·L^{-1}), and high sulfate content (~6800 mg·L^{-1}). On the other hand, the determinations made for flotation wastewater showed, among others, a pH of approximately 12 and ca. 780 mg·L^{-1} of sulfates. AMD and flotation wastewater neutralization by the waste rock was shown to be possible and efficient. However, in both cases, the final solution contained elevated concentrations of metals and sulfates. Premixing AMD with alkaline flotation wastewater in the first step and then neutralizing the obtained mixture with the waste rock was considered the best solution. The produced solution had a circumneutral pH. However, the obtained solution does not meet the legislative requirements but could be further treated by, for example, passive treatment systems. It is noteworthy that the proposed approach is low cost and does not require any chemical reagents.

Keywords: acid mine drainage; alkaline flotation wastewater; waste rock; mining waste material

1. Introduction

The advancement of the mining industry and its improved efficiency entail an increase in the generation of various types of waste. Lottermoser divided them into mining wastes (waste rocks, mine waters, mine drainage sludges), processing wastes (tailings), and metallurgical wastes (bauxite red mud, historical base metal smelting slags, phosphogypsum) [1,2]. The wastes most frequently produced by processing plants result from mineral extraction, physical and chemical processing of metal ores, physical and chemical processing of minerals other than metal ores, drilling muds, and other drilling wastes. Mine wastes are estimated to be one of the largest waste streams in the world [3,4]. Existing waste management technologies are still not efficient enough and/or require large financial outlays, therefore these solid, flotation wastewater or slurry are most often deposited at or near the mine, and since they usually contain high concentrations of elements and compounds, they can harm the environment. This can happen not only in the vicinity of the processing plant but also over a much larger area, e.g., contamination of rivers streams, and lakes, the devastation of the landscape [5–9].

It seems there is a pressing need to develop and apply solutions that are not only environmentally friendly and efficient but also economically viable [10].

AMD is an issue that mines have been struggling with for decades. Its origin can be twofold. In areas where minerals containing sulfides are present in the geological structure, the formation of AMD is the result of chemical, physical, and biological complex processes. As a consequence of the impact of water and air on the iron-bearing sulfides, mainly pyrite and pyrrhotite (oxidizing conditions), the resultant processes, often including microbial activity, lead to AMD formation [11,12]. Pyrite oxidation, ferrous sulfur, and sulfuric acid are formed first, followed by reddish-orange ferric oxyhydroxide with additional sulfuric acid. The resulting conditions allow for the further dissolution of other minerals containing hazardous and toxic elements including arsenic, cadmium, lead, and zinc [13–15]. Alongside pyrite and pyrrhotite, other sulfide-bearing minerals such as sphalerite, galena, arsenopyrite, gersdorffite, or chalcopyrite contribute to the formation of AMD [16,17]. Acid mine drainage can also be caused by human activity as a result of mining and processing works [18]. When the mine is in operation, it is necessary to maintain a sufficiently low level of groundwater. The cessation of mining activities causes the abandoned excavations to slowly fill up with water. High water oxygenation promotes the oxidation of sulfide minerals, especially pyrite. As a consequence of the reactions, sulfates are formed and metals are released. Commonly fine-grained waste rocks, which are exposed to the atmosphere, are stored in piles. Due to precipitation, geochemical weathering of metal-containing primary minerals takes place. The result is an outflow of watercourses containing significant loads of metals [17,19–21]. Over the years, many methods of AMD treatment have been developed [11,22–27]. Generally, they can be divided into two categories: passive and active types of acid mine drainage clean-up. One of the most well-studied and employed methods is based on chemical neutralization with the use of limestone [22], which is characterized by its low cost and wide availability. The result is a pH increase and, by precipitation, metal removal [17,28]. However, this method has some limitations due to its tendency to form an external coating caused by secondary mineral precipitation [29]. This may result in a significant reduction of the concentration of such elements, e.g., Cu, Ni, or Zn. In turn, in the presence of aqueous sulfate a gypsum sediment precipitates [30]. What is more, such a solution would be effective enough for AMD treatment only when the formation of AMD was temporary or short-term [15].

Problems with mine leachate treatment relate not only to AMD but also to other effluents generated during operation and deposit processing. Many mines also transform the ore into the final product. The most frequent ore dressing method is flotation. Briefly, pre-prepared rock material is subjected to chemical processes and air. In this manner, a particular metal concentrate is obtained (solid/solid separation). As usual, this type of process produces waste containing off-balance amounts of metals, including heavy metals with a strongly acidic or strongly alkaline reaction. In recent years, many attempts have been made to conduct post-flotation wastewater treatment [31–34]. Despite the many available solutions, due to the specific nature of flotation, it is necessary to adapt them to the type of waste being treated. Thus, storage in sedimentary tailings ponds is standard practice. Of course, there exist no longer used tailings containing waste that have been deposited for decades, whose treatment can be beneficial not only for the environment but also economically feasible. The technologies used in the past were less efficient, and, therefore, settling ponds contain significant amounts of metals that might be recycled [35,36]. Nonetheless, in the event of a breakdown or failure, a leak of hundreds of Mt of waste may pose an unimaginable risk to the environment, even over an extended period [35,37].

The zinc and lead mine Gradir in Sula, Montenegro, is an example of a site that still operates as a mine and which faces significant environmental issues. This mine has been operating for decades. Former and present activity contributes to the generation of four types of waste (AMD, flotation wastewater, and two types of waste rock). Traces of past activity can be seen in the form of closed, more or less collapsed adits. Most of them are filled with acid mine drainage. The presence of AMD is the result of both previous human activity and natural processes. As mentioned above, the formation of AMD is based on sulfide-containing minerals, and Montenegro is rich in them [38]. The production of lead and zinc concentrate is carried out by a flotation, the by-product of which is liquid waste. It is pumped into flotation tailings. To minimize waste emissions and energy consumption, the mine operates in a circular loop: that is, part of the leachate, after the suspension has settled, is reused in production activity. At present, mining is carried out using the open-pit method. A variety of rock material is produced during mining (blasting) operations. Based on the decision of the geologist the material is then separated. Limestone, which is the overburden of the deposit, is used both inside and outside the mine. The rock poor in the deposit are considered waste and are stored on the slopes of the mountain in the form of different size rock coarse and fine rock material. The material stored on the slopes is exposed to weather conditions, thus can contribute to the formation of AMD. The rock rich in Pb and Zn ore is crushed in mills, Appropriate chemical reagents are added and then hydro-transported to the hall where the flotation takes place. The final product is Pb and Zn concentrates. The wastewaters remaining after the process are discharged into the pond.

In this paper, an attempt was made to use waste materials produced in the mining of zinc and lead in the treatment and neutralization of acid and alkaline wastewater before discharge into the environment. We investigated whether the waste material remaining after the extraction of zinc and lead ore can be used to neutralize and stabilize the pH of acid mine waters flowing out of ancient adits and alkaline wastewater generated by flotation of Zn and Pb ore. The study of wastewater treatment was preceded by a detailed geochemical description of the raw material in terms of an environmental risk assessment associated with the uncontrolled release of toxic elements. In turn, acid mine drainage and flotation wastewater were monitored for 12 months to analyze seasonal variations in pollutant concentrations.

2. Materials and Methods

2.1. Site Description

In 1948, at the site of the current Gradir Montenegro mine, the search for zinc- and lead-bearing veins began. During the four-year investigation, the presence of deposits was documented, allowing the underground mine to be exploited and 250,000 tons of ore to be floated. The dominant minerals were sphalerite and galena. Over 50,000 meters of underground adits were drilled. Mining ceased in 1987 for two reasons: low metal prices and a lack of reserves of the vein type of ore. The mine resumed operations in 1997, though as an open-pit mine. The war in the Balkans and the economic sanctions imposed once again resulted in the closure of the mine. The mine was shut down from 2000 to 2008. Current mining is based on the exploitation of the remaining massive sulfide ore veins and disseminated sulfide stockwerk mineralization. The outline of the geology of the area is presented in Figure 1. Postvolcanic sulfide ores are related to Triassic bimodal effusive volcanism including basalts and rhyolites and later dioritic intrusions which were the source of hydrothermal ore-bearing fluids. The host rock for the ores is highly altered andesite. The metasomatic alteration includes chloritization, sericitization, silification, and pyritisation processes. Annual production reaches 1,000,000 tons of ore, resulting in 350,000 tons of pre-concentrate after the flotation process (more than 17,000 t of Zn and 5000 t of Pb). The Gradir Montenegro mine is located in the NW part of Montenegro at an elevation of 1180–1470 meters above sea level. Winters are mostly snowy (November–March) with several degrees of frost. In summer, the average daily temperature is about 20°, although the maximum reaches 30°. Most rainfall occurs from October to November. However, April has the most days with rain, and July to September have the least.

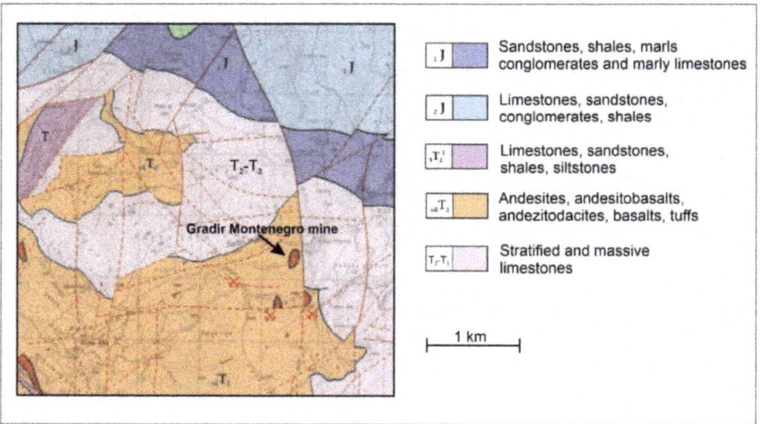

Figure 1. Geological map of the exploration area with a marked Gradir Montenegro mine [39].

2.2. Waste Materials

2.2.1. Acid Mine Drainage

The cessation of the use of adits for mining works has caused their gradual destruction. Currently, most of them have collapsed. Water seeping through the rocks is not pumped out, and it fills the corridors and, as a result, flows out naturally. The conditions inside (air, temperature, microbial activity, etc.) are conducive to acid mine drainage formation [4,40,41]. For the purposes of this publication, AMD samples were taken from one of the adits. The initial characteristics observed confirmed the low pH and high content of sulfate ions.

2.2.2. Alkaline Flotation Wastewater

The main activity of the Gradir mine (Sula, Montenegro) is based on the production of lead and zinc concentrates. After crushing and initial enrichment, the excavated deposit is directed to the flotation. It is a multi-stage process, and during its course, various chemical reagents (collectors, frothers, depressants, activants, etc.) are dosed [42,43]. Apart from the final product, liquid waste is produced. Its composition depends on the chemical reagents used in the process. The highly alkaline wastewater that is generated, about 70 cubic meters per hour, is pumped into a tailings pond nearby the site. The pond itself is a geomembrane-lined basin with strengthened banks. As a result, potential leaks and environmental threats are minimized. Most of the waste is deposited, and some of it is returned to the mine to be used for additional processes that are carried out.

2.2.3. Mine Waste Rocks

Among the minerals of most interest in the opencast area are pyrite, sphalerite, galena, and chalcopyrite. Of course, they are accompanied by other minerals not used in the exploitation of zinc and lead ores. A variety of rock material is produced during mining (blasting) operations. Then, according to the assessment of the mining geologist, the material is separated. Limestone, which is the overburden of the deposit, is not treated as waste material as it is used both inside and outside the mine. Barren rocks are considered waste and are stored on the slopes of the mountain in the form of different size rock coarse and fine rock material. One use for the rocks is the reconstruction of the hill, which is destroyed as a result of the opencast mining of rock material. The material stored on the slopes is exposed to weather conditions. In this work, an attempt has been made to check whether there is any difference in the properties of waste material that has just been acquired and that has been stored for some time.

2.3. Sample Collection and Data Acquisition

Since February 2018, flotation wastewater and acid mine drainage content have been monitored in order to measure their quality and determine the seasonal variability of discharged effluents. Each month, samples (in triplicates) were collected. All were filtered with the use of a syringe filter (0.45 µm). For anions analysis, samples were filled to full capacity. Cation samples were preserved with nitric acid. Then, samples were transported (at 4 °C) to the laboratory. At the site, pH and conductivity were measured with the use of a portable pH/conductometer. During the research, the capacity of the settling pond was reached. A new one was launched. Therefore, subsequent samples were taken from it. AMD was taken from a partially collapsed adit. The tunnel is located near the flotation building and filled with water, which flows at an almost constant speed.

2.3.1. Chemical Analysis of Wastewaters (AMD and Flotation Wastewater)

A whole range of analyses was performed, beginning with pH and conductivity (portable pH/conductometer with temperature compensation, CPC-411, Elmetron, Zabrze, Poland). The metal composition was determined by Inductively Coupled Plasma Atomic Emission Spectroscopy (ICP-OES, iCAP 6500 DUO, Thermo Fisher Scientific, Waltham, MA, USA), while anions concentration was assessed with chromatography (HPLC with conductivity and photodiode array detectors, Alliance, Waters). The of limits quantification of selected techniques are given in the tables below (Tables 1 and 2). Changes in the samples matrix caused multiple dilutions, thus increasing the limits of quantification. These are given in brackets.

Table 1. Limits of Quantification (LOQ) of the measurement elements by ICP-OES.

Element	LOQ (mg·L^{-1})	Element	LOQ (mg·L^{-1})
Al	0.01	Mg	0.1 (1.00)
Ba	0.001 (0.1)	Mn	0.001 (0.01)
Ca	0.1	Na	0.5 (50.0)
Cd	0.001 (0.01)	Ni	0.005 (0.5)
Co	0.002 (0.2)	Pb	0.01
Cu	0.002 (0.02)	Si (as SiO_2)	0.1
Fe	0.01 (0.1)	Sr	0.002
K	0.5 (50.0)	Zn	0.003

In brackets increased limit because of matrix.

Table 2. Limits of Quantification (LOQ) of the measurement anions by HPLC.

Element	LOQ (mg·L^{-1})
Cl^-	0.5 (15.0)
NO_3^-	0.01 (3.00)
SO_4^-	0.5

In brackets increased limit because of matrix.

2.3.2. Geochemical Analysis of Waste Rocks

Chemical composition of the waste material was analyzed using X-ray fluorescence spectroscopy. The Rigaku ZSX Primus II WDS spectrometer (Tokyo, Japan) equipped with an Rh lamp was used. The mineral composition was characterized by scanning electron microscopy and X-ray diffractometry. Scanning electron microscopy observations were carried out in a low vacuum mode, using an FEI 200 Quanta FEG microscope equipped with an EDS/EDAX spectrometer (Thermo Fisher Scientific, Waltham, MA, USA). XRD patterns were collected using a Rigaku SmartLab diffractometer equipped with a graphite monochromator and rotation Cu anode in a recording range of 2–75°2θ at a 0.05° step size, and a counting time of 1 s per step.

2.3.3. Titration with Mine Wastewater

Acid mine drainage from the abandoned adit and the alkaline flotation wastewater was used for the research on neutralization. The samples were collected twice: in February and April. For the titration, 25 mL of flotation effluent was used, to which acidic mine drainage was added. During the experiment, the pH was constantly monitored. Titration was carried out in five independent series.

2.3.4. Mine Water Neutralization by Waste Rocks

Two types of experiments were conducted. For the first one, two types of waste rock were selected: short- (less than one month of storage) and long-term deposited (more than three months of storage). Water, acid mine drainage, and flotation wastewater were used as a medium. The quantities of the used materials were 1, 4, 6, and 10 g. The volume of the medium was 10 mL. The neutralization was performed in PP tubes shaken at 120 rpm for 84 h. Values of pH and conductivity were measured at selected times. For the second experiment, only long-term deposited waste rock was used. Solid material (20 g) was mixed with AMD, flotation wastewater, and water as a control. In this test, we also used AMD and flotation wastewater mixture (with a final pH of 9). Each time, 20 mL of medium was used. The test was carried out in the same manner, except it took 96 h.

3. Results

3.1. Characterization of Waste Rock

The analyzed waste material comes from the Zn-Pb pre-enrichment process and is a quartz-feldspar rock of likely hydrothermal origin. The X-ray diffraction pattern (Figure 2) shows that phyllosilicates (chlorite and mica) and sulfides (chiefly pyrite) constitute the rock besides quartz and potassium feldspar (orthoclase). The presence of minor dolomite-type carbonate, as well as traces of gypsum, are also apparent.

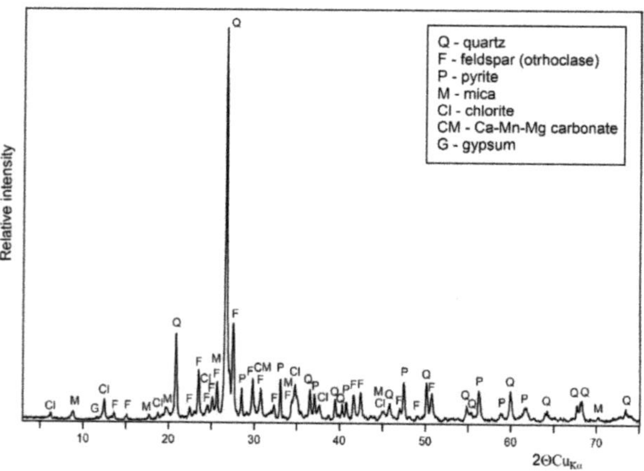

Figure 2. X-ray diffraction pattern of the long-term stored waste rock.

The dominance of quartz and potassium feldspar is reflected by chemical composition, with SiO_2, Al_2O_3, and K_2O being the main constituents (Table 3). On the other hand, the presence of ore minerals results in high iron, sulfur, zinc, and lead contents. The sulfides are disseminated within the rock mass (Figure 3A) but may form more dense accumulations in places (Figure 3B). Euhedral or subhedral cubic pyrite crystals up to 0.3–0.4 mm in size do not show distinct weathering alterations. Minor sulfides (sphalerite and galena) usually form subhedral grains and vein infillings (Figure 3D), which are also apparently unaltered. Electron microscopic observations revealed the presence of accessory minerals:

apatite, Ti oxides, dolomite, and Ca-Mn carbonate (Figure 3C), as well as manganese oxides. Traces of monazite were also encountered.

Table 3. Chemical composition of the long-term stored waste rock used in the experiments.

Main Oxides (wt. %)		Trace Elements (mg kg^{-1})	
SiO$_2$	54.5	As	202
TiO$_2$	0.60	Br	bd
Al$_2$O$_3$	16.2	Cl	85
Fe$_2$O$_3$	8.22	Co	84
CaO	0.92	Cr	73
MgO	2.41	Cu	95
MnO	0.34	Ni	46
SrO	0.02	Pb	2995
BaO	0.51	Sn	bd
Na$_2$O	0.31	Zn	7044
K$_2$O	7.50		
P$_2$O$_5$	0.12		
SO$_3$	7.05		
ZrO$_2$	0.04		
LOI	0.25		

bd—below limit of detection; LOI—loss on ignition.

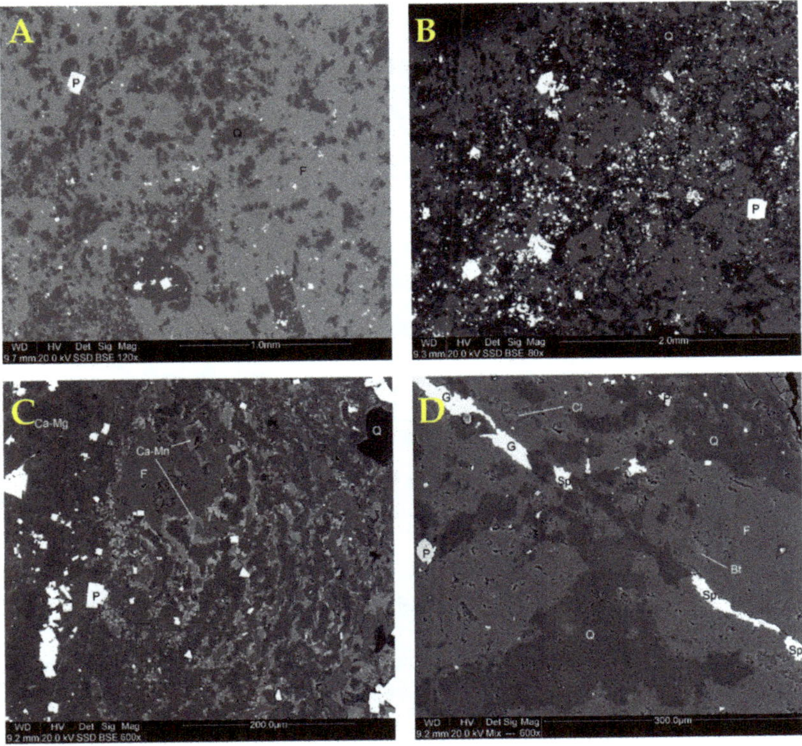

Figure 3. Back-scattered electron images (BSE) of the waste rock. (**A**) structure of a typical pyrite-poor area, (**B**) structure of pyrite-rich area, (**C**) carbonate-rich area, (**D**) typical vein infilling with minor Zn and Pb sulfides. Explanations: Q—quartz, F—potassium feldspar, P—pyrite, Ca-Mg—calcium-magnesium carbonate (dolomite), Ca-Mn—calcium-manganese carbonate, Bt—biotite (mica), Cl—chlorite, Sp—sphalerite, G—galena.

3.2. Characterization of Mine Wastewater

In the area where the Gradir mine operates today, zinc and lead ores have been mined since the mid-20th century. Until the late 1990s, the deposit was mined through drilled tunnels. When mining ceased, the abandoned drifts began to be destroyed. Unused mining galleries began to be filled with water seeping through the rock mass. As a result of leaching rocks containing, among others, pyrite, the water has the character of acidic mine water (Table 4). Low pH, high concentrations of sulfate ions, and significant amounts of zinc, cadmium, or aluminum were recorded. It was observed that the outflow of water from the adit is almost constant during the year. It goes directly to the creek (Mjedenički), flowing through the mine area and then to the nearby river (Cehotina).

Table 4. General characteristics of acid mine drainage (AMD) and flotation wastewater. Twelve-month average compared with the Standards for the industrial water discharge (Službeni list Crne Gore) [44].

Parameter	Unit	Flotation Wastewater		AMD		Montenegro Standards
		Average	SD	Average	SD	
pH		12.10	0.40	3.11	0.09	6.0–8.5
Conductivity	mS cm^{-1}	2.69	1.15	6.22	2.26	-
Cl$^-$		9.40	2.89	23.0	20.1	-
NO$_3$$^-$		3.04	1.14	7.10	38.6	50
SO$_4$$^{2-}$		774	208	6769	2312	20
Al		0.22	0.13	88.0	36.1	3.0
Ca		469	111	386	19.2	-
Cd		<0.01	-	0.90	0.31	0.01
Co		<0.02	-	0.51	0.15	1.0
Cu		<0.02	-	1.25	1.13	0.5
Fe	mg L^{-1}	<0.1	-	107	40.0	2.0
K		31.3	7.03	<50.0	-	-
Mg		<1.00	-	880	320	-
Mn		0.01	0.02	199	84.8	2.5
Na		57.9	15.8	<50.0	-	-
Ni		<0.05	-	0.86	0.23	1.25
Pb		0.55	0.94	0.24	0.06	0.5
SiO$_2$		2.67	1.25	66.4	17.2	-
Sr		0.79	0.21	1.77	0.05	-
Zn		0.06	0.11	746	303	1

< below limits of quantification.

As mentioned earlier, after extraction, crushing, and enrichment, zinc- and lead-bearing ore undergoes a flotation, which requires the use of significant amounts of various chemical compounds. The final product is a zinc and lead concentrate. The by-product is an alkaline flotation effluent. It contains off-balance amounts of zinc and lead and significant sulfate content (Table 4). The wastewater in the amount of 70 cubic meters per hour is discharged to a nearby pond where it is deposited. It is not subjected to further treatment.

The obtained results of samples taken each month over the year are shown in Tables S1 and S2. Due to heavy snowfall, it was not possible to take samples on the scheduled date in March 2018.

Changes in the pH of acid mine drainage were not significant. The minimum was 3.00, the maximum 3.26, and the average 3.11 (Figure 4). However, an upward trend is noticeable. The variation in conductivity was much greater. The maximum (9.64 mS·cm^{-1}) was in April and the minimum (3.20 mS·cm^{-1}) in December 2018. Throughout the year, it averages 6.22 mS·cm^{-1}. Sulfate concentrations were changing. They reached their maximum (10,333 mg·L^{-1}) at the beginning of the third quarter of 2018, then decreased in the following months until November (minimum 3700 mg·L^{-1}). A similar trend can be seen in the case of most metals (Figure 4). Maximum concentrations or close to maximum concentrations were reached in July 2018. The following months brought about a decrease.

From December 2018 onwards, the concentrations increased. The changes in the chemical composition of acid mine wastewater flowing from the abandoned mine adit are not expected to be a result of mine operations. The relationships of the determined parameters with average rainfall and average temperature were compared (Figure 4 and Figure S1). The amount of rainfall was observed to influence such parameters as conductivity and concentrations of sulfates and metal cations. As the amount of rainfall decreases, the mentioned above parameters also decrease, but with a one-, two-month delay. In the case of pH, this effect was not observed. Furthermore, in the temperature, its influence on pH is not observed, whereas the influence on other parameters is visible. However, the change in the values is seen about a month ahead of the temperature change. In June 2018, part of the tunnel collapsed. Unfortunately, the technical conditions of the adit did not allow for safe penetration. Undoubtedly, this incident could have affected the results obtained.

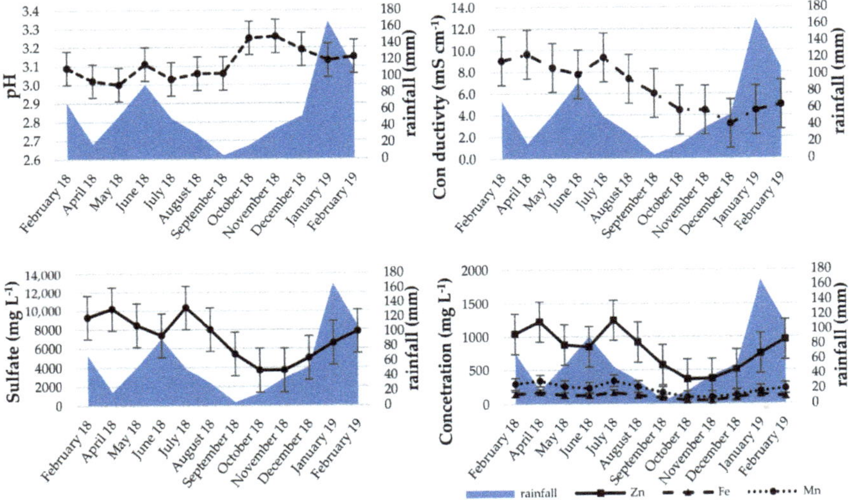

Figure 4. Evolution of selected parameters in AMD with rainfall data. Metrological seasons: Spring: 1 March to 31 May; Summer: 1 June to 31 August; Autumn: 1 September to 30 November; Winter: 1 December to 28 February.

It can be noted that for flotation wastewater, the pH varies from 11.32 to 12.56, with an average of 12.10. There was no correlation observed between pH and conductivity (or other parameters), which ranged from 0.99 to 5.04 mS·cm^{-1} (average 2.69 mS·cm^{-1}) (Figure 5). It can be assumed that the physicochemical parameters of flotation wastewater do not depend on seasonal variations. They could be dependent on the processes taking place in the tailing pond. High summer temperatures are conducive to the evaporation of water and consequently impact the concentration of the solution. On the other hand, rainfall and spring melting of snow can cause dilution. It is also possible that some of the elements, together with the sediment, are bound in the lower parts of the settling tank. However, it seems that changes in the flotation, such as the amount of reagent added, shutdowns, failures, etc., have the greatest impact on the chemical composition of flotation waters. The content of lead varied from 0.03 to 3.48 mg·L^{-1}. The calculated average was 0.68 mg·L^{-1}. Differences are visible in individual months. The maximum was determined in May 2018, and the minimum was observed in December 2018 and January 2019 when the new tailing pond was launched. The sulfate content also varied: the minimum amount was 440 and the maximum 1100 mg·L^{-1}. The average concentration was determined at the level of 774 mg·L^{-1}.

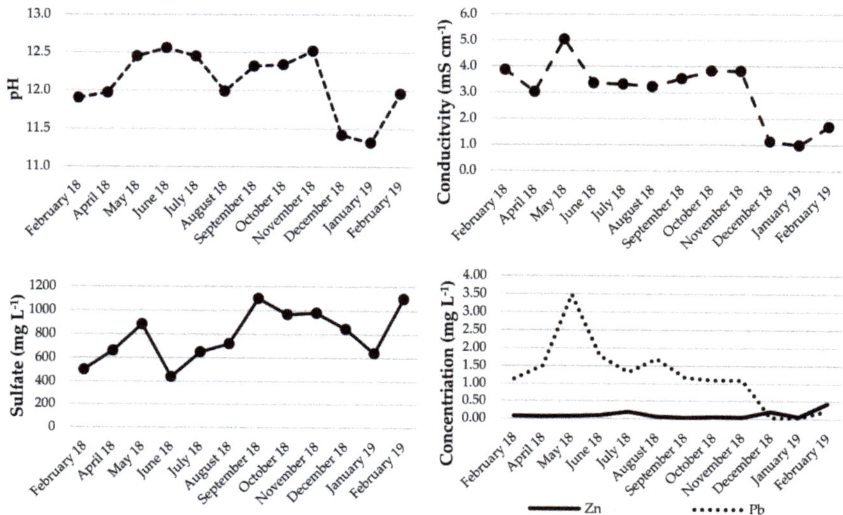

Figure 5. Evolution of selected parameters in flotation wastewater.

3.3. Flotation Wastewater Neutralization by Acid Mine Drainage

The essence of this test was to determine the scope to neutralize flotation wastewater by acid mine drainage. As presented above, the flotation wastewater had a high pH of ~12, and AMD had a low pH of ~3, while the water discharged into the river should have a pH of 7–8. Thus, acid mine waters were used to neutralize alkaline flotation wastewater. For titration, 25 mL of post-flotation effluent was used, to which acidic mine waters were added, and pH was monitored with pH-meter. The titration was carried out in five independent series, and the averaged results are presented in Table S3 and Figure 6.

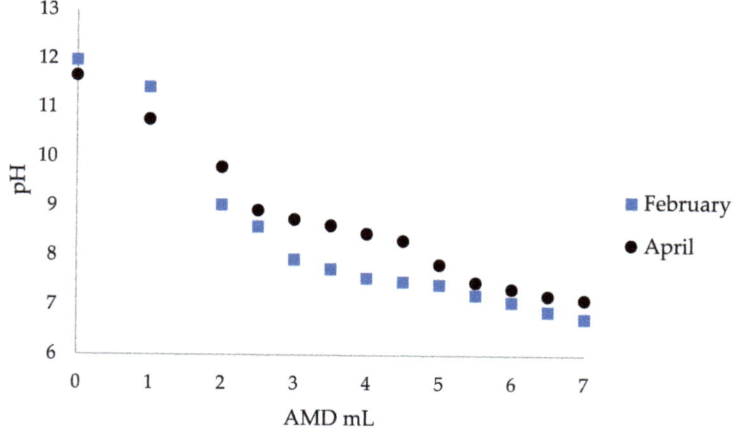

Figure 6. Titration of flotation wastewater with acid mine drainage from the adit collected in February and April.

Based on the results obtained, it can be concluded that four times less acid mine drainage from the adit (1:0.24 *v/v*) should be used to neutralize the flotation wastewater water to pH 7. It is also possible to reduce sulfate content from circa 10,000 to less than 2500 mg·L^{-1} (Tables S1 and S3). The experiment was conducted to observe the influence of the aggregate on the physicochemical composition of the medium. When acidic water is mixed with alkaline water, a colloidal mixture is formed, which then settles down. The sediments precipitated at pH 7 had a creamy color. An attempt was made to assess the structure of the sediment. As can be seen on the XRD pattern below (Figure 7), the crystalline iron (both ferrous and ferric), zinc and manganese sulfate hydrates as well as calcium and magnesium carbonates dominated the precipitate composition with an admixture of aluminum hydroxide.

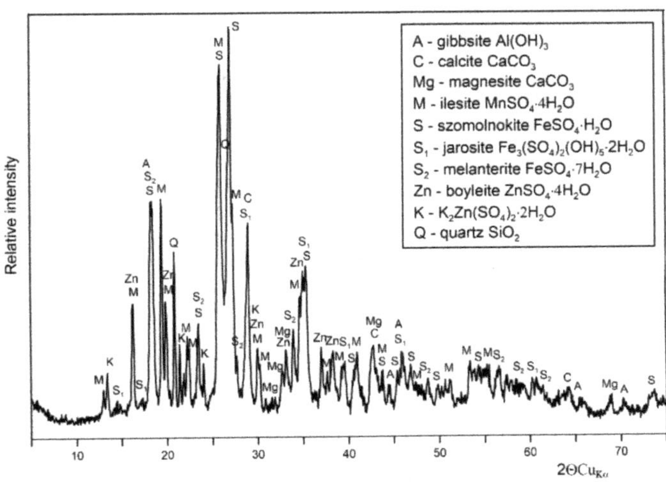

Figure 7. X-ray diffraction pattern of the sediment formed after mixing AMD and flotation wastewater.

3.4. Mine Wastewater Neutralization by Waste Rock—Batch Test 1

In order to verify which solid material (short- or long-term stored) should be used for the neutralization of mine wastewater, an initial 84-h experiment was carried out. A batch adsorption test was conducted by mixing a known weight of long-term and short-term stored waste rock with 10 mL of water, flotation wastewater, and AMD. The mixture was shaken in a mechanical shaker for 84 h. After a specified time, the pH and conductivity were measured.

Water as a medium was used to verify how the length of storage of waste rock affects pH and conductivity. A pH increase could be observed, especially in the case of small weights (Table 6). While larger weights showed a slight increase in pH, when compared to pure water, conductivity rose significantly. On the other hand, small weights had a minor influence on the conductivity value. The pH of the water remained close to neutral with bigger weights in waste rock stored for a shorter period (Table 5). The conductivity, however, increased similarly when solid waste stored for a long time was used. In the case of small weights, an increase in both pH and conductivity was observed.

In order to neutralize flotation wastewater, it is necessary to use bigger weights. A significant drop in pH was visible for a 10-gram weighting. For the long-term stored waste rock, values close to neutral were obtained (Table 5). The short-term stored waste rock showed instability, which was expressed by a definite difference in conductivity. Short-term stored waste rocks are unoxidized and are highly reactive; its properties change significantly in a short time. This effect was also observed in the case of the long-term stored material, where the values for the respective weightings were similar (1.71–1.82 mS·cm^{-1}).

Table 5. Waste rocks stored for a short and long amount of time treated with water, flotation wastewater, and AMD. The pH and conductivity were measured after 84 h of mixing.

	Water		Flotation Wastewater		Acid Mine Drainage	
Weight (g)	pH	Conductivity ($mS \cdot cm^{-1}$)	pH	Conductivity ($mS \cdot cm^{-1}$)	pH	Conductivity ($mS \cdot cm^{-1}$)
Short-term stored waste rocks						
0	7.00	0.795	12.37	2.06	2.83	5.21
1	7.99	0.859	11.85	1.66	3.57	5.33
4	7.35	2.14	9.14	1.15	4.57	5.14
6	7.27	1.42	8.96	1.26	5.44	5.07
10	7.41	2.33	7.86	1.62	5.70	5.06
Long-term stored waste rocks						
0	7.00	0.795	12.37	2.06	2.83	5.21
1	7.60	1.20	10.10	1.81	3.33	5.39
4	7.31	2.09	8.22	1.77	6.14	4.68
6	7.14	2.12	7.45	1.71	6.03	4.57
10	7.07	2.21	7.14	1.82	6.24	4.15

The influence of short- and long-term storage of waste rock on the changes in acid mine drainage parameters was also tested within 84 h. The use of 10 grams of long-term stored solid waste was also much more effective here. In this case, it was possible to increase the pH from 2.83 up to 6.24 (5.70 in the case of the short-term stored solid waste) (Table 5). Longer contact time may encourage a further decrease in pH. Short-term stored solid waste did not demonstrate any significant changes in acid mine drainage conductivity. The use of long-term stored solid waste caused only a slight reduction in conductivity: from 5.21 to 4.15.

It can be concluded from the conducted studies that long-term storage of waste rock is more effective in neutralizing both alkaline and acidic leachate. It is worth noting that the storage time of the waste rock may be critical due to the influence of atmospheric conditions.

3.5. Mine Wastewater Neutralization by Waste Rock—Batch Test 2

Long-term stored waste rock and four liquids were selected for the study. This time, a mixture of AMD and flotation wastewater, alongside AMD and flotation wastewater, was used. Distilled water was used as a control. As noted in previous experiments, it is possible to neutralize flotation wastewater using AMD. Therefore, considering that waste rock can lower the pH, AMD and flotation wastewater were mixed to reach a pH of 9 only. An appropriate weight (20 g) was placed in vessels with 20 mL of each medium. Then, the whole sample was shaken for 96 h. After a selected time, pH and conductivity were measured. At the beginning and end of the experiment, the sulfate concentration was determined. The results are given in Table 6.

Table 6. Neutralization of four selected liquids by long-term stored waste rocks 20 g of the waste rocks was mixed with 20 mL of the liquid.

Medium	Parameter	Time (h)				
		0	24	48	72	96
AMD + waste rocks	pH	2.85	6.08	6.24	6.43	6.41
	SD	0.03	0.17	0.10	0.15	0.17
	Conductivity (mS·cm^{-1})	8.15	6.43	6.56	6.17	6.23
	SD	0.09	0.13	0.25	0.27	0.30
	SO$_4{}^{2-}$ (mg·L^{-1})	8110	-	-	-	5627
	SD	59	-	-	-	105
Flotation wastewater + waste rocks	pH	11.14	6.60	6.74	6.64	6.61
	SD	0.10	0.08	0.06	0.05	0.08
	Conductivity (mS·cm^{-1})	2.20	2.72	2.78	2.69	2.78
	SD	0.02	0.05	0.02	0.06	0.06
	SO$_4{}^{2-}$ (mg·L^{-1})	991	-	-	-	1634
	SD	48	-	-	-	61
Flotation wastewater + AMD + waste rocks	pH	9.02	6.56	6.72	6.70	6.68
	SD	0.03	0.02	0.02	0.09	0.05
	Conductivity (mS·cm^{-1})	2.14	2.77	2.86	2.74	2.84
	SD	0.05	0.04	0.07	0.02	0.15
	SO$_4{}^{2-}$ (mg·L^{-1})	1204	-	-	-	1713
	SD	58	-	-	-	29
Water + waste rocks	pH	6.65	6.58	6.85	6.75	6.78
	SD	0.03	0.04	0.04	0.06	0.06
	Conductivity (mS·cm^{-1})	0.0055	2.54	2.62	2.54	2.63
	SD	0.0001	0.04	0.02	0.01	0.02
	SO$_4{}^{2-}$ (mg·L^{-1})	2	-	-	-	1505
	SD	0.03	-	-	-	16

Analyzing the results, it can be concluded that by using long-term stored waste rock, it is possible to neutralize AMD, flotation wastewater, and their mixtures. As shown in Table 6, a sharp change in pH occurred after 24 h. If the contact time is extended, the reaction was stabilized and the measurements in the following days brought similar results (respectively). Furthermore, as the control test showed, a 96 h contact time did not fundamentally change the pH of the water. The use of long-term stored waste rock in leachate treatment poses the risk of increasing the load of undesired components. As a result of the processes and reactions occurring on the surface of rocks, elements released from the used aggregate may be released into solution. The chemical results of the tests carried out after the 96-h experiment are presented in Table 7. They are compared with the initial results for AMD and flotasation wastewater used in the experiment. In the case of AMD neutralization with waste rock, a slight increase in lead concentration was observed. However, an essential decrease in zinc concentration was visible. Furthermore, the sulfate content decreased by over 1000 mg·L^{-1}. Using waste rock for flotation wastewater neutralization had the opposite effect. Both zinc and sulfate concentrations increased. As the results of the experiment show, to utilize AMD, flotation wastewater, and waste rock, it seems optimal to initially lower the flotation wastewater pH by mixing with AMD and then adding waste rock. Such a solution leads to pH neutralization and a slight increase in conductivity. It must be admitted that the concentration of Zn and sulfates increased, though in a range similar to other variants, and even less than in the use of AMD to neutralize flotation wastewater.

Table 7. Comparison of the selected results of physicochemical parameters of flotation wastewater, AMD, and mixtures of long-term stored waste rocks, and proposed medium.

Parameter	Unit	AMD		Flotation Wastewater		AMD + Waste Rocks		Flotation Wastewater + Waste Rocks		Flotation Wastewater + AMD + Waste Rocks		Water + Waste Rocks	
		Average	SD	Average	SD	Average	SD	Average	SD	Average	SD	Average	SD
pH		2.85	0.03	11.14	0.10	6.41	0.17	6.61	0.08	6.68	0.05	6.78	0.06
Conductivity	$mS \cdot cm^{-1}$	8.15	0.09	2.20	0.02	6.23	0.30	2.78	0.06	2.83	0.15	2.63	0.02
Al		127	31.8	0.12	0.03	<0.1	-	<0.02	-	<0.02	-	<0.02	-
Ca		384	96.0	341	85	446	116	571	143	574	144	575	144
Cd		1.41	0.35	<0.01	-	0.48	0.12	0.05	0.01	0.06	0.02	0.06	0.02
Co		0.63	0.16	<0.02	-	0.66	0.17	0.09	0.02	0.09	0.02	0.10	0.03
Cu		3.49	0.87	<0.02	-	<0.05	-	0.01	0.003	0.01	0.003	0.01	0.003
K		<50	-	40.5	10.1	25.2	6.30	34.5	8.63	34.0	8.50	19.2	4.80
Mg		964	241	<1	-	956	239	57.9	14.5	81.5	20.4	51.9	13.0
Mn	$mg \cdot L^{-1}$	224	56.0	<0.01	-	220	55.0	33.8	8.45	35.5	8.88	29.4	7.35
Mo		<0.3	-	<0.03	-	<0.05	-	0.01	0.003	0.01	0.00	0.01	0.00
Na		<50	-	66.0	16.5	8.03	2.01	52.7	13.2	51.0	12.8	2.23	0.56
Ni		0.95	0.24	<0.05	-	0.60	0.15	0.07	0.02	0.08	0.02	0.08	0.02
Pb		0.19	0.05	0.23	0.06	0.46	0.12	0.08	0.02	0.06	0.02	0.05	0.01
SiO_2		76.5	19.1	4.80	1.20	19.8	4.95	13.1	3.28	15.1	3.78	14.3	3.58
Sr		1.83	0.46	1.13	0.28	1.56	0.39	1.10	0.28	1.14	0.29	0.75	0.19
Zn		957	239	0.43	0.11	204	51.0	8.15	2.04	7.99	2.00	11.5	2.88
SO_4^{2-}		8110	59	991	48	5626	105	1633	61.0	1713	29.00	1505	16.0

< below limits of quantification.

4. Discussion

Regardless of the many efforts, the majority of the mining waste generated still goes to waste disposal sites [2,45]. Increasing emphasis on protection against climate change and concern for environmental protection have resulted in the mining industry beginning to change its approach to waste management [46]. In the presented case, the waste includes AMD flowing from abandoned adits, slurry from the processing of zinc and lead ore stored in the tailings storage facility, and finally, waste rocks stored on the slopes of the mountain. The XRD analysis shows that alongside quartz and potassium feldspar, the rock material contains phyllosilicates as well as sulfides iron, zinc, lead. What is characteristic for rocks containing sulfide minerals where besides pyrite, there are also other sulfide minerals containing valuable metals such as gold, silver, copper, lead, and zinc [14,47]. As Lefebvre [48] noted, oxidation of sulfide minerals occurs gradually. Once the pyrite close to the surface of the mineral is oxidized, the oxidant has to penetrate deep into the rock. This is following the observations made during the experiments. The rock material deposited no more than a month (short-term stored) reacted more strongly than that which has been longer (long-term stored) in contact with the atmospheric conditions. Mining waste containing sulfate-bearing minerals exposed to atmospheric oxygen and water causes the uncontrolled formation of AMD [49]. Moreover, the accurate prediction of the volume of released AMD is difficult not only due to the mentioned gradual oxidation but also dependent on co-occurring minerals. The presence, among others, of carbonates and silicates may cause neutralization of the leachate [36,50,51].

The solubility of carbonates, depending on the pH, increases the amount of carbonate ions in the solution, thus increasing the potential to neutralize the solution. As it progresses, the pH increases and the calcite may precipitate as a secondary mineral. The dissolution of silicates is much slower than carbonates. The feldspar weathering process depends mainly on pH, silica concentration, Na, K, and Ca. It results in the formation of secondary minerals such as muscovite, kaolinite, and gibbsite [52]. The experiments carried out have shown that waste rocks are very reactive. The scope to change the pH values of the leaching medium, even in a short time, is significant. By using acid mine wastewater as a leaching solution after 24 h, a pH above 6 was obtained (at the beginning, 2.85). However, there is a danger that the acidification of the solution may occur again over time. Still, this solution brings a 31% reduction in sulfate and a 79% reduction in zinc concentration. This is probably due to the precipitation of sulfate hydrates typical of the AMD environment. Unfortunately, a 2.5 increase in lead concentration was observed. The effect of using solid waste to neutralize flotation wastewater appeared to be unsuccessful. Admittedly, a sharp drop in pH and lead but with the simultaneous growth of zinc and sulfate concentrations.

Abandoned underground mines are quite a challenging task. Mining shafts very often are collapsed and filled with water. Access to them is very difficult and it is usually impossible to reach the precise plans [53]. The Gradir Montenegro mine is no different. Most of the over 50,000 meters of underground adits are already inundated and destroyed. Flooding out water is characterized by very low pH values (avg. 3.11) and substantial sulfate content (avg. 6769 mg·L^{-1}). It was observed that the chemical composition depended on weather conditions. As the temperature and the amount of rainfall decreased, the concentrations of sulfates and metal cations also decreased. Considering the significantly elevated content of metals, including heavy metals, the acidic leachate may have an irreversible impact on the environment [54–58]. Alkaline substances are often used to neutralize AMD [18,59–61]. Their use allows to reduce the metal content almost completely with a 60% reduction of sulfates [62]. Cost estimates and limitations of the use of alkaline substances were undertaken by Taylor et al. [63]. One of the most well studied and employed methods is based on chemical neutralization with the use of limestone [22], which is characterized by its low cost and wide availability. The result is a pH increase and, by precipitation, metal removal [17,28]. However, this method has some limitations due to its tendency to form an external coating caused by secondary mineral precipitation [29]. In turn, in the presence of aqueous sulfate a gypsum sediment precipitates [30]. What is more, such a solution would be effective enough for AMD treatment only when the formation of AMD was temporary or short-term [15]. The process creates significant amounts of metal hydroxide sediment. As described by Sibrell and Watten [64], the neutralization of 50 metric tons of AMD generated 450 metric tons of sludge, the disposal of which was costly. In the case of the mine described here, it is estimated that from only one abandoned adit, about 30 cubic meters per hour flows out. Thus, almost 263,000 cubic meters are required to be neutralized during the year.

The extraction of metals from sulfide ores results in significant amounts of waste. The amount of waste generated by the mines might be close to the volume of processed raw material [65]. As it has been proven, the tailing content is related to a specific process, its modifications, failures, etc. used in the mine. Still, the most common method is depositing it into sedimentary ponds without any treatment [33,66,67]. Tailings no longer used are protected by a layer of soil. However, without a protective plant cover, wind and water erosion is possible, resulting in the release of pollutants into the environment [68–70]. Some solutions have been proposed using different plants or microorganisms not only for stabilization but also for pond remediation [32,71–74]. As mentioned, these are only applicable to tailings no longer in use. Attempts have been made to treatment alkaline wastewater from an operating mine by O'Sulivan et al. [75] using Surface-Flow Wetlands. The pilot-scale installation has been in operation for two years. The chemical composition of the slurry was quite similar to that of the one being treated in this paper—sulfates 900 mg·L^{-1}, lead 0.15 mg·L^{-1}, and zinc 2.0 mg·L^{-1}. Unfortunately, the authors do not mention a pH value. At the flow of 1.5 L min^{-1}, it was possible to reduce sulfate and lead concentrations by about 30% and 74% of zinc.

The authors propose the use of acidic mine drainage to neutralize alkaline flotation leachate. The conducted experiments indicated that it is possible to reduce the pH from ~12 to values close to neutral by using a volume of acidic mine wastewater four times smaller than that of flotation wastewater. After mixing both streams, colloidal sediment was formed, constituting about 1/6 of the volume. Besides the iron, zinc, and manganese sulfate hydrates calcium and magnesium carbonates prevailed in the precipitate. As expected, gibbsite was observed among the secondary minerals. The final solution contained four times less sulfate than the flotation wastewater and AMD combined. However, the value of the mentioned parameter does not allow for safe discharge into the environment. Still, it is worth considering the use of the AMD flowing from the abandoned adit for the neutralization of all flotation wastewater (~70 cubic meters per hour) instead of deposition in the tailing pond. The received solution could then be treated using a passive treatment system, for example, constructed wetlands.

The experiences described above have led to the proposal to use three waste materials for mutual treatment. In the first step, mixing the flotation wastewater with AMD to pH 9 causes fewer secondary minerals to be precipitated and thus formation less sludge. Simultaneously, the sulfate concentration decreased. In the next step, mixing the resulting mixture with the rock material results in neutralization of the pH, reduction of metal and sulfate concentrations (compared to a total of three waste materials). The management of the three waste materials produced in the mining of zinc and lead does not require the use of any additional reagents (e.g., limestone) and although the parameters of the obtained solution do not meet the legislative requirements, their further treatment seemed to be relatively easy.

5. Conclusions

The rock waste material stored on the hillsides is a quartz-feldspar rock still containing certain amounts of zinc and lead. When exposed to weather conditions, it can be a potential source of acidic mine water. The flotation wastewater stored in the sedimentary tailings is characterized by a high pH (about 12), elevated sulfate, and low metal concentrations. Acid mine drainage filling abandoned adits have an acidic pH (about 3). They also have a significant amount of sulfates and metals. The chemical composition of these two solutions appear to change over time. In the case of flotation wastewater, these changes result from changes in the flotation. However, AMD has a rather seasonal background. The performed experiments proved that AMD can be used to neutralize flotation wastewater. Positive results indicate that four times less AMD is sufficient for the pH neutralization of the final solution. Furthermore, the possibility of using both liquid and waste rocks was proven. Premixing alkaline flotation wastewater with AMD and then flowing through waste rock neutralized the pH and reduced the total charge of metals and sulfates. The proposed solution does not require the use of additional chemical reagents and minimizes the sediments needed for storage.

Supplementary Materials: The following are available online at http://www.mdpi.com/2075-163X/10/12/1061/s1, Figure S1: Evolution of selected parameters in acid mine drainage (AMD) with temperature data. Table S1: Monitoring results of AMD, Table S2: Monitoring results of tailings, Table S3: Titration of tailings using acid mine drainage (collected in February and April) from the adit.

Author Contributions: Conceptualization, J.R., and L.D.; methodology, J.R., G.R., T.B.; validation, J.R., T.B., G.R.; formal analysis, J.R., L.D. investigation, J.R., G.R., T.B.; writing—original draft preparation, J.R.; writing—review and editing, L.D.; visualization, J.R., T.B., G.R.; project administration, L.D.; funding acquisition, L.D. All authors have read and agreed to the published version of the manuscript.

Funding: This research was funded by Team Tech programme of Foundation for Polish Science No. POIR.04.04-00-00-2053/16 (TEAM TECH 2016-2/9) as a part of Measure 4.4 of the 2014–2020 Smart Growth Operational Programme, EU.

Conflicts of Interest: The authors declare no conflict of interest.

References

1. Lottermoser, B.G. *Mine Wastes: Characterization, Treatment and Environmental Impacts*, 3rd ed.; Springer: Berlin/Heidelberg, Germany, 2010. [CrossRef]
2. Lottermoser, B.G. Recycling, reuse and rehabilitation of mine wastes. *Elements* **2011**, *7*, 405–410. [CrossRef]
3. Hudson-Edwards, K.A.; Jamieson, H.E.; Lottermoser, B.G. Mine Wastes: Past, Present, Future. *Elements* **2011**, *7*, 375–380. [CrossRef]
4. Bian, Z.; Miao, X.; Lei, S.; Chen, S.; Wang, W.; Struthers, S. The Challenges of Reusing Mining and Mineral-Processing Wastes. *Science* **2012**, *337*, 702–703. [CrossRef] [PubMed]
5. Sarmiento, A.M.; DelValls, A.; Nieto, J.M.; Salamanca, M.J.; Carballo, M.A. Toxicity and potential risk assessment of a river polluted by acid mine drainage in the Iberian Pyrite Belt (SW Spain). *Sci. Total Environ.* **2001**, *409*, 4763–4771. [CrossRef] [PubMed]
6. Grande, J.A.; Santisteban, M.; de la Torre, M.L.; Davila, J.M.; Perez-Ostale, E. Map of impact by acid mine drainage in the river network of The Iberian Pyrite Belt (Sw Spain). *Chemosphere* **2018**, *199*, 269–277. [CrossRef]
7. Byrne, P.; Wood, P.J.; Reid, I. The impairment of river systems by metal mine contamination: A review including remediation options. *Crit. Rev. Environ Sci. Technol.* **2012**, *42*, 2017–2077. [CrossRef]
8. Iribar, V.; Izco, F.; Tames, P.; Antiguedad, I.; da Silva, A. Water contamination and remedial measures at the Troya abandoned Pb-Zn mine (The Basque Country, Northern Spain). *Environ. Geol.* **2000**, *39*, 800–806. [CrossRef]
9. Lebre, E.; Corder, G.D.; Golev, A. Sustainable practices in the management of mining waste: A focus on the mineral resource. *Miner. Eng.* **2017**, *107*, 34–42. [CrossRef]
10. Durucan, S.; Korre, A.; Munoz-Melendez, G. Mining life cycle modelling: A cradle-to-gate approach to environmental management in the minerals industry. *J. Clean. Prod.* **2006**, *14*, 1057–1070. [CrossRef]
11. Simate, G.S.; Ndlovu, S. Acid mine drainage: Challenges and opportunities. *J. Environ. Chem. Eng.* **2014**, *2*, 1785–1803. [CrossRef]
12. Dold, B. Evolution of Acid Mine Drainage Formation in Sulphidic Mine Tailings. *Minerals* **2014**, *4*, 621–641. [CrossRef]
13. Yuniati, M.D.; Kitagawa, K.; Hirajima, T.; Miki, H.; Okibe, N.; Sasaki, K. Suppression of pyrite oxidation in acid mine drainage by carrier microencapsulation using liquid product of hydrothermal treatment of low-rank coal, and electrochemical behavior of resultant encapsulating coatings. *Hydrometallurgy* **2015**, *158*, 83–93. [CrossRef]
14. Li, X.; Hiroyoshi, N.; Tabelin, C.B.; Naruwa, K.; Harada, C.; Ito, M. Suppressive effects of ferric-catecholate complexes on pyrite oxidation. *Chemosphere* **2019**, *214*, 70–78. [CrossRef]
15. Igarashi, T.; Herrera, P.S.; Uchiyama, H.; Miyamae, H.; Iyatomi, N.; Hashimoto, K.; Tabelin, C.B. The two-step neutralization ferrite-formation process for sustainable acid mine drainage treatment: Removal of copper, zinc and arsenic, and the influence of coexisting ions on ferritization. *Sci. Total Environ.* **2020**, *715*, 136877. [CrossRef]

16. Chopard, A.; Benzaazoua, M.; Bouzahzah, H.; Plante, B.; Marion, P. A contribution to improve the calculation of the acid generating potential of mining wastes. *Chemosphere* **2017**, *175*, 97–107. [CrossRef]

17. Park, I.; Tabelin, C.B.; Jeon, S.; Li, X.; Seno, K.; Ito, M.; Hiroyoshi, N. A review of recent strategies for acid mine drainage prevention and mine tailings recycling. *Chemosphere* **2019**, *219*, 588–606. [CrossRef]

18. Akcil, A.; Koldas, S. Acid Mine Drainage (AMD): Causes, treatment and case studies. *J. Clean. Prod.* **2006**, *14*, 1139–1145. [CrossRef]

19. Schaider, L.A.; Senn, D.B.; Estes, E.R.; Brabander, D.J.; Shine, J.P. Sources and fates of heavy metals in a mining-impacted stream: Temporal variability and the role of iron oxides. *Sci. Total Environ.* **2014**, *490*, 456–466. [CrossRef]

20. Soni, A.K.; Mishra, B.; Singh, S. Pit lakes as an end use of mining: A review. *J. Min. Environ.* **2014**, *5*, 99–111. [CrossRef]

21. Tomiyama, S.; Igarashi, T.; Tabelin, C.B.; Tangviroon, P.; Li, H. Acid mine drainage sources and hydrogeochemistry at the Yatani mine, Yamagata, Japan: A geochemical and isotopic study. *J. Contam. Hydrol.* **2019**, *225*, 103502. [CrossRef]

22. Kefeni, K.K.; Msagati, T.A.M.; Mamba, B.B. Acid mine drainage: Prevention, treatment options, and resource recovery: A review. *J. Clean. Prod.* **2017**, *151*, 475–493. [CrossRef]

23. Pyrbot, W.; Shabong, L.; Singh, O.P. Neutralization of acid mine drainage contaminated water and ecorestoration of stream in a coal mining area of east Jaintia Hills, Meghalaya. *Mine Water Environ.* **2019**, *38*, 551–555. [CrossRef]

24. Fosso-Kankeu, E.; Mittal, H.; Waanders, F.; Ray, S.S. Thermodynamic properties and adsorption behavior of hydrogel nanocomposites for cadmium removal from mine effluents. *J. Ind. Eng. Chem.* **2017**, *48*, 151–161. [CrossRef]

25. Kaartinen, T.; Laine-Ylijoki, J.; Ahoranta, S.; Korhonen, T.; Neitola, R. Arsenic removal from mine waters with sorption techniques. *Mine Water Environ.* **2017**, *36*, 199–208. [CrossRef]

26. Garcia, V.; Hayrynen, P.; Landaburu-Aguirre, J.; Pirila, M.; Keiski, R.L.; Urtiaga, A. Purification techniques for the recovery of valuable compounds from acid mine drainage and cyanide tailings: Application of green engineering principles. *J. Chem. Technol. Biotechnol.* **2013**, *89*, 803–813. [CrossRef]

27. Wingenfelder, U.; Hansen, C.; Furrer, G.; Schulin, R. Removal of heavy metals from mine waters by natural zeolites. *Environ. Sci. Technol.* **2005**, *39*, 4606–4613. [CrossRef]

28. Tabelin, C.B.; Igarashi, T.; Villacorte-Tabelin, M.; Park, I.; Opiso, E.M.; Ito, M.; Hiroyoshi, N. Arsenic, selenium, boron, lead, cadmium, copper, and zinc in naturally contaminated rocks: A review of their sources, modes of enrichment, mechanisms of release, and mitigation strategies. *Sci. Total Environ.* **2018**, *645*, 1522–1553. [CrossRef]

29. Jacobs, J.A.; Lehr, J.H.; Testa, S.M. (Eds.) *Acid Mine Drainage, Rock Drainage, and Acid Sulfate Soils. Causes, Assessment, Prediction, Prevention, and Remediation*; John Wiley & Sons, Inc.: Hoboken, NJ, USA, 2014. [CrossRef]

30. Offeddu, F.G.; Cama, J.; Solera, J.M.; Dávila, G.; McDowell, A.; Craciunescu, T.; Tiseanu, I. Processes affecting the efficiency of limestone in passive treatments for AMD: Column experiments. *J. Environ. Chem. Eng.* **2015**, *3*, 304–316. [CrossRef]

31. Othmani, M.A.; Souissi, F.; Bouzahzah, H.; Bussiere, B.; Silva, E.F.; Benzaazoua, M. The flotation tailings of the former Pb-Zn mine of Touiref (NW Tunisia): Mineralogy, mine drainage prediction, base-metal speciation assessment and geochemical modeling. *Environ. Sci. Pollut. Res.* **2015**, *22*, 2877–2890. [CrossRef]

32. Ciarkowska, K.; Hanus-Fajerska, E.; Gambus, F.; Muszynska, E.; Czech, T. Phytostabilization of Zn-Pb ore flotation tailings with Dianthus carthusianorum and Biscutella laevigata after amending with mineral fertilizers or sewage sludge. *J. Environ. Manag.* **2017**, *189*, 75–83. [CrossRef]

33. Stankovic, V.; Milosevic, V.; Milicevic, D.; Gorgievski, M.; Bogdanovic, G. Reprocessing of the old flotation tailings deposited on the RTB Bor tailings pond—A case study. *Chem. Ind. Chem. Eng. Q.* **2018**, *24*. [CrossRef]

34. Khalil, A.; Argane, R.; Benzaazoua, M.; Bouzahzah, H.; Taha, Y.; Hakkou, R. Pb–Zn mine tailings reprocessing using centrifugal dense media separation. *Miner. Eng.* **2019**, *131*, 28–37. [CrossRef]

35. Antonijevic, M.M.; Dimitrijevic, M.D.; Stevanovic, Z.O.; Serbula, S.M.; Bogdanovic, G.D. Investigation of the possibility of copper recovery from the flotation tailings by acid leaching. *J. Hazard. Mater.* **2008**, *158*, 23–34. [CrossRef] [PubMed]

36. Kuhn, K.; Meima, J.A. Characterization and Economic Potential of Historic Tailings from Gravity Separation: Implications from a Mine Waste Dump (Pb-Ag) in the Harz Mountains Mining District, Germany. *Minerals* **2019**, *9*, 303. [CrossRef]

37. Shadrunova, I.V.; Orekhova, N.N. A Process for Advanced Recycling of Water Originating from Mining Operations, with Metal Recovery. *Mine Water Environ.* **2015**, *34*, 478–484. [CrossRef]

38. Wolkersdorfer, C.; Bowell, R. Contemporary Reviews of Mine Water Studies in Europe, Part 2. *Mine Water Environ.* **2005**, *24*, 2–37. [CrossRef]

39. Geological Map of the Exploration Area. Available online: http://videolectures.net/outbursts_herlec_sshm/ (accessed on 17 October 2020).

40. Gutierrez, M.; Mickus, K.; Camacho, L.M. Abandoned Pb-Zn mining wastes and their mobility as proxy to toxicity: A review. *Sci. Total Environ.* **2016**, *565*, 393–400. [CrossRef]

41. Naidu, G.; Ryu, S.; Thiruvenkatachari, R.; Choi, Y.; Jeong, S.; Vigneswaran, S. A critical review on remediation, reuse, and resource recovery from acid mine drainage. *Environ. Pollut.* **2019**, *247*, 1110–1124. [CrossRef]

42. Jennett, J.C.; Wixson, B.G. Problems in lead mining waste control. *J. Water Pollut. Control Fed.* **1972**, *44*, 2103–2110.

43. Wang, L.K.; Shammas, N.K.; Selke, W.A.; Aulenbach, D.B. (Eds.) *Flotation Technology*; Humana Press: London, UK, 2010. [CrossRef]

44. Službeni List Crne Gore, broj 26/2012 od 24.05.2012. Available online: http://www.sluzbenilist.me/pregled-dokumenta/?id=\{E16B3C65-E56B-4812-A15D-022796224105\} (accessed on 17 October 2020).

45. Tayebi-Khorami, M.; Edraki, M.; Corder, G.; Golev, A. Re-Thinking Mining Waste through an Integrative Approach Led by Circular Economy Aspirations. *Minerals* **2019**, *9*, 286. [CrossRef]

46. Guerin, T.F. Heavy equipment maintenance wastes and environmental management in the mining industry. *J. Environ. Manag.* **2002**, *66*, 185–199. [CrossRef] [PubMed]

47. Bonnissel-Gissinger, P.; Alnot, M.; Ehrhardt, J.J.; Behera, P. Surface Oxidation of Pyrite as a Function of pH. *Environ. Sci. Technol.* **1998**, *32*, 2839–2845. [CrossRef]

48. Lefebvre, R.; Gelinas, P.J. Numerical modeling of AMD production in waste rock dumps. In Proceedings of the Sudbury '95, Conference on Mining and the Environment, Sudbury, ON, Canada, 28 May–1 June 1995.

49. Jamieson, H.E. Geochemistry and Mineralogy of Solid Mine Waste: Essential Knowledge for Predicting Environmental Impact. *Elements* **2011**, *7*, 381–386. [CrossRef]

50. Becker, M.; Dyantyi, N.; Broadhurst, J.L.; Harrison, S.T.L.; Franzidis, J.P. A mineralogical approach to evaluating laboratory scale acid rock drainage characterisation tests. *Miner. Eng.* **2015**, *80*, 33–36. [CrossRef]

51. Karlsson, T.; Raisanen, M.L.; Lehtonen, M.; Alakangas, L. Comparison of static and mineralogical ARD prediction methods in the Nordic environment. *Environ. Monit. Assess.* **2018**, *190*, 719. [CrossRef]

52. Dold, B. Basic concepts in environmental geochemistry of sulfidic mine-waste management. In *Waste Management*; Kumar, S.E., Ed.; IntechOpen: London, UK, 2010; ISBN 978-953-7619-84-8.

53. Skousen, J.G.; Ziemkiewicz, P.F.; McDonald, L.M. Acid mine drainage formation, control and treatment: Approaches and strategies. *Extr. Ind. Soc.* **2019**, *6*, 241–249. [CrossRef]

54. Rodriguez, L.; Ruiz, E.; Alonso-Azcarate, J.; Rincon, J. Heavy metal distribution and chemical speciation in tailings and soils around a Pb–Zn mine in Spain. *J. Environ. Manag.* **2009**, *90*, 1106–1116. [CrossRef]

55. Casiot, C.; Egal, M.; Elbaz-Poulichet, F.; Bruneel, O.; Bancon-Montigny, C.; Cordier, M.-A.; Gomez, E.; Aliaume, C. Hydrological and geochemical control of metals and arsenic in a Mediterranean river contaminated by acid mine drainage (the Amous River, France); preliminary assessment of impacts on fish (Leuciscus cephalus). *Appl. Geochem.* **2009**, *24*, 787–799. [CrossRef]

56. Strosnider, W.H.J.; Llanos Lopez, F.S.; Nairn, R.W. Acid mine drainage at Cerro Rico de Potosí II: Severe degradation of the Upper Rio Pilcomayo watershed. *Environ. Earth Sci.* **2011**, *64*, 911–923. [CrossRef]

57. Cherry, D.S.; Currie, R.J.; Soucel, D.J.; Latimer, H.A.; Trent, G.C. An integrative assessment of a watershed impacted by abandoned mined land discharges. *Environ. Pollut.* **2001**, *111*, 377–388. [CrossRef]

58. Uster, B.; O'Sullivan, A.D.; Ko, S.Y.; Evans, A.; Pope, J.; Trumm, D.; Caruso, B. The use of mussel shells in upward-flow sulfate-reducing bioreactors treating acid mine drainage. *Mine Water Environ.* **2015**, *34*, 442–454. [CrossRef]

59. Doye, I.; Duchesne, J. Neutralisation of acid mine drainage with alkaline industrial residues: Laboratory investigation using batch-leaching tests. *Appl. Geochem.* **2003**, *18*, 1197–1213. [CrossRef]

60. Iakovleva, E.; Makila, E.; Salonen, J.; Sitarz, M.; Wang, S.; Sillanpaa, M. Acid mine drainage (AMD) treatment: Neutralization and toxic elements removal with unmodified and modified limestone. *Ecol. Eng.* **2015**, *81*, 30–40. [CrossRef]

61. Garcia-Valero, A.; Martinez-Martinez, S.; Faz, A.; Rivera, J.; Acosta, J.A. Environmentally sustainable acid mine drainage remediation: Use of natural alkaline material. *J. Water Process. Eng.* **2020**, *33*, 101064. [CrossRef]

62. Tolonen, E.T.; Sarpola, A.; Hu, T.; Ramo, J.; Lassi, U. Acid mine drainage treatment using by-products from quicklime manufacturing as neutralization chemicals. *Chemosphere* **2014**, *117*, 419–424. [CrossRef] [PubMed]

63. Taylor, J.; Pape, S.; Murphy, N. A Summary of Passive and Active Treatment Technologies for Acid and Metalliferous Drainage (AMD). In Proceedings of the Fifth Australian Workshop on Acid Drainage, Fremantle, Australia, 29–31 August 2005.

64. Sibrell, P.L.; Watten, B.J. Evaluation of sludge produced by limestone neutralization of AMD at the Friendship Hill National Historic Site. In Proceedings of the 2003 National Meeting of the American Society of Mining and Reclamation and the 9th Billings Land Reclamation Symposium Billings MT, ASMR, Lexington, KY, USA, 3–6 June 2003; pp. 1151–1169. [CrossRef]

65. Adiansyah, J.S.; Rosano, M.; Vink, S.; Keir, G. A framework for a sustainable approach to mine tailings management: Disposal strategies. *J. Clean. Prod.* **2015**, *108*, 1050–1062. [CrossRef]

66. Harrison, J.; Heijnis, H.; Caprarelli, G. Historical pollution variability from abandoned mine sites, Greater Blue Mountains World Heritage Area, New South Wales, Australia. *Environ. Geol.* **2003**, *43*, 680–687. [CrossRef]

67. Bauerek, A.; Cabala, J.; Smieja-Krol, B. Mineralogical alterations of Zn-Pb flotation wastes of the Mississippi Valley type ores (Southern Poland) and their impact on contamination of rain water runoff. *Pol. J. Environ. Stud.* **2009**, *18*, 781–788.

68. Zhang, S.; Li, T.; Huang, H.; Zou, T.; Zhang, X.; Yu, H.; Zheng, Z.; Wang, Y. Cd accumulation and phytostabilization potential of dominant plants surrounding mining tailings. *Environ. Sci. Pollut. Res.* **2012**, *19*, 3879–3888. [CrossRef]

69. Lekovski, R.; Mikic, M.; Krzanovic, D. Impact of the flotation tailing dumps on the living environment of Bor and protective measures. *Min. Metall. Eng. Bor* **2013**, *2*, 97–116. [CrossRef]

70. Pajak, M.; Gasiorek, M.; Cygan, A.; Wanic, T. Concentrations of Cd, Pb and Zn in the top layer of soil and needles of Scots Pine (Pinus Sylvestris L.); A case study of two extremely different conditions of the forest environment in Poland. *Fresenius Environ. Bull.* **2015**, *24*, 71–76.

71. Krzaklewski, W.; Pietrzykowski, M. Selected physico-chemical properties of zinc and lead ore tailings and their biological stabilization. *Water Air Soil Pollut.* **2002**, *141*, 125–142. [CrossRef]

72. Parraga-Aguado, I.; Querejeta, J.-I.; Gonzalez-Alcaraz, M.-N.; Jimenez-Carceles, F.J.; Conesa, H.M. Usefulness of pioneer vegetation for the phytomanagement of metal(loid)s enriched tailings: Grasses vs. shrubs vs. trees. *J. Environ. Manag.* **2014**, *133*, 51–58. [CrossRef] [PubMed]

73. Ye, Z.H.; Wong, J.W.C.; Wong, M.H.; Baker, A.J.M.; Shu, W.S.; Lan, C.Y. Revegetation of Pb/Zn Mine Tailings, Guangdong Province, China. *Restor. Ecol.* **2001**, *8*, 87–92. [CrossRef]

74. Ye, M.; Li, G.; Yan, P.; Ren, J.; Zheng, L.; Han, D.; Sun, S.; Huan, S.; Zhong, Y. Removal of metals from lead-zinc mine tailings using bioleaching and followed by sulfide precipitation. *Chemosphere* **2017**, *185*, 1189–1196. [CrossRef]

75. O'Sullivan, A.D.; Murray, D.A.; Otte, M.L. Removal of Sulfate, Zinc, and Lead from Alkaline Mine Wastewater Using Pilot-scale Surface-Flow Wetlands at Tara Mines, Ireland. *Mine Water Environ.* **2004**, *23*, 58–65. [CrossRef]

Publisher's Note: MDPI stays neutral with regard to jurisdictional claims in published maps and institutional affiliations.

 *minerals*

Article

Co-Disposal of Coal Gangue and Red Mud for Prevention of Acid Mine Drainage Generation from Self-Heating Gangue Dumps

Zhou Ran [1,2], **Yongtai Pan** [1,2] **and Wenli Liu** [1,*]

1   School of Chemical and Environmental Engineering, China University of Mining and Technology,
    Beijing 100083, China; rzhou9112@163.com (Z.R.); panyongtai@cumtb.edu.cn (Y.P.)
2   Engineering Research Center for Mine and Municipal Solid Waste Recycling, China University of Mining
    and Technology, Beijing 100083, China
*   Correspondence: liuwenli08@163.com; Tel.: +86-10-6233-9883

Received: 9 October 2020; Accepted: 30 November 2020; Published: 2 December 2020

Abstract: The seepage and diffusion of acid mine drainage (AMD) generated from self-heating coal gangue tailings caused acid pollution to the surrounding soil and groundwater. Red mud derived from the alumina smelting process has a high alkali content. To explore the feasibility of co-disposal of coal gangue and red mud for prevention of AMD, coal gangue and red mud were sampled from Yangquan (Shanxi Province, China), and dynamic leaching tests were carried out through the automatic temperature-controlled leaching system under the conditions of different temperatures, mass ratios, and storage methods. Our findings indicated that the heating temperature had a significant effect on the release characteristics of acidic pollutants derived from coal gangue, and that the fastest rate of acid production corresponding to temperature was 150 °C. The co-disposal dynamic leaching tests indicated that red mud not only significantly alleviated the release of AMD but also that it had a long-term effect on the treatment of acid pollution. The mass ratio and stacking method were selected to be 12:1 (coal gangue: red mud) and one layer was alternated (coal gangue covered with red mud), respectively, to ensure that the acid-base pollution indices of leachate reached the WHO drinking-water quality for long-term discharge. The results of this study provided a theoretical basis and data support for the industrial field application of solid waste co-treatment.

Keywords: coal gangue; red mud; prevention; acid mine drainage; dynamic leaching

1. Introduction

Coal gangue is a by-product of coal mining and washing and its discharge amount accounts for approximately 10–15% of total coal production [1,2]. The coal gangue dump is formed by the direct storage of coal gangue and occupies a great deal of farmland and arable land resources [1,3]. The active components contained in coal gangue, such as combustible carbon, pyrite, and heavy metal sulphide, may induce spontaneous combustion in the weathering process. The four stages of spontaneous combustion of coal gangue are: oxygen combining with coal gangue through physical and chemical adsorption; the combined oxygen and pyrite in coal gangue undergoing chemical and microbial catalytic (*Thiobacillus ferrooxidans*) oxidation and emitting heat; the combustible carbon being rapidly oxidized and emitting heat at an accumulation temperature of 80–90 °C; and combustion occurring when the temperature reaches the fire point of the combustible carbon. The critical factors for causing the spontaneous combustion of coal gangue are combustible substance, oxygen, and thermal storage environment. The spontaneous combustion of coal gangue usually occurs at a depth of 2 m from the surface of the coal gangue and causes the release of a large quantity of toxic and harmful gases, like CO, H_2S, SO_2, and NO_x into the atmosphere [4–7]. The soluble salt in coal gangue and the soluble secondary sulfate formed by natural weathering are dissolved and leached under the action of precipitation and leaching to generate acid mine drainage (AMD), which is harmful to the soil, river, and groundwater systems. In particular, a considerable amount of toxic metalloids (arsenic (As), selenium (Se), and antimony (Sb)) and a minute amount of heavy metals (chromium (Cr), copper (Cu), lead (Pb), and zinc (Zn)) are transformed from sulfide form to soluble sulfate form by oxidation and contaminate soils and groundwater [8–14].

Many studies have been conducted on the prevention and control of environmental pollution caused by separate storage of coal gangue. The main options for remediating AMD are divided into active and passive treatments [6,15–22]. The active remediation, that is, adding the neutralizer, adsorbents, and sulfate-reducing bacteria into the collected wastewater, both increases the pH value of the wastewater through acid-base neutralization and reduces the content of sulfates by the adsorption and precipitation of heavy metals [23–31]. Another treatment is to flow the AMD through the artificial wetland, which effectively and continuously improves the quality of AMD by adsorption, filtration, and precipitation [32,33]. In addition, the recycling of wastewater, resource recovery of metals and rare earth resources from AMD are obtained by conducting electrodialysis, membrane separation, and nanofiltration, respectively [34,35]. However, the active treatment of AMD derived from coal gangue needs to meet two requirements: small water flow, and tiny acidic fluctuation. A multi-stage drainage treatment process should be adopted when the AMD has a high concentration of dissolved heavy metals. Moreover, the treatment produces large volumes of sludge.

The principal aim of passive treatment is to prevent the AMD produced by the oxidation of pyrite in coal gangue, and it can be obtained by removing sulfide, isolating water and air, directly passivating pyrite, and adding antibacterial agents and alkaline components [36]. The risk of AMD leaching was decreased from the source by removing pyrite from fine-grained coal gangue during the bubble flotation process [37–39]. Meanwhile, the low-sulfur clean coal was separated from coal gangue by the oil gathering process [40,41]. By covering coal gangue with soil, desulfurized tailings, activated sludge, and oxygen-consuming organic matter, or establishing the gangue tailing pond underwater, the coal gangue was prevented from contact with air, which weakened the chemical oxidation [18,37,42–44]. The carrier-microencapsulation was proposed to passivate pyrite by forming a barrier on the mineral surface [45,46]. In addition, the chemical and microbial catalytic oxidation (*Thiobacillus ferrooxidans*) of pyrite in coal gangue was weakened by mixing with fly ash, quicklime, and limestone, or spraying antibacterial agents on coal gangue [47–50]. The prevention process requires neither continuous alkali addition nor sludge retreatment, thus, it shortens the time of the treatment cycle and provides more economic benefits compared with active treatment [51].

Red mud, as a type of red silt-like solid waste, is generated from alumina production, in which bauxite is the raw material. Typically, about 1 to 2 tons of red mud are generated from 1 t of alumina production, and the annual discharge of red mud worldwide is approximately 150 Mt [52,53]. The comprehensive utilization rate of red mud is less than 15%, and the remaining part of red mud is disposed of by deep-sea dumping or dam storage [54,55]. The particle size of red mud particles is very small, of which more than 90% of particles have a size range from 5 to 75 μm. Red mud generates a large amount of fugitive dust after being air-dried and disintegrated, which increases the content of inhalable particulate matter in the atmosphere [56]. Red mud is a highly alkaline (pH = 10–12.5) and intensely sodic (exchangeable sodium percentage, or ESP, from 70% to 90%) material [57]. Alkaline components in red mud include the adsorption alkali and bound alkali. Among them, the adsorption alkali (free alkali) exists in the forms of $Na_2O \cdot Al_2O_3$, Na_2CO_3, $NaHCO_3$, $NaAl(OH)_4$, $NaAlO_2$, Na_2SO_4 and $KOH(H_2O)_4$, which can be directly dissolved and leached, while the bound alkali is composed mainly of calcite ($CaCO_3$), sodalite ($Na_6Al_6Si_6O_{24} \cdot 2Na_2X$), hydrated garnet ($Ca_3Al_2(SiO_4)_x(OH)_{12-4x}$) and tricalcium aluminate ($Ca_3Al_2(OH)_{12}$), which is dissolved through a chemical reaction [58–61]. Leaching of bauxite residues not only results in alkaline pollution and excessive fluoride content in groundwater but causes salinization and consolidation of the soil [62].

Therefore, red mud added to prevent acid pollution caused by coal gangue could be a potential method for solid waste's co-treatment projects. The present study focused on the release characteristics of the acid contaminant generated from self-heating oxidation of coal gangue. Meanwhile, during the co-disposal of red mud and coal gangue, the leaching regularities of acid pollutants were researched through the automatic temperature-controlled leaching system, which was designed to simulate the conditions of rapid oxidation precipitation, and leaching. Moreover, an optimal co-disposal program of red mud and coal gangue was proposed to ensure that water meets the quality standards so that it could be discharged outward.

2. Materials and Methods

2.1. Samples Preparation

The coal gangue and red mud samples were collected from the No.1 Gangue Dump of Yangquan Coal Industry and the red mud Reservoir of Zhaofeng Alumina Plant, respectively, both in Shanxi Province, China. The fresh coal gangue samples were the discharge of raw coal after washing, and their original particle size composition was >25 mm 61.26 wt%, 13–25 mm 23.51 wt%, and <13 mm 15.23 wt%, respectively. The pyrite in coal gangue presented in the single crystal, disseminated, and stripped states. When used directly without any pretreatment, the original sample would have an immense effect on the comparative analysis of results for the huge fluctuation of the sample composition. Therefore, the original coal gangue samples were ground into leaching samples, whose diameter was 1 mm, while the red mud samples obtained from the settled and thickened alumina tailings slurry through a quick-opening filter press contained 39.84 wt% water. To facilitate the co-disposal test, the red mud samples derived from the filter process were air-dried, crushed, and sieved into agglomerated samples with a diameter of 1 mm, which had the same physical and chemical properties as the original red mud samples [63].

2.2. Mineralogical Determination

The proximate analysis (M, Ash, VM, and FC) of the coal gangue samples (1 mm) was conducted according to ASTM D 3173-03 (Moisture, M), ASTM D 3174-04 (Ash), ASTM D 3175-07 (Volatile matter, VM), and the Fixed carbon (FC) was determined by subtracting the sum of M, Ash, and VM from 100%. The ultimate analysis was performed on an elemental analyzer (Vario MACRO CHNS, Elementar) with an oxygen kit. The mineralogical compositions of the samples were studied using a D/max-2550 X-ray diffractometer (Rigaku Corporation, Tokyo, Japan) with Cu-K$_{\alpha 1}$ radiation operating at 40 kV and 40 mA. The samples were scanned over 3–90° 2-theta with a step size of 0.02° 2-theta and a counting time of 1 s. The mineral content of the samples semi-quantitatively analyzed by the reference intensity ratio (RIR) method was obtained with the help of MDI Jade6.5 software. An XRF-1800 X-ray fluorescence spectrometer (Shimadzu Corporation, Kyoto, Japan) with Rh radiation and 60 kV(Max) X-ray tube pressure was used to measure the chemical composition of the samples. The extraction characteristics of the samples were analyzed according to ASTM D3987-12. The extractant was the reagent water, of which the volume equalled in milliliters to 20 times the mass in grams of the samples. Besides, the particle sizes of red mud samples (1 mm) were determined by laser particle analyzer (LS-POP(VI), Zhuhai, China). All of the mineralogical indicators were measured 3 times, and the data was analyzed for standard deviation (SD).

2.3. Leaching Experimental Device

To truly simulate the physical and chemical conditions of the self-heating coal gangue dumps' precipitation, an automatic temperature-controlled leaching system was designed for this research, as is shown in Figure 1.

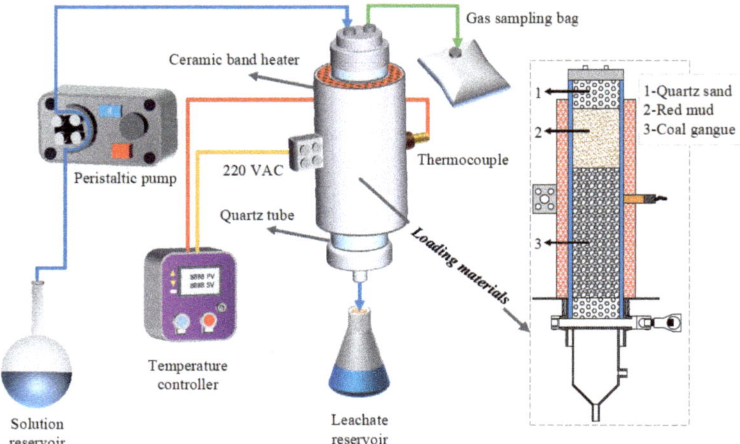

Figure 1. Schematic diagram of the automatic temperature-controlled leaching system and material loading.

The system was mainly composed of the quartz tube, ceramic band heater, temperature controller, peristaltic pump, gas sampling bag, solution reservoir, and leachate reservoir. The size of the quartz tube was Φ 70 mm × 5 mm, h = 300 mm, and the quartz tube and its inside samples were heated by the ceramic band heater tightly clamped outside. The temperature controller with a control range of 0–300 °C was used to detect and control the working temperature of the ceramic band heater. The spray liquid in the solution reservoir was sent to the quartz tube by the peristaltic pump. The gas sampling bag collected toxic and harmful gases generated from the heating process, and the leaching solution was collected by the leachate reservoir.

2.4. Dynamic Heating and Leaching

To explore the prevention effect of red mud on the acid pollution derived from coal gangue dumps in a self-heating state, it was necessary to grasp the release characteristics of the coal gangue acid contaminant at different temperatures. Under the condition of the fastest release rate of acidic components in coal gangue, dynamic leaching tests were performed on the co-disposal of red mud and coal gangue to research the co-disposal effect.

Firstly, the releasing characteristics of acidic pollutants derived from coal gangue were studied. The coal gangue samples (1 mm) were separately packed into the quartz tube according to the loading method shown in Figure 1. The samples were filled into the quartz tube by gravity accumulation without external pressure. The mass of the coal gangue sample in the middle layer was set as 600 g, and the height of the coal gangue layer was 60 mm. The upper and lower layers were paved with 10–20 mesh quartz sand to ensure the spray solution's uniform infiltration and filter out the coarse particles in the leaching solution. The spray intensity of the peristaltic pump was set to be 6 mL·min⁻¹ according to the rain capacity in Yangquan area (Shanxi Province, China), and the spray liquid was deionized water. At t = 0 h, the materials in the quartz tube were initially wetted, and the wetting process was stopped when a 250 mL leaching solution had been collected. Then, temperatures of the ceramic band heater were set at 50 °C, 100 °C, 150 °C, and 200 °C to heat the coal gangue sample. The quartz tube was heated at intervals of 10 h for spraying operation until 250 mL leaching solution had been collected, and the cumulative heating time of single factor test (e.g., 50 °C) was 160 h. The pH value, electrical conductivity (EC) value, sulfate (SO_4^{2-}) concentration, oxidation-reduction potential (ORP), total dissolved solids content (TDS), and acid neutralization potential (ANP) of the leachate were detected immediately. These indicators of the leachate were repeatedly tested three times, and the average value was taken to reduce the detection error. The apparatus and methods are shown in Table 1, and the dynamic heating and leaching process is shown in Figure 2.

Table 1. The apparatus and methods for detecting pollution indices of leachate.

Parameter	Apparatus and Methods
pH	FE28-standard pH meter (Mettler Toledo, Küsnacht, Switzerland)
EC	DDS-11A electrical conductivity meter (Lei Ci, Shanghai, China)
SO_4^{2-}	Water quality-Determination of sulfate-Gravimetric method (GB11899-89)
ORP	Orion 3 Star mV Detector (Thermo Scientific, Waltham, MA, USA)
TDS	TDS-5 m (Greensky, Hangzhou, China)
ANP	Acid-base neutralization titration
Mineral composition	D/max-2550 X-ray diffractometer (Rigaku Corporation, Tokyo, Japan)

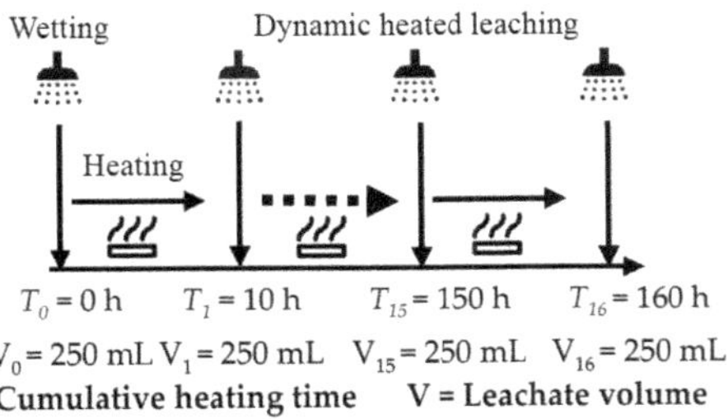

Figure 2. Schematic diagram of dynamic heating and leaching process.

Additionally, at the temperature corresponding to the fastest release rate of acidic components of coal gangue, the dynamic leaching tests of the co-disposal of the coal gangue and red mud were carried out under the conditions of different mass ratios (12:1, 8:1, 5:1, and 3:1) and storage methods (alternating 1 layer, 2 layers, 5 layers, and uniform mixing). In the mass ratios study, 600 g coal gangue (thickness: 60 mm) was used as the mass benchmark and covered with the red mud of different mass, and the filled red mud masses were 50 g (40 mm), 75 g (60 mm), 120 g (96 mm), and 200 g (160 mm), respectively, corresponding to the mass ratios of 12:1, 8:1, 5:1, and 3:1. Besides, in the storage method study, the mass ratio of coal gangue to red mud remained unchanged (e.g., 12:1 or 600 g:50 g), and the certain mass of coal gangue and red mud was equally divided into 1, 2 and 5 parts, respectively, corresponding to the storage method of alternating 1 layer, 2 layers, and 5 layers. Each part was filled into the quartz tube alternately in the order of coal gangue in the lower layer and red mud in the upper layer. As for the storage method of uniform mixing, the coal gangue and red mud samples were uniformly mixed by the mixer prototype (MR2L, France). The samples were filled into the quartz tube by gravity accumulation without external pressure, and the co-disposal dynamic heating and leaching mechanism was the same as the research method of the release characteristics of the acidic pollutant derived from coal gangue.

3. Results and Discussion

3.1. Properties of Materials

The proximate and ultimate analysis of coal gangue samples are shown in Table 2.

Table 2. Proximate and ultimate analysis of coal gangue (wt%, mean ± SD, n = 3).

Proximate Analysis		Ultimate Analysis	
Moisture, M_{ad} [1]	1.07 ± 0.02	Carbon, C_{ad}	17.60 ± 0.06
Ash, A_{ad}	72.70 ± 0.5	Hydrogen, H_{ad}	1.36 ± 0.05
Volatile matter, VM_{ad}	9.25 ± 0.02	Nitrogen, N_{ad}	0.41 ± 0.03
Fixed carbon, FC_{ad}	16.98 ± 0.41	Sulfur, S_{ad}	4.50 ± 0.16

[1] ad: Air-dried statement.

The coal gangue had great potential of self-heating and spontaneous combustion, and the content of the FC_{ad} and C_{ad} accounted for 16.98 ± 0.41 wt% and 17.60 ± 0.06 wt%, respectively [64]. The potential acid production of coal gangue was 137.64 kg $CaCO_3 \cdot t^{-1}$ calculated by the standard Sobek acid-base counting test based on the 4.50 ± 0.16 wt% S_{ad} [65]. The coal gangue sample was medium-sulfuric and had a strong potential for acid production.

XRD pattern of coal gangue revealed that the main minerals consisted of 31.88 ± 0.14% kaolinite, 26.78 ± 0.27% illite, 22.96 ± 0.20% quartz, 9.80 ± 0.37% pyrite and, 8.58 ± 0.09% calcite (Figure 3a). The content of the clay minerals (kaolinite, illite, and montmorillonite) exceeded 50%. This indicates that the sample is clay salt coal gangue and had a certain degree of acid neutralization potential, which could neutralize acidic components generated by coal gangue itself. As is shown in Figure 3b, the mineral composition of the red mud sample included 51.74 ± 0.35% hydrated andradite, 21.71 ± 0.21% calcite, 21.46 ± 0.05% hydrated sodalite, and 5.09 ± 0.05% hematite. The hydrated sodalites is the main alkaline component that accounts for the great potential of alkali production in red mud [66].

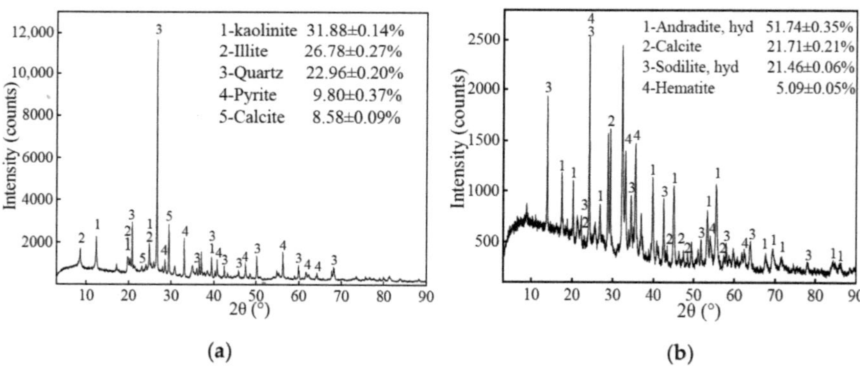

(a)   (b)

Figure 3. The XRD patterns of (**a**) coal gangue, (**b**) red mud, and the mineralogical compositions (%, mean ± SD, n = 3).

According to Table 3, the chemical composition of coal gangue was mainly SiO_2 and Al_2O_3 with contents of 49.81 ± 0.06% and 24.82 ± 0.13%, respectively. It showed that the main compound in the coal gangue sample was aluminosilicate. The total content of Al_2O_3 and Fe_2O_3 in red mud exceeded the total mass of SiO_2 and CaO, and the content of Na_2O was 12.05 ± 0.07%. This indicated that the red mud sample was the bauxite residue of the Bayer process and was strongly alkaline.

Table 3. Chemical compositions of coal gangue and red mud (wt%, mean ± SD, n = 3).

Material	Oxides											Loss
	SiO_2	Al_2O_3	Fe_2O_3	CaO	SO_3	Na_2O	Cr_2O_3	NiO	CuO	ZnO	PbO	
Coal gangue	49.81 ± 0.06	24.82 ± 0.13	9.82 ± 0.05	6.39 ± 0.04	5.52 ± 0.03	0.45 ± 0.01	0.03 ± 0.002	0.01 ± 0.00	0.01 ± 0.00	0.02 ± 0.00	0.01 ± 0.00	27.2 ± 0.12
Red mud	20.43 ± 0.16	25.92 ± 0.09	14.62 ± 0.18	17.22 ± 0.08	2.61 ± 0.02	12.05 ± 0.07	0.06 ± 0.003	0.08 ± 0.002	0.009 ± 0.00	0.01 ± 0.00	0.01 ± 0.00	7.10 ± 0.16

The shake extraction indices of samples are shown in Table 4. The pH value and ANP of coal gangue were 7.25 ± 0.04 and −0.071 ± 0.003 g (CaCO$_3$)·L^{-1}, respectively, and it indicated that the sample presented as weakly alkaline during the shake extraction process. The coal gangue that has not been weathered and oxidized does not cause acid pollution to the environment. However, the red mud presented as strongly alkaline, in which the alkaline components were leached in a large amount during the extraction process, and the pH value was 11.11 ± 0.05. The EC values of coal gangue and red mud were 0.715 ± 0.009 and 0. 852 ± 0.006 mS·cm^{-1}, respectively, and the TDS were 581 ± 5 and 509 ± 4 mV. Since both the EC value and TDS reflected the concentration of soluble ions in the solution (natural water EC value, 0.102–2.079 mS·cm^{-1}), the result showed that both extracted solutions of the samples had little effect on the hardness of water [67]. Moreover, the extracted solutions of coal gangue and red mud presented as weakly oxidative for the ORP of 141.3 ± 4.2 and 8.52 ± 0.6 mV, respectively. Besides, the average particle diameter ($D_{(4,\,3)}$) of the red mud sample was 5.74 ± 0.02 μm and the D_{90} was 8.59 ± 0.02 μm, indicating that the sample had an extremely fine particle size (Figure 4). The fine particles cause the red mud to have uniformity of physical and chemical properties and low permeability.

Table 4. The shake extraction indices of coal gangue and red mud (mean ± SD, n = 3).

Parameter	Unit	Material	
		Coal Gangue	Red Mud
pH		7.25 ± 0.04	11.11 ± 0.05
EC	mS·cm^{-1}	0.715 ± 0.009	0.852 ± 0.006
ORP	mV	141.3 ± 4.2	8.5 ± 0.6
TDS	ppm	581 ± 5	509 ± 4
ANP	g(CaCO$_3$)·L^{-1}	−0.071 ± 0.003 *	−0.52 ± 0.02

* A positive value indicates that the acidic solution consumes alkali, while a negative value indicates that the solution is alkaline.

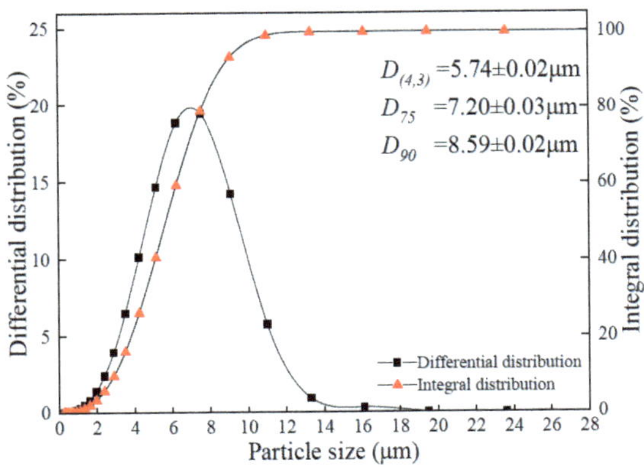

Figure 4. The particle size (mean ± SD, n = 3) distribution of red mud.

3.2. Release Characteristics of Acidic Contamination

The pH value and ANP of leaching solution derived from coal gangue corresponding to different dynamic heating and leaching tests varied as the cumulative heating time increased, as shown in Figure 5.

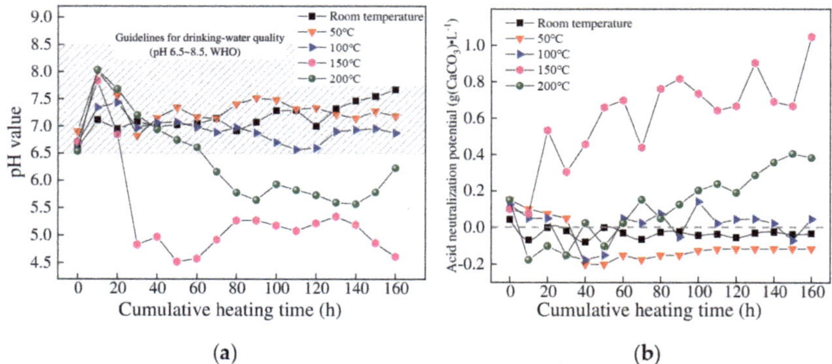

Figure 5. Variations of (**a**) pH value and (**b**) ANP of coal gangue leaching solution at different temperatures.

As is shown in Figure 5a, the variations of the pH value of coal gangue's leachate under different temperature conditions were similar as the cumulative heating time went by, that was, the pH value increased in the beginning, then decreased and eventually stabilized. When the pH value presented as stable ($t > 80$ h), the steady-state pH value decreased at first and then increased as the temperature increased, during which the minimum pH value was about 4.61 at 150 °C (T_c). The minimum pH value appeared to be strongly acidic and was out of the limit of the WHO guidelines for drinking-water quality (pH 6.5–8.5). This means that the acidic contamination of coal gangue was fully released under this temperature condition.

The main acid producing component in coal gangue is the pyrite; the related Equations (1)–(6) chemical reaction equations during the heating and filtration processes can be seen as follows.

The pyrite produces acids directly by oxidation under the condition of air and water [11]:

$$2FeS_2(s) + 7O_2(aq) + 2H_2O \rightarrow 2Fe^{2+}(aq) + 4SO_4^{2-}(aq) + 4H^+(aq) \tag{1}$$

$$FeS_2(s) + 14Fe^{3+}(aq) + 8H_2O \rightarrow 15Fe^{2+}(aq) + 2SO_4^{2-}(aq) + 16H^+(aq) \tag{2}$$

$$4Fe^{2+}(aq) + O_2(aq) + 4H^+(aq) \rightarrow 4Fe^{3+}(aq) + 2H_2O \tag{3}$$

The pyrite is oxidized by air without water:

$$FeS_2(s) + 11O_2(g) \rightarrow 2Fe_2O_3(s) + 8SO_2(g) \tag{4}$$

$$2SO_2(g) + O_2(g) \rightarrow 2SO_3(g) \tag{5}$$

$$4FeS_2(s) + 3O_2(g) \rightarrow 2Fe_2O_3(s) + 8S(s) \tag{6}$$

The rate of pyrite's oxidizing and production acid accelerates with the increase in external temperature under the condition of water [68,69]. However, when the temperature reaches a higher state ($T > 200$ °C), a large amount of water inside the coal gangue is evaporated, and it curbs the direct oxidization and acid production of pyrite. The main products of the pyrite oxidized by the air are sulfur oxides (SO_x) and sulfur (S), which means there is a decrease in the acid-soluble components [70–72].

In addition, the calcite and silicate minerals contained in the coal gangue sample can neutralize AMD derived from the coal gangue itself, and this causes the pH value to gradually increase in the early stage of dynamic leaching tests. The related Equations (7)–(9) chemical reaction equations during the AMD neutralization processes present as follows.

$$CaCO_3(s) + 2H^+(aq) \Leftrightarrow Ca^{2+}(aq) + CO_2(g) + H_2O \tag{7}$$

$$NaAlSi_3O_8(s) + 4H^+(aq) + 4H_2O \rightarrow Na^+(aq) + Al^{3+}(aq) + 3H_4SiO_4(aq) \tag{8}$$

$$KAlSi_3O_8(s) + 4H^+(aq) + 4H_2O \rightarrow K^+(aq) + Al^{3+}(aq) + 3H_4SiO_4(aq) \tag{9}$$

Figure 5b shows that the leaching solution's ANP was small under a low-temperature condition (room temperature–100 °C), which fluctuated in the range of −0.2–0.18 g (CaCO$_3$)·L^{-1} and had little potential to cause acid pollution, under a high-temperature condition (from 150 °C to 200 °C), however, the ANP gradually increased as the cumulative heating time increased. In the temperature range of this study (room temperature–200 °C), the ANP reached its highest level at 150 °C·(T_c), indicating that the pyrite in the coal gangue produced the largest quantities of acidic components at T_c, which accords with the pH analysis results in Figure 5a. Meanwhile, it also indicated that the coal gangue sample continuously caused acid pollution for a long time under a suitable temperature condition.

The mineral composition of coal gangue samples had been analyzed after being heated separately. The X-ray diffraction pattern is shown in Figure 6.

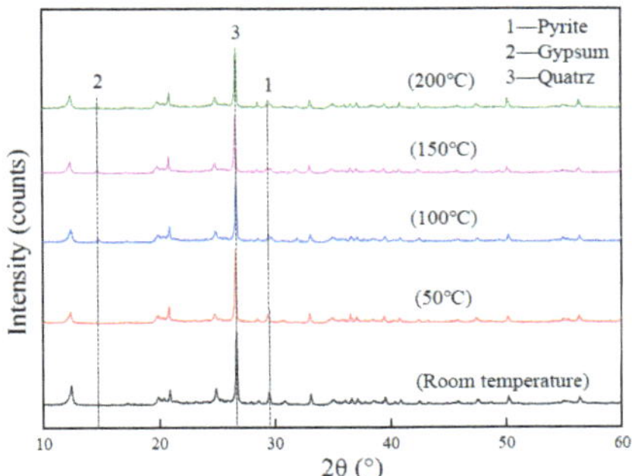

Figure 6. XRD patterns of the remaining coal gangue at different temperatures.

As the temperature increased, the intensity of the diffraction peaks of the pyrite in the remaining coal gangue weakened first and then strengthened, while that of the gypsum first strengthened and then weakened, and the change in quartz was not significant. The content of the pyrite was negatively correlated with the gypsum. The reason is that the sulfide in the coal gangue is heated and oxidized to produce sulfate, whereas part of the gypsum component in the remaining coal gangue is produced by the oxidation of pyrite. The pyrite in coal gangue has strong chemical activity. However, the generated gypsum has good chemical stability, and its coverage on the pyrite can prevent further oxidation and acid production.

3.3. Co-Disposal Prevention

Based on the research results of the acid pollution release characteristics of coal gangue in Section 3.2, the temperature of 150 °C (T_c) was chosen to carry out the dynamic leaching test through the co-disposal of red mud and coal gangue to prevent the acid pollution caused by the coal gangue. The tests were conducted under the condition of different mass ratios and storage methods to find a suitable storage solution.

3.3.1. Mass Ratio

pH Value

The coal gangue covered with the red mud was stacked in layers with different mass ratios (12:1~3:1) and then dynamically heated and leached. The pH value of the leaching solution changed as the cumulative heating time increased, as shown in Figure 7.

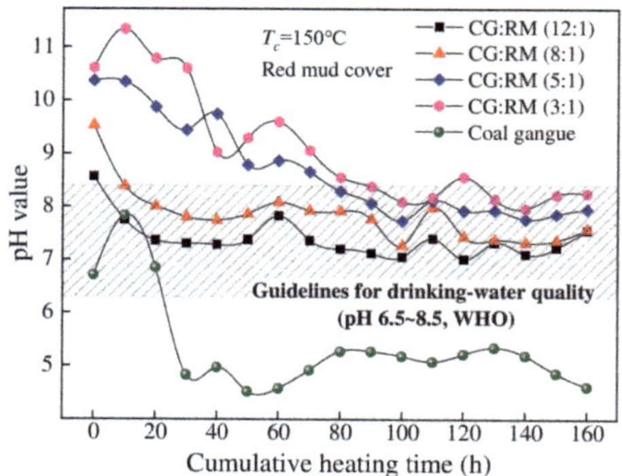

Figure 7. Variations of pH value of co-disposal leaching solution with different mass ratios.

The pH values of the leaching solution corresponding to different mass ratios (12:1~3:1) changed similarly as the cumulative heating time increased, that is, they gradually decreased and then remained stable. In the late stage of leaching (80–160 h), the steady-state pH value of the leaching solution was within the range of the drinking-water quality. This indicates that the addition of red mud can effectively alleviate the AMD generated from the coal gangue. The AMD is neutralized by the alkaline component in red mud as described by Equations (10)–(17) [66].

AMD is neutralized by the adsorption alkali:

$$NaHCO_3(s) + H^+(aq) \rightarrow Na^+(aq) + CO_2(g) + H_2O \tag{10}$$

$$Na_2CO_3(s) + 2H^+(aq) \rightarrow 2Na^+(aq) + CO_2(g) + H_2O \tag{11}$$

$$NaAl(OH)_4(s) + H^+(aq) \rightarrow Na^+(aq) + Al(OH)_3(s) + H_2O \tag{12}$$

$$KOH(H_2O)_4(s) + H^+(aq) \rightarrow K^+(aq) + 5H_2O \tag{13}$$

AMD is neutralized by the bound alkali:

$$CaCO_3(s) + 2H^+(aq) \rightarrow Ca^{2+}(aq) + CO_2(g) + H_2O \tag{14}$$

$$[Na_6Al_6Si_6O_{24}]\cdot[2NaX, Na_2X](s) \rightarrow 8Na^+(aq) + 6H_4SiO_4(aq) + 6Al(OH)_3(s) + 8X(OH^-, HCO_3^{2-})(aq) \tag{15}$$

$$2H_4SiO_4(aq) + 4H^+ \rightarrow Si^{4+}(aq) + H_2SiO_3(aq) + 5H_2O \tag{16}$$

$$Al(OH)_3(s) + 3H^+(aq) \rightarrow Al^{3+}(aq) + 3H_2O \tag{17}$$

At t = 0 h, the initial wetting solutions corresponding to each mass ratio were alkaline (pH > 8.5), and the initial pH value increased as the mass ratio decreased. This indicates that the more bauxite tailings are added, the stronger alkaline of the initial wetting solution would be. As shown in Equations (10)–(13), the adsorption alkali component in the red mud is particularly soluble and can be dissolved and leached in a large amount under the action of wetting spray, of which the leaching amount increases with the addition of red mud. How to effectively control the rapid dissolution of adsorption alkali in the early stage of leaching is a problem to be solved urgently.

During the entire process, the lower the mass ratio was (12:1→3:1), the higher the pH value of the corresponding leaching solution was. The pH value corresponding to the low mass ratio (5:1~3:1) was higher than that of the index limit of the drinking-water quality (pH 6.5–8.5) in the early stage of leaching (0–80 h). By contrast, the pH value for the high mass ratio (12:1~8:1) was within the range of the drinking-water quality during the entire process. The cumulative heating time required for the pH value to reach a stable state increased as the mass ratio decreased, in which the pH value corresponding to the mass ratio of 3:1 was 100 h. The more red mud was added into the stack, the longer the leaching time that was required for the complete leaching of the adsorption alkali component. In the late stage of leaching (80–160 h), the steady-state pH value increased slowly with the decrease in the mass ratio for the leaching component of red mud due mainly to the bound alkali. Acid-base neutralizing between the acidic component produced by coal gangue and the bound alkali component derived from red mud takes place mainly by the displacement reaction, catalytic reaction, and hydration reaction as shown in Equations (14)–(17), of which the degree of the acid-base neutralization depends on the content of the acidic components produced by coal gangue.

ANP

The ANP of the leaching solution obtained from the co-disposal of red mud and coal gangue according to different mass ratios changed as the cumulative heating time increased, as shown in Figure 8.

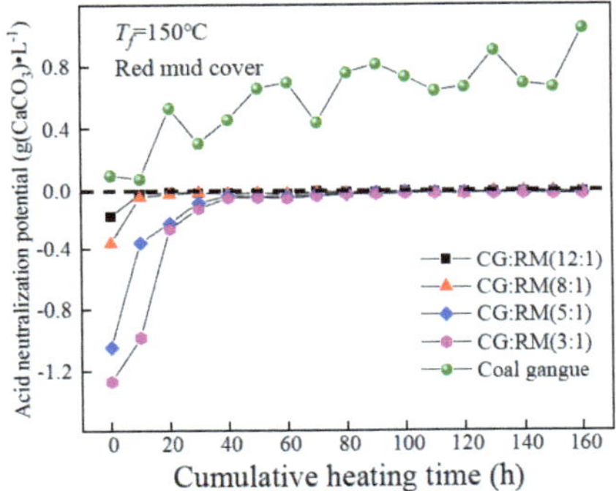

Figure 8. Variations of ANP of co-disposal leaching solution with different mass ratios.

Throughout the entire heating and leaching process, the ANP values corresponding to different mass ratios were all negative, and the leaching solution presented as alkaline. Additionally, the ANP increased rapidly and eventually approached 0 g (CaCO$_3$)·L^{-1}. This further demonstrates that the red mud could prevent the acid pollution caused by the self-heating oxidation of the separately-stored coal gangue. At $t = 0$ h, the ANP of the initial wetting solution increased with the increase in mass ratio (3:1→12:1). The ANP of the coal gangue wetting solution was 0.10 g(CaCO$_3$)·L^{-1}, while the mass ratio of 3:1 corresponded to the ANP of −1.27 g(CaCO$_3$)·L^{-1}. This indicates that the addition of red mud has a significant impact on the ANP of the leaching solution. The possible reason is that the rapid dissolution of adsorption alkali (free alkali) components in the red mud layer results in the strong alkalinity in the leaching solution.

In the late stage of leaching (40–160 h), the mass ratio had little effect on the ANP of the leaching solution, as the ANP of different mass ratios steadily converged towards 0 g(CaCO$_3$)·L^{-1}. Meanwhile, adding a small amount of the red mud was enough to decrease the ANP of the coal gangue leaching solution, for it was the bound alkali that neutralized the acidic product produced by coal gangue and the releasing amount of the bound alkali depended on the quantity of the acid product from coal gangue. Compared with the pH index of the leaching solution, the cumulative heating time required for the ANP to reach the steady-state is less than that of the pH value. This indicates that the complex composition of alkaline minerals in the red mud and its strong anti-acidity capacity could delay the decrease in the leaching solution's pH value [73].

EC Value

The EC value of the leaching solution obtained by the co-disposal of red mud and coal gangue in different mass ratios varied with the cumulative heating time, as shown in Figure 9.

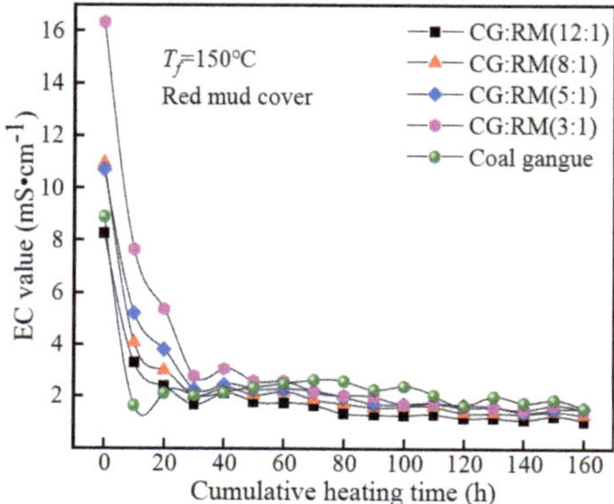

Figure 9. Variations of EC value of co-disposal leaching solution with different mass ratios.

The EC values of the leaching solutions corresponding to different mass ratios had a similar trend in the entire heating and leaching process, that was, they first decreased sharply, and then stabilized at about 2 mS·cm^{-1} for a long time, which was higher than the range of general natural water's EC value (0.102–2.079 mS·cm^{-1}) [74]. This indicated that the co-disposal of red mud and coal gangue would increase the hardness of the surface water systems and the groundwater bodies after being leached. The EC value of the solution depends on the concentration of soluble ions and their salt content. The soluble mineral components contained in the coal gangue and red mud samples were relatively high and quickly dissolved and released in the early stage of leaching (0–30 h), while they were significantly reduced in the late stage of leaching (30–160 h). At $t = 0$ h, the EC values of the initial wetting solutions corresponding to different mass ratios all reached the maximum, and the maximum value increased as the mass ratio decreased (12:1→3:1). When the mass ratio decreased from 8:1 to 3:1, the wetting solution was regarded as highly mineralized water (EC > 10 mS·cm^{-1}). The coal gangue contains a large number of clay minerals (kaolinite, montmorillonite, and illite), meanwhile, the red mud contains various forms of adsorption alkali. The clay minerals and the adsorption alkali components are quickly dissolved and leached under the action of initial wetting and spraying. The co-disposal of red mud and coal gangue can significantly increase the water hardness.

In the late stage of dynamic leaching (30–160 h), the EC values corresponding to the different mass ratios reached a stable state. The mass ratio had no significant effect on the steady-state EC value, for the concentration of soluble ions in the leaching solution depended on the acidic components produced by the coal gangue and the bound alkali component derived from the red mud. The total leaching amount of bound alkali depended on the content of acidic components produced by the coal gangue. Therefore, the mass ratio of coal gangue to red mud does not have a significant impact on the hardening water quality caused by long-term storage.

Therefore, the mass ratios of the co-disposal of coal gangue and red mud should be 12:1, for the pH of the leaching solution was within the range of drinking-water quality for a long time and the leaching solution had the least impact on the hardness of groundwater.

3.3.2. Storage Method

pH Value

The mass ratio of coal gangue to red mud was designed as 12:1, and the dynamic leaching tests were carried out under the condition of 150 °C. The pH value of leaching solution derived from the co-disposal of coal gangue and red mud based on the different storage methods (alternating 1, 2, 5 layers, and uniform mixing) varied with the accumulated heating time, as shown in Figure 10.

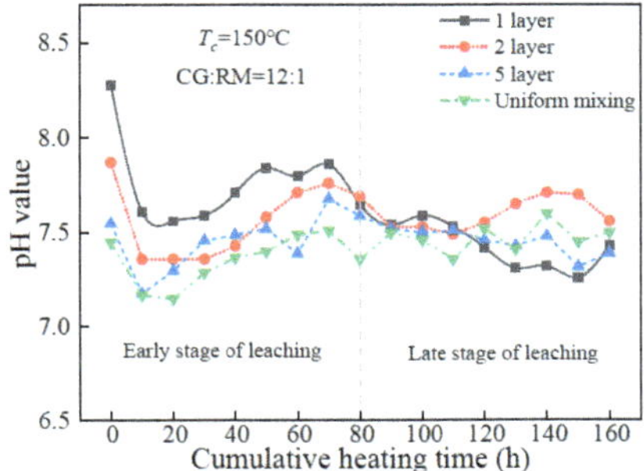

Figure 10. Variations of pH value of co-disposal leaching solution with different storage methods.

As shown in Figure 10, the storage method of coal gangue and red mud had no significant impact on the co-treatment of acid-base pollution throughout the entire heating and leaching process. The pH values of the leaching solution corresponding to the different alternate layers underwent similar changes, that was, they first decreased, then increased, and finally stabilized. The pH values did not exceed the drinking-water quality (pH 6.5–8.5) during the entire process, and the steady-state pH value approached 7.5 in the late stage of leaching (80–160 h).

At $t = 0$ h, the pH values of the initial wetting solutions reached the maximum, and the maximum values of pH decreased with the increase in the number of the alternate accumulation layers. It was the increasing number of the alternate accumulation layer that promoted the coal gangue being in contact with the red mud to strengthen the acid-base neutralization.

In the early stage of leaching (10–80 h), the pH value of the leaching solution corresponding to the various storage methods gradually increased. This indicated that the total alkaline concentration of red mud was higher compared with the acid concentration of coal gangue, for the leaching of the alkali in the red mud sample contained the adsorption alkali and the bound alkali at this stage. The bound alkali has a strong buffering capacity, which can fully react with the acidic components generated from coal gangue by displacement, catalytic, and hydration reactions. In the late stage of leaching (90–160 h), the pH value tended to be stable, and the steady-state pH value was not significantly affected by the number of alternate accumulation layers, for the residual content of adsorption alkali in the red mud sample was extremely small. Besides, the leaching alkali solution mainly contained the bound alkali in the late stage, and the amount of bound alkali leaching depended on the total amount of acid produced by the oxidation of coal gangue.

ANP

The ANP of the leaching solution varied with the cumulative heating time, as shown in Figure 11.

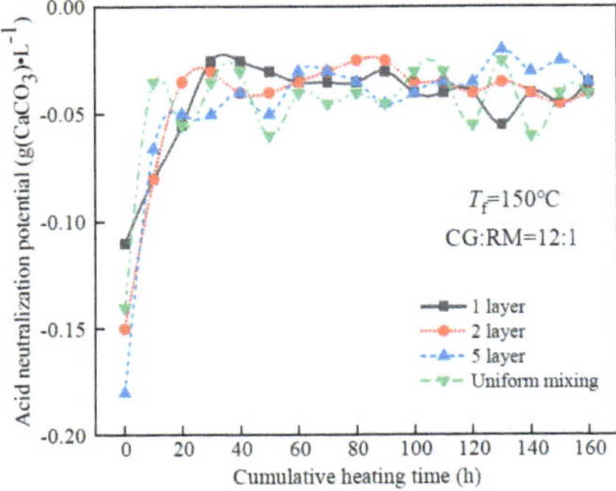

Figure 11. Variations of ANP of co-disposal leaching solution with different storage methods.

Figure 11 shows that the ANPs of the leaching solution corresponding to various stacking methods had a similar change during the entire heating and leaching process. The ANP reached a stable state after a rapid increase and the steady-state value approached −0.04 g $(CaCO_3)\cdot L^{-1}$, which did not cause acid-base pollution to the surrounding water bodies. At $t = 0$ h, the ANP of the initial wetting solution was the smallest while alkalinity was the strongest. Meanwhile, the ANP decreased with the increase in the alternating stacking layers but the uniform mixing one increased exceptionally, for the ANP was determined by the acidic components of coal gangue, the alkaline components of coal gangue, and the adsorption alkali of red mud at the early stage of leaching. The alkaline components of coal gangue are mainly clay minerals, which leached in large amounts during the initial wetting, and the amount of leaching increases with the increase in the number of alternate accumulation layers of coal gangue and red mud.

During the late stage of leaching (40–160 h), the ANP of the leaching solution tended to be stable and was not significantly affected by the storage methods of coal gangue and red mud. The soluble clay minerals in the coal gangue and the adsorption alkali in the red mud were eluted in the early stage of leaching. However, in the late stage of leaching, the ANP of the leaching solution was mainly determined by the acidic components produced by the oxidation of pyrite and the bound alkali derived from red mud. The amount of bound alkali leaching depended on the amount of acid produced by the coal gangue oxidation. Therefore, there was no significant difference in the ANP of the leaching solution for the different storage methods.

EC Value

The EC value of the leaching solution varied with the cumulative heating time, as shown in Figure 12.

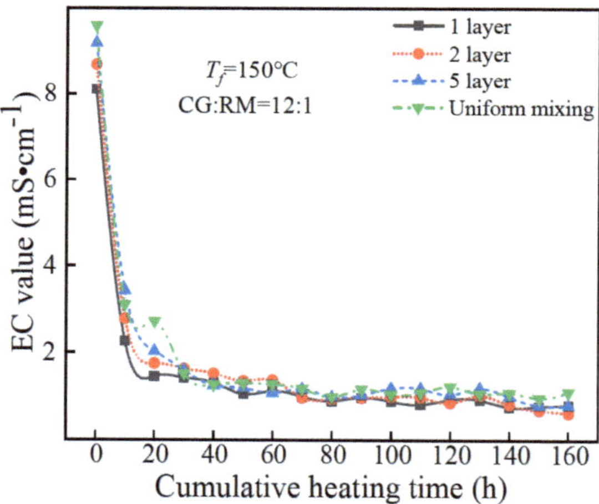

Figure 12. EC value variations of the co-disposal leaching solutions corresponding to various storage methods.

Figure 12 shows that the EC value of the leaching solution corresponding to different storage methods had a similar change with the cumulative heating time, that was, it first decreased sharply and then remained stable, and the steady-state value was lower than 1.5 mS/cm. The variation of ANP indicated that the soluble minerals in the co-disposal of coal gangue and red mud under the different storage methods leached rapidly in the early stage.

At $t = 0$ h, the EC value of the initial wetting solution corresponding to the different storage methods reached the maximum, and the maximum value increased as the alternate accumulation layers increased. It also indicated that the increased contact area between coal gangue and red mud helped to improve the leaching of the clay minerals in the coal gangue and the adsorption alkali in the red mud.

In the late stage of leaching (30–160 h), the different storage methods had no significant effect on the steady-state EC value, for the soluble salt content of coal gangue and red mud had decreased, and the soluble minerals leached were mainly the acidic components obtained from coal gangue and the bound alkali components of red mud. The leaching amount of bound alkali depended on the amount of acid derived from coal gangue. Meanwhile, the acid produced by the oxidation of coal gangue was significantly affected by the sulfur content of coal gangue, the external temperature, and the degree of being contact with the air but had nothing to do with the co-disposal method of coal gangue and red mud.

Considering the operability of the co-disposal of coal gangue and red mud to prevent and control the acid pollution caused by coal gangue, it is proper to adopt the alternative single layer by the coal gangue covered with the red mud during the stacking process.

4. Conclusions

(1) The acid production rate of coal gangue is significantly affected by the self-heating temperature. The oxidation degree of pyrite increases as the self-heating temperature increases. However, excessive temperature leads to the reduction in soluble acidic products. In the temperature range of this study, coal gangues had the fastest rate of acid production at 150 °C, the leaching solution presented as strongly acidic and the acid pollution to the environment lasted for a long time.

(2) The addition of red mud can significantly alleviate the acid pollution caused by coal gangue and has long-term effectiveness in the treatment of acid pollution. The mass ratio of coal gangue to red mud has a significant impact on the co-treatment of acid-base pollution. The most appropriate mass ratio was 12:1, and it ensured that the acid pollution index of the leaching solution reached the standard for long-term discharge. However, the co-treatment effect on acid-base pollution was not significantly affected by the storage method. It was more appropriate to adopt the alternate one layer stacking method to simplify the co-disposal of red mud and coal gangue.

(3) The physical and chemical process of covering red mud to prevent acid pollution derived from coal gangue is complex, which involves the gas-solid-liquid three-phase complex chemical environment. Various components are involved in direct and indirect chemical reactions, microbial catalytic oxidation, and displacement reaction. Mutual transformation occurs under the action of chemical reaction, and migration occurs under the action of thermal stress and seepage force field. Thus, the mechanism of co-disposal of coal gangue and red mud used for alleviating the acid-base pollution, and the transformation states of pollutant components, need to be further studied.

Author Contributions: Conceptualization, Z.R. and Y.P.; methodology, Z.R. and W.L.; software, Z.R.; validation, Z.R., Y.P. and W.L.; investigation, Z.R., Y.P. and W.L.; data curation, Z.R. and W.L.; writing—original draft preparation, Z.R.; writing—review and editing, Z.R. and W.L.; visualization, Z.R. All authors have read and agreed to the published version of the manuscript.

Funding: This research was funded by the Science and Technology Development Program of Yangquan Coal Industry (Group) Co. Ltd., Grant No. YM12066 and the National Natural Science Foundation of China (51974324).

Conflicts of Interest: The authors declare no conflict of interest.

References

1. Li, J.; Wang, J. Comprehensive utilization and environmental risks of coal gangue: A review. *J. Clean. Prod.* **2019**, *239*, 117946. [CrossRef]

2. Li, M.S. Ecological restoration of mineland with particular reference to the metalliferous mine wasteland in China: A review of research and practice. *Sci. Total Environ.* **2006**, *357*, 38–53. [CrossRef]

3. Zhang, Y.; Ling, T. Reactivity activation of waste coal gangue and its impact on the properties of cement-based materials—A review. *Constr. Build. Mater.* **2020**, *234*, 117424. [CrossRef]

4. Pone, J.D.N.; Hein, K.A.A.; Stracher, G.B.; Annegarn, H.J.; Finkleman, R.B.; Blake, D.R.; McCormack, J.K.; Schroeder, P. The spontaneous combustion of coal and its by-products in the Witbank and Sasolburg coalfields of South Africa. *Int. J. Coal Geol.* **2007**, *72*, 124–140. [CrossRef]

5. Querol, X.; Izquierdo, M.; Monfort, E.; Alvarez, E.; Font, O.; Moreno, T.; Alastuey, A.; Zhuang, X.; Lu, W.; Wang, Y. Environmental characterization of burnt coal gangue banks at Yangquan, Shanxi Province, China. *Int. J. Coal Geol.* **2008**, *75*, 93–104. [CrossRef]

6. Zhai, X.; Wu, S.; Wang, K.; Drebenstedt, C.; Zhao, J. Environment influences and extinguish technology of spontaneous combustion of coal gangue heap of Baijigou coal mine in China. *Energy Procedia* **2017**, *136*, 66–72. [CrossRef]

7. Wang, H.; Cheng, C.; Chen, C. Characteristics of polycyclic aromatic hydrocarbon release during spontaneous combustion of coal and gangue in the same coal seam. *J. Loss Prev. Proc.* **2018**, *55*, 392–399. [CrossRef]

8. Fu, T.; Wu, Y.; Ou, L.; Yang, G.; Liang, T. Effects of thin covers on the release of coal gangue contaminants. *Energy Procedia* **2012**, *16*, 327–333. [CrossRef]

9. Baker, B.J.; Banfield, J.F. Microbial communities in acid mine drainage. *FEMS Microbiol. Ecol.* **2003**, *44*, 139–152. [CrossRef]

10. Campbell, R.N.; Lindsay, P.; Clemens, A.H. Acid generating potential of waste rock and coal ash in New Zealand coal mines. *Int. J. Coal Geol.* **2001**, *45*, 163–179. [CrossRef]

11. Tabelin, C.B.; Veerawattananun, S.; Ito, M.; Hiroyoshi, N.; Igarashi, T. Pyrite oxidation in the presence of hematite and alumina: I. Batch leaching experiments and kinetic modeling calculations. *Sci. Total Environ.* **2017**, *580*, 687–698. [CrossRef]

12. Yu, J.; Heo, B.; Choi, I.; Cho, J.; Chang, H. Apparent solubilities of schwertmannite and ferrihydrite in natural stream waters polluted by mine drainage. *Geochim. Cosmochim. Acta* **1999**, *63*, 3407–3416. [CrossRef]

13. Acharya, B.S.; Kharel, G. Acid mine drainage from coal mining in the United States—An overview. *J. Hydrol.* **2020**, *588*, 125061. [CrossRef]

14. Tabelin, C.B.; Corpuz, R.D.; Igarashi, T.; Villacorte-Tabelin, M.; Alorro, R.D.; Yoo, K.; Raval, S.; Ito, M.; Hiroyoshi, N. Acid mine drainage formation and arsenic mobility under strongly acidic conditions: Importance of soluble phases, iron oxyhydroxides/oxides and nature of oxidation layer on pyrite. *J. Hazard. Mater.* **2020**, *399*, 122844. [CrossRef] [PubMed]

15. Kaksonen, A.H.; Puhakka, J.A. Sulfate reduction based bioprocesses for the treatment of acid mine drainage and the recovery of metals. *Eng. Life Sci.* **2007**, *7*, 541–564. [CrossRef]

16. Taylor, J.S.P.; Murphy, N. A Summary of Passive and Active Treatment Technologies for Acid and Metalliferous Drainage (AMD). In Proceedings of the 5th Australian Workshop on Acid Drainage, Fremantle, Australia, 29–31 August 2005.

17. Johnson, D.B.; Hallberg, K.B. Acid mine drainage remediation options: A review. *Sci. Total Environ.* **2005**, *338*, 3–14. [CrossRef]

18. Liu, B.; Tang, Z.; Dong, S.; Wang, L.; Liu, D. Vegetation recovery and groundwater pollution control of coal gangue field in a semi-arid area for a field application. *Int. Biodeter. Biodegr.* **2018**, *128*, 134–140. [CrossRef]

19. Tang, Y.; Wang, H. Development of a novel bentonite–acrylamide superabsorbent hydrogel for extinguishing gangue fire hazard. *Powder Technol.* **2018**, *323*, 486–494. [CrossRef]

20. Wu, Y.; Yu, X.; Hu, S.; Shao, H.; Liao, Q.; Fan, Y. Experimental study of the effects of stacking modes on the spontaneous combustion of coal gangue. *Process. Saf. Evniron.* **2019**, *123*, 39–47. [CrossRef]

21. Demers, I.; Mbonimpa, M.; Benzaazoua, M.; Bouda, M.; Awoh, S.; Lortie, S.; Gagnon, M. Use of acid mine drainage treatment sludge by combination with a natural soil as an oxygen barrier cover for mine waste reclamation: Laboratory column tests and intermediate scale field tests. *Miner. Eng.* **2017**, *107*, 43–52. [CrossRef]

22. Igarashi, T.; Herrera, P.S.; Uchiyama, H.; Miyamae, H.; Iyatomi, N.; Hashimoto, K.; Tabelin, C.B. The two-step neutralization ferrite-formation process for sustainable acid mine drainage treatment: Removal of copper, zinc and arsenic, and the influence of coexisting ions on ferritization. *Sci. Total Evniron.* **2020**, *715*, 136877. [CrossRef]

23. Wang, X.; Jiang, H.; Fang, D.; Liang, J.; Zhou, L. A novel approach to rapidly purify acid mine drainage through chemically forming schwertmannite followed by lime neutralization. *Water Res.* **2019**, *151*, 515–522. [CrossRef]

24. Hong, S.; Cannon, F.S.; Hou, P.; Byrne, T.; Nieto-Delgado, C. Adsorptive removal of sulfate from acid mine drainage by polypyrrole modified activated carbons: Effects of polypyrrole deposition protocols and activated carbon source. *Chemosphere* **2017**, *184*, 429–437. [CrossRef]

25. Kaur, G.; Couperthwaite, S.J.; Millar, G.J. Performance of bauxite refinery residues for treating acid mine drainage. *J. Water Process. Eng.* **2018**, *26*, 28–37. [CrossRef]

26. Kefeni, K.K.; Msagati, T.A.M.; Nkambule, T.T.I.; Mamba, B.B. Synthesis and application of hematite nanoparticles for acid mine drainage treatment. *J. Evniron. Chem. Eng.* **2018**, *6*, 1865–1874. [CrossRef]

27. Lee, G.; Cui, M.; Yoon, Y.; Khim, J.; Jang, M. Passive treatment of arsenic and heavy metals contaminated circumneutral mine drainage using granular polyurethane impregnated by coal mine drainage sludge. *J. Clean. Prod.* **2018**, *186*, 282–292. [CrossRef]

28. Masindi, V.; Gitari, M.W.; Tutu, H.; DeBeer, M. Efficiency of ball milled South African bentonite clay for remediation of acid mine drainage. *J. Water Process Eng.* **2015**, *8*, 227–240. [CrossRef]

29. Masindi, V.; Madzivire, G.; Tekere, M. Reclamation of water and the synthesis of gypsum and limestone from acid mine drainage treatment process using a combination of pre-treated magnesite nanosheets, lime, and CO_2 bubbling. *Water Resour. Ind.* **2018**, *20*, 1–14. [CrossRef]

30. Núñez-Gómez, D.; Rodrigues, C.; Lapolli, F.R.; Lobo-Recio, M.Á. Adsorption of heavy metals from coal acid mine drainage by shrimp shell waste: Isotherm and continuous-flow studies. *J. Environ. Chem. Eng.* **2019**, *7*, 102787. [CrossRef]

31. Gibert, O.; de Pablo, J.; Cortina, J.L.; Ayora, C. Treatment of acid mine drainage by sulphate-reducing bacteria using permeable reactive barriers: A review from laboratory to full-scale experiments. *Rev. Environ. Sci. Biotechnol.* **2002**, *1*, 327–333. [CrossRef]

32. Whitehead, P.G.; Prior, H. Bioremediation of acid mine drainage: An introduction to the Wheal Jane wetlands project. *Sci. Total Evniron.* **2005**, *338*, 15–21. [CrossRef]

33. Sheoran, A.S.; Sheoran, V. Heavy metal removal mechanism of acid mine drainage in wetlands: A critical review. *Miner. Eng.* **2006**, *19*, 105–116. [CrossRef]

34. Naidu, G.; Ryu, S.; Thiruvenkatachari, R.; Choi, Y.; Jeong, S.; Vigneswaran, S. A critical review on remediation, reuse, and resource recovery from acid mine drainage. *Evniron. Pollut.* **2019**, *247*, 1110–1124. [CrossRef]

35. Wadekar, S.S.; Hayes, T.; Lokare, O.R.; Mittal, D.; Vidic, R.D. Laboratory and Pilot-Scale Nanofiltration Treatment of Abandoned Mine Drainage for the Recovery of Products Suitable for Industrial Reuse. *Ind. Eng. Chem. Res.* **2017**, *56*, 7355–7364. [CrossRef]

36. Doye, I.; Duchesne, J. Neutralisation of acid mine drainage with alkaline industrial residues: Laboratory investigation using batch-leaching tests. *Appl. Geochem.* **2003**, *18*, 1197–1213. [CrossRef]

37. Bois, D.; Poirier, P.; Benzaazoua, M.; Bussière, B. A feasibility study on the use of desulphurized tailings to control acid mine drainage. *CIM Bull.* **2004**, *98*, 361–380.

38. Kazadi, M.C.; Harrison, S.T.L.; Franzidis, J.P.; Broadhurst, J.L. Mitigating Acid Rock Drainage Risks while Recovering Low-Sulfur Coal from Ultrafine Colliery Wastes Using Froth Flotation. *Miner. Eng.* **2012**, *29*, 13–21. [CrossRef]

39. Jha, R.K.T.; Satur, J.; Hiroyoshi, N.; Ito, M.; Tsunekawa, M. Suppression of floatability of pyrite in coal processing by carrier microencapsulation. *Fuel Process. Technol.* **2011**, *92*, 1032–1036. [CrossRef]

40. Yaşar, Ö.; Uslu, T.; Şahinoğlu, E. Fine coal recovery from washery tailings in Turkey by oil agglomeration. *Powder Technol.* **2018**, *327*, 29–42. [CrossRef]

41. Sahinoglu, E.; Uslu, T. Effect of particle size on cleaning of high-sulphur fine coal by oil agglomeration. *Fuel Process Technol.* **2014**, *128*, 211–219. [CrossRef]

42. Beauchemin, S.; Clemente, J.S.; Thibault, Y.; Langley, S.; Gregorich, E.G.; Tisch, B. Geochemical stability of acid-generating pyrrhotite tailings 4 to 5 years after addition of oxygen-consuming organic covers. *Sci. Total Evniron.* **2018**, *645*, 1643–1655. [CrossRef]

43. Querol, X.; Zhuang, X.; Font, O.; Izquierdo, M.; Alastuey, A.; Castro, I.; van Drooge, B.L.; Moreno, T.; Grimalt, J.O.; Elvira, J.; et al. Influence of soil cover on reducing the environmental impact of spontaneous coal combustion in coal waste gobs: A review and new experimental data. *Int. J. Coal Geol.* **2011**, *85*, 2–22. [CrossRef]

44. Kotsiopoulos, A.; Harrison, S.T.L. Application of fine desulfurised coal tailings as neutralising barriers in the prevention of acid rock drainage. *Hydrometallurgy* **2017**, *168*, 159–166. [CrossRef]

45. Li, X.; Gao, M.; Hiroyoshi, N.; Tabelin, C.B.; Taketsugu, T.; Ito, M. Suppression of pyrite oxidation by ferric-catecholate complexes: An electrochemical study. *Miner. Eng.* **2019**, *138*, 226–237. [CrossRef]

46. Li, X.; Hiroyoshi, N.; Tabelin, C.B.; Naruwa, K.; Harada, C.; Ito, M. Suppressive effects of ferric-catecholate complexes on pyrite oxidation. *Chemosphere* **2019**, *214*, 70–78. [CrossRef]

47. Lottermoser, B.G. *Mine Wastes: Characterization, Treatment and Environmental Impacts*; Springer: Berlin/Heidelberg, Germany, 2003.

48. Hu, Z.; Zhang, M.; Ma, B.; Wang, P.; Kang, J. Fly ash for control pollution of acid and heavy metals from coal refuse. *Meitan Xuebao/J. China Coal Soc.* **2009**, *34*, 79–83.

49. Mylona, E.; Xenidis, A.; Paspaliaris, I. Inhibition of acid generation from sulphidic wastes by the addition of small amounts of limestone. *Miner. Eng.* **2000**, *13*, 1161–1175. [CrossRef]

50. Zhang, M.; Wang, H. Utilization of Bactericide Technology for Pollution Control of Acidic Coal Mine Waste. In Proceedings of the 2017 6th International Conference on Energy, Environment and Sustainable Development, Zhuhai, China, 11–12 March 2017.

51. Park, I.; Tabelin, C.B.; Jeon, S.; Li, X.; Seno, K.; Ito, M.; Hiroyoshi, N. A review of recent strategies for acid mine drainage prevention and mine tailings recycling. *Chemosphere* **2019**, *219*, 588–606. [CrossRef]

52. Paramguru, R.K.; Rath, P.C.; Misra, V.N. Trends in red mud utilization—A review. *Min. Proc. Ext. Met. Rev.* **2005**, *26*, 1–29. [CrossRef]

53. Zhu, X.; Li, W.; Guan, X. An active dealkalization of red mud with roasting and water leaching. *J. Hazard. Mater.* **2015**, *286*, 85–91. [CrossRef]

54. Jones, B.E.H.; Haynes, R.J.; Phillips, I.R. Addition of an organic amendment and/or residue mud to bauxite residue sand in order to improve its properties as a growth medium. *J. Evniron. Manag.* **2012**, *95*, 29–38. [CrossRef]

55. Evans, K. The History, Challenges, and New Developments in the Management and Use of Bauxite Residue. *J. Sustain. Metall.* **2016**, *2*, 316–331. [CrossRef]

56. Khairul, M.A.; Zanganeh, J.; Moghtaderi, B. The composition, recycling and utilisation of Bayer red mud. *Resour. Conserv. Recycl.* **2019**, *141*, 483–498. [CrossRef]

57. Jones, B.E.H.; Haynes, R.J. Bauxite processing residue: A critical review of its formation, properties, storage, and revegetation. *Crit. Rev. Environ. Sci. Technol.* **2011**, *44*, 271–315. [CrossRef]

58. Li, X.; Ye, Y.; Xue, S.; Jiang, J.; Wu, C.; Kong, X.; Hartley, W.; LI, Y. Leaching optimization and dissolution behavior of alkaline anions in bauxite residue. *Trans. Nonferrous Met. Soc.* **2018**, *28*, 1248–1255. [CrossRef]

59. Lyu, F.; Hu, Y.; Wang, L.; Sun, W. Dealkalization processes of bauxite residue: A comprehensive review. *J. Hazard. Mater.* **2021**, *403*, 123671. [CrossRef]

60. Wu, Y.; Li, M.; Zhu, F.; Hartley, W.; Liao, J.; An, W.; Xue, S.; Jiang, J. Variation on leaching behavior of caustic compounds in bauxite residue during dealkalization process. *J. Evniron. Sci.-China* **2020**, *92*, 141–150. [CrossRef]

61. Xue, S.; Kong, X.; Zhu, F.; Hartley, W.; Li, X.; Li, Y. Proposal for management and alkalinity transformation of bauxite residue in China. *Evniron. Sci. Pollut. Res.* **2016**, *23*, 12822–12834. [CrossRef]

62. Wang, W.; Pranolo, Y.; Cheng, C.Y. Recovery of scandium from synthetic red mud leach solutions by solvent extraction with D2EHPA. *Sep. Purif. Technol.* **2013**, *108*, 96–102. [CrossRef]

63. Zhang, J.; Li, S.; Li, Z.; Liu, C.; Gao, Y.; Qi, Y. Properties of red mud blended with magnesium phosphate cement paste: Feasibility of grouting material preparation. *Constr. Build. Mater.* **2020**, *260*, 119704. [CrossRef]

64. Yang, Z.; Zhang, Y.; Liu, L.; Wang, X.; Zhang, Z. Environmental investigation on co-combustion of sewage sludge and coal gangue: SO_2, NO_x and trace elements emissions. *Waste Manag.* **2016**, *50*, 213–221. [CrossRef]

65. Sobek, A.A.; Shuller, W.A.; Freeman, J.R.; Smith, R.M. *Field and Laboratory Methods Applicable to Overburdens and Minesoils*; EPA 600/2-78-054; U.S. Environmental Protection Agency: Washington, DC, USA, 1978.

66. Hu, G.; Lyu, F.; Khoso, S.A.; Zeng, H.; Sun, W.; Tang, H.; Wang, L. Staged leaching behavior of red mud during dealkalization with mild acid. *Hydrometallurgy* **2020**, *196*, 105422. [CrossRef]

67. Alam, S.; Das, S.K.; Rao, B.H. Characterization of coarse fraction of red mud as a civil engineering construction material. *J. Clean. Prod.* **2017**, *168*, 679–691. [CrossRef]

68. Nicholson, R.V.; Gillham, R.W.; Reardon, E.J. Pyrite oxidation in carbonate-buffered solution: 1. Experimental kinetics. *Geochim. Cosmochim. Acta* **1988**, *52*, 1077–1085. [CrossRef]

69. Wang, H.; Dowd, P.A.; Xu, C. A reaction rate model for pyrite oxidation considering the influence of water content and temperature. *Miner. Eng.* **2019**, *134*, 345–355. [CrossRef]

70. Chandra, A.P.; Gerson, A.R. The mechanisms of pyrite oxidation and leaching: A fundamental perspective. *Surf. Sci. Rep.* **2010**, *65*, 293–315. [CrossRef]

71. Long, H.; Dixon, D.G. Pressure oxidation of pyrite in sulfuric acid media: A kinetic study. *Hydrometallurgy* **2004**, *73*, 335–349. [CrossRef]

72. Hu, G.; Dam-Johansen, K.; Wedel, S.; Hansen, J.P. Decomposition and oxidation of pyrite. *Prog. Energy Combust. Sci.* **2006**, *32*, 295–314. [CrossRef]

73. Kishida, M.; Harato, T.; Tokoro, C.; Owada, S. In situ remediation of bauxite residue by sulfuric acid leaching and bipolar-membrane electrodialysis. *Hydrometallurgy* **2017**, *170*, 58–67. [CrossRef]

74. McCleskey, R.B.; Nordstrom, D.K.; Ryan, J.N. Comparison of electrical conductivity calculation methods for natural waters. *Limnol. Oceanogr. Methods* **2012**, *10*, 952–967. [CrossRef]

Publisher's Note: MDPI stays neutral with regard to jurisdictional claims in published maps and institutional affiliations.

Article

Evaluation of Efficiencies of Locally Available Neutralizing Agents for Passive Treatment of Acid Mine Drainage

Casey Oliver A. Turingan [1],*, Giulio B. Singson [1], Bernadette T. Melchor [1], Richard D. Alorro [2], Arnel B. Beltran [1] and Aileen H. Orbecido [1]

[1] Chemical Engineering Department, De La Salle University, Manila 1004, Philippines; giulio_singson@dlsu.edu.ph (G.B.S.); bernadette_melchor@dlsu.edu.ph (B.T.M.); arnel.beltran@dlsu.edu.ph (A.B.B.); aileen.orbecido@dlsu.edu.ph (A.H.O.)

[2] Western Australia School of Mines: Minerals, Energy and Chemical Engineering, Curtin University, Kalgoorlie Campus, Kalgoorlie, WA 6430, Australia; richard.alorro@curtin.edu.au

* Correspondence: casey_oliver_turingan@dlsu.edu.ph

Received: 11 August 2020; Accepted: 22 September 2020; Published: 24 September 2020

Abstract: Acid mine drainage (AMD) generated from the mining industry elevates environmental concerns due to the pollution and contamination it causes to bodies of water. Over the years, passive treatment of AMD using alkalinity-generating materials have been widely studied with pH neutralization as its commonly observed mechanism. During the treatment process, heavy metal removal is also promoted by precipitation due to pH change or through adsorption facilitated by the mineral component of the materials. In this study, four materials were used and investigated: (1) a low grade ore (LGO) made up of goethite, calcium oxide, and manganese aluminum oxide (2–3) limestone and concrete aggregates (CA) composed of calcite, and (4) fly ash consisting of quartz, hematite, and magnetite. The performance of each alkalinity-generating agent at varying AMD/media ratios was based on the change in pH, total dissolved solids (TDS), oxidation reduction potential (E_H); and heavy metals (Fe, Ni, and Al) removal and sulfate concentration reduction. Concrete aggregate displayed the most significant effect in treating AMD after raising the pH to 12.42 and removing 99% Fe, 99% Ni, 96% Al, and 57% sulfates. Afterwards, the efficiency of CA at various particle sizes were evaluated over 1 h. The smallest range at 2.00–3.35mm was observed to be most effective after 60 min, raising the pH to 6.78 and reducing 94% Fe, 78% Ni, and 92% Al, but only 28% sulfates. Larger particles of CA were able to remove higher amounts of sulfate up to 57%, similar to the jar test. Overall, CA is an effective treatment media for neutralization; however, its performance can be complemented by a second media for heavy metal and sulfate removal.

Keywords: iron hydroxide; iron oxyhydroxide; acid mine drainage; nickel ore; fly ash; concrete aggregate; calcite; goethite; hematite; magnetite; quartz

1. Introduction

Acid mine drainage (AMD) is known as one of the worst environmental problems worldwide related to mining, mineral processing operations, and other large-scale excavations [1,2]. It is characterized by low pH and high concentrations of sulfate, heavy metals, and other toxic elements, which cause negative effects to the surrounding areas [3,4]. In the United States, there is an estimated 20,000 to 50,000 mines capable of impacting up to 10,000 miles of water bodies, since 90% of the mines are abandoned [5]. Other countries with a large number of mines include Australia (52,534), Canada (10,139), South Africa (>6000), Japan (5487), China (>5000), the UK (>2000), and South Korea (1692) [6]. In the Philippines, the mining industry has significantly contributed to the economy with an estimated \$840 billion worth of mineral wealth that had not been extracted as of 2012 [7]. In 2019, the industry contributed 0.6% to GDP and 6% of the total exports, which amounted to approximately \$2 billion and \$4.3 billion, respectively [8]. However, generation of AMD from several AMD-generating mine sites has posed a detrimental effect on the environment [4].

One of the main and most common minerals responsible for AMD is pyrite (FeS_2) [9]. It is typically found in sediments, ore deposits, mineralized veins, hydrothermally altered rocks, and soils [3,10]. As mining activities expose pyrite and other sulfide-bearing ores to oxygen and water, AMD production accelerates [11]. Pyrite oxidation is facilitated by various reactions that occur in different situations as shown in Equations (1)–(6) [9,12].

$$FeS_2 + 15/2O_2 + 7/2H_2O \rightarrow Fe(OH)_3 + 2SO_4^{2-} + 4H^+ \tag{1}$$

$$FeS_2 + 14Fe^{3+} + 8H_2O \rightarrow 15Fe^{2+} + 2SO_4^{2-} + 16H^+ \tag{2}$$

$$4Fe^{2+} + O_{2\,(aq)} + 4H^+ \rightarrow 4Fe^{3+} + 2H_2O \tag{3}$$

$$FeS_2 + 8H_2O \rightarrow Fe^{2+} + 2SO_4^{2-} + 16H^+ + 14e^- \tag{4}$$

$$O_{2\,(aq)} + 4H^+ + 4e^- \rightarrow 2H_2O \tag{5}$$

$$Fe^{3+} + e^- \rightarrow Fe^{2+} \tag{6}$$

$$FeS_2 + 15/4O_2 + 1/2H_2O \rightarrow Fe^{3+} + 2SO_4^{2-} + H^+ \tag{7}$$

Equation (1) describes pyrite oxidation when exposed to oxygen and excess water at a neutral pH. On the other hand, Equation (2) utilizes Fe^{3+} as the oxidizing agent with excess water and proceeds at a faster rate compared to that of Equation (1) [13]. In this case, Fe^{3+} is generated by Equation (3) or through iron-oxidizing bacteria [12]. Equation (4) shows the electrochemical reaction which occurs in aqueous systems. Dissolved oxygen and Fe^{3+} are reduced in their respective cathodic sites through the electrons (Equations (5) and (6)) transferred from the anodic sites of the pyrite crystal until its dissociation [14]. Equation (7) represents the complete oxidation of pyrite in low water content. This reaction is accompanied by the odd formation of goethite (α-FeOOH) or schwertmannite ($Fe_8O_8(OH)_6 \cdot nH_2O$) [15]. As the oxidation process occurs, acidic water interacts with rocks containing various minerals and dissolves toxic metals along the way. Fe, Al, sulfate, and bicarbonate are the most common ions dissolved in AMD [2]. Moreover, pyrite alone is known to be incorporated with As, Pb, Cu, Zn, Se, and B, depending on its source [3].

The pollution caused by AMD in bodies of water poses a serious threat to surrounding flora and fauna [4]. Thus, proper treatment of acid mine drainage is necessary during and after operations. Generally, not all mine sites have the same AMD potential, and thus different active and passive treatments are employed in treating produced AMD from various mine sites [11,16,17]. Active treatments include using alkaline metals to precipitate metals, adsorption, ion exchange, and membrane technology, among others [9]. These methods are typically used to treat AMD with very high levels of acidity while being capable of adjusting to the varying geochemical properties. However, active treatment is limited by its cost and sludge generation, making it unsustainable in the long run [6]. In comparison, passive treatments are considered to be more cost-effective to use in a closed and abandoned mine site due to the stable chemistry at these mine sites as well as the accessible land for remediation systems [16,18]. As a result of several closed and abandoned mine sites in the Philippines, passive treatment is employed in this study [19].

In the passive treatment of AMD, some of the materials used that generate alkalinity are limestone ($CaCO_3$), lime (CaO), hydrated lime ($Ca(OH)_2$), pebble quicklime (CaO), periclase (MgO), magnesite ($MgCO_3$), brucite ($Mg(OH)_2$), dolomite ($MgCO_3 \cdot CaCO_3$), fly ash, soda ash (Na_2CO_3), caustic soda ($NaOH$), and ammonia (NH_3) [16,20,21]. AMD is treated through pH neutralization as the materials generate alkalinity through the reaction of oxide minerals present. Limestone is the main material in some passive treatment systems such as open limestone channels (OLCs), open limestone drains (OLDs), and limestone leaching beds (LLBs) [22–24]. However, a major drawback encountered in these systems is the armoring with hydroxides when heavy metal concentrations, specifically Fe, are too high [25]. In addition, these systems are unable to treat AMD with pH < 2 and flows greater than 50 L/s. Moreover, effluents would only reach up to pH 8, rendering the treatment ineffective against some heavy metals with higher solubilities [6]. In addition, waste by-products were also used as an alkalinity-generating material. These waste by-products are eggshell waste [26], wood ash [27], phosphatic waste rock [28], and concrete aggregate [29]. Another material used in recent studies was serpentinite ($(Mg,Fe,Ni,Al,Zn,Mn)_{2-3}(Si,Al,Fe)_2O_5(OH)_4$) found in mining waste rock that belongs to the serpentine group of minerals considered as alkaline-rich [30,31]. After a series of batch experiments, serpentinite was able to raise the pH of the AMD to 4.5, 5.5, and 6.5–8.5 for the first, second, and third series of experiments, respectively [32]. In terms of metal removal, the mineral was able to remove 99.95–100% Fe, 85.7–98.9% Al, and 27.3–28.8% SO_4 from the synthetic acid mine drainage.

In the Philippines, nickel mines are ubiquitous and account for 29 out of 104 operational mines in the country [8]. However, only those containing >1.4% Ni are eligible for exportation, with the remaining material considered as low-grade ore (LGO). This is combined with deposits that have relatively high purity to meet the standard; otherwise, the LGO will be stockpiled in mine sites. The stock would then act as a source of pollution when its runoff flows to nearby streams. In order to repurpose LGO, it was characterized and its potential to treat AMD was evaluated. Goethite was identified as a major component. Moreover, LGO was able to remove >99% Fe and Al, 94% Ni, and 93% of sulfate despite only raising pH from 2.21 to 5.36 [33]. This shows the potential of LGO to complement an alkaline-generating material in treating AMD to make up permeable reactive barriers. Permeable reactive barriers are an example of an in situ passive treatment wherein the contaminants are immobilized via adsorption, precipitation process and ion exchange. Some examples of these are iron oxyhydroxides (FeOOH), iron oxides (Fe_xO_y), and carbonates (CO_3^{2-}). Iron oxyhydroxides naturally occurs as the following minerals: α-FeOOH (goethite), β-FeOOH (akageneite), γ-FeOOH (lepidocrocite), and δ-FeOOH (feroxyhyte). Among the polymorphs of FeOOH, its most common form occurs as goethite and lepidocrocite. Past studies made use of these minerals in the adsorption of cadmium (Cd), copper (Cu), zinc (Zn), lead (Pb), selenium (Se), and chromate (CrO_4^{2-}) [34–36].

Even though limestone has been widely studied, its limitations generated from previous studies due to its low solubility above pH 7 and armoring have led the researchers of this study to look for alternative alkalinity-generating materials [2]. Concrete aggregate, fly ash, and low-grade nickel ore are waste materials in abundant supply locally and are considered to be excellent alternatives. Concrete aggregate makes up as much as 50% of municipal solid waste while 1.4 MMT of fly ash is generated annually [37,38]. In addition, these materials have varying composition compared to those used in previous studies, and thus, performance in AMD treatment would also differ. Hence, this study is conducted to evaluate the efficiencies of low-grade nickel ore, fly ash, and concrete aggregate in comparison with limestone in passive treatment of AMD. The treatment efficiencies were evaluated in terms of the geochemical properties and the heavy metal and sulfate concentration of the treated AMD.

2. Materials and Methods

2.1. Materials and Reagents

Low-grade nickel ore (LGO), limestone, fly ash, and concrete aggregates were the four locally available neutralizing agents compared based on their efficiencies in treating acid mine drainage (AMD). LGO and limestone were both collected from a mine site in Mindanao, Philippines. Fly ash was obtained from a coal-fired power plant situated in Bataan, Philippines; while concrete aggregates were collected from a local construction site in Manila, Philippines. Concrete aggregate and limestone were crushed and sieved through a mesh size 20 (<841 μm) to obtain their powder form. LGO and fly ash were left as they were since particle size analysis was not possible for these during the study.

2.2. Synthetic Acid Mine Drainage Preparation

The treatment efficiencies of the neutralizing agents were evaluated with a synthetic solution of acid mine drainage (AMD) based on Bernier (2005) [32]. The synthetic AMD was prepared with analytical grade reagents diluted to 7 L of distilled water, as shown in Table 1. Then 1.5 M sulfuric acid was carefully added dropwise to adjust the acidity of the synthetic solution. The geochemical properties of the synthetic solution, including the metals (Fe, Ni, and Al) and sulfates content are immediately measured, which are summarized in Table 2.

Table 1. Reagents needed for synthetic acid mine drainage (AMD) preparation.

Reagents	Mass (g)
$FeSO_4 \cdot 7H_2O$	65.46
$Al_2(SO_4)_3 \cdot 18H_2O$	17.28
$NiSO_4 \cdot 6H_2O$	9.39
$CuSO_4 \cdot 5H_2O$	1.4
$MnSO_4 \cdot H_2O$	1.4
$MgSO_4 \cdot 7H_2O$	1.4
1.5 M H_2SO_4	60 mL

Table 2. Geochemical properties and metal content of synthetic AMD.

Parameter	Measurement
pH	2.44
Oxidation-Reduction Potential (E_H)	453.6 mV
Conductivity	6.98 mS/cm
Total Dissolved Solids (TDS)	3420.5 ppm
Fe	1647.48 ppm
Ni	302.49 ppm
Al	690 ppm
Sulfates	30,350 ppm

2.3. Characterization of Neutralizing Agents

The four types of neutralizing agents were analyzed for their whole-rock chemistry and mineralogy. The chemistry of the materials was determined using X-ray Fluorescence (Horiba Scientific XGT-7200 X-ray Analytical Microscope, HORIBA Ltd., Kyoto, Japan). The samples were dried to 105 °C to remove the free moisture content before undergoing actual XRF procedures. Meanwhile, X-ray Diffraction was employed to analyze their mineralogy using Shimadzu LabX XRD-6100 X-ray Diffractometer (Shimadzu Scientific Instruments, Columbia, MD, USA) at step size = 0.02° over a 5–60° 2θ range. Furthermore, limestone and concrete aggregates underwent mechanical crushing and sieving in order to obtain fine powder particles (<0.841 mm) similar to the other two samples of neutralizing agents (LGO and fly ash).

2.4. Jar Test Using Different Alkaline-Generating Materials

For the first batch of experiment, an overhead stirrer (Lovibond Floc Tester, Tintometer Inc., Dortmund, Germany) was employed to evaluate the efficiency of each neutralizing agents in treating synthetic AMD with various AMD/Media ratios of 0.75, 1.0, 1.5, and 2.0 mL/g. Constant mixing at 180 rpm was done for 1 h. Afterwards, the mixtures settled for 30 min before collecting 50 mL of the supernatant for analysis.

2.5. Batch Test Using Various Particle Sizes

In order to determine the effect of varying particle sizes, concrete aggregate, the most efficient neutralizing agent from the jar test, was utilized. The material underwent crushing and mechanical sieving to obtain particle sizes of 4.75 mm, 3.35 mm, and 2.00 mm. Similarly, the AMD/Media ratio of 0.75 mL/g wherein the four neutralizing agents performed best was used in this set of runs. In each run, enough AMD is poured into the reactor containing concrete aggregate to achieve the desired AMD/media ratio. Samples are obtained at specific time intervals through a spout located at the bottom of the reactor (Figure 1). Analysis of geochemical properties, heavy metals, and sulfates content were done for all treated samples from both batch experiments. Benchtop meters were used to measure geochemical properties of the treated samples, specifically: pH, reduction-oxidation potential (E_H), conductivity, and total dissolved solids (TDS). On the other hand, heavy metals and sulfate content were analyzed through atomic absorption spectrometry (AAS) (Shimadzu AA-6300, Shimadzu Scientific Instruments, Columbia, MD, USA) with the following detection limits: Fe, 0.06–15 ppm; Ni, 0.1–20 ppm; Al, minimum DL = 0.01 ppm; sulfates, minimum DL = 1 ppm.

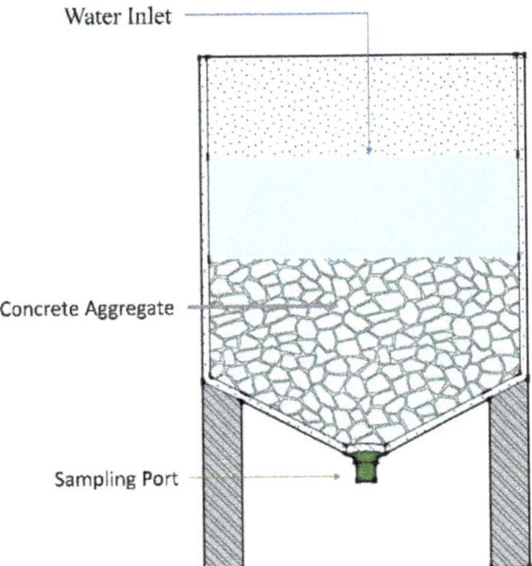

Figure 1. Schematic diagram of the batch test reactor.

2.6. Geochemical Modeling

Geochemical models were developed to complement the experimental results and allow further insight into the processes during the treatment of AMD. PHREEQC Interactive v.3.6.2 (USGS, Reston, VA, USA) [39] was used to calculate the saturation indices of potential precipitates using the THERMODEM database from the BGRM Institute, and the data obtained from AAS analysis. Ca and HCO_3^- were counted in the effluent solution to consider the dissolution of calcite from limestone and concrete aggregates. On the other hand, Ca was included for the fly ash-treated effluent. These ions were incorporated to show the precipitates that may form when AMD is treated with these materials.

3. Results

3.1. Whole Rock Chemistry

The whole rock chemistry of the materials in terms of their oxide composition is presented in Table 3. Furthermore, oxide composition of the concrete aggregate provided was taken from literature [40–42] since no actual analysis was available.

Table 3. Percentage of oxide composition of neutralizing agents.

Element	Material			
	LGO	Limestone	Fly Ash	Concrete Aggregate [40–42]
Al_2O_3	4.19	0.93	8.55	4.71 ± 1.50
CaO	4.56	85.66	16.04	58.57 ± 11.68
Fe_2O_3	54.07	7.59	44.63	4.11 ± 1.88
MgO	7.96	0.02	n/a	1.38 ± 1.01
SiO_2	23.87	1.78	24.65	26.63 ± 9.75
NiO	2.13	0.76	n/a	n/a
Others	3.22	3.26	6.13	4.03 ± 1.87

n/a = metal oxide was not included in the analysis.

Table 3 shows that LGO is mainly comprised of iron oxide in the form of Fe_2O_3 (54.07%), followed by silicon dioxide (23.87%). Other oxides present are aluminum (4.19%), calcium (4.56%), magnesium (7.96%), and nickel (2.13%). Basic oxides in the form of CaO, MgO, and NiO are only less than 10% in total which suggests that LGO would raise the pH but not to a neutral pH. Additionally, SiO_2, which makes up for 23.87% of the sample is considered a very weak acidic oxide and would mean that it could be factor in decreasing the pH even for a little amount.

The limestone used consisted of 85.66% CaO with less than 10% other metal oxides. CaO, MgO, and NiO are among the basic oxides that exist in limestone and would suggest that it would be able to successfully neutralize the AMD. On the other hand, fly ash was mainly composed of Fe_2O_3 (44.63%) and SiO_2 (24.65%), similar to LGO. Although fly ash is comparable to LGO, its performance can be varied as its CaO content is greater. Its SiO_2 content, however, may also limit its capability in neutralizing the solution.

Lastly, numerous studies show that concrete aggregate [40–42] has CaO and SiO_2 as its main components. The presence of CaO in this material suggests that it has the potential to increase pH of acidic solutions, in this case, AMD.

3.2. Mineralogy

Figure 2 shows the presence of goethite in LGO along with calcium oxide and manganese aluminum oxide. Goethite is a form of an iron oxyhydroxide, and in their poorly crystalline form holds higher surface area and smaller site density [43,44]. The presence of higher surface area and smaller site density of iron oxyhydroxides makes its sorption capabilities more extensive [45].

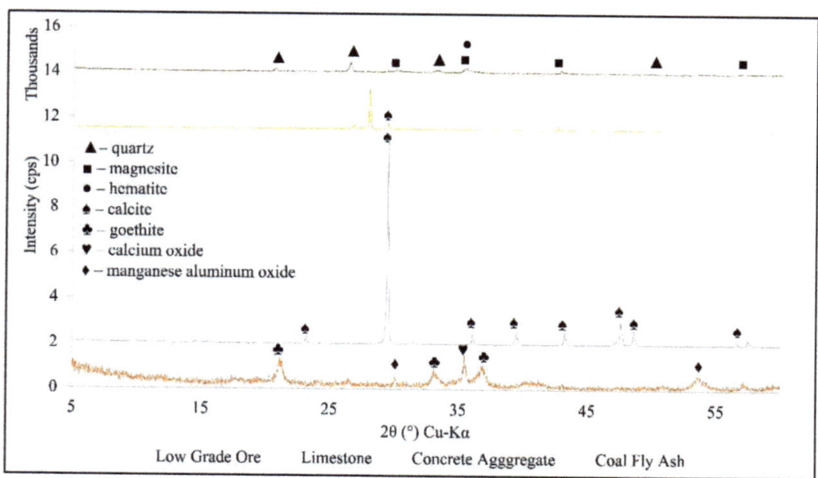

Figure 2. X-ray diffraction of raw low-grade ore, limestone, concrete aggregate, and coal fly ash.

Limestone, on the other hand, is mostly composed of the mineral calcite ($CaCO_3$), which is also the same with the XRD results from past studies [46,47]. Whereas results for fly ash show that quartz, magnetite, and hematite are the minerals present in the sample. Lastly, concrete aggregate had similar result with limestone having calcite on its peaks.

3.3. Jar Test Using Various Neutralizing Agents

The geochemical properties of AMD treated using the four neutralizing agents are shown in Figures 3–6. The highest pH recorded was 11.53 using concrete aggregates at 0.75 mL/g AMD/Media ratio. Furthermore, all four materials had similar trends wherein the highest pH attained was observed at the lowest ratio (0.75 mL/g). Among the four, LGO appeared to have the minimum neutralizing capability as it only raised the pH to 4.66; while limestone was able to raise the pH of the solution to a value of 6.62 whereas fly ash was capable of neutralizing the solution and raising the pH up to 9.34. Finally, the trend for the change in pH, expected to be based on their alkaline component, CaO, was not followed.

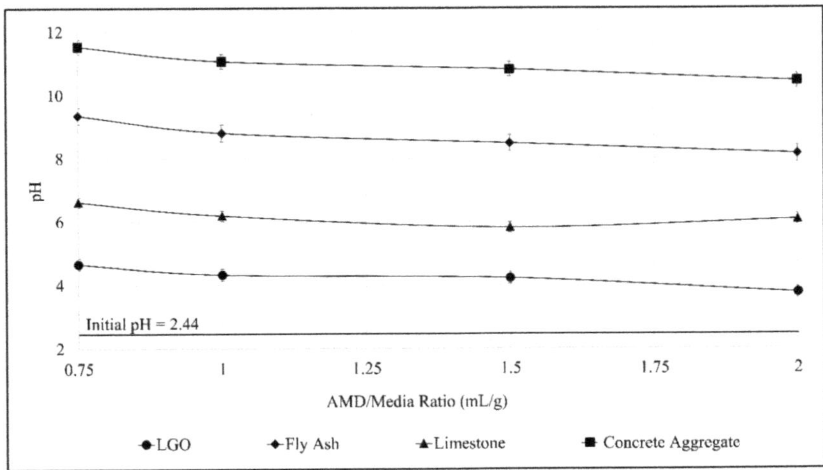

Figure 3. Effect of AMD/Media ratio on pH after 1-h mixing.

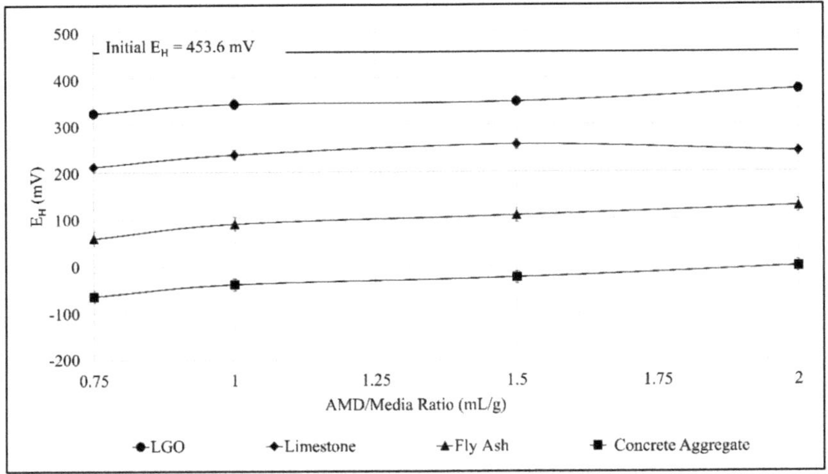

Figure 4. Effect of AMD/Media ratio on E_H after 1-h mixing.

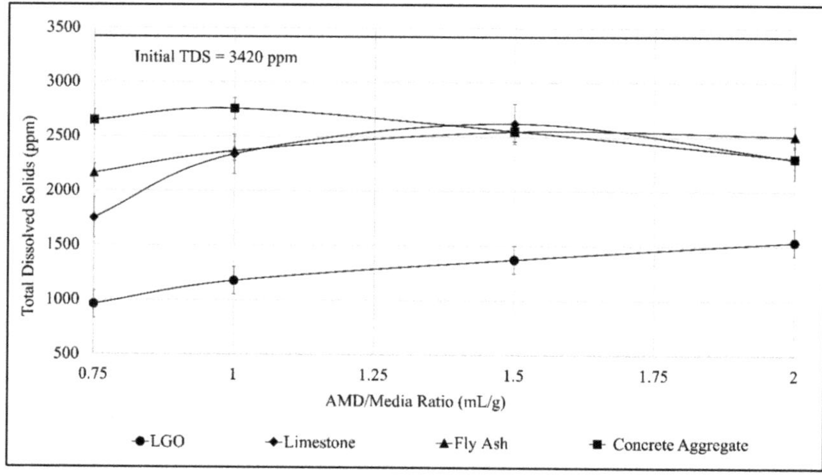

Figure 5. Effect of AMD/Media ratio on total dissolved solids (TDS) after 1-h mixing.

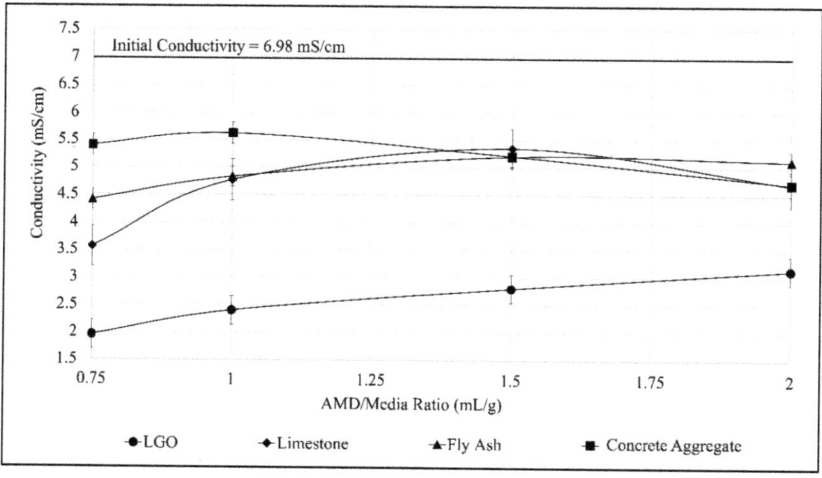

Figure 6. Effect of AMD/Media ratio on conductivity after 1-h mixing.

The initial E_H reading of the synthetic solution was 254 mV. After conducting jar test, the AMD treated with concrete aggregate and fly ash caused the redox potential of the solution to be negative. The AMD treated with LGO still had a high redox potential close to the initial value since the solution was still essentially acidic as it was not able to get neutralized significantly.

The AMD treated using a ratio of 0.75 mL/g had the lowest value of total dissolved solids (TDS) and conductivity for all neutralizing agents. After jar test, LGO-treated AMD at 0.75 mL/g ratio had the lowest TDS and conductivity suggesting that low-grade ore was able to remove the most amount of heavy metals from the synthetic solution. Meanwhile, TDS and conductivity results from the solution treated by limestone had a significant increase when ratio of 1.0 and 1.5 were used, then had a decrease when the ratio of 2.0 was employed. Heavy metal ions were removed through adsorption onto goethite present in LGO while precipitation is the main suspect in the removal of heavy metal ions in the other three materials which resulted to a more alkaline effluent [33,40,48]. In relation to its pH results, ratio 1.5 for limestone also had the lowest pH attained which could suggest that heavy metals in AMD were not able to precipitate as much at that pH level as compared to the other ratios.

The comparison for the metals (Fe, Ni, and Al) and sulfate concentrations of the initial and treated synthetic AMD is shown in Figure 7. The trend of removal was expected wherein the lowest AMD/Media ratio corresponds to the highest removal of metals and sulfates for each material used. For iron, the removal was fairly consistent at 99% at ratio of 0.75 up to 1.5 mL/g for LGO and all ratios using fly ash and concrete aggregate. Meanwhile for nickel, removal was only at most 80% using LGO and 70% for limestone; but it attained almost complete removal when fly ash and concrete aggregate were used with any of the ratios employed. Aluminum also reached complete removal as high as 98% using LGO at 0.75 mL/g, while only achieving 82% at 2.0 mL/g, which was the highest removal for aluminum among the four materials. Nevertheless, out of the four metals and sulfates, iron was removed most efficiently wherein complete removal was almost achieved at >99% for all neutralizing agents.

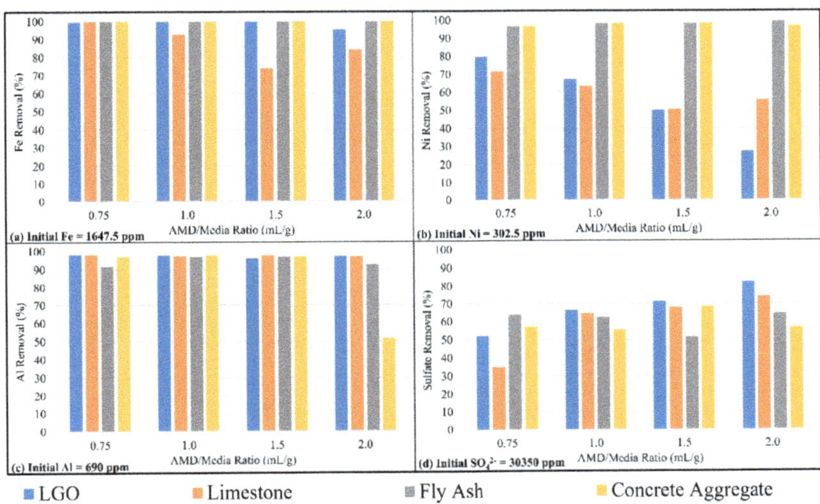

Figure 7. Effect of AMD/Media ratio on the removal of (**a**) iron; (**b**) nickel; (**c**) aluminum; and (**d**) sulfates.

3.4. Effect of Particle Size on AMD Treatment

After the treatment of AMD, the highest pH measured was 6.78 at the 60-min interval of 3.35 mm–2.00 mm size range; whereas pH obtained upon using 4.75 mm–3.35 mm was at 6.42; and pH 6.16 was obtained using 4.75 mm (Figure 8a). The results do not reflect a similar behavior of the concrete aggregate material as seen in the previous experiment. This may be due to the slow dissolution rate of Ca-based compounds which is slower compared to that of other chemicals, and thus there is a need for mechanical mixing to increase its neutralizing capacity [49].

In comparison with the result obtained from using concrete aggregate in the jar test which resulted to 99.64% removal for Fe and 99.69% removal for Ni, percent removals acquired for the batch experiment (Figure 9) have significant difference which may be accounted to the improved optimal surface exposure provided by the constant mechanical mixing, as compared to the undisturbed contact between the concrete aggregate and AMD without mechanical mixing. Additionally, only little amount of Ni was removed from the AMD solution after 1-h treatment. Again, the smallest particle size was still able to remove more metals than compared to the other two sizes as it had the largest total surface area reacting with the synthetic solution.

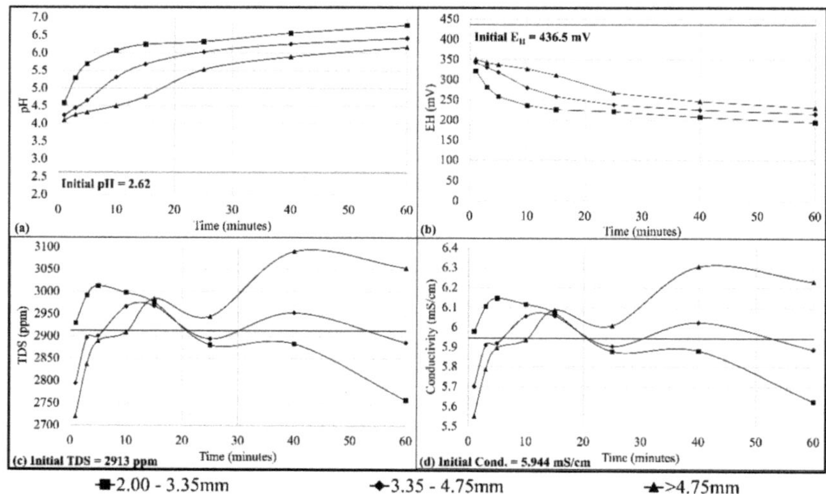

Figure 8. Effect of particle size on (**a**) pH; (**b**) E_H; (**c**) TDS; and (**d**) conductivity after 1-h contact.

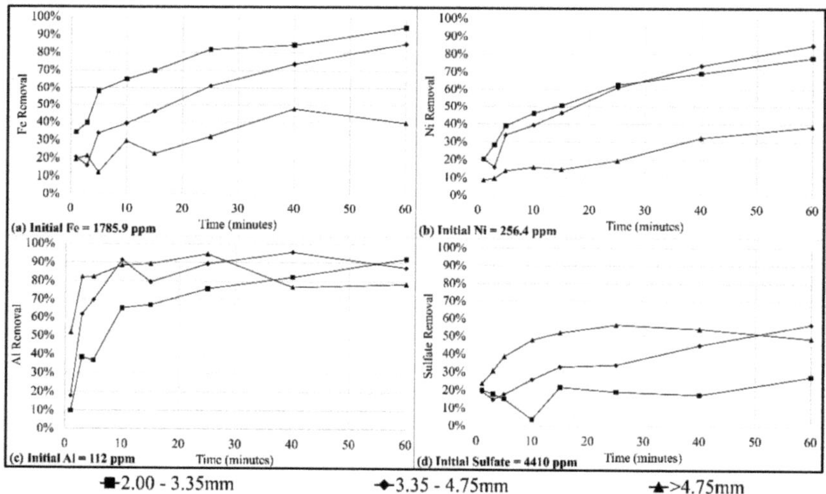

Figure 9. Effect of particle size on the removal of (**a**) iron; (**b**) nickel; (**c**) aluminum; and (**d**) sulfates.

4. Discussion

4.1. Comparison of the Treatment Efficiencies of Various Neutralizing Agents

Low grade ore (LGO) caused a slight change in pH of synthetic AMD from 2.44 to 4.46 at the 0.75 ratio. However, this material caused the largest decrease in total dissolved solids and conductivity, suggesting that it was able to remove the largest amount of metals and sulfates from the solution. Its performance can be attributed to the capabilities of goethite to adsorb metals such as nickel, copper, manganese, and aluminum [50], including sulfates [51]. Such capabilities of goethite hold true as it is often poorly crystalline and rich in impurities that makes it have high specific surface area with good surface activity. Furthermore, the material's surface area and site density are important factors for determining its capabilities as a good adsorbent [50]. Aluminum may not only be adsorbed onto the surface of goethite but also substitute some of the Fe on the surface by as much as 33 mol% to form

Al-FeOOH [52]. This compound has shown a higher adsorption rate of Ni in comparison to goethite through inner-sphere surface complexation [53]. PHREEQC suggests that some precipitates formed, such as anhydrite ($CaSO_4$), gypsum ($CaSO_4 \cdot 2H_2O$), bassanite ($CaSO_4 \cdot 0.5H_2O$), diaspore ($AlO(OH)$), boehmite ($AlO(OH)$), and gibbsite ($Al(OH)_3$), in decreasing saturation indices (SIs). During their formation, Ni and sulfate ions co-precipitate with the aluminum hydroxides, resulting in a higher sulfate removal at a higher ratio when using LGO as treatment media [54].

Limestone was not able to successfully neutralize the synthetic AMD, thus revealing its limitation in neutralization that may be due to its low solubility and armoring. The latter occurs after the initial neutralization process when Fe precipitates on the surface of the limestone, slowing down the rate of its dissolution [21]. Based on geochemical modeling, these are magnetite (Fe_2O_3), goethite ($FeOOH$), lepidocrocite ($FeOOH$), and ferrihydrite ($Fe(OH)_3$), and siderite ($FeCO_3$). Other precipitates were diaspore, boehmite, gibbsite, gypsum, and bassanite. Nonetheless, results obtained from treating AMD with limestone were expected to be and were indeed consistent in comparison to past studies, as it has a neutralizing capability ranging from pH 6.00 to 7.5 [55]. Nickel is still quite soluble at this range, suggesting that its removal may have been mainly due to co-precipitation and cation sorption on the surface of the Fe and Al precipitates [56].

Fly ash performed better in increasing the pH of AMD despite containing 16.04% CaO only. This shows that the CaO component in fly ash is more readily dissolved relative to that of limestone. Other studies have observed a similar change in pH and attributed it to the dissolution of the fly ash [43,57,58]. At the pH level reached using this material, PHREEQC suggested the precipitation of the following: dolomite ($CaMg(CO)_3$), hydrotalcite ($Mg_4Al_2O_7 \cdot 10H_2O$), magnesite, diaspore, boehmite, gibbsite, $NiCO_3$, nesquehonite ($MgCO_3 \cdot 3H_2O$), lansfordite ($MgCO_3 \cdot 5H_2O$), artinite ($Mg_2(OH)_2(CO_3) \cdot 3H_2O$, spinel ($MgAl_2O_4$), gypsum, and anhydrite. Concrete aggregates had the highest effluent pH = 11.75. This allowed the precipitation of Ni compounds, specifically $NiCO_3$ and $Ni(OH)_2$. PHREEQC also showed the formation of ettringite ($Ca_6Al_2(SO_4)_3(OH)_{12} \cdot 26H_2O$), magnetite, goethite, lepidocrocite, ferrihydrite, diaspore, boehmite, gibbsite, gypsum, and anhydrite. Iron compounds had the highest Sis, followed by those containing nickel, then aluminum, and sulfates.

Overall, oxyhydroxides had relatively high SIs followed by hydroxides, gibbsite ($Al(OH)_3$), and ferrihydrite ($Fe(OH)_3$), except for LGO, wherein gypsum had the highest SI. The precipitates from the use of concrete aggregates also followed a different trend, dictated by the metal component rather than the presence of hydroxides and oxyhydroxides.

4.2. Effect of Particle Size on Acid Mine Drainage Treatment

The trend of pH shown in Figure 8a for this batch experiment was based from the amount of material exposed to the AMD in contrast to the jar test which is significantly affected by the amount of alkalinity generating component from each neutralizing agent. The trend was consistent with the size ranges being tested as the highest pH was obtained from the small particle size, which is attributed to the total surface area exposed and reacted with the AMD. However, only minute differences between the obtained pH was observed from each particle size range which may be accounted to the minor size differences concrete aggregate; thus, a more significant variation of sizes may be employed in order to further understand their effects.

As for the removal of Fe, majority was removed using the smallest particle size which attained a pH value of 6.78. Its successful removal can be attributed to its supposed precipitation at a pH range of 3.0–6.0 and would dissolve again at pH 7.0 and above. Since the pH of the AMD had almost reached a value of 7.0, it would suggest that this may have caused the leftover dissolved Fe concentration of around 180 ppm. On the other hand, Al removal is more efficient with larger particles. This may be due to the rate of pH change in the treated AMD. The longer neutralization time needed by larger particles may benefit the formation of Al precipitates, which translates to a higher Al removal from AMD. This trend is also observed for sulfates which may be highly influenced by the speciation of Al over time and similarly affected by the rate of pH change. However, the limited data available from the experiment need to be expanded to verify this claim.

5. Conclusions

This study was conducted to determine the AMD treatment efficiency of low-grade nickel ore, limestone, fly ash, and concrete aggregates as alternative alkaline-generating materials at varying AMD/Media ratio following a 1-h contact time. After which, the effect of particle size was evaluated using the material that displayed the most significant result from the first batch of experiment. The optimal AMD/Media ratio was used for the second batch of experiment while maintaining the residence time to 1 h and varying grain size from 4.75 mm to 2.00 mm.

Results from the first batch of experiment showed that concrete aggregate provided the most significant result among the four neutralizing agents that were used. Although fly ash was able to elevate the pH of AMD to 9.34, AMD treated with concrete aggregates displayed the most improvement based on geochemical properties and heavy metal concentrations. Alternatively, despite a high percentage removal of sulfate (82.22%) obtained using LGO, the treated AMD had a pH of 4.66 which is very far from the acceptable pH range of 6.5–9.0. Limestone, on the other hand, had the least percentage removal of sulfates (32.47%) despite it having a significant percentage removal for heavy metals. It was also not able to neutralize the AMD to at least neutral level.

The second batch of experiment showed that the smallest particle size range of concrete aggregates had the best performance in neutralizing AMD at the optimal pH of 6.78. The differences in pH for each size range were recorded at minimal values due to the minute difference between the size ranges used. Consequently, it is recommended for future studies to use varying particle sizes with significant size differences. In addition to expanding the difference in particle size, methods that can employ mechanical mixing during the treatment can be developed to improve the neutralizing capability of concrete aggregates.

Author Contributions: Conceptualization, A.H.O. and A.B.B.; methodology, A.H.O. and C.O.A.T.; validation, C.O.A.T., A.H.O. and R.D.A.; writing—original draft preparation, G.B.S. and B.T.M.; writing—review and editing, C.O.A.T. and A.H.O.; visualization, G.B.S. and B.T.M.; supervision, A.H.O.; project administration, A.H.O.; and funding acquisition, A.H.O. All authors have read and agreed to the published version of the manuscript.

Funding: This research was funded by the Department of Science and Technology—Philippine Council for Industry, Energy, and Emerging Technology Research and Development (DOST-PCIEERD), Project No. 04660.

Acknowledgments: The researchers would like to acknowledge Agata Mining Ventures Inc. (AMVI, Agusan Del Norte, Philippines) for their support in the procurement of materials used in this research, and Hirofumi Hinode of Tokyo Institute of Technology for his assistance in the XRF and XRD analysis of the low-grade nickel ore and limestone samples.

Conflicts of Interest: The authors declare no conflict of interest.

References

1. Igarashi, T.; Herrera, P.; Uchiyama, H.; Miyamae, H.; Iyatomi, N.; Hashimoto, K.; Tabelin, C. The two-step neutralization ferrite-formation process for sustainable acid mine drainage treatment: Removal of copper, zinc and arsenic, and the influence of coexisting ions on ferritization. *Sci. Total Environ.* **2020**, *715*, 136877. [CrossRef] [PubMed]

2. Skousen, J.; Ziemkiewicz, P.; Mcdonald, L. Acid mine drainage formation, control and treatment: Approaches and strategies. *Extr. Ind. Soc.* **2019**, *6*, 241–249. [CrossRef]

3. Tabelin, C.; Igarashi, T.; Villacorte-Tabelin, M.; Park, I.; Opiso, E.; Ito, M.; Hiroyoshi, N. Arsenic, selenium, boron, lead, cadmium, copper, and zinc in naturally contaminated rocks: A review of their sources, modes of enrichment, mechanisms of release, and mitigation strategies. *Sci. Total Environ.* **2018**, *645*, 1522–1553. [CrossRef] [PubMed]

4. Simate, G.S.; Ndlovu, S. Acid mine drainage: Challenges and opportunities. *J. Environ. Chem. Eng.* **2014**, *2*, 1785–1803. [CrossRef]

5.   Berk, B.; Pemmasani, N.; Phagura, S. Acid Mine Drainage Treatment Public Policy Report. 2015. Available online: https://www.cmu.edu/epp/people/faculty/course-reports/NTCReport2015-AcidMineDrainage.pdf (accessed on 7 September 2020).

6.   Park, I.; Tabelin, C.; Jeon, S.; Li, X.; Seno, K.; Ito, M.; Hiroyoshi, N. A review of recent strategies for acid mine drainage prevention and mine tailings recycling. *Chemosphere* **2019**, *219*, 588–606. [CrossRef]

7.   Quintans, J. Mining industry in the Philippines. *The Manila Times*. 2017. Available online: https://www.manilatimes.net/2017/09/04/supplements/mining-industry-philippines/348610/#:~{}:text=THE%20Philippines%20is%20the%20fifth,mineral%20wealth%2C%20as%20of%202012 (accessed on 7 September 2020).

8.   Mines & Geosciences Bureau. Mining Industry Statistics. 2020. Available online: https://mgb.gov.ph/images/Mineral_Statistics/MIS_Q12020_May292020.pdf (accessed on 7 September 2020).

9.   Kefeni, K.; Msagati, T.; Mamba, B. Acid mine drainage: Prevention, treatment options, and resource recovery: A review. *J. Clean. Prod.* **2017**. [CrossRef]

10.   Huyen, D.; Tabelin, C.; Thuan, H.; Dang, D.; Truong, P.; Vonghuthone, B.; Kobayashi, M.; Igarashi, T. The solid-phase partitioning of arsenic in unconsolidated sediments of the Mekong Delta, Vietnam and its modes of release under various conditions. *Chemosphere* **2019**, *233*, 512–523. [CrossRef]

11.   Akcil, A.; Koldas, S. Acid mine drainage (AMD): Causes, treatment and case studies. *J. Clean. Prod.* **2006**, *14*, 1139–1145. [CrossRef]

12.   Li, X.; Gao, M.; Hiroyoshi, N.; Tabelin, C.; Taketsugu, T.; Ito, M. Suppression of pyrite oxidation by ferric-catecholate complexes: An electrochemical study. *Miner. Eng.* **2019**, *138*, 226–237. [CrossRef]

13.   Pierre Louis, A.; Yu, H.; Shumlas, S.; Van Aken, B.; Schoonen, M.; Strongin, D. Effect of Phospholipid on Pyrite Oxidation and Microbial Communities under Simulated Acid Mine Drainage (AMD) Conditions. *Environ. Sci. Technol.* **2015**, *49*, 7701–7708. [CrossRef]

14.   Holmes, P.; Crundwell, F. The kinetics of the oxidation of pyrite by ferric ions and dissolved oxygen: An electrochemical study. *Geochim. Cosmochim. Acta* **2000**, *64*, 263–274. [CrossRef]

15.   Chen, M.; Lu, G.; Guo, C.; Yang, C.; Wu, J.; Huang, W.; Yee, N.; Dang, Z. Sulfate migration in a river affected by acid mine drainage from the Dabaoshan mining area, South China. *Chemosphere* **2015**, *119*, 734–743. [CrossRef] [PubMed]

16.   Skousen, J.G.; Sexstone, A.; Ziemkiewicz, P. Acid mine drainage control and treatment. *Reclam. Drastically Disturb. Lands* **2000**. [CrossRef]

17.   Waters, J.C.; Santomartino, S. Acid rock drainage treatment technologies: Identifying appropriate solutions. In Proceedings of the 6th International Conference on Acid Rock Drainage, Cairns, QLD, Australia, 14–17 July 2003.

18.   Skousen, J.; Ziemkiewicz, P. Performance of 116 passive treatment systems for acid mine drainage. In Proceedings of the 2005 National Meeting of the American Society of Mining and Reclamation, Breckenridge, CO, USA, 19–23 June 2005. [CrossRef]

19.   Morallo, A. DENR to Close 21 Mining Firms. *The Philippine Star*. 2017. Available online: https://www.philstar.com/business/2017/02/02/1668370/denr-close-21-mining-firms (accessed on 31 May 2020).

20.   Masindi, V.; Akinwekomi, V. Comparison of mine water neutralisation efficiencies of different alkaline generating agents. *J. Environ. Chem. Eng.* **2017**, *5*, 3903–3913. [CrossRef]

21.   Potgieter-Vermaak, S.S.; Potgieter, J.H. Comparison of limestone, dolomite and fly ash as pre-treatment agents for acid mine drainage. *Miner. Eng.* **2005**, *19*, 454–462. [CrossRef]

22.   Ziemkiewicz, P.F.; Skousen, J.G.; Lovett, R. Open limestone channels for treating acid mine drainage: A new look at an old idea. *Green Lands* **1994**, *24*, 36–41.

23.   Cravotta, C.A., III; Trahan, M.K. Limestone drains to increase pH and remove dissolved metals from acidic mine drainage. *Appl. Geochem.* **1999**, *14*, 581–606. [CrossRef]

24.   Arnold, D.E. Diversion wells—A low-cost approach to treatment of acid mine drainage. In Proceedings of the Twelfth West Virginia Surface Mine Drainage Task Force Symposium, Morgantown, WV, USA, 3–4 April 1991.

25.   Hedin, R.; Narin, R.; Kleinmann, R. *The Passive Treatment of Coal Mine Drainage*; Bureau of Mines IC 9389; U.S. Department of the Interior: Washington, DC, USA, 1994.

26.   Muliwa, A.; Leswifi, T. Performance evaluation of eggshell waste material for remediation of acid mine drainage drom coal dump leachate. *Miner. Eng.* **2018**, *122*, 241–250. [CrossRef]

27. Heviánková, S.; Bestová, I. The application of wood ash as a reagent in acid mine drainage treatment. *Miner. Eng.* **2014**, *56*, 109–111. [CrossRef]

28. Ouakibi, O.; Hakkou, R. Phosphate carbonated wastes used as drains for acidic mine drainage passive treatment. *Procedia Eng.* **2014**, *83*, 407–414. [CrossRef]

29. Kamal, N.M.; Sulaiman, S.K. Bench-Scale study of acid mine drainage treatment using local neutralisation agents. *Malays. J. Fundam. Appl. Sci.* **2014**, *10*, 150–153. [CrossRef]

30. Cowan, C.E.; Zachara, J.M.; Resch, C.T. Cadmium adsorption on iron oxides in the presence of alkaline-earth elements. *Environ. Sci. Technol.* **1991**, *25*, 437–446. [CrossRef]

31. The Editors of Encyclopaedia Britannica. 2017. Carbonate Mineral. Available online: https://www.britannica.com/science/carbonate-mineral (accessed on 23 June 2020).

32. Bernier, L.R. The potential use of serpentinite in the passive treatment of acid mine drainage: Batch experiments. *Environ. Geol.* **2005**, *47*, 670–684. [CrossRef]

33. Turingan, C.; Singson, G.; Melchor, B.; Alorro, R.; Beltran, A.; Orbecido, A. A comparison of acid mine drainage (AMD) neutralization potential of low grade nickel laterite and other alkaline-generating materials. *J. Mater. Sci. Chem. Eng.* **2020**. [CrossRef]

34. Benjamin, M.M.; Leckie, J.O. Multiple-Site adsorption of Cd, Cu, Zn, and Pb on amorphous iron oxyhydroxide. *J. Colloid Interface Sci.* **1981**, *79*, 209–221. [CrossRef]

35. Balistrieri, L.S.; Chao, T. Adsorption of selenium by amorphous iron oxyhydroxide and manganese dioxide. *Geochim. Cosmochim. Acta* **1990**, *54*, 739–751. [CrossRef]

36. Zachara, J.M.; Girvin, D.C.; Schmidt, R.L.; Resch, C.T. Chromate adsorption on amorphous iron oxyhydroxide in the presence of major groundwater ions. *Environm. Sci. Technol.* **1987**, *21*, 589–594. [CrossRef]

37. Holcim, G.T.Z. Reuse and Recycling of Construction and Demolition Waste. 2007. Available online: https://pdfs.semanticscholar.org/5963/263e9892bc081f55928f45fe9ff09eaa51bb.pdf (accessed on 4 September 2020).

38. Kalaw, M.; Culaba, A.; Hinode, H.; Kurniawan, W.; Gallardo, S.; Promentilla, M. Optimizing and Characterizing Geopolymers from Ternary Blend of Philippine Coal Fly Ash, Coal Bottom Ash and Rice Hull Ash. *Materials* **2016**, *9*, 580. [CrossRef]

39. Parkhurst, D.L.; Appelo, C.A.J. Description and Input and Examples for PHREEQC Version 3—A Computer Program for Speciation, Batch-Reaction, One-Dimensional Transport, and Inverse Geochemical Calculations. In *USGS Techniques and Methods 6-A43*; USGS: Denver, CO, USA, 2013.

40. Jones, S.; Cetin, B. Evaluation of waste materials for acid mine drainage remediation. *Fuel* **2017**, *188*, 294–309. [CrossRef]

41. Parisatto, M.; Dalconi, M.; Valentini, L.; Artioli, G.; Rack, A.; Tucoulou, R.; Cruciani, G.; Ferrari, G. Examining microstructural evolution of Portland cements by in-situ synchrotron micro-tomography. *J. Mater. Sci.* **2014**, *50*, 1805–1817. [CrossRef]

42. Habeeb, G.; Mahmud, H. Study on Properties of Rice Husk Ash and Its Use as Cement Replacement Material. *Mater. Res.* **2010**, *13*. [CrossRef]

43. Henmi, T.; Wells, N. Poorly-Ordered iron-rich precipitates from springs and streams on andesitic volcanoes. *Geochim. Cosmochim. Acta* **1980**, *44*, 365–372. [CrossRef]

44. Jambor, J.L.; Dutrizac, J.E. Occurrence and constitution of natural and synthetic ferrihydrite, a widespread iron oxyhydroxide. *Chem. Rev.* **1998**, *98*, 2549–2586. [CrossRef] [PubMed]

45. Koeppenkastrop, D.; De Carlo, E.H. Sorption of rare-earth elements from seawater onto synthetic mineral particles: An experimental approach. *Chem. Geol.* **1992**, *95*, 251–263. [CrossRef]

46. Soler, J.M.; Boi, M. The passivation of calcite by acid mine water: Column experiments with ferric sulfate and ferric chloride solutions at pH 2. *Appl. Geochem.* **2008**, *23*, 3579–3588. [CrossRef]

47. Offeddu, F.; Cama, J. Processes affecting the efficiency of limestone in passive treatments for AMD: Column experiments. *J. Environ. Chem. Eng.* **2015**, *3*, 304–316. [CrossRef]

48. Ya, Z.; Zhou, L. High efficiency of heavy metal removal in mine water by limestone. *Chin. J. Geochem.* **2009**, *28*, 293–298. [CrossRef]

49. The Editors of Encyclopaedia Britannica. Hydroxide. Britannica. 2017. Available online: https://www.britannica.com/science/hydroxide (accessed on 31 May 2020).

50. Liu, H.; Chen, T. An overview of the role of goethite surfaces in the environment. *Chemosphere* **2014**, *103*, 1–11. [CrossRef]

51. Rietra, R.P.; Hiemstra, T. Sulfate adsorption on goethite. *J. Colloid Interface Sci.* **1999**, *218*, 511–521. [CrossRef]
52. Cornell, R.M.; Schwertmann, U. *The Iron Oxides: Structure, Properties, Reactions, Occurrence and Uses*; VCH: Weinheim, Germany, 2003; Volume 2, p. 15.
53. Ma, M.; Gao, H.; Sun, Y.; Huang, M. The adsorption and desorption of Ni(II) on Al substituted goethite. *J. Mol. Liq.* **2015**, *201*, 30–35. [CrossRef]
54. Silva, R.; Cadorin, L.; Rubio, J. Sulphate ions removal from an aqueous solution: I. Co-precipitation with hydrolysed aluminum-bearing salts. *Miner. Eng.* **2010**, *23*, 1220–1226. [CrossRef]
55. Skousen, J.; Politan, K. Acid mine drainage treatment systems: Chemicals and costs. *Green Lands* **1990**, *20*, 31–37.
56. Miller, A.; Figueroa, L.; Wildeman, T. Zinc and nickel removal in simulated limestone treatment of mining influenced water. *Appl. Geochem.* **2011**, *26*, 125–132. [CrossRef]
57. Vadapalli, V.; Klink, M.; Etchebers, O.; Petrik, L.; Gitari, W.; White, R.; Key, D.; Iwuoha, E. Neutralization of Acid Mine Drainage using Fly Ash and Strength Development of the Resulting Solid Residues. *South Afr. J. Sci.* **2008**, *104*, 317–322.
58. Golab, A.; Peterson, M.; Indraratna, B. Selection of potential reactive materials for a permeable reactive barrier for remediating acidic groundwater in acid sulphate soil terrains. *Fac. Eng. Pap.* **2006**, *39*. [CrossRef]

 *minerals*

Article

Effectiveness of Fly Ash and Red Mud as Strategies for Sustainable Acid Mine Drainage Management

Viktoria Keller [1], Srećko Stopić [1,*], Buhle Xakalashe [2], Yiqian Ma [1], Sehliselo Ndlovu [3] , Brian Mwewa [3], Geoffrey S. Simate [3] and Bernd Friedrich [1]

[1]   IME Process Metallurgy and Metal Recycling, RWTH Aachen University, 52056 Aachen, Germany; viktoria.keller@rwth-aachen.de (V.K.); yma@ime-aachen.de (Y.M.); bfriedrich@ime-aachen.de (B.F.)
[2]   Pyrometallurgy Division, MINTEK, Private Bag X3015, Randburg 2125, South Africa; buhlex@mintek.co.za
[3]   School of Chemical and Metallurgical Engineering, University of the Witwatersrand, Johannesburg 2000, South Africa; Sehliselo.Ndlovu@wits.ac.za (S.N.); sirbhimself@yahoo.co.uk (B.M.); Geoffrey.Simate@wits.ac.za (G.S.S.)
*   Correspondence: sstopic@ime-aachen.de; Tel.: +49-176-7826-1674

Received: 7 June 2020; Accepted: 6 August 2020; Published: 10 August 2020

Abstract: Acid mine drainage (AMD), red mud (RM) and coal fly ash (CFA) are potential high environmental pollution problems due to their acidity, toxic metals and sulphate contents. Treatment of acidic mine water requires the generation of enough alkalinity to neutralize the excess acidity. Therefore, red mud types from Germany and Greece were chosen for the neutralization of AMD from South Africa, where this problem is notorious. Because of the high alkalinity, German red mud is the most promising precipitation agent achieving the highest pH-values. CFA is less efficient for a neutralization and precipitation process. An increase in temperature increases the adsorption kinetics. The maximum pH-value of 6.0 can be reached by the addition of 100 g German red mud at 20 °C to AMD-water with an initial pH value of 1.9. German red mud removes 99% of the aluminium as aluminium hydroxide at pH 5.0. The rare earth elements (yttrium and cerium) are adsorbed by Greek red mud with an efficiency of 50% and 80% at 60 °C in 5 min, respectively.

Keywords: acid mine drainage; secondary materials; management; absorption; precipitation

---

1. Introduction

AMD is a term for wastewaters from mining processes. AMD contains a high concentration of dissolved metals from ores in a sulphate solution [1,2]. While mining, the mass of the rocks is fragmented. This leads to an increase of the surface area contacting the atmosphere which results in higher acid production rate [3]. As mentioned by Tabelin et al. [4] the contaminated excavation debris/spoils/mucks, loosely referred to as "naturally contaminated rocks", contain various hazardous and toxic inorganic elements like arsenic (As), selenium (Se), boron (B), and heavy metals like lead (Pb), cadmium (Cd), copper (Cu), and zinc (Zn). If left untreated, these naturally contaminated rocks could pose very serious problems not only to the surrounding ecosystem but also to people living around the construction and disposal sites.

Park et al. [5] mentioned that remediation options like neutralization, adsorption, ion exchange, membrane technology, biological mediation, and electrochemical approach have been developed to reduce the negative environmental impacts of AMD on ecological systems and human health. However, these techniques require a continuous supply of chemicals and energy, expensive maintenance and labor cost, and long-term monitoring of affected ecosystems until AMD generation stops.

As mentioned by Igarashi et al. [6], AMD containing Zn, Cu and As was treated by using a laboratory-type continuous ferrite process flow setup. Cu and As were removed in the first sludge, which are stable in standard leaching tests. Magnetic magnesioferrites and magnetite were generated when dissolved silicon (Si) was low. However, in the study by Igarashi et al., information about the recovery of critical metals such as rare earth elements is missing.

AMD also originates from sulphide conglomerates stored on deposits where the rain rinses the acid and metals such as uranium (U) out of the dumps [3]. Finally, the acid infiltrates the groundwater and the environment such as rivers in the affected region around the deposits which contaminates the aquatic life and the soil in the surroundings [7]. Additionally, South Africa is a water-poor country with an average rainfall of under 450 mm per year [7]. Acid mine drainage containing around 3500 mg/L [3] of sulphate has a pH value between 2 and 3 which increases the solubility of certain metals such as heavy metals. Along the river pathways, the iron (Fe) in a sulphide form oxidizes and precipitates, leading to a bright orange trail. The formation of AMD is the result of oxidation of pyrite, FeS_2, with oxygen and ferric iron, Fe^{3+}, as shown in Equations (1)–(3) [8]. Consequently, sulphuric acid is formed during this reaction. At this pH, bacteria like *Thiobacillus ferrooxidans* can survive [7] and accelerate the geochemical reactions.

$$FeS_{2(s)} + \frac{7}{2}O_{2(aq)} + H_2O \rightarrow Fe^{2+} + 2SO_4{}^{2-} + 2H^+ \tag{1}$$

$$Fe^{2+} + \frac{1}{4}O_{2(aq)} + H^+ \rightarrow Fe^{3+} + \frac{1}{2}H_2O \tag{2}$$

$$Fe^{3+} + 3H_2O \rightarrow Fe^{3+} + 3OH^- + 3H^+ \rightarrow Fe(OH)_{3(s)} + 3H^+ \tag{3}$$

AMD can be classified by the content of acid and dissolved metals with Ficklin diagram [9]. At the surface, the AMD stream is diluted by rain or surface water which leads to an increase of the pH values with white precipitation products in the AMD stream. The precipitation products contain mainly crystalline and amorphous aluminium phases [10] which adsorb other metals. Heavy metals are generally removed by iron precipitation. Aluminum precipitation only becomes important when iron content is low, but this is rarely the case.

In South Africa, coal is extracted by underground mining or open-pit mining with little surface dumping [3]. FeS_2 is contained in the host rock but is more abundant in the coal layers. When mining terminates, the mine will be closed by collapsing the upper layers. Water fills in the voids of the fractured rocks and reacts with the FeS_2 [3]. Rain penetrates through the soil and becomes acidic and influences the natural groundwater. In the dams around coal mines in Middelburg and Witbank in South Africa, the salinity and sulphate concentration amount to about 200 mg/L [3], which is the limit for domestic use.

Regarding the problem of growing deposits, the branch of research on wastewater treatment is of high importance. The primary aim of wastewater treatment is to recover valuable metals and neutralize the acidic solution [11]. An advantage is the higher content of valuable metals in the wastewater compared to the ores used in primary production.

There are further possibilities for water purification like the SAVMIN™ process. This is the simplest technology for reducing high sulphate concentrations is lime precipitation developed by MINTEK in South Africa [9]. Precipitation of metal hydroxides and sulphides was performed using sodium hydroxide, calcium carbonate and hydrogen sulphide. Regarding the costs, calcium carbonate is most suitable for the neutralization experiments. Because the neutralization is exothermic, room temperature was chosen for the removal of metals from waste solution. Pre-oxidation with hydrogen peroxide transforms the bivalent Fe to the trivalent Fe which precipitates at lower pH values [12] than the bivalent Fe. Analogously, As is oxidised from the trivalent state to the pentavalent state to enable hardly dissolving oxides [13,14]. Removal of As(III) and As(V) can occur during neutralization simultaneously. Additionally, because of the high reactivity of sulphides with heavy metal ions, they can be applied to achieve a larger decrease in metal concentration after neutralization with hydroxides [13]. Trivalent Fe cannot be removed by sulphides precipitation since they do not form stable sulphides under wastewater conditions [13]. Instead, the trivalent will be reduced to the bivalent state which forms iron(II) sulphide (FeS) [14]. Another possibility for selective removal of metals is cementation with nanoscaled zerovalent powders which have a large reaction surface area [15]. Cementation at room temperature and reductive precipitation took a short time in contrast to adsorption. There is a tendency that metals are adsorbed in this sequence: Fe(II)/Fe(III) > Cu(II) > Mn(II) > Zn(II) [16] following the Langmuir model which states that all surfaces of a given solid have the same affinity to adsorb metals [17]. Addition of brine can decrease the kinetics of neutralization, so there is a long-term effect which avoids a too-high pH value [18]. Precipitation of dissolved metals is achieved by the introduction of chemical additives such as sodium carbonate and oxalic acid to mostly affect changes in pH and/or ligand concentration at room temperature.

High standards are applied for drinking water which is chosen as a better reference in comparison to the purified water. Pure water is clear, colour- and tasteless and low in microbiological contamination [14]. Some metals, like Fe^{3+}, change the colour of water. Fe^{3+} turns the water into a reddish-brown colour [19]. The optimum pH for drinking water ranges between 6.5–8.5 [19]. The taste threshold for salts depends on the cation (sodium, potassium) and the anion (chloride, sulphate) concentrations. For sodium chloride, the taste threshold concentration is 200–300 mg/L [19]. Turbidity indicates any kind of contamination which can be caused by inorganic and organic matter in the form of suspended particles [19]. Microorganisms are attached to particles [14]; therefore, filtration of particles can reduce their population in treated water. Turbidity of AMD can also be a sign of natural precipitation of iron. In Germany, the most important laws for wastewater treatment are:

- Water Resources Act (WHG, Wasserhaushaltsgesetz) [1]
- Wastewater Ordinance (AbwV, Abwasserverordnung) [20]
- Drinking Water Ordinance (TrinkwV, Trinkwasserverordnung) [21]

The European and German standards are based on the recommendations of the World Health Organization (WHO) [14], which has set up health-based guidelines for chemicals in drinking water [19]. The effects of contaminants on taste, odour and appearance are cited by the WHO, while the most important effects such as toxicity are not mentioned [19]. Mostly, because the concentration in treated water is too low, it is not possible to determine guideline values for some metals such as silver (Ag) and gold (Au). In the case of aluminium, the guideline value is nine-fold higher than the value achievable by current technology [19]. The TrinkwV restricts the concentration of aluminium since there could be a correlation between aluminium consumption and Alzheimer's diseases [14]. The pH value of drinking water is between 6.5 and 9.5. Water with carbonic acid (H_2CO_3) can have a lower pH value than 6.5 [21].

Purification of wastewaters is a high-complex process. The process line depends on the origin of the wastewater and therefore the contaminations as well as the subsequent use of the purified water. For use as drinking water, the water needs to fulfil special regulations, for example, the TrinkwV [21]. The purification process contains several steps such as mechanical separation of solids and suspended particles (e.g., flotation, sedimentation, filtering) [14,22]; oxidation of dissolved metals to solids [14]; ion exchange processes to neutralize the water from salts and metals [23]; disinfection of water by oxidation media (ClO_2, O_3, hyper chlorites) or UV-light [24] and biological treatment (aerobic processes and anaerobic processes) [11].

The by-products used in this paper for treating AMD are red mud and CFA. Red mud comes from the Bayer process [25,26], whereas coal fly ash is the inorganic residue from coal combustion. In red mud, there are mainly hematite, other crystalline aluminium silicate phases, silica, titanium oxide, rare earth oxides and un-leached residual alumina [27]. Coal fly ash contains approximately 50% of crystalline phases, mostly quartz and mullite [28].

The neutralization process enables the transformation of several waste streams into valuable products regarding the zero waste guidelines in technology. There are several water treatment methods based on red mud and CFA [29] like acid and base treated powder [30–33] or zeolites as special aluminium silicates [34,35], but not all of them have yet been tested in acid mine drainage. Transforming CFA to zeolites can increase the adsorption of dissolved metals [36].

Our main aim is to perform a neutralization of AMD using secondary materials such as red mud and CFA and to discuss the possibility of the recovery of valuable metals such as Al, Zn, Mn and rare earth elements. The neutralization of AMD in this study is a preparation of wastewater for return into downstream processing or releasing to the environment. Acidic water can damage the plants by corroding or damaging the environment. Another aim of neutralization with alkaline material is the preparation of solid waste materials for further metal winning processes. The final aim of this study is to develop a process for water purification which fulfils the following criteria: economical, good water quality after clarification, resources of materials used for neutralization are as regionally available (<500 km) as possible and waste products are environmentally friendly.

2. Materials and Experimental Procedure

2.1. Characterisation of the Studied Materials

The AMD sample was collected from Mpumalanga, South Africa. All sampling and laboratory analysis is performed in accordance with recognized global standards such as the International Standards organization (ISO). After sampling and laboratory analysis in South Africa, all samples were sent to Germany. The AMD water was characterized by using ICP-OES analysis (SPECTRO ARCOS, SPECTRO Analytical Instruments GmbH, Kleve, Germany) and the solid samples by X-ray fluorescence, (Axios FAST, Malvern Panalytical GmbH, Kassel, Germany). The AMD was first filtrated in order to remove the formed precipitate, but AMD was not acidified. The solid samples were ground up before the X-ray diffraction analysis (XRD) analysis. Additionally, characterization of the red mud and fly ash to the X-ray powder diffraction XRD (Bruker AXS, Karlsruhe, Germany) was operated. Bauxite residue, employed during AMD-treatment as the main raw material, was provided by Aluminum of Greece plant, Metallurgy Business Unit, Mytilineos S.A. (AoG). The sample was first homogenized by using laboratory sampling procedures (riffling method) and then a representative sample was dried in a static furnace at 105 °C for 24 h. Subsequently, the material was milled using a vibratory disc mill and the sample was fully characterized.

Chemical analyses of major and minor elements were executed via the fusion method (1000 °C for 1 h with a mixture of $Li_2B_4O_7/KNO_3$ followed by direct dissolution in 10% HNO_3 solution) through a Perkin Elmer 2100 Atomic Absorption Spectrometer (AAS) (Waltham, MA, USA), a Spector Xepos Energy Dispersive X-ray Fluorescence Spectroscope (ED-XRF) (SPECTRO, Kleve, Germany), a Thermo Fisher Scientific X-series 2 Inductively Coupled Plasma Mass Spectrometer (ICP-MS) (Waltham, MA, USA) and a Perkin Elmer Optima 8000 Inductively Coupled Plasma Optical Emission Spectrometer (ICP-OES) (Waltham, MA, USA), whereas the loss of ignition (LOI) of the sample was provided by differential thermal analysis (DTA), using a SETARAM TG Labys-DS-C (Caluire, France) system in the temperature range of 25–1000 °C with a 10 °C/min-heating rate, in air atmosphere.

Mineralogical phases were detected by XRD using a Bruker D8 Focus powder diffractometer with nickel-filtered CuKα radiation (λ = 1.5405 Å) coupled with XDB Powder Diffraction Phase Analytical System version 3.107 which evaluated the quantification of mineral phases via profile fitting specifically for bauxite ore and bauxite residues.

2.1.1. Acid Mine Drainage from South Africa

The pH of the AMD wastewater ranges around 2.0. The ICP-OES analysis given in Table 1 shows the composition from the wastewater which has a dark red–yellow colour as shown in Figure 1 and their limits in the TrinkwV. According to the Ficklin diagram from [9], the AMD sample is defined as highly acidic and as having a high metal content inn the water. The total concentration of the relevant metals (mg/L), amounted to 18.903 mg. The single concentration of elements in AMD (mg/L) amounted to 14.2 Zn, 2.09 Ni, 1.94 Co, 0.65 Cu, 0.02 Cd, and 0.003 Pb. Comparison of the metal concentration with the TrinkwV shows that the critical metals were cadmium, chromium and nickel, iron manganese and the sulphate. The pH of 2.0 was too low (interval is 6.5–9.5).

Table 1. Metal containing wastewater (three significant figures).

Substance	$Fe^{2+/3+}$	Ca^{2+}	Mg^{2+}	Al^{3+}	Mn
Concentration [mg/L]	2970	503	435	370	82
Limit (TrinkwV) [mg/L]	0.2	-	-	0.2	0.05
Substance	Na^{2+}	Si^{4+}	Zn^{3+}	Ce^{3+}	Ni^{2+}
Concentration in [mg/L]	52.4	44.6	14.3	5.51	2.09
Limit (TrinkwV) [mg/L]	-	-	-	-	0.02
Substance	Y^{3+}	Co^{2+}	Nd^{3+}	La^{3+}	Cu^{2+}
Concentration in [mg/L]	2.05	1.94	1.31	0.85	0.65
Limit (TrinkwV) [mg/L]	-	-	-	-	2
Substance	Cr^{3+}	Sc^{3+}	Cd^{2+}	Pb^{2+}	SO_4^{2-}
Concentration in [mg/L]	0.15	0.11	0.02	0.003	10,200
Limit (TrinkwV) [mg/L]	0.05	-	0.003	0.01	250

Figure 1. Mine drainage water from South Africa.

2.1.2. Red Mud from Germany and Greece

For the neutralisation experiments, wto sorts of red mud were used. In Figure 2, the red mud shown on the left originates from Germany (Stade, Lower Saxony, Germany), and the right one was from Greece (Aluminium of Greece, Viotia). The Greek red mud is darker due to its higher content of iron as shown in Table 2. This table shows that the Greek red mud contains more iron oxide, alumina and lime than the sample from Germany. In contrast, the German sample contains more sodium oxide, titanium dioxide and silica. Red mud is very alkaline due to the sodium hydroxide from the Bayer process (pH < 12). For further experiments, the s/L-rates of 1:10 (100 g/L AMD) and 1:5 (200 g/L AMD) were chosen.

Table 2. Content of oxides in the red mud samples wt.% or ppm (Sc).

Origin	Fe_2O_3	Al_2O_3	CaO	SiO_2	TiO_2	Na_2O	Cr_2O_3	Sc
Germany	35.3	15.7	6.7	14.0	11.4	8.9	0.2	86
Greek	44.0	23.0	10.2	5.5	5.6	1.8	0.3	122

Figure 2. Red mud samples, (**left**) from Germany, (**right**) from Greece.

The XRD analysis of Greek Red Mud is presented in Table 3.

Table 3. X-ray diffraction analysis (XRD) analysis of Greek red mud (d = 1.87 µm).

Mineralogical Phase	(wt.%)
Cancrinite [$Na_6Ca_{1.5}Al_6Si_6O_{24}(CO_3)_{1.6}$]:	15.0
Perovskite [$CaTiO_3$]	4.5
Hematite [Fe_2O_3]:	30.0
Boehmite [$AlO(OH)$]:	3.0
Goethite [$FeO(OH)$]:	9.0
Anatase [TiO_2]:	0.5
Andradite [Ca–Fe–Al–Si oxides]:	17.0
Quartz [SiO_2]:	2.0
Rutile [TiO_2]:	0.5
Calcite [$CaCO_3$]:	4.0
Chamosite [$(Fe^{2+},Mg)_5Al(AlSi_3O_{10})(OH)_8$]:	4.0
Diaspore [$AlO(OH)$]:	9.0
Gibbsite [$Al(OH)_3$]:	2.0
Total	98.5

The XRF analysis of the rare earth elements in Greek red mud is shown in Table 4.

Table 4. Composition of rare earth elements in Greek red mud.

Elements	Y	La	Ce	Pr	Nd	Sm	Eu	Gd
(g/t)	76	114	368	28	99	21	5	22
Elements	Dy	Ho	Er	Tm	Yb	Lu	Y	Tb
(g/t)	17	4	13	2	14	2	80	3

Mineralogical analysis of the German red mud was explained by Kaussen and Friedrich [37]. The missing 1.5% accounts for unaccounted, unknown or amorphous content. There is no sufficient reference data (known crystallographic measurements) to characterize 100% of bauxite residue mineralogical composition. In the current XDB full profile fitting mineral phase quantification it was not possible to quantify amorphous content. Amorphous content can be determined in phase quantification when a known quantity of an internal standard such as corundum is added to the sample.

2.1.3. Coal Fly Ash from South Africa

The coal fly ash used in this study was obtained from an Eskom thermal power station, Mpumalanga, South Africa. The XRD analyses of fly ashes have been measured with a Bruker D4 powder diffractometer in Bragg–Brentano geometry. The qualitative evaluation was done with the program EVA and the PDF 2 file from ICSD. The quantitative determination was carried out using a Rietveldt refinement with the program TOPAS from Bruker. The analyses were carried out on powders <60 μm which were prepared as backloading compacts. SEM-analysis was performed using GEMINI SEM 300, Car Zeiss Microscopy GmbH; Oberkochen, Germany.

As shown in Table 5, the main components of fly ashes are silica, alumina, lime and hematite with a sum of 96.1 wt.%, and Y_2O_3 in small amounts (0.013). X-ray diffraction shows that 49% of the phases are amorphous phases, and the proportion of crystalline phase is (in %): 0.3 CaO, 0.7 Fe_2O_3, 38 Al_4SiO_8, 12 SiO_2, and 0.1 TiO_2. Most of the other oxides are present in traces. The coal fly ash in Figure 3 has a light grey colour (left) and spherical submicron particles (right).

Table 5. X-ray fluorescence (XRF) analysis of the content of compounds in the coal fly ash (wt.%).

SiO_2	Al_2O_3	CaO	Fe_2O_3	TiO_2	MgO	Na_2O
56.2	32.1	4.4	3.4	1.68	1.1	0.352
Cr_2O_3	MnO	NiO	PbO	CuO	Y_2O_3	ZnO
0.057	0.031	0.019	0.017	0.015	0.013	0.010

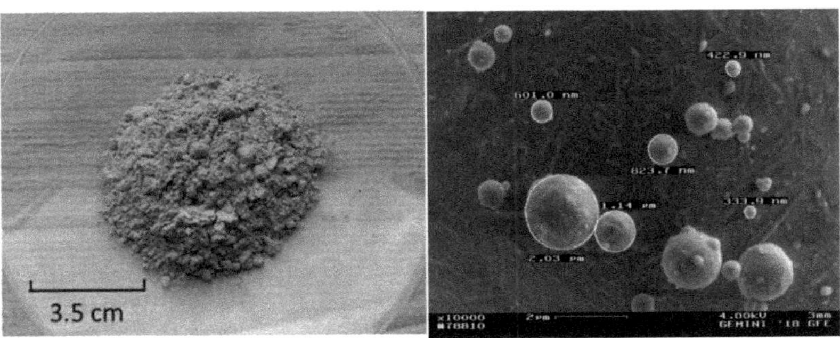

Figure 3. Fly ash powder (**left**) and SEM analysis (**right**).

2.2. Parameters for Neutralization

For neutralization, there are four important parameters:

- Experimental time (min): 5, 10, 20, 40, 60, 120.
- Precipitation media: coal fly ash (CFA), German red mud (GerRM), Greek red mud (GrRM)
- solid/liquid (s/L)-Ratio: 1:10, 1:5, 1:2
- Temperature (°C): 20 (room temperature), and 60 (after Heating)

The experiments, see Table 6, were designed for each media separately because coal fly ash is less alkaline, but contains yttrium (0.01%), and it is very stable for leaching. Every experiment was conducted twice to reproduce the results.

Table 6. Design of the experiment.

Factor	Low	High
s/L-rate	1:10 (RM), 1:5 (CFA)	1:5 (RM), 1:2 (CFA)
Temperature	Room temperature, 20 °C	60 °C

2.3. Setup and Test Performance

The aim of the following experiments is to prove the necessity of by-products to neutralize 500 mL of metallic wastewater. The quantity of the substances is:

- 500 mL or 1 L of defiled wastewater
- up to 200 g red mud and 250 g fly ash

The acid mine drainage was heated up to between 20 °C and 60 °C and agitated for 120 min up to 350 rpm. After precipitation, 6 samples with approximately 10 mL (from 500 mL samples) or 50 mL (from 1 L samples) were taken out of the suspension. The filtered 10 mL samples were diluted with deionised water, from a 5 mL solution samples to 50 mL, without new dilution. The solid residues were analysed after precipitation in order to determine the metal content and the absorption kinetics regarding the consumed red mud and fly ash. The precipitation and absorption fraction were calculated using Equations (4) and (5), respectively.

3. Results and Discussion

With respect to iron ions as the greatest fraction in the AMD wastewater, white and yellow flakes can be found in the samples after 5 and 10 min. In the samples with the red muds, there are white flakes, as seen Figure 4. In the samples with coal fly ash there is a yellow suspension due to a lower pH than 3 as seen in Figure 5. Yellow-orange powders are left after a second filtration of the solution. These yellow–orange flakes indicate that iron has precipitated as iron(III) hydroxide from the solution. In the red mud samples, the iron hydroxide is already removed from the red mud. The reformation of $Fe(OH)_3$ in the filtrate of coal fly ash indicates that the formation of $Fe(OH)_3$ at constant pH has a long term kinetic of several minutes.

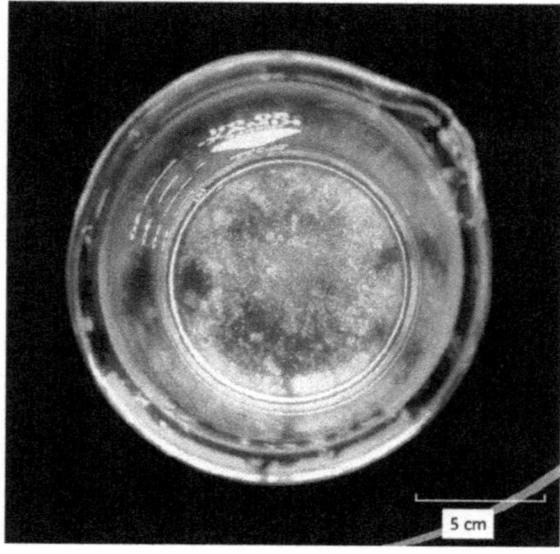

Figure 4. Typical flakes in the 5 min and 10 min samples after treatment with red mud.

Figure 5. Suspension in the 5, 10 and 20 min samples after treatment with coal fly ash.

The XRF analysis showed that the yellow powder after treatment with coal fly ash, filtration and drying mainly contains approximately

- 30.94 ± 0.01% iron
- 11.66 ± 0.01% sulphur
- 4.577 ± 0.010% silicium
- 3.783 ± 0.015% aluminium
- 1.416 ± 0.003% calcium

We supposed that iron and aluminium precipitate as iron(III) and aluminium(III) hydroxide, respectively. Additionally, some compounds precipitate as sulphates, e.g., $CaSO_4$, because it is slightly soluble, what is presented in Figure 6. Additionally, the presence of aluminium oxide in the product was revealed.

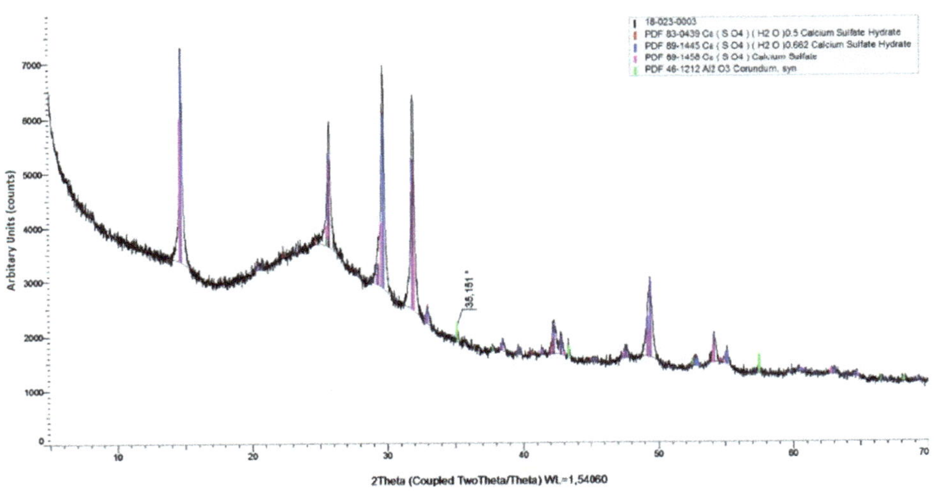

Figure 6. XRD analysis of the precipitation product.

3.1. pH Values as Function of the Time

The initial pH value of the AMD water was 1.973 ± 0.031 (the calculated relative error ΔpH/pH = 15.7%) for all 12 experiments. After the measurement of pH-Value, AMD solution was added in a glass reactor and heated to 60 °C. Red mud and fly ash (50–100 g) were added at 60 °C in the solution of AMD. The pH-Values were measured in time of 120 min. The same procedure was performed at room temperature and results were compared with ones obtained at 60 °C.

The pH was plotted as a function of time as seen in Figure 7. All samples have a strong pH increase in the first 5 min. The maximum pH of 6 can be reached by 100 g German red mud at 20 °C. The pH values from the Greek red mud tend to be lower than the ones from the German red mud, the maximum pH being 5.9, which is in the same range as the German red mud. The samples with coal fly ash indicate that in the first 5 min, the average pH was 2.78 ± 0.14 (ΔpH/pH = 5%) and most of the iron ions were still dissolved in the solution. After 10 min, the average pH was 2.93 ± 0.11 (ΔpH/pH = 4%) and there was still some iron(III) dissolved. For full iron(III) removal, the pH needs to be at least higher than three, which was mentioned by Stiefel [11].

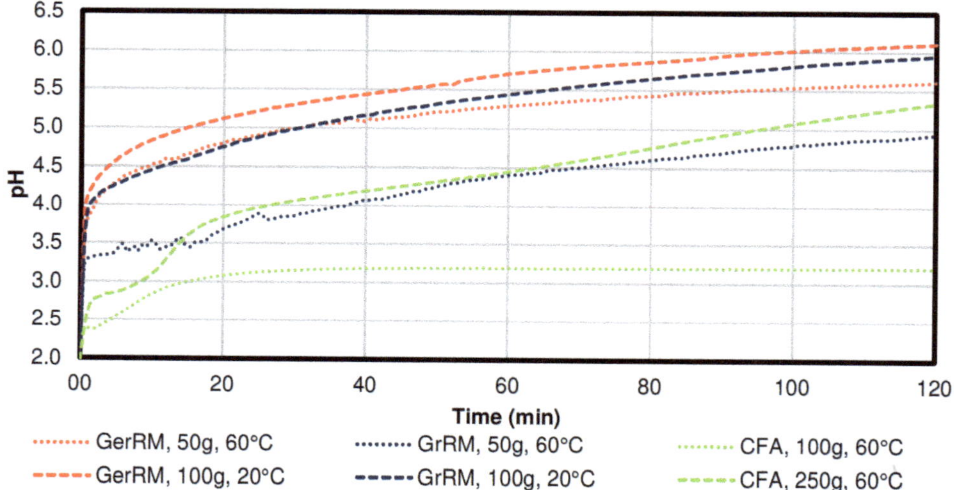

Figure 7. pH values in time (GerRM—German red mud; GrRM—Greek red mud; CFA—coal flying ash).

Coal fly ash had pH-values between 8 and 11 depending on the quantity of the added sample, in comparison to red mud (pH > 12). Especially because of 49% of amorph structure and 38% of very stable mullite in this structure, as shown at Figure 3, the fly ash has smaller neutralisation efficiency in comparison to used red mud. The temperature is very important in regards to pH measurements. As the pH value changes with the change in temperature, the new measured pH value is technically the true pH value. Under laboratory conditions, a note of the temperature and pH value should be made together. There is only one major temperature effect in pH measurement that can cause errors in readings. It is the only reasonably predictable error due to changes in temperature, and is the only temperature-related factor that a pH meter with temperature compensation can correct. This temperature error is very close to 0.003 pH/°. Generally in our experiments, Because of an influence of the diffusion of the acidic solution, an increase of temperature from room temperature to 60 °C increases the neutralization efficiency of AMD during the addition of fly ash and red mud. It seems that adding FA and RM can increase pH of AMD at the scale of 120 min. Long-term neutralization efficiency is required, as confirmed in subsequent work, and in the work of Paradis et al. [38].

3.2. Comparison of Precipitation Efficiency per Material

Regarding the literature [8], the critical pH for iron(III) precipitation using CFA can be assumed as approximately 3.0. At this point, 99% of the dissolved iron is precipitated. The maximum concentration of dissolved iron is 44.2 mg/L. However, it must be considered that the iron concentration in the solution was higher direct after the filtration since iron(III) hydroxide has precipitated after filtration. The following six figures (Figures 7–12) show the precipitation efficiency as a function of the pH of aluminium, manganese and zinc. The symbols represent the time of sampling (5, 10, 20, 40, 60, 120 min). The precipitation efficiency x_P was calculated by Equation (4). Negative values for x_P indicate leaching effects and enrichment in the acid mine drainage.

$$x_P = 1 - \frac{\text{Concentration of metal at time } t}{\text{Initial concentration of metal in AMD}} = 1 - \frac{c_t}{c_0} \tag{4}$$

3.2.1. Manganese

In Figure 8, by using the German red mud, manganese had a lower precipitation efficiency at a higher temperature (60 °C) than the sample with more mass at 20 °C at the same pH-value. There is a tendency of reaching a maximum precipitation efficiency of approximately 75% as the sample with 100 g at 20 °C, which shows a light equilibrium state. Using a Greek 50 g red mud results in leaching in acid mine drainage up to 10% at a higher temperature. Greek red mud with higher mass at room temperature leads to the precipitation of manganese up to 30%. Manganese tends to be leached in both coal fly ash experiments. In the 100 g sample at 60 °C it is leached from 10% to 22%. In the 250 g lower heated experiment, manganese was leached to 20% in the first 5 min and then precipitated until it was at a value of 6% of the initial concentration. It follows that manganese may precipitate after pH 4 by using red mud.

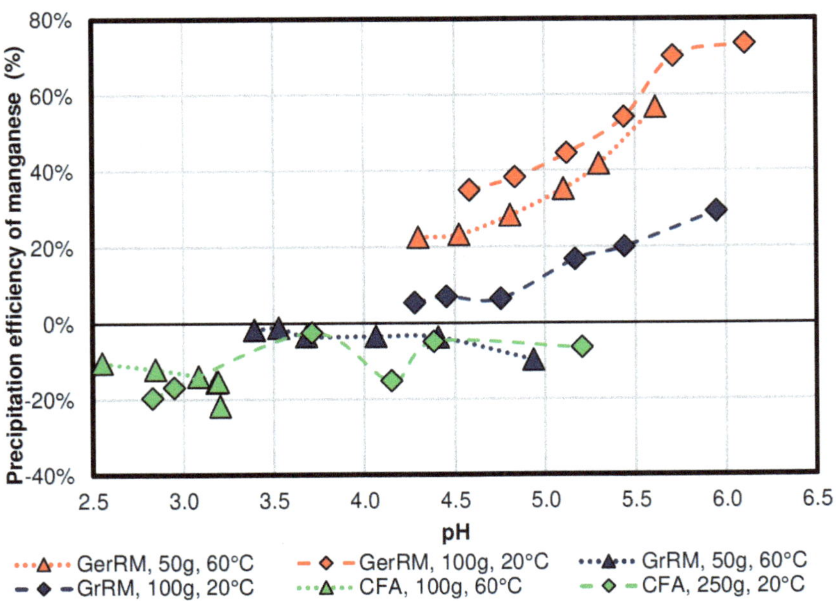

Figure 8. Participation efficiency of manganese.

3.2.2. Aluminium

Figure 9 shows the precipitation efficiency of aluminium for all precipitation media compared to the initial concentration of 370 mg/L. German red mud removes 99% of the aluminium as aluminium hydroxide at pH 5.0. Higher temperature leads to an earlier precipitation pH and therefore, to higher efficiency at the same pH compared to the experiment with higher mass at 20 °C. As a contrast, higher mass results in faster pH increase so that the pH is higher after the same time. Experiments with Greek red mud showed a similar tendency, although the precipitation efficiency of the sample in heated acid mine drainage was just 14% after 5 min. In all coal fly ash samples, the aluminium was not precipitated under pH 3.5. In the experiment with heated acid mine drainage, a leaching from coal fly ash up to 89% at pH 2.8 occurred at first, then re-precipitation. As a conclusion, all samples showed similar correlation of pH and precipitation efficiency, like an s-curve with a strong increase between pH 3.0–3.6 in all experiments, see the red circle. Positive precipitation efficiency starts at pH 3.3–3.5 until pH-Value of 5. After the pH of 5.0, precipitation of aluminum is complete.

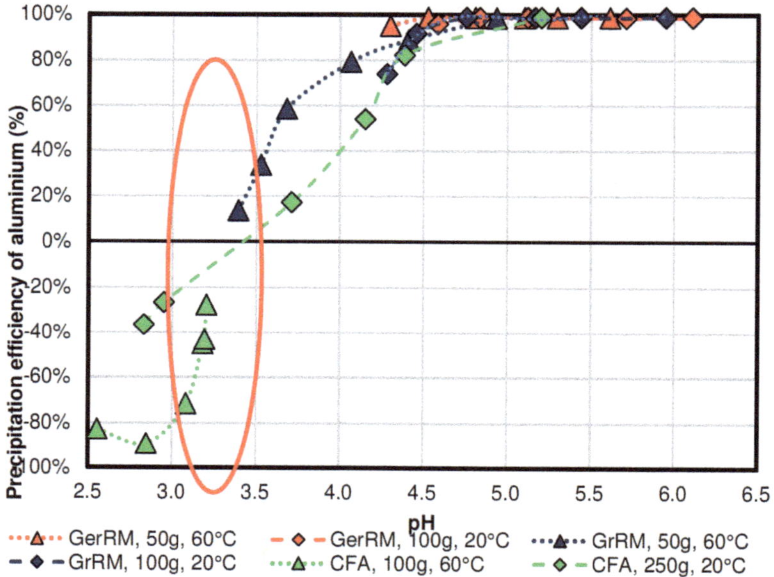

Figure 9. Participation efficiency of aluminium.

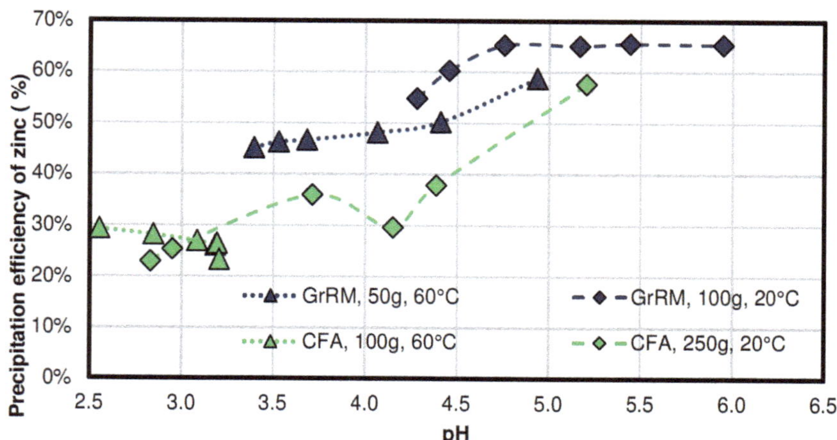

Figure 10. Participation efficiency of zinc.

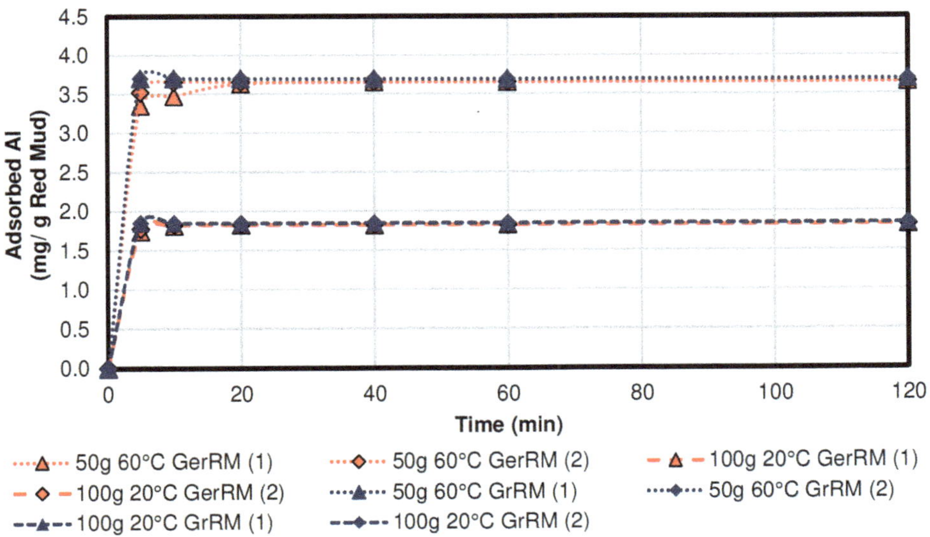

Figure 11. Kinetics of aluminium adsorption for both red muds [mg/g].

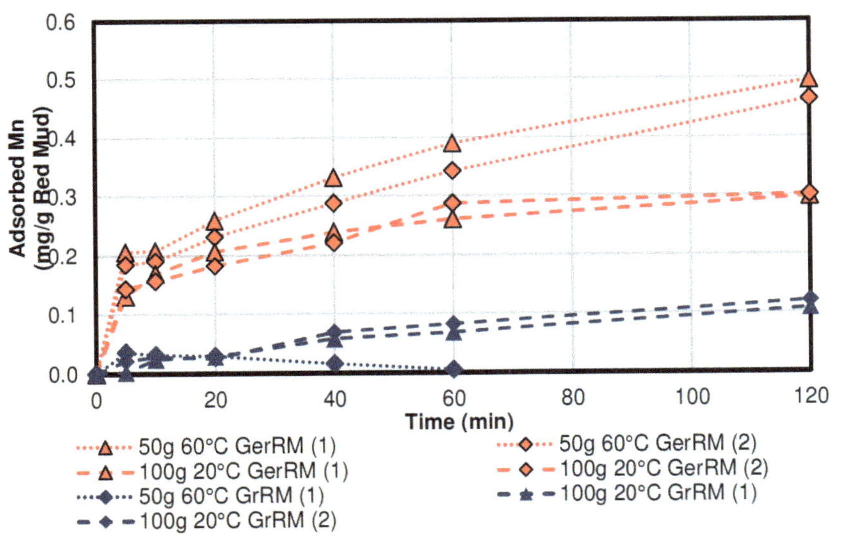

Figure 12. Kinetics of manganese adsorption for both types of red muds [mg/g].

3.2.3. Zinc

In German red mud, the zinc is completely precipitated, to under the detection limit. Zinc, as shown in Figure 10, can reach a minimum precipitation of 45% in both combinations of Greek red mud. The maximum precipitation efficiency can be at least 65% for the sample at 20 °C with 100 g and 59% for the pre-heated acid mine drainage experiment with 50 g Greek red mud. In coal fly ash, zinc was precipitated between 20% and 60% in both cases, whereas the non-heated sample had a better efficiency of 60% after 120 min. There is a linear correlation between pH and precipitation efficiency. As a conclusion, pre-heated acid mine drainage results in leaching of all metals, which is also expected of zinc but only for the first 40 min, longer neutralization leads to stronger zinc back-dissociation.

3.3. Adsorption Kinetics of Adsorbed Metals

The adsorption kinetics can be compared by using the rate of adsorbed material per gram precipitation media, see Equation (5).

$$q_t = \frac{(c_0 - c_t)_{Me} \cdot V_{AMD}}{m_{RM/CFA}} \tag{5}$$

with:

q_t rate of adsorbed material [mg/g]
c_0, c_t concentration of the metal ions [mg/L] (c_0: initial, c_t: after t min)
V_{AMD} volume of the tested AMD water [L]
$m_{RM/CFA}$ mass of the used precipitation media [g]

3.3.1. Aluminium

Figure 11 shows the adsorption kinetics for the aluminium in the acid mine drainage precipitated by the both types of red muds. The coal fly ash samples were missing due to enrichment in the AMD. The aluminium in the ash may be a soluble phase. The concentration of aluminium in the solution was 370 mg/L, therefore there were 185 mg dissolved in 500 mL. It can be recognised that the adsorption potential was higher using red mud at 60 °C than at 20 °C. This indicates that the aluminium adsorption is a temperature-activated process and therefore, it is more like chemisorption than physisorption. The increase to 60 °C doubles the adsorption potential per gram red mud from 1.8 mg/g (0.9% of the total amount of dissolved aluminium in solution) to 3.7 mg/g (2%). The difference between the use of the German red mud and the Greek red mud was relatively small, Greek red mud was a little bit better than the German type. In case of the 50 g 60 °C sample, the mass of the adsorbed part was 185 mg, which was the maximum, whereas the 100 g 20 °C adsorbed only 180 mg. In the case of aluminium, both combinations achieved equal precipitation efficiency of 100%. After 20 min (average pH = 4.751 ± 0.495), the precipitation can be assumed as completed.

3.3.2. Manganese

The concentration of manganese was 82 mg/L which means 41 mg in 500 mL acid mine drainage water. The adsorption kinetics for manganese in Figure 12 show that the German red mud had better adsorption potential than the Greek one. Because of enrichment, the 50 g sample at 20 °C with the Greek red mud is missing, see Equation (6).

$$MnO \rightarrow Mn^{2+} + O^{2-} \tag{6}$$

None of the samples have achieved the equilibrium state. The German red mud 100 g and 20 °C sample could have an equilibrium at 0.31 mg/g manganese. One gram can adsorb 0.7% of the dissolved manganese which is in total 30 mg (73% in total). For the German red mud 50 g at 60 °C sample, it can be assumed that the equilibrium is approximately 0.6 mg/g red mud which means an adsorption grade of 1.4% and therefore, 30 mg in total (also 73% in total). This means that both combinations can work equally for manganese, see Table 7. The equilibrium for the 50 g Greek red mud sample at 60 °C was negative due to leaching effects. Therefore, the 120 min adsorption rate cannot be determined exactly. Under recognition of instrumental errors, the equilibrium can be near zero.

Table 7. Adsorption kinetics of manganese by red mud with the percentage at assumed equilibrium.

Sample	GerRM 50/60	GerRM 100/20	GrRM 50/60	GrRM 100/20
rate in mg/g	~0.6	~0.31	~0	~0.15
% per gram	~1.4	~0.7	~0	~0.4
mass [mg]	~30	~30	~0	~15
% in total	~73	~73	~0	~37

3.3.3. Zinc

Figure 13 shows the sorption kinetic of zinc. The concentration of dissolved zinc was 14.3 mg/L, therefore, there were 7.15 mg dissolved in acid mine drainage. The equilibrium of zinc for coal fly ash was between 0.01 and 0.02 mg/g which were at a maximum 0.28% of the total amount of the dissolved zinc per gram in coal fly ash. The 100 g of coal fly ash could adsorb a maximum 2 mg which is 28% of the total amount and 250 g maximum 5 mg zinc (nearly 70% of the total zinc). For Greek red mud, the equilibrium at 20 °C was approximately 0.047 mg/g which is a grade of 0.66%. A 100 g portion of this sample can adsorb 4.7 mg zinc which is 66% of the total zinc in solution. The best sample with 50 g and 60 °C had a short equilibrium 0.067 mg/g for 15 min which was 0.9% adsorption per gram with total adsorption of 3.35 mg (47%) zinc. After 20 min, an increase of adsorption can be seen. This indicates that there could be another mechanism of adsorption, like precipitation. After 40 min, the mixture achieved a pH of 4.1. Regarding the results, the Greek red mud had a better adsorption potential than the coal fly ash but overall, the 250 g coal fly ash sample at 20 °C can be as useful as the 100 g of Greek red mud at 20 °C, as shown in Table 8. The precipitation follows Equation (7).

$$Zn^{2+} + 2OH^- \rightarrow Zn(OH)_2 \tag{7}$$

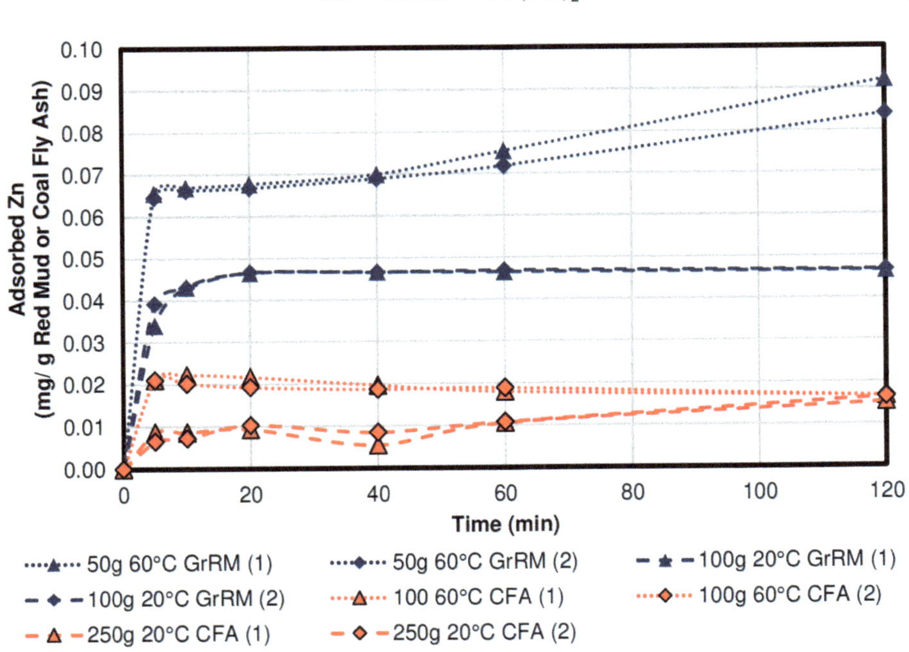

Figure 13. Kinetics of zinc adsorption for Greek red mud and coal fly ash [mg/g].

Table 8. Kinetics of zinc adsorption for Greek red mud and coal fly ash.

Sample	GrRM 50/60	GrRM 100/20	CFA 100/60	CFA 250/20
rate [mg/g]	>0.067	0.047	~0.02	~0.02
% per gram	>0.94	0.66	~0.28	~0.28
mass [mg]	>3.35	4.7	~2	~5
% in total	>47	66	~28	~70

3.4. Leaching Kinetics

Analogously, the leaching kinetics can be calculated with Equation (8) as well as the adsorption kinetics in the previous chapter.

$$q_t = \frac{(c_t - c_0)_{Me} V_{AMD}}{m_{RM/CFA}} \tag{8}$$

Sodium

Due to a high concentration of sodium hydroxide, it is the main component for neutralization of acid mine drainage. The consumption of sodium from coal fly ash for neutralization of AMD was not very high, only approximately 0.05 mg/g coal fly ash since the content of sodium in coal fly ash was just 0.35 wt.% (<1 mg/g), whereas the German red mud contains 8.9 wt.% (66 mg/g) and the Greek red mud 1.8 wt.% (13 mg/g). Regarding the average leaching kinetics of sodium in Figure A1 in Appendix A, German red mud at 60 °C has the highest leaching ability with 50 mg/g sodium which is 76% of total sodium in the sample as seen in Table 9 and 20 mg/g in the non-heated sample which was 30%. The heated Greek red mud had a similar leaching ability like the non-heated German red mud with 20 mg/g. The non-heated Greek red mud had a leaching ability of 10 mg/g which was 77% of the total sodium in the sample. As a conclusion, the non-heated German red mud and the heated Greek red mud sample have a similar ability to achieve the same neutralization pH per gram.

$$NaOH \rightarrow Na^+ + OH^- \tag{9}$$

Table 9. Content of sodium at the beginning and after leaching in the types of red mud.

Sample	GerRM 50/60	GerRM 100/20	GrRM 50/60	GrRM 100/20
Sodium in solid [mg/g]	66	66	13	13
Leached sodium [mg/g]	50	20	20	10
Percentage [%]	76	30	(153)	77

3.5. Rare Earth Concentration in the One Litre Experiments

The initial concentration of cerium is 5.51 mg/L and of yttrium was 2.05 mg/L. Figure 14 and Table 10 shows the concentration of the rare earth elements yttrium and cerium during the experiments. The detection limit for the rare earths was 1.0 mg/L, which marks the start of the y-axis. The rare earth elements (yttrium and cerium) were adsorbed by Greek red mud with an efficiency of 50% and 80% at 60 °C in 5 min, respectively. After 5 min, using German red mud can reduce the concentration of the rare earth to nearly under the detection limit (Ce: 1.12 mg/L, Y: <1 mg/L). Using Greek red mud reduces cerium to under the detection limit after 60 min, yttrium after 40 min. Using coal fly ash can result in re-dissolving into acid mine drainage which indicates an equilibrium of the concentration in the solid and in the acid mine drainage. After 60 min, the concentration of the rare earths does not change much. There is a steady-state in concentration.

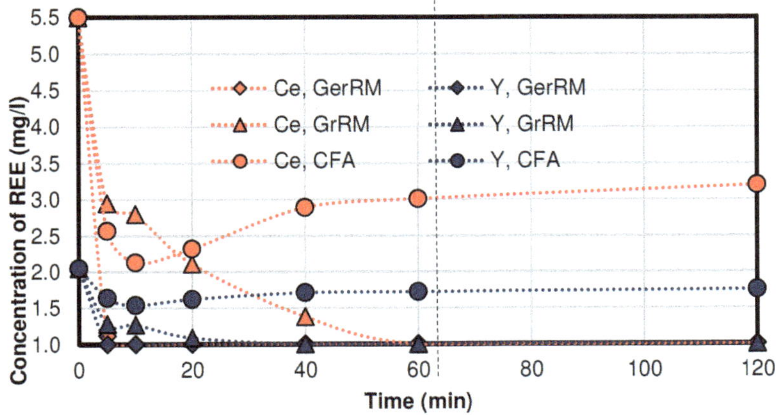

Figure 14. Concentration of yttrium and cerium in time.

Table 10. Yttrium and cerium in precipitation experiment [mg/L].

Time [min]	5	10	20	40	60	120
Ce, GerRM	1.12	<1	<1	<1	<1	<1
Y, GerRM	<1	<1	<1	<1	<1	<1
Ce, GrRM	2.95	2.80	2.11	1.38	<1	<1
Y, GrRM	1.28	1.27	1.09	< 1	<1	<1
Ce, CFA	2.57	2.13	2.32	2.89	3.01	3.20
Y, CFA	1.64	1.54	1.62	1.71	1.72	1.75

3.6. Sodium Concentration Regarding the Guidelines

The initial concentration of the sodium in acid mine drainage is 52.4 mg/L. The guideline for sodium in drink water is 200 mg/L [21], see the red line in Figure A2 in Appendix A. The German red mud leads to a final concentration of 4–5 g/L which is twenty-five times that of the recommended concentration. The final concentration of the Greek red mud was approximately 2 g/L which is tenfold of the recommended concentration. Only the coal fly ash sample fit into the guideline, with a final concentration of approximately 70 mg/g. Therefore, for the red mud samples, it is necessary to distinguish if the water is used as drink water after treatment.

With respect to the results, there is a possibility of purifying the acid mine drainage by precipitation with red mud and coal fly ash. Overall iron is removed nearly completely, and aluminium also precipitates to 99% in most samples. Same samples as Greek red mud and coal fly ash have a potential for selective removal of metals which can be useful for multistage process designing. In all red mud samples, the concentration of sodium has increased strongly.

Comparison of the adsorption kinetics shows that in some cases, the combinations of higher mass or higher temperature could give similar results like those in the experiment with German red mud and manganese (both 73% of total) or the one with both kinds of red muds and aluminium. Additionally, for zinc precipitation, the 100 g, 20 °C Greek red mud and the 250 g, 20 °C coal fly ash have equal precipitation efficiency. For economical use, higher mass and low temperature are preferred. Additionally, the red mud and the coal fly ash can replace the soil, which enables outside application for neutralisation. The neutralised red mud and coal fly ash can be used for further metal-winning treatment.

The main cause for the increase of pH is the sodium hydroxide. Since the red mud can release up to 80% of the sodium ions into acid mine drainage, a pH increase up to 6 is possible. A combination of higher mass and higher temperature can achieve the neutral water state. The coal fly ash causes pH increase by dissociation of metal oxides in acids, but the pH increase is not as high as using red mud since coal fly ash does not contain sodium oxide. The oxides release oxygen to acid mine drainage, which influences the potential of acid mine drainage, resulting in the formation of hydroxyl ions and shifts the precipitation pH to lower values.

The total amount of the rare earth elements was 9.83 mg/L, which can be reduced after neutralization under 4 mg/L using German red mud.

The German red mud performs the best results of all media. The Greek red mud also has good results, but they were not as good as the German ones. The precipitation efficiency of coal fly ash was low in some samples. Coal fly ash is usable when iron, aluminium and zinc needs to be precipitated but manganese would be dissolved in solution.

4. Conclusions

The following conclusions are found in this work:

- In all samples, the iron is removed after a pH of 3.0 (20 min) from acid mine drainage, and aluminum after a pH of 5.0.
- The required pH for drinking water (pH > 6.5) can be nearly reached.
- Higher mass of red mud and flying ash leads to higher final pH.
- In all experiments, the German red mud achieved the highest pH values.
- Because of it's high alkaline nature (pH > 12), German red mud is the best precipitation media for iron, aluminium, cerium and yttrium.
- The red muds removed the metals but enriches the dissolved sodium.
- Sample with coal fly ash enriches the acid mine drainage with sodium and zinc since sodium is contained in low concentrations.
- Increase of temperature can increase adsorption kinetics.
- Under a pH of 3, yellow flakes have precipitated in the filtrate. The flakes contain iron(III) hydroxide.
- White flakes indicate the precipitation of aluminum(III) hydroxide.

Kinetic study shows that contrast combinations of mass and temperature can achieve the same precipitation adsorption. Increasing of temperature to 60 °C can double the adsorption capacity of material like aluminium, which is equivalent to the twofold mass of material. The alternative with the higher amount of material should be preferred because of the higher pH increase due to easily-soluble sodium hydroxide. As the results indicated, there is much less metal removal from solution using the adsorption method.

Author Contributions: Conceptualization, B.X., V.K. and S.S.; Funding acquisition, S.N. and B.F.; Investigation, V.K., Y.M. and B.M.; Methodology, V.K., S.S. and B.X.; Supervision, S.N., G.S.S. and B.F.; Writing—original draft, S.S., B.X., S.N., B.M., G.S.S. and B.F. All authors have read and agreed to the published version of the manuscript.

Funding: This research was funded by the International Office of the BMBF in Germany, grant number. 01DG17024, and by NRF in South Africa (grant number: GERM160705176077). The APC was funded by the International Office of the BMBF in Germany, grant number. 01DG17024.

Conflicts of Interest: The authors declare no conflict of interest.

Appendix A

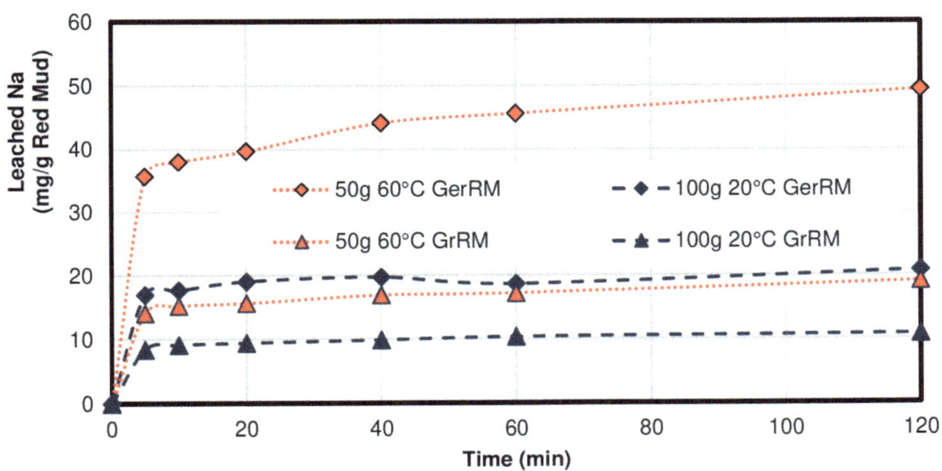

Figure A1. Kinetics of sodium adsorption for German and Greek red mud [mg/g].

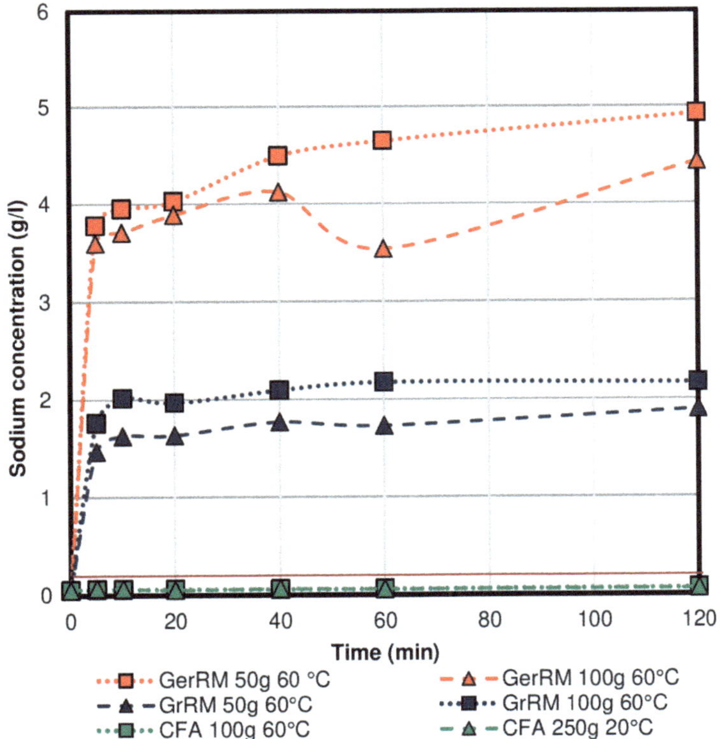

Figure A2. The concentration of sodium in treated acid mine drainage regarding the guidelines [18].

References

1. WHG. *Gesetz zur Ordnung des Wasserhaushalts*; German Federal Ministry of Justice and Consumer Protection: Berlin, Germany, 2009.
2. Knight, J.; Rogerson, C.M. *The Geography of South Africa*, 1st ed.; Springer: Cham, Switzerland, 2019; pp. 27–31.
3. Mccarthy, T.S. The impact of acid mine drainage in South Africa. *S. Afr. J. Sci.* **2011**, *107*, 5–6. [CrossRef]
4. Tabelin, C.B.; Igarashi, T.; Tabelin, M.V.; Park, I.; Opiso, E.M.; Ito, M. Arsenic, selenium, boron, lead, cadmium, copper, and zinc in naturally contaminated rocks: A review of their sources, modes of enrichment, mechanisms of release, and mitigation strategies. *Sci. Total Environ.* **2018**, *645*, 1522–1553. [CrossRef] [PubMed]
5. Park, I.; Tabelin, C.B.; Jeon, S.; Li, X.; Seno, K.; Ito, M.; Hiroyoshi, N. A review of recent strategies for acid mine drainage prevention and mine tailings recycling. *Chemosphere* **2019**, *219*, 588–606. [CrossRef] [PubMed]
6. Igarashi, T.; Herrera, P.S.; Uchiyama, H.; Miyamae, H.; Iyatomi, N.; Hashimoto, K.; Baltazar, C.B.T. The two-step neutralization ferrite-formation process for sustainable acid mine drainage treatment: Removal of copper, zinc and arsenic, and the influence of coexisting ions on ferritization. *Sci. Total Environ.* **2020**, *715*, 136877. [CrossRef] [PubMed]
7. Bwapwa, J.K. A Review of Acid Mine Drainage in a Water-Scarce Country: Case of South Africa. *Environ. Manag. Sustain. Dev.* **2018**, *7*, 1–20. [CrossRef]
8. Singer, P.C.; Stumm, W. Acidic Mine Drainage: The Rate-Determining Step. *Science* **1970**, *167*, 1121–1123. [CrossRef]
9. Plumlee, G.S.; Smith, K.S.; Montour, M.R.; Ficklin, W.H.; Mosier, E.L. Geologic Controls on the Composition of Natural Waters and Mine Waters Draining Diverse Mineral-Deposit Types. In *The Environmental Geochemistry of Mineral Deposits. Reviews in Economic Geology*; Society of Economic Geologists: Littleton, CO, USA, 1999; Chapter 19; Volume 6, pp. 373–435.
10. Lim, J.; Yu, J.; Wang, L.; Jeong, Y.; Shin, J.H. Heavy Metal Contamination Index Using Spectral Variables for White Precipitates Induced by Acid Mine Drainage: A Case Study of Soro Creek. *IEEE Trans. Geosci. Remote Sens.* **2019**, *57*, 1–19. [CrossRef]
11. Stiefel, R. *Abwasserrecycling: Technologien und Prozesswassermanagement: Das Konzept Prozesswasserautarkie*, 1st ed.; Springer: Wiesbaden, Germany, 2017; p. 104.
12. Smit, J.P. *The Treatment of Polluted Mine Water, Presented at the Mine, Water and Environment*; International Mine Water Association: Sevilla, Spain, 1999.
13. Dietrich, G. *Hartinger Handbuch Abwasser und Recyclingtechnik*, 3rd ed.; Carl Hanser Verlag: München, Germany, 2017; pp. 57–68, 102, 113–118.
14. Baur, A.; Fritsch, P.; Hoch, W.; Merkl, G.; Rautenberg, J.; Weiß, M.; Wricke, B. *Mutschmann/Stimmelmayr Taschenbuch der Wasserversorgung*, 17th ed.; Springer: Wiesbaden, Germany, 2019; pp. 5, 236, 252, 258, 286–300, 312–317, 327–332.
15. Crane, R.A.; Sapsford, D.J. Selective formation of copper nanoparticles from acid mine drainage using nanoscale zerovalent iron particles. *J. Hazard. Mater.* **2018**, *347*, 252–265. [CrossRef]
16. Asokbunyarat, V.; van Hullebusch, E.D.; Lens, P.N.L.; Annachhatre, A.P. Immobilization of Metal Ions from Acid Mine Drainage by Coal Bottom Ash. *Water Air Soil Pollut.* **2017**, *228*, 328. [CrossRef]
17. Sawyer, C.N. *Chemistry for Environmental Engineering and Science*, 5th ed.; McGraw-Hill: New York, NY, USA, 2003; p. 99.
18. Paradis, M.; Duchesne, J.; Lamontagne, A.; Isabel, D. Long-term neutralisation potential of red mud bauxite with brine amendment for the neutralisation of acidic mine tailings. *Appl. Geochem.* **2007**, *22*, 2326–2333. [CrossRef]
19. World Health Organization. *Guidelines for Drinking-Water Quality*, 4th ed.; World Health Organization: Geneva, Switzerland, 2011; pp. 223–228.
20. German Federal Ministry of Justice and Consumer Protection. *Verordnung über Anforderungen an das Einleiten von Abwasser in Gewässer*; AbwV, German Federal Ministry of Justice and Consumer Protection: Berlin, Germany, 1997.
21. German Federal Ministry of Justice and Consumer Protection. *Verordnung über die Qualität von Wasser für den Menschlichen Gebrauch*; TrinkwV, German Federal Ministry of Justice and Consumer Protection: Berlin, Germany, 2001.

22. Christen, D.S. *Praxiswissen der chemischen Verfahrenstechnik: Handbuch für Chemiker und Verfahrensingenieure*, 2nd ed.; Springer: Berlin/Heidelberg, Germany, 2010; pp. 333–366.

23. Kurzweil, P. *Chemie: Grundlagen, Aufbauwissen, Anwendungen und Experimente*, 10th ed.; Springer: Wiesbaden, Germany, 2015; pp. 177–178.

24. German Federal Environment Agency. *Bekanntmachung der Liste der Aufbereitungsstoffe und Desinfektionsverfahren gemäß § 11 der Trinkwasserverordnung—20. Änderung—(Stand: Dezember 2018)*; German Federal Environment Agency: Dessau-Roßlau, Germany, 2018.

25. Mortimer, C.E.; Müller, U. *Chemie*, 9th ed.; Georg Thieme Verlag KG: Stuttgart, Germany, 2007; p. 475.

26. Kammer, C. *Aluminium Taschenbuch Band 1: Grundlagen und Werkstoffe*, 16th ed.; Aluminium: Düsseldorf, Germany, 2002; pp. 20–25.

27. Castaldi, P.; Silvetti, M.; Santona, L.; Enzo, S.; Melis, P. XRD, FTIR, and Thermal Analysis of Bauxite Ore-Processing Waste (Red Mud) Exchanged with Heavy Metals. *Clays Clay Miner.* **2008**, *56*, 461–469. [CrossRef]

28. Kutchko, B.G.; Kim, A.G. Fly ash characterization by SEM–EDS. *Fuel* **2006**, *85*, 2537–2544. [CrossRef]

29. Ahmed, M.J.K.; Ahmaruzzaman, M. A review on potential usage of industrial waste materials for binding heavy metal ions from aqueous solutions. *J. Water Process. Eng.* **2016**, *10*, 39–47. [CrossRef]

30. Smičiklas, I.; Smilijanic, S.; Peric, A.-G.; Sljivic, M.-I.; Mitric, M.; Antonovic, D. Effect of acid treatment on red mud properties with implications on Ni(II) sorption and stability. *Chem. Eng. J.* **2014**, *242*, 27–35. [CrossRef]

31. Sahu, M.K.; Mandal, S.; Dash, S.S.; Badhai, P.; Patel, R.K. Removal of Pb(II) from aqueous solution by acid activated red mud. *J. Environ. Chem. Eng.* **2013**, *1*, 1315–1324. [CrossRef]

32. Nadaroglu, H.; Kalkan, E.; Demir, N. Removal of copper from aqueous solution using red mud. *Desalination* **2010**, *251*, 90–95. [CrossRef]

33. Visa, M.; Isac, L.; Duta, A. Fly ash adsorbents for multi-cation wastewater treatment. *Appl. Surf. Sci.* **2012**, *258*, 6345–6352. [CrossRef]

34. Qiu, W.; Zheng, Y. Removal of lead, copper, nickel, cobalt, and zinc from water by a cancrinite-type zeolite synthesized from fly ash. *Chem. Eng. J.* **2009**, *145*, 483–488. [CrossRef]

35. Smart, L.E.; Moore, E.A. *Solid State Chemistry—An Introduction*, 3rd ed.; Taylor & Francis Group: Boca Raton, FL, USA, 2005; p. 301.

36. Prasad, B.; Kumar, H. Treatment of Lignite Mine Water with Lignite Fly Ash and Its Zeolite. *Mine Water Environ.* **2019**, *38*, 24–29. [CrossRef]

37. Kaussen, F.; Friedrich, B. Phase characterization and thermochemical simulation of (landfilled) bauxite residue ("red mud") in different alkaline processes optimized for aluminum recovery. *Hydrometallurgy* **2018**, *176*, 49–61. [CrossRef]

38. Paradis, M.; Duchesne, J.; Lamontagne, A.; Isabel, D. Using red mud bauxite for the neutralization of acid mine tailings: A column leaching test. *Can. Geotech. J.* **2006**, *43*, 1167–1179. [CrossRef]

 *minerals*

Article

Effects of Backfilling Excavated Underground Space on Reducing Acid Mine Drainage in an Abandoned Mine

Kohei Yamaguchi [1,2,*], Shingo Tomiyama [3], Toshifumi Igarashi [3], Saburo Yamagata [4], Masanori Ebato [5] and Masatoshi Sakoda [6]

[1] Division of Sustainable Resources Engineering, Graduate School of Engineering, Hokkaido University, Sapporo 060-8628, Japan

[2] Mitsubishi Materials Corporation, 1-600, Kitabukuro-cho, Omiya-ku, Saitama-shi, Saitama 330-8508, Japan

[3] Faculty of Engineering, Hokkaido University, Sapporo 060-8628, Japan; tomiyama@mmc.co.jp (S.T.); tosifumi@eng.hokudai.ac.jp (T.I.)

[4] Mitsubishi Materials Corporation, 3-2-3, Marunouchi, Chiyoda-ku, Tokyo 100-8117, Japan; s-yamaga@mmc.co.jp

[5] Oyo Corporation, 2-10-9, Daitakubo, Minami-ku, Saitama-shi, Saitama 336-0015, Japan; ebato-masanori@oyonet.oyo.co.jp

[6] Japan Oil, Gas and Metals National Corporation (JOGMEC), 2-10-1, Toranomon, Minato-ku, Tokyo 105-0001, Japan; sakoda-masatoshi@jogmec.go.jp

* Correspondence: kyama@mmc.co.jp; Tel.: +81-48-641-5691

Received: 14 June 2020; Accepted: 28 August 2020; Published: 31 August 2020

Abstract: Three-dimensional groundwater flow around an abandoned mine was simulated to evaluate the effects of backfilling the excavated underground space of the mine on reducing the acid mine drainage (AMD). The conceptual model of the groundwater flow consists of not only variable geological formations but also vertical shafts, horizontal drifts, and the other excavated underground space. The steady-state groundwater flow in both days with high and little rainfall was calculated to calibrate the model. The calculated groundwater levels and flow rate of the AMD agreed with the measured ones by calibrating the hydraulic conductivity of the host rock, which was sensitive to groundwater flow in the mine. This validated model was applied to predict the flow rate of the AMD when backfilling the excavated underground space. The results showed that the flow rate of the AMD decreased by 5% to 30%. This indicates that backfilling the excavated space is one of the effective methods to reduce AMD of abandoned mines.

Keywords: abandoned mine; groundwater flow analysis; acid mine drainage; backfilling

---

1. Introduction

Acid mine drainage (AMD) is generated at many active, closed, and abandoned mines throughout the world. AMD is a serious environmental issue in the mining industry [1–9], which is generally characterized by low pH and high concentrations of sulfate, heavy metals (e.g., copper (Cu), lead (Pb), zinc (Zn), and cadmium (Cd)) [10,11], and metalloids (e.g., arsenic (As), [12]). Low pH is caused by oxidation of sulfide minerals and dissolves heavy metals in the host rock. Moreover, the contribution of bacteria is important under in situ conditions for AMD formation [13,14]. The chemical reactions (1)–(4) [15] causes low pH when the oxygenated rainwater contacts with pyrite in the unsaturated zone. Backfilling the excavated underground space may reduce the amount of AMD due to the decrease in hydraulic gradient [16,17]. In addition, the contact area between oxygenated rainwater and pyrite decreases due to the rise of groundwater levels.

$$FeS_2(s) + 7/2O_2 + H_2O = Fe^{2+} + 2SO_4^{2-} + 2H^+ \tag{1}$$

$$Fe^{2+} + 1/4O_2 + H^+ = Fe^{3+} + 1/2H_2O \tag{2}$$

$$Fe^{3+} + 3H_2O = Fe(OH)_3(s) + 3H^+ \tag{3}$$

$$FeS_2(s) + 14Fe^{3+} + 8H_2O = 15Fe^{2+} + 2SO_4^{2-} + 16H^+ \tag{4}$$

Sulfide minerals in the host rock exposed to shallow groundwater or rainwater are oxidized during the operation of the mine. In addition, AMD is continuously generated for more than several decades after closing or abandoning the mines [18].

In Japan, there are over 5500 closed or abandoned non-ferrous metal mines and 79 mines of these are treating AMD. AMD from closed or abandoned mines is commonly treated by neutralization with hydrated lime or sodium hydroxide. In this application, most toxic heavy metals are precipitated and removed [19], and then the treated water is released into nearby rivers. The treatment of AMD induces a large load from an economic perspective although lime neutralization has effectively been used over the last 40 years in Japan. Total subsidiary aid cost for about 50 years is ~70 billion JPY (~650 million USD) in Japan [20].

Active treatments by adding alkaline reagents are costly and should last for decades. On the other hand, passive treatments are expected to be applied to mines with a relatively lower flow rate of AMD [21]. For both treatments, it is necessary to reduce the amount of produced AMD. Thus, it is important to clarify the processes of generating AMD. There have been a variety of studies of AMD monitoring and characterization in not only Japan (e.g., [22–24]) but also other countries [25–34]. Some studies pointed out that groundwater flow patterns in and around mines should be evaluated to purpose countermeasures against AMD reduction. There are several methods of the mitigation, such as covering of ground surface with low-permeable layers and impoundment of land subsidence [35–41]. In this study, the effects of backfilling the underground space in a mine on produced AMD were examined by 3-D groundwater flow model to reduce the AMD produced because the mine selected has a huge volume of underground space already excavated.

2. Geology and Mining of Study Area

The selected mining area is located in valley terrain at an altitude of 300 to 400 m (Figure 1). The basement rock mostly consists of pre-Neogene granite. Neogene tuff, andesite, and rhyolite were deposited on the granite basement. They erupted and deposited on the granite in the marine. The type of mine is a vein-type deposit formed in faults and fractures in the Neogene strata. Distribution of the mined area was recorded in the pit map created during the operating period when drifts were excavated at a depth of 60, 90, and 150 m from the ground surface (−2 L, −3 L, and −5 L levels, respectively; Figure 1). Among them, the −5 L level drift is used as drainage of mine water to the mine mouth. Mining was carried out by the shrinkage method at the beginning of the operation because the host rock of the Neogene strata was solid and because the veins in the Neogene strata were inclined with slopes of 70° to 90°. Since the mine is producing sludge generated from AMD neutralization, backfilling the formed sludge mixed with cement (sandy slime) into excavated underground space was adopted to prevent collapses of the space. This is because the host rock to be excavated became crumblier with the progress of operation.

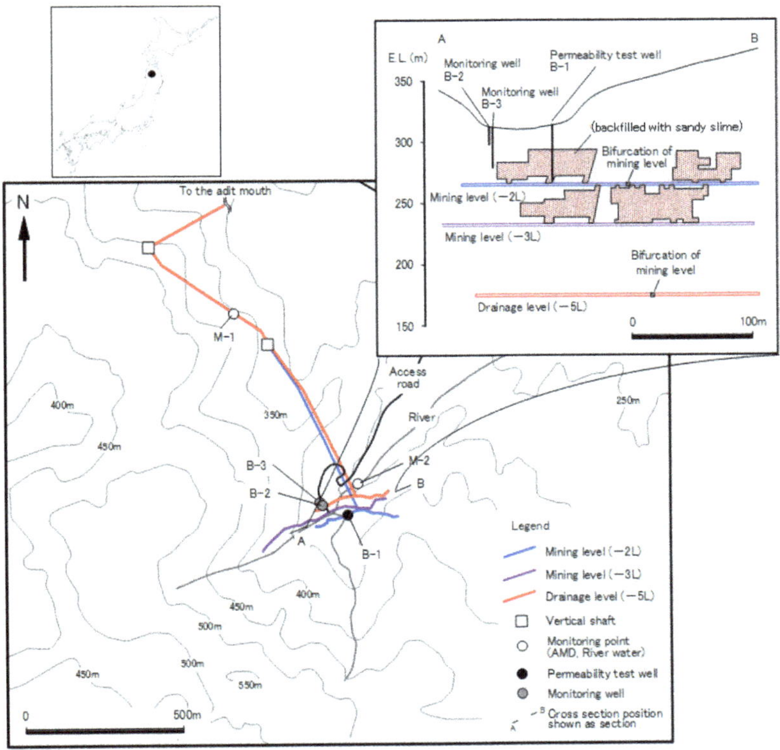

Figure 1. Distribution of old mining levels and excavated underground space of the study area.

3. Conceptual Model of Groundwater Flow

Groundwater flows through aquifers consisting of surface soil and weathered layers at this mine site. Groundwater in the aquifers flows to mining areas through faults and veins as shown in Figure 2 [42]. Groundwater in the mined area flows through the mining levels (−2 L, −3 L, and −5 L) to the adit mouth of the drainage level (−5 L). Thus, the total head of groundwater decreases from the ground surface to the deeper underground, and thereby duplicate aquifers, a shallower aquifer, and a deeper aquifer, are formed.

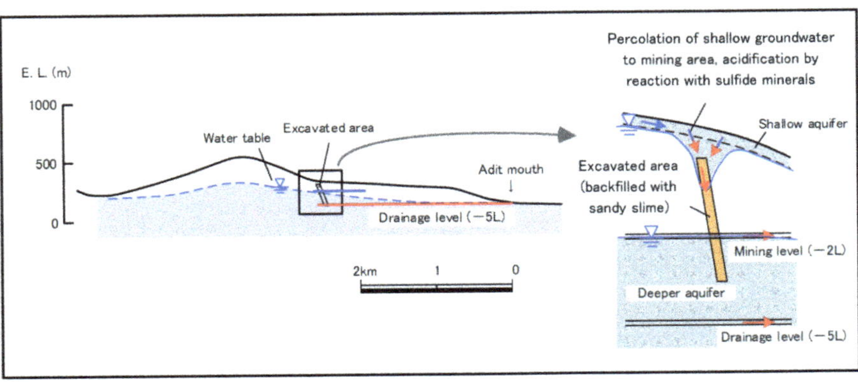

Figure 2. Conceptual model of groundwater flow and its discharge.

The flow rate of AMD is about 4 to 10 m^3/min from this mine. The AMD from the adit mouth of -5 L level accounts for about 0.04 to 0.16 m^3/min.

4. Methods

4.1. In Situ Survey

Groundwater levels were continuously measured at B-2 and B-3 from August 4, 2014 to December 17, 2016 at intervals of 60 min. The strainer pipes were installed from 15.0 to 35.0 m deep of B-2 well and from 3.0 to 15.0 m deep of B-3 well. B-2 is located less than 10 m away from B-3 and both wells are at the same ground level. This means that B-2 is to monitor the deeper groundwater level whereas B-3 is to monitor the shallower groundwater level. The location of these boreholes corresponds to upstream of groundwater into underground space, sensitive to the AMD production.

The flow rate of AMD was measured at M1 in Figure 1 using triangular weirs installed in the drain of the tunnel. The flow rate of AMD was measured from 4 August 2014 to 17 December 2016 at intervals of 60 min. The daily precipitation was calculated by accumulating hourly precipitations measured at the observatory 2 km away from the study area.

4.2. Theoretical Equation

Saturated–unsaturated groundwater flow analysis was applied to evaluate groundwater flow using Dtransu-3D-EL software [42,43] together with G-TRAN/3D pre- and post-processing softwares for Dtransu-3D [36,40]. Dtransu-3D-EL software solves the equation for saturated–unsaturated groundwater flow derived from the mass preservation and Darcy's equation, which can be written as

$$\rho_f \theta \gamma \frac{\partial c}{\partial t} + \rho\{\beta Ss + Cs(\theta)\}\frac{\partial \varphi}{\partial t} = \frac{\partial}{\partial x_i}\{\rho K_{ij}^S K_r(\theta)\frac{\partial \varphi}{\partial x_j} + \rho K_{i3}^S K_r(\theta)\rho_r\} \tag{5}$$

where φ is the pressure head, θ is the (volumetric) water content, Ss is the specific storage, $Cs(\theta)$ is the specific water capacity, K^S_{ij} is the directional components of the saturated hydraulic conductivity function, K_r is the relative hydraulic conductivity, t is time, ρ_f is the density of solvent, ρ is the density of fluid, ρ_r is the ratio of ρ_f to ρ, $\beta = 1$ is the saturated zone, $\beta = 0$ is the unsaturated zone, and γ is the solute density ratio [43].

4.3. Numerical Model

Basic configuration of the numerical model is shown in Table 1. The model domain had an area of 0.58 km^2 with a total elevation of 520 m and bounded by topographic ridges as shown in Figure 3. The mined area above the mining levels (-2 L, -3 L, and -5 L levels) was assumed to be in an unsaturated zone. The ground surface topography was reproduced by using the digital elevation model (DEM) created based on numerical maps by aerial survey. Ore body, mining levels (-2 L, -3 L, and -5 L) and excavated area (backfilled with sandy slime) were constructed in the mesh diagram of the numerical model based on the handwritten ore map and level map with 1:3000 scale. First, the ground plane was divided into squares with a side of 20 m. Next, a two-dimensional square on the ground surface were extended in the underground direction to form quadrangular prisms. Quadrangular prisms were added downward to create a three-dimensional mesh model.

Table 1. Basic configuration of the numerical model.

Items		Configurations
Boundary conditions	Side and bottom	Impermeable
	Ground surface	Infiltration: 5.0 mm/day (days with high rainfall) 0.47 mm/day (days with little rainfall)
	River	Pressure head: 0 m
	Mining levels (−2 L, −3 L, and −5 L)	Seepage
Finite element method grid	Number of elements	19,890
	Number of layers	12
	Basic element size	x = 20 m; y = 0.5–140 m; z = 20 m
Method of analysis		Saturated-unsaturated three-dimensional seepage analysis

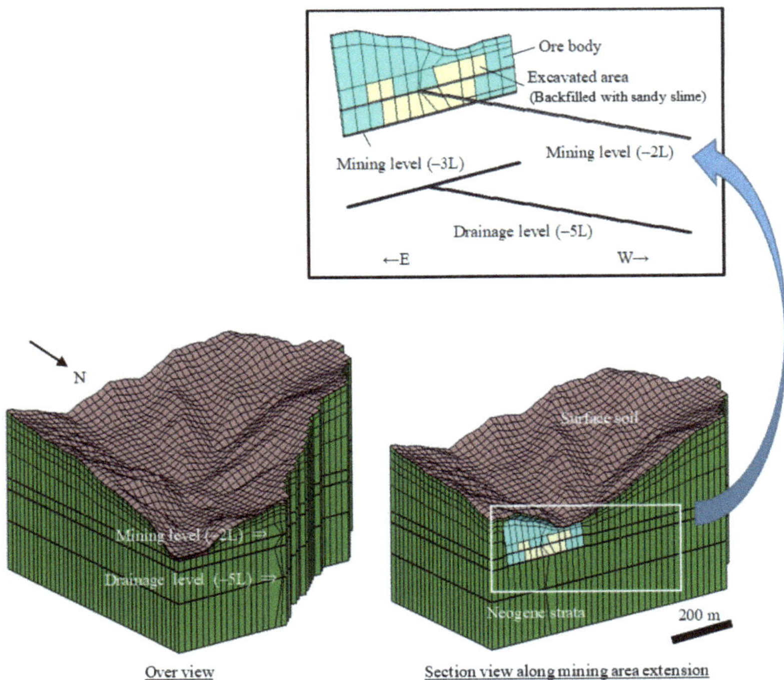

Figure 3. Simulation model of three-dimensional groundwater flow.

The hydraulic conductivities (K) and unsaturated properties of numerical blocks representing the surface soil, embankment, and Neogene strata as well as the excavated areas were based on either in situ measured results or estimated from the results of other papers as listed in Table 2. Hydraulic conductivities were obtained from in situ permeability tests or estimated from rock permeability data collected throughout Japan [44].

The properties of various geological strata under unsaturated conditions were obtained by the van Genuchten model [45], which are given by

$$S_e(\varphi) = \frac{\theta - \theta_r}{\theta_s - \theta_r} = \left[\frac{1}{1 + |a\varphi|^n}\right]^m \tag{6}$$

$$K_r\{S_e(\varphi)\} = S_e{}^l[1 - (1 - S_e{}^{1/m})^m]^2 \qquad (7)$$

where S_e is effective water saturation, θ_s is the saturated water content, θ_r is the residual water content, a, m, and n are the van Genuchten parameters [45], and l is a parameter representing the degree of pore connectivity (no unit). The van Genuchten parameters used for the current simulations are presented in Table 3, and the corresponding functions are depicted in Figure 4.

Table 2. Hydraulic conductivities of different geological strata.

Type of Geological Stratum	Hydraulic Conductivity K (m/s)	Reference
Surface soil	1.1×10^{-5}	[44]
Embankment	7.0×10^{-7}	[46]
Neogene strata	$1.0 \times 10^{-9}, 9.2 \times 10^{-9},$ $1.9 \times 10^{-8}, 9.4 \times 10^{-8}$	in situ measurements
Ore body	1.9×10^{-6}	in situ measurements
Mining levels (−2 L, −3 L, and −5 L)	1.0×10^{-1}	[40]
Excavated area (backfilled with sandy slime)	7.3×10^{-6}	in situ measurements

Table 3. Residual water content θ_r, saturated water content θ_s, and van Genuchten parameters a, l, and n [45].

	θ_r (-)	θ_s (-)	a (m^{-1})	l (-)	n (-)	Reference
Tailings	0.00	0.42	0.15	0.5	1.87	[46]
Sand slime	0.01	0.29	3	0.5	3.72	[47]
Andesite	0.00	0.104	0.06	0.5	3.57	[48]

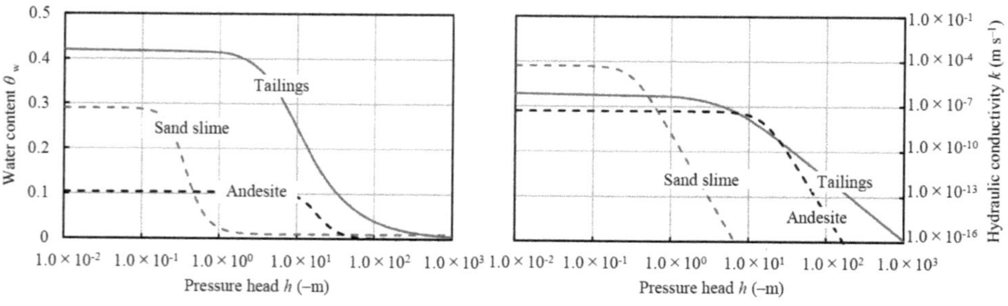

Figure 4. Water content and hydraulic conductivity curves using the van Genuchten's model for sandy slime, tailings, and andesite. See Table 3 for the corresponding parameters of the van Genuchten model [45].

The distribution of mined areas and tunnels (−2 L, −3 L, and −5 L levels) in the numerical model was based on the records of the operating period. At that time, the thickness of the mining area was set at 3 m based on the actual space size. Similarly, the height and width of mining levels were set at 2 m. The average rainfall in August 2014 was 13.5 mm/day (hereafter, days with high rainfall) and the average in August 2015 was 3.5 mm/day (hereafter, days with little rainfall). The infiltration rate was calculated based on the water balance analysis, days with high rainfall: 5.0 mm/day (recharge rate 0.37) and days with little rainfall: 0.47 mm/day), (recharge rate 0.13). Both mining level (−2 L level) and drainage level (−5 L level) were assumed to be seepage boundary conditions.

The boundary conditions of the numerical simulations are as follows:

- The river was assumed to be connected to the groundwater surface and was set as a fixed head boundary condition (pressure head 0 m).
- The ground surface (except that of the river) was set as infiltration boundary condition (Table 1).

The flow chart of the simulation is the same as Nishigaki (1995) [49].

5. Results and Discussion

5.1. Monitored Data

The groundwater level of B-2 in the deeper aquifer was almost constant at the elevation of about 287 m while that of the B-3 in the shallow aquifer was changed in response to rainfall. This means that the groundwater level of B-3 is sensitive to rain whereas that of B-2 is not so sensitive to rain.

The AMD flow rate is shown in Figure 5. The maximum value was 0.16 m^3/min in April during the snowmelt season. Although the AMD flow rate tended to increase during the snowmelt season, the AMD during the summer varied from year to year, high in 2014 and low in 2015. This means that the infiltration of rainwater directly affects the flow rate of the AMD.

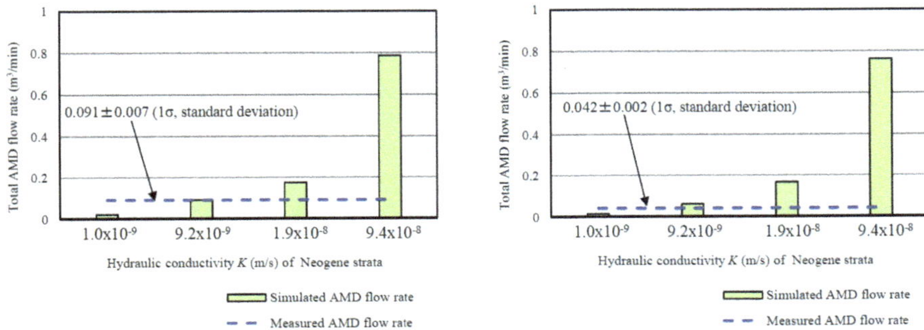

Figure 5. Comparison between the simulated and measured acid mine drainage (AMD) flow rate of days with high (**right**) and little (**left**) rainfall.

5.2. Calibration of Hydraulic Conductivity of the Neogene Strata

The study area is snowy in winter and snowmelt increases the amount of AMD in spring (March to May). However, the rain depends on the year during summer. Thus, the AMD flow rates in days with high rainfall (August 2014) and days with little rainfall (August 2015) were selected for calibration. In order to evaluate seasonal variations of AMD flow rates, analysis of steady state groundwater flow was performed for both days with high and little rainfall.

The simulated AMD flow rates by changing the hydraulic conductivity of the Neogene strata ranging from 1.0×10^{-9} m/s to 9.4×10^{-8} m/s are compared with those measured rates as shown in Figures 5 and 6. Since measured hydraulic conductivity of the Neogene strata ranged from 1.0×10^{-9} m/s to 9.4×10^{-8} m/s, the hydraulic conductivity was parametrically changed in the simulation. The calculated AMD flow rate strongly depended on the hydraulic conductivity of the Neogene strata. The AMD flow rate was the highest in case of the highest hydraulic conductivity of 9.4×10^{-8} m/s, and lowest in case of the lowest hydraulic conductivity of 1.0×10^{-9} m/s. When the hydraulic conductivity of the Neogene strata was 9.2×10^{-9} m/s, the flow rate of AMD agreed with the measured ones during both days with high and little rainfall. This means that the same hydraulic conductivity of the Neogene strata is applicable to any season.

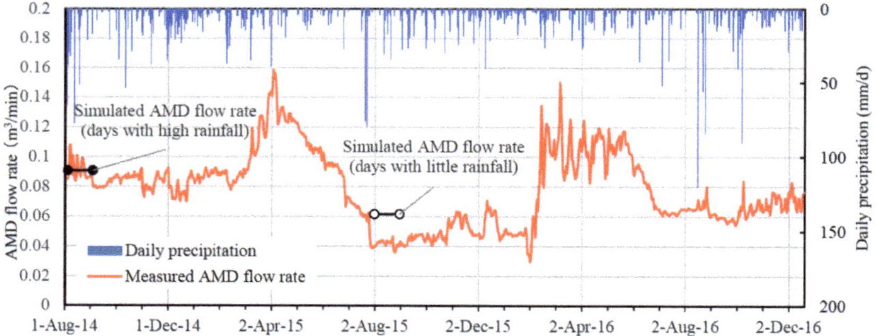

Figure 6. Change of precipitation and AMD flow rate.

The sensitivity analysis of hydraulic conductivity of the ore body was also conducted to investigate whether the ore body distributed on the upper part of −5 L affected the flow rate of AMD. When the hydraulic conductivity of the ore body was reduced to 1.9×10^{-5} m/s, the AMD increased by 20%, and when the hydraulic conductivity was increased to 1.3×10^{-5} m/s, the AMD decreased by 10%. This indicates that the effect of hydraulic conductivity of the Neogene strata on AMD flow rate was more significant than that of the ore body. It was also found that the amount of AMD from the deeper −5 L decreased when the shallower ore body collected more groundwater.

Calculated groundwater levels are compared with observed ones in boreholes as shown in Figure 7. The measured groundwater levels agreed with the calculated ones, irrespective of seasons when the hydraulic conductivity of the Neogene strata of 9.2×10^{-9} m/s was used. The measured and simulated levels of B-3 increased during the days with high rainfall and decreased during the little rain days. The simulated groundwater level of B-2 had a difference of about 4 m between the days with high and little rainfall. However, the measured values of B-2 were constant during both seasons. This indicates that the groundwater flow model can simulate the shallower groundwater flow and not deeper groundwater flow well.

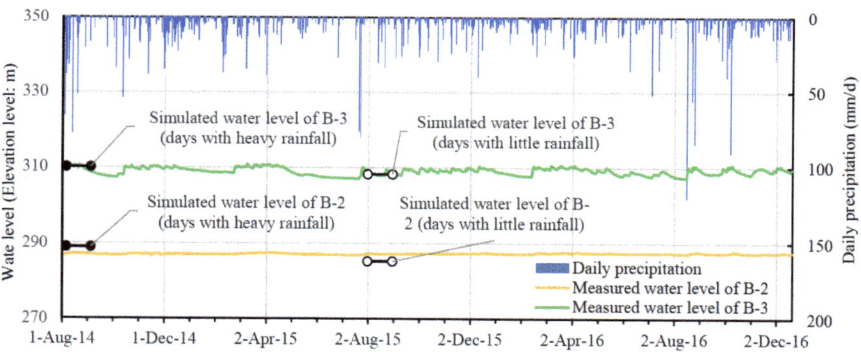

Figure 7. Change of precipitation and groundwater levels at B-2 and B-3.

Figure 7 shows changes in groundwater levels of monitoring wells of B-2 and B-3. The simulated results agreed with the observed ones. This implies that the calibrated model is effective in evaluating groundwater levels of both rainwater-sensitive and rainwater-insensitive wells.

The simulated value of B-2 responds to the increase/decrease in rainfall with a water level difference of 3.8 m between the days with high rainfall and little rainfall. Even if the amount of rainfall is large, if it runs off the surface layer, it may not contribute to the rise in water level. Therefore, we conducted a sensitivity analysis in which the hydraulic conductivity of the surface soil was increased by 10%. As a result, the water level difference between days with high rainfall and little rainfall was 3.7 m. Although it had the effect of suppressing the increase/decrease in water level difference, it was small.

5.3. Prediction of the Effect of Backfilling

Figure 8 shows the vertical distribution of total head and Darcy velocity along the drainage tunnel simulated by 3-D groundwater flow model. Before backfilling the underground space, the total head decreased from the ground surface to the mining level (−2 L level) as shown in Figure 8b. Both groundwater surfaces of the shallower and lower levels around the mining level (−2 L level) were clearly observed in the simulation (Figure 8b). The AMD flow rate exceeded 1.0 m/day as Darcy flow velocity around the mining level (−2 L) (Figure 8c). This flow rate corresponds to 0.091 m^3/min of AMD produced. After backfilling the mining levels (−2 L, −3 L, and −5 L), AMD flow rate was calculated at 0.059 m^3/min, 64.9% of AMD flow rate before backfilling, during days with high rainfall (35.1% reduction). In days with little rainfall, AMD flow rate was calculated at 0.059 m^3/min after backfilling, which was 95.1% of the flow rate before backfilling (4.9% reduction). This indicates that the AMD flow rate was reduced after backfilling whether it was the rainy season or not. In particular, backfilling the underground space was more effective during the rainy season. On the other hand, the groundwater level rose after the mined area was backfilled (Figure 8b). In addition, the groundwater flow around the −2 L level was reduced after the backfilling (Figure 8c). Higher groundwater level could mitigate the reaction between atmospheric oxygen and sulfide minerals in the underground. Thus, the rise in groundwater level may prevent AMD formation. If DO (Dissolved Oxygen) is consumed in the saturated zone, there is no new supply of oxygen and the oxidation reaction of pyrite does not occur. The geochemical modelling like previous theoretical model and experiments [50–52] is necessary to estimate the amount of acidification.

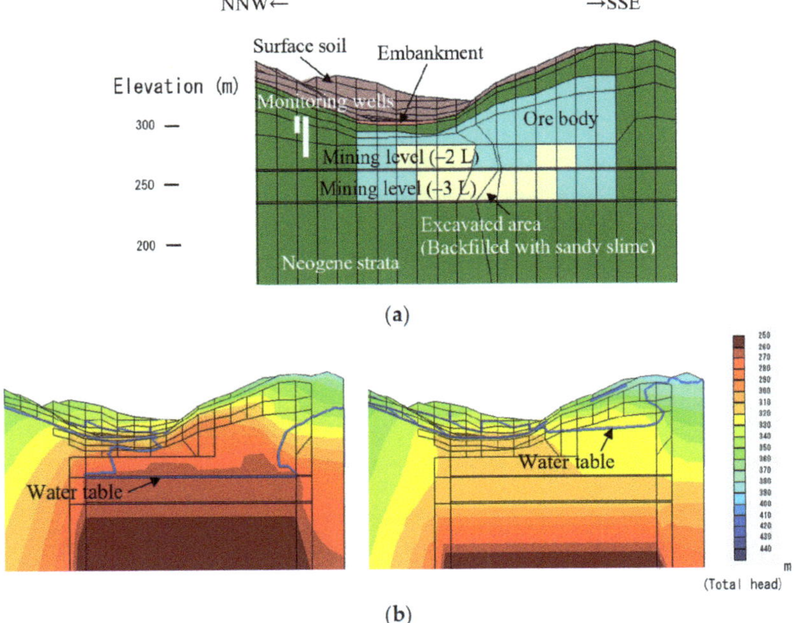

Figure 8. *Cont.*

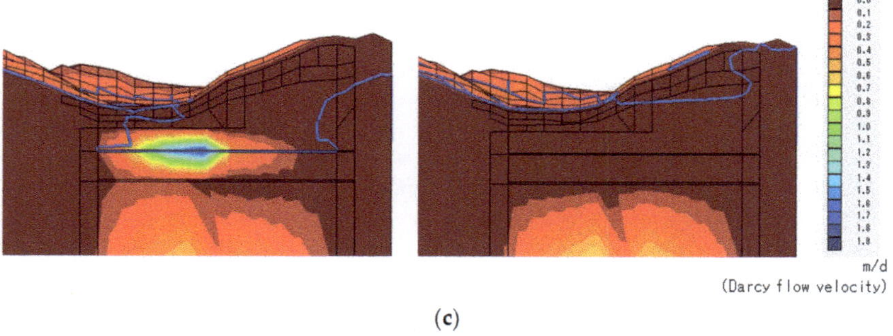

(**c**)

Figure 8. Vertical section of simulation model and simulated results of groundwater flow. (**a**) Vertical section of simulation model. (**b**) Distribution of total heads before (left) and after backfilling (right). (**c**) Distribution of Darcy flow velocity before (left) and after backfilling (right).

The measured amounts of AMD showed a difference of about twice in the days with high rainfall and the days with little rainfall as shown in Figure 8 and 1.5 times in the simulation. The amount of AMD in the days with high rainfall and that in the days with little rainfall were almost the same after backfilling because AMD was not significantly affected by the infiltration of rainwater.

The obtained results imply that backfilling underground space is effective in reducing the flow rate of AMD and preventing AMD formation is also expected. The backfilling method is promising when inclined water-conductive zones, such as excavated underground space and fractures, are located in the mine. Based on the outcome of this study, an appropriate design of backfilling underground space can be performed to mitigate the AMD in closed and abandoned mines. The effects of backfilling underground space on AMD production depend on the geology and hydrogeology of mine sites, so applying this method to a variety of sites is required to quantitatively evaluate the performance.

6. Conclusions

3-D numerical model consisting of a variety of geological formations and underground tunnels was constructed. By using the calibrated model, both groundwater levels around the mine and flow rates of AMD agreed with the measured values, irrespective of the season. In this mine, AMD after backfilling the underground space was reduced to more than 30% in days with high rainfall and to 5% in days with little rainfall. In addition, the acidification of groundwater may be expected due to the rise in groundwater levels. These results imply that backfilling the underground space is effective in reducing AMD in this mine.

Author Contributions: Conceptualization, K.Y. and S.T.; software, K.Y. and S.T.; validation, K.Y., S.T. and S.Y.; investigation, M.E.; writing—original draft preparation, K.Y. and S.T.; writing—review and editing, T.I.; visualization, K.Y. and S.T.; supervision, M.S. All authors have read and agreed to the published version of the manuscript.

Funding: This work was supported in part by Development of advanced technology for mine drainage treatment from 2012 to 2014 of Ministry of Economy, Trade, and Industry.

Acknowledgments: The authors wish to thank the editor and anonymous reviewers for their constructive comments that improved this manuscript. We would also like to thank the staffs of Mitsubishi Materials Corporation, Eco-Management Corporation, and Mitsubishi Materials Techno Corporation for their help, advice, and cooperation during this study.

Conflicts of Interest: The authors declare no conflict of interest. The funders had no role in the design of the study; in the collection, analyses, or interpretation of data; in the writing of the manuscript, or in the decision to publish the results.

References

1. Younger, P.L. Mine water pollution in Scotland: Nature, extent and preventative strategies. *Sci. Total Environ.* **2001**, *265*, 309–326. [CrossRef]
2. Gault, A.G.; Cooke, D.R.; Townsend, A.T.; Charnock, J.M.; Polya, D.A. Mechanisms of arsenic attenuation in acid mine drainage from Mount Bischoff, western Tasmania. *Sci. Total Environ.* **2005**, *345*, 219–228. [CrossRef] [PubMed]
3. Molson, J.W.; Fala, O.; Aubertin, M.; Bussière, B. Numerical simulations of pyrite oxidation and acid mine drainage in unsaturated waste rock piles. *J. Contam. Hydrol.* **2005**, *78*, 343–371. [CrossRef] [PubMed]
4. Boularbah, A.; Schwartz, C.; Bitton, G.; Morel, J.L. Heavy metal contamination from mining sites in south Morocco: I. Use of a biotest to assess metal toxicity of tailings and soils. *Chemosphere* **2006**, *63*, 802–810. [CrossRef] [PubMed]
5. Lim, H.S.; Lee, J.S.; Chon, H.T.; Sager, M. Heavy metal contamination and health risk assessment in the vicinity of the abandoned Songcheon Au–Ag mine in Korea. *J. Geochem. Explor.* **2008**, *96*, 223–230. [CrossRef]
6. Hien, N.T.T.; Yoneda, M.; Matsui, A.; Hai, H.T.; Pho, N.V.; Quang, N.H. Environmental contamination of arsenic and heavy metals around Cho Dien lead and zinc mine, Vietnam. *J. Water Environ. Technol.* **2012**, *10*, 253–265. [CrossRef]
7. Hierro, A.; Olías, M.; Cánovas, C.R.; Martín, J.E.; Bolivar, J.P. Trace metal partitioning over a tidal cycle in an estuary affected by acid mine drainage (Tinto estuary, SW Spain). *Sci. Total Environ.* **2014**, *497–498*, 18–28. [CrossRef]
8. Skierszkan, E.K.; Mayer, K.U.; Weis, D.; Beckie, R.D. Molybdenum and zinc stable isotope variation in mining waste rock drainage and waste rock at the Antamina mine, Peru. *Sci. Total Environ.* **2016**, *550*, 103–113. [CrossRef]
9. Zhao, Q.; Guo, F.; Zhang, Y.; Ma, S.; Jia, X.; Meng, W. How sulfate-rich mine drainage affected aquatic ecosystem degradation in northeastern China, and potential ecological risk. *Sci. Total Environ.* **2017**, *609*, 1093–1102. [CrossRef]
10. Tabelin, C.B.; Igarashi, T.; Villacorte-Tabelin, M.; Park, I.; Opisod, E.M.; Ito, M.; Hiroyoshi, N. Arsenic, selenium, boron, lead, cadmium, copper, and zinc in naturally contaminated rocks: A review of their sources, modes of enrichment, mechanisms of release, and mitigation strategies. *Sci. Total Environ.* **2018**, *645*, 1522–1553. [CrossRef]
11. Igarashi, T.; Herrera, P.S.; Uchiyama, H.; Miyamae, H.; Iyatomi, N.; Hashimoto, K.; Tabelin, C.B. The two-step neutralization ferrite-formation process for sustainable acid mine drainage treatment: Removal of copper, zinc and arsenic, and the influence of coexisting ions on ferritization. *Sci. Total Environ.* **2020**, *715*, 136877. [CrossRef] [PubMed]
12. Migaszewski, Z.M.; Gałuszka, A.; Dołęgowska, S. Arsenic in the Wiśniówka acid mine drainage area (south-central Poland)—Mineralogy, hydrogeochemistry, remediation. *Chem. Geol.* **2018**, *493*, 491–503. [CrossRef]
13. Tabelin, C.B.; Veerawattananun, S.; Ito, M.; Hiroyoshi, N.; Igarashi, T. Pyrite oxidation in the presence of hematite and alumina: I. Batch leaching experiments and kinetic modeling calculations. *Sci. Total Environ.* **2017**, *580*, 687–698. [CrossRef] [PubMed]
14. Tabelin, C.B.; Veerawattananun, S.; Ito, M.; Hiroyoshi, N.; Igarashi, T. Pyrite oxidation in the presence of hematite and alumina: II. Effects on the cathodic and anodic half-cell reactions. *Sci. Total Environ.* **2017**, *581–582*, 126–135. [CrossRef] [PubMed]
15. Stumm, W.; Morgan, J.J. *Aquatic Chemistry*, 3rd ed.; Wiley: New York, NY, USA, 1995; pp. 690–691.
16. Park, I.; Tabelin, C.B.; Jeon, S.; Li, X.; Seno, K.; Ito, M.; Hiroyoshi, N. A review of recent strategies for acid mine drainage prevention and mine tailings recycling. *Chemosphere* **2019**, *219*, 588–606. [CrossRef] [PubMed]
17. Tabelin, C.B.; Corpuz, R.D.; Igarashi, T.; Villacorte-Tabelin, M.; Alorro, R.D.; Yoo, K.; Raval, S.; Ito, M.; Hiroyoshi, N. Acid mine drainage formation and arsenic mobility under strongly acidic conditions: Importance of soluble phases, iron oxyhydroxides/oxides and nature of oxidation layer on pyrite. *J. Hazard. Mater.* **2020**, *399*, 122844. [CrossRef]
18. Tabelin, C.; Sasaki, A.; Igarashi, T.; Tomiyama, S.; Villacorte-Tabelin, M.; Ito, M.; Hiroyoshi, N. Prediction of acid mine drainage formation and zinc migration in the tailings dam of a closed mine, and possible countermeasures. *MATEC Web Conf. RSCE* **2019**, *268*, 06003. [CrossRef]

19. Kalin, M.; Fyson, A.; Wheeler, W.N. The chemistry of conventional and alternative treatment systems for the neutralization of acid mine drainage. *Sci. Total Environ.* **2006**, *366*, 395–408. [CrossRef]

20. Kawabe, N. The significance of the soil environment, aqueous environment and planting trees measures in mine pollution control. In Proceedings of the MMIJ and EARTH 2017, Sapporo, Japan, 26–28 September 2017; pp. 89–90.

21. Skousen, J.; Zipper, C.E.; Rose, A.; Ziemkiewicz, P.F.; Nairn, R.; McDonald, L.M.; Kleinmann, R.L. Review of passive systems for acid mine drainage treatment. *Mine Water Environ.* **2017**, *36*, 133–153. [CrossRef]

22. Iwatsuki, T.; Yoshida, H. Groundwater chemistry and fracture mineralogy in the basement granitic rock in the Tono uranium mine area, Gifu Prefecture, Japan: Groundwater composition, Eh evolution analysis by fracture filling minerals. *Geochem. J.* **1999**, *33*, 19–32. [CrossRef]

23. Okumura, M. Hydrogeochemical study of the 3M level adit drainage at the old Matsuo mine. *Resour. Geol.* **2003**, *53*, 173–182.

24. Mahara, Y.; Nakata, E.; Oyama, T.; Miyakawa, K.; Igarashi, T.; Ichihara, Y.; Matsumoto, K. Proposal for the methods to characterize fossil seawater: Distribution of anions, cations and stable isotopes, and estimation on the groundwater residence time by measuring ^{36}Cl at the Taiheiyou Coal Mine. *J. Groundw. Hyd.* **2006**, *48*, 17–33. [CrossRef]

25. Hazen, J.M.; Milliams, M.W.; Stover, B.; Wireman, M. Characterization of acid mine drainage using a combination of hydrometric, chemical and isotopic analyses, Mary Murphy Mine, Colorado. *Environ. Geochem. Health* **2002**, *24*, 1–22. [CrossRef]

26. Sracek, O.; Choquette, M.; Gélinas, P.; Lefebvre, R.; Nicholson, R.V. Geochemical characterization of acid mine drainage from a waste rock pile, Mine Doyon, Quebec, Canada. *J. Contam. Hydrol.* **2004**, *69*, 45–71. [CrossRef]

27. Lee, J.S.; Chon, H.T. Hydrogeochemical characteristics of acid mine drainage in the vicinity of an abandoned mine, Daduk Creek, Korea. *J. Geochem. Explor.* **2006**, *88*, 37–40. [CrossRef]

28. Leybonrne, M.L.; Clark, I.D.; Goodfellow, W.D. Stable isotope geochemistry of ground and surface waters associated with undisturbed massive sulfide deposits; Constraints on origin of waters and water-rock reactions. *Chem. Geol.* **2006**, *231*, 300–325. [CrossRef]

29. Hubbard, C.G.; Black, S.; Coleman, L. Aqueous geochemistry and oxygen isotope compositions of acid mine drainage from the Río Tinto, SW Spain, highlight inconsistencies in current models. *Chem. Geol.* **2009**, *265*, 321–334. [CrossRef]

30. Gammons, C.H.; Duaime, T.E.; Parker, S.R.; Poulson, S.R.; Kennelly, P. Geochemistry and stable isotope investigation of acid mine drainage associated with abandoned coal mines in central Montana, USA. *Chem. Geol.* **2010**, *269*, 100–112. [CrossRef]

31. Bethune, J.; Randell, J.; Runkel, R.I.; Singha, K. Non-invasive flow path characterization in a mining-impacted wetland. *J. Contam. Hydrol.* **2015**, *183*, 29–39. [CrossRef]

32. Cánovas, C.R.; Macías, F.; Olías, M.; Pérez-López, R. Metal and acidity fluxes controlled by precipitation/dissolution cycles of sulfate salts in an anthropogenic mine aquifer. *J. Contam. Hydrol.* **2016**, *188*, 29–43. [CrossRef]

33. Galván, L.; Olías, M.; Cánovas, C.R.; Sarmiento, A.M.; Nieto, J.M. Hydrological modeling of a watershed affected by acid mine drainage (Odiel River, SW Spain). Assessment of the pollutant contributing areas. *J. Hydrol.* **2016**, *540*, 196–206. [CrossRef]

34. Cánovas, C.R.; Macías, F.; Olías, M.; Pérez-López, R.; Nieto, J.M. Metal-fluxes characterization at a catchment scale: Study of mixing processes and end-member analysis in the Meca River watershed (SW Spain). *J. Hydrol.* **2017**, *550*, 590–602. [CrossRef]

35. Razowska, L. Changes of groundwater chemistry caused by the flooding of iron mines (Czestochowa Region, Southern Poland). *J. Hydrol.* **2001**, *244*, 17–32. [CrossRef]

36. Tomiyama, S.; Ueda, A.; Ii, H.; Nakamura, Y.; Koizumi, Y.; Saito, K. Sources and flow system of groundwater in the Hosokura mine, Miyagi prefecture, using geochemical method and numerical simulation. *J. MMIJ* **2010**, *126*, 31–37. [CrossRef]

37. Tomiyama, S.; Igarashi, T.; Ii, H.; Takano, H. Sources and flow system of groundwater in the Shimokawa mine, north Hokkaido, using geochemical method and numerical simulation. *J. MMIJ* **2016**, *132*, 80–88. [CrossRef]

38. Tomiyama, S.; Igarashi, T.; Tabelin, C.B.; Tangviroon, P.; Ii, H. Acid mine drainage sources and hydrogeochemistry at the Yatani mine, Yamagata, Japan: A geochemical and isotopic study. *J. Contam. Hydrol.* **2019**, *225*, 103502. [CrossRef]

39. Tomiyama, S.; Igarashi, T.; Tabelin, C.B.; Tangviroon, P.; Ii, H. Modeling of the groundwater flow system in excavated areas of an abandoned mine. *J. Contam. Hydrol.* **2020**, *230*, 103617. [CrossRef]

40. Tomiyama, S.; Ii, H.; Koizumi, Y.; Metsugi, H. Modeling of groundwater recharge and discharge in Tomitaka mine, Miyazaki Prefecture. *J. Groundw. Hydol.* **2010**, *52*, 261–274. [CrossRef]

41. Tokoro, C.; Fukaki, K.; Kadokura, M.; Fuchida, S. Forecast of AMD quantity by a series tank model in three stages: Case studies in two closed Japanese mines. *Minerals* **2020**, *10*, 430. [CrossRef]

42. Yamaguchi, K.; Tomiyama, S.; Metsugi, H.; Ii, H.; Ueda, A. Flow and geochemical modeling of drainage from Tomitaka mine, Miyazaki, Japan. *J. Environ. Sci.* **2015**, *36*, 130–143. [CrossRef]

43. Hishiya, T.; Nishigaki, M.; Hashimoto, N. The three dimensional numerical analysis method for density dependent groundwater flow with mass transport. *J. Constr. Man. Eng.* **1999**, *638*, 59–69.

44. Umeda, K.; Yanagisawa, K.; Yoneda, S. Creation of permeability coefficient database for Japanese soils. *J. Groundw. Hydol.* **1995**, *37*, 69–77.

45. Van Genuchten, M.T. A closed-form equation for predicting the hydraulic conductivity of unsaturated soils. *Soil Sci. Soc. Am. J.* **1980**, *44*, 892–898. [CrossRef]

46. Ethier, M.-P.; Bussière, B.; Broda, S.; Aubertin, M. Three-dimensional hydrogeological modeling to assess the elevated-water-table technique for controlling acid generation from an abandoned tailings site in Quebec, Canada. *Hydrogeol. J.* **2018**, *26*, 1201–1219. [CrossRef]

47. Fala, O.; Molson, J.; Aubertin, M.; Bussière, B. Numerical modelling of flow and capillary barrier effects in unsaturated waste rock piles. *Mine Water Environ.* **2005**, *24*, 172–185. [CrossRef]

48. Imai, H.; Amemiya, K.; Matsui, H.; Sato, T.; Saegusa, H.; Watanabe, K. The proposals relevant to seepage flow simulation in rockmass around tunnel under unsaturated condition. Method for estimating unsaturated seepage parameters of stones and setting of boundary condition on tunnel wall. *J. Jpn. Soc. Civ. Eng. C Geotech.* **2013**, *69*, 285–296.

49. Nishigaki, M.; Hishiya, T.; Hashimoto, N.; Kohno, I. The numerical method for saturated-unsaturated fluid density dependent groundwater flow with mass transport. *J. Jpn. Soc. Civ. Eng.* **1995**, *511*, 135–144.

50. Tabelin, C.B.; Igarashi, T. Mechanisms of arsenic and lead release from hydrothermally altered rock. *J. Hazard. Mater.* **2009**, *169*, 980–990. [CrossRef]

51. Tabelin, C.B.; Sasaki, R.; Igarashi, T.; Park, I.; Tamoto, S.; Arima, T.; Ito, M.; Hiroyoshi, N. Simultaneous leaching of arsenite, arsenate, selenite and selenate, and their migration in tunnel-excavated sedimentary rocks: I. Column experiments under intermittent and unsaturated flow. *Chemosphere* **2017**, *186*, 558–569. [CrossRef]

52. Tabelin, C.B.; Sasaki, R.; Igarashi, T.; Park, I.; Tamoto, S.; Arima, T.; Ito, M.; Hiroyoshi, N. Simultaneous leaching of arsenite, arsenate, selenite and selenate, and their migration in tunnel-excavated sedimentary rocks: II. Kinetic and reactive transport modeling. *Chemosphere* **2017**, *188*, 444–454. [CrossRef]

Article

Chemical Treatment of Highly Toxic Acid Mine Drainage at A Gold Mining Site in Southwestern Siberia, Russia

Svetlana Bortnikova [1], Olga Gaskova [2,3] , Nataliya Yurkevich [1,*], Olga Saeva [1] and Natalya Abrosimova [1]

[1] Trofimuk Institute of Petroleum Geology and Geophysics, Siberian Branch of the Russian Academy of Sciences, Koptug ave. 3, 630090 Novosibirsk, Russia; BortnikovaSB@ipgg.sbras.ru (S.B.); SaevaOP@ipgg.sbras.ru (O.S.); AbrosimovaNA@ipgg.sbras.ru (N.A.)

[2] Sobolev Institute of Geology and Mineralogy, Siberian Branch of the Russian Academy of Sciences, Koptug ave. 3, 630090 Novosibirsk, Russia; gaskova@igm.nsc.ru

[3] Geological and Geophysical Faculty, Novosibirsk State University, Pirogova str. 2, 630090 Novosibirsk, Russia

* Correspondence: YurkevichNV@ipgg.sbras.ru; Tel.: +7-(383)-363-91-94; Fax: +7-(383)-330-28-07

Received: 21 August 2020; Accepted: 29 September 2020; Published: 30 September 2020

Abstract: The critical environmental situation in the region of southwestern Siberia (Komsomolsk settlement, Kemerovo region) is the result of the intentional displacement of mine tailings with high sulfide concentrations. During storage, ponds of acidic water with incredibly high arsenic (up to 4 g/L) and metals formed on the tailings. The application of chemical methods to treat these extremely toxic waters is implemented: milk of lime $Ca(OH)_2$, sodium sulfide Na_2S, and sodium hydroxide NaOH. Field experiments were carried out by sequential adding pre-weighed reagents to the solutions with control of the physicochemical parameters and element concentrations for each solution/reagent ratio. In the experiment with $Ca(OH)_2$, the pH increased to neutral values most slowly, which is contrary to the results from the experiment with NaOH. When neutralizing solutions with NaOH, arsenic-containing phases are formed most actively, arsenate chalcophyllite $Cu_{18}Al_2(AsO_4)_4(SO_4)_3(OH)_{24}\cdot36H_2O$, a hydrated iron arsenate scorodite, kaatialaite $FeAs_3O_9\cdot8H_2O$ and $Mg(H_2AsO_4)_2$. A common specificity of the neutralization processes is the rapid precipitation of Fe hydroxides and gypsum, then the reverse release of pollutants under alkaline conditions. The chemistry of the processes is described using thermodynamic modeling. The main species of arsenic in the solutions are iron-arsenate complexes; at the end of the experiments with $Ca(OH)_2$, Na_2S, and NaOH, the main species of arsenic is $CaAsO_4^-$, the most toxic acid H_3AsO_3 and AsO_4^{3-}, respectively. It is recommended that full-scale experiments should use NaOH in the first stages and then $Ca(OH)_2$ for the subsequent neutralization.

Keywords: mine water treatment; milk of lime; sodium sulfide; sodium hydroxide; arsenic-containing tailings

---

1. Introduction

Minimizing the influence of toxic components in acid mine drainage (AMD) and acid rock drainage (ARD) has been a widely discussed topic in the scientific literature since the 1990s [1–9]. Along with the use of natural materials as sorbents, including zeolites [10,11], clay minerals [12,13], plant materials [14], charcoal ash [15] and iron and aluminum oxides [16], to extract various elements from solutions are offered. Other various modified and induced sorbents include aluminosilicates, ferrocyanide sorbents based on hydrated titanium dioxide, resins, organosilicon ion-exchange and complex-forming sorbents and cementitious materials [17–20]. Different approaches for the precipitation of metals and metalloids from AMD solutions have been applied, such as the use of zero-valent iron [21], a pulsed limestone bed treatment system minimizing armor formation [22], and alkaline industrial wastes [23]. Currently, the focus is on two directions of the tailings problem: (1) prevention methods; and (2) secondary mineral processing of mine waste [24].

Highly mineralized acid drainage solutions are extremely toxic for the environment; they have to be treated, making a neutral pH and the removal of metal and metalloids necessary to reduce potential hazards [25]. Nevertheless, the critical environmental situation in the industrial region of southwestern Siberia (Kemerovo region) is the result of the intentional displacement (for reprocessing) and uncontrolled storage for 16 years (due to inexpediency) of the cyanide leaching tailings with high sulfide and arsenic concentrations. Ponds of acidic water with extremely high concentrations of arsenic (up to 4 g/L) and metals are formed on the surface of the solid tailings.

Limestone or portlandite has been used for the neutralization of drainage solutions of different composition and acidity [25]. These lime neutralizers in case of H_2SO_4 precipitate gypsum ($CaSO_4$ $2H_2O$) and passivate the surface. This indicates that alternatives could be needed.

Igarashi et al. [26] assert that AMD or ARD neutralization is effective but unsustainable in the long term. Estimation of the number of years required for metals in AMD from abandoned tailings dams to decrease below maximum permissible concentrations (MPC) has been done [27]. The authors of the reactive-transport model have estimated that the formation of AMD and release of Zn will persist for a thousand years. Igarashi et al. [27] introduced a technique for AMD management, where acid solution with high concentrations of Zn, Cu, and As was treated using a laboratory setup of ferrite flow.

An alternative AMD processing method is the alkaline barium calcium desalination. Both sulfate and metals were reduced below the MPC. The interesting thing about this technology is that low levels of sludge are disposed after useful chemicals are recovered from AMD [28,29].

The precipitation of metal sulfides, the solubility and stability of the sulfide ion, and its complexes with metals were reported by Lewis and Hille [30]. A great attention has been paid to the precipitation of metal sulfides from AMD and saturated leach solutions [31]. These authors conclude, "notwithstanding and irrespective of the source of sulfide, metallic sulfide precipitation has many challenges. Further research is needed to address these issues".

A fractional precipitation process was conducted to precipitate metals from the AMD [32]. With the help of four-step precipitation, AMD was treated to World Health Organization (WHO) requirements. Detoxification of zinc plant leach residues from Kabwe, Zambia has been performed by removing Pb, using a coupled extraction-cementation method in chloride media [33].

Special attention is devoted to the removal of arsenic from acidic and alkaline drainage, due to its high toxicity and mobility [34–38]. The formation of ferric arsenate, or a Fe/Al arsenate phase, as well as strong adsorption of As to Fe-oxyhydroxides/oxides have limited its mobility at low pH conditions [3]. The initial step in the extraction of As from solution is the oxidation of trivalent arsenic to pentavalent arsenic to remove it in one stable form [39,40]. To do this, they use hydrochemical neutralization with lime, resulting in the formation of svabite $Ca_5(AsO_4)_3F$ [41,42], sulfide precipitation [43,44], coprecipitation with iron and formation of scorodite $FeAsO_4·2H_2O$ [45,46], bacterial deposition [47,48] and various sorbents [49–51]. As(V) in acidic solutions (pH 3–4) can be removed effectively by synthesized schwertmannite [52], biomineralization [53] and scorodite precipitation [54,55]. Besides, As mobility can be decrease via sorption reactions with carbonates and/or gypsum [56,57]. However, the arsenic problem is far from being resolved now.

For example, Yuan et al. [51] investigated the effect of pH on fast As removal from AMD containing high arsenic. The authors studied Fe/As molar ratio, oxygen flow rate, temperature, initial As concentration and the action of reagents (NaOH vs. $Ca(OH)_2$). The mechanisms of solid precipitation, including As removal, were deeply discussed. This study provides further evidence on the speciation of As and its distribution in Fe(II/III)-As(III/VI)-S(II/VI) aqueous and solid systems, which affects the fate of arsenic. To date, there is no better summary of arsenic mineralogy than [58].

In this article, we presented a case study for both acidic solutions neutralization and metal precipitation from a highly mineralized tailings pond water, based on the addition of $Ca(OH)_2$, Na_2S, and NaOH in the field. From a geochemical point of view, according to the mechanism, the reagents act as precipitating $Ca(OH)_2$, reducing Na_2S, and neutralizing NaOH barriers. Actually, $Ca(OH)_2$ is also a neutralizing reagent, and NaOH may trigger metal precipitation as well; therefore, this difference is conditional for convenience. The above references show the relevance of this case study for a broader international audience. This means that current techniques like chemical neutralization should be improved to facilitate the better and more sustainable management of AMD solutions. In our study, we focus on (1) the optimal S/R ratio for the precipitation of elements during a sequential decrease of its variable; (2) identification of the mineral species of elements formed during neutralization; and (3) description of the chemistry of the water-rock interactions by thermodynamic modeling. Our task was to develop a quick and cheap way to prevent in situ the influence of toxic solutions, within the limits of the population's habitat.

2. Materials and Methods

2.1. Study Area

The object of the study was to find highly mineralized solutions from ponds on the surface of displaced tailings after the cyanide leaching of the sulfide flotation concentrate from the Berikul gold extracting plant (BGEP), Figure 1.

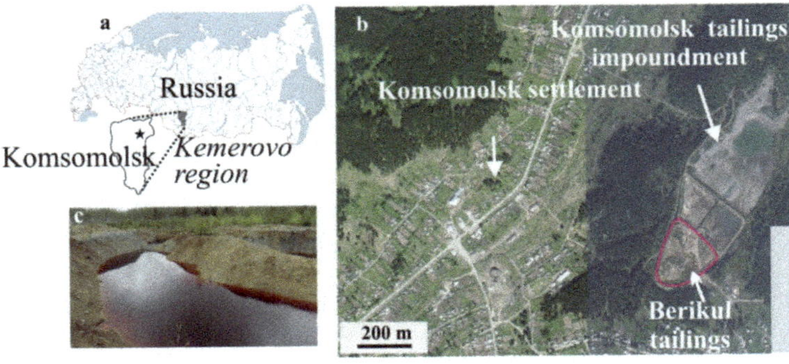

Figure 1. (**a**) Geographic position; (**b**) sample site at the Berikul tailings; and (**c**) photograph of acid ponds.

In 2004, the tailings were transported and stored in the southern part of the Komsomolsk tailings impoundment [59]. After 2004, ponds were formed by seasonal precipitation on top of the displaced Berikul tailings. The volume and configuration of the ponds changed seasonally. The pond water has a brownish-black to light red color. Interactions with highly oxidized tailings led to the formation of the current hydrochemical composition of the solutions. The solutions of the ponds are acidic, and ultra-acidic solutions with extremely high concentrations of many chemical elements, such as arsenic, are the main danger. The arsenic maximum concentration reached 4 g/L. A detailed description of the compositions of the ponds and solutions is provided in a previous article [59]. Based on the measured volume of the ponds and the mean concentration of arsenic in the solutions, the calculated amount of dissolved arsenic is 80–120 kg. The existence of open ponds with highly toxic solutions within the village boundaries raised concerns about their neutralization and precipitation of elements. Since there are no sorbents applicable for the extraction of elements with such high mineralization (Table 1), the precipitation of elements was carried out using chemical bonding technology, with three common reagents: $Ca(OH)_2$, Na_2S and NaOH.

Table 1. Chemical composition of the Berikul pond water formed on the top of mine wastes, Eh in mV, electrical conductivity (EC) in mSm/cm, element concentration in mg/L.

	Ber-1/0		**Ber-1/0**
pH	2.10	Zn	140
Eh	657	Pb	6.4
EC	11	Cd	3.4
SO_4^{2-}	27,000	Ba	0.32
Ca	630	Rb	0.0072
Mg	420	Sr	1.9
Na	22	As	1300
K	6.20	Sb	0.18
Al	520	Bi	1.7
Fe	8600	P	34
Mn	18	Mo	0.052
Cr	3.2	Sn	0.028
Co	4.2	In	0.054
Ni	4.8	Ag	0.0064
Cu	32		

Chemical treatment is used as a method for the deep purification of industrial wastewater. The processes of the binding of elements and their precipitation in the form of newly formed phases were studied by field experiments, which did not require the transportation and storage of the solutions.

2.2. Methods and Used Reagents

Field batch tests on the interaction of mine drainage with the three chemical barriers were carried out in the same regime under ambient conditions. To select a solution for the experiments, samples were taken from different ponds. The electrical conductivity and pH were measured in situ, and the most acidic saline solution, Ber-1 (pH 2.1), was chosen for the experiment. To allow the initial composition of the solution to be identical in all experiments, a larger aliquot for the experiments was taken with a polyethylene bucket, which was previously rinsed three times at the sampling site. Then, a sample for analysis (Ber-1/0) was taken from the bucket, and 1 L of the solution was poured into polyethylene bottles and covered with lids. Different amounts of reagents (Table 2) were added sequentially to 1 L of this solution to fix the alkaline conditions.

Table 2. The weight of the portion (PW, g), the total weight of the reagent (TW, g) and the ratios of the solution and the reagent (S/R) during the experiments.

Sample	$Ca(OH)_2$			Sample	Na_2S			Sample	NaOH		
	PW, g	TW, g	S/R		PW, g	TW, g	S/R		PW, g	TW, g	S/R
B-1/1/1	0.20	0.20	5000	B-1/2/1	0.20	0.20	5000	B-1/3/1	0.20	0.20	5000
B-1/1/2	0.20	0.40	2500	B-1/2/2	0.20	0.40	2500	B-1/3/2	0.20	0.40	2500
B-1/1/3	5.0	5.4	185	B-1/2/3	1.0	1.4	714	B-1/3/3	1.0	1.4	714
B-1/1/4	5.0	10	96	B-1/2/4	1.0	2.4	417	B-1/3/4	1.0	2.4	417
B-1/1/5	5.0	15	65	B-1/2/5	1.0	3.4	294	B-1/3/5	1.0	3.4	294
B-1/1/6	5.0	20	49	B-1/2/6	1.0	4.4	227	B-1/3/6	1.0	4.4	227
B-1/1/7	10	30	33	B-1/2/7	2.0	6.4	156	B-1/3/7	2.0	6.4	156
B-1/1/8	10	40	25	B-1/2/8	2.0	8.4	119	B-1/3/8	2.0	8.4	119
B-1/1/9	10	50	20	B-1/2/9	2.0	10	96	B-1/3/9	2.0	10	96
B-1/1/10	10	60	17	B-1/2/10	4.0	14	69	B-1/3/10	2.0	12	81
				B-1/2/11	4.0	18	54	B-1/3/11	2.0	14	69
				B-1/2/12	8.0	26	38	B-1/3/12	2.0	16	61
				B-1/2/13	16	42	24				

We used the following reagents: 1) industrial milk of lime ($Ca(OH)_2$; JSC "Iskitimcement", Iskitim, Novosibirsk region, Russia); 2) chemically pure dry sodium sulfide (Na_2S; JSC "Lenreaktiv",

Sankt-Petersburg, Russia); and 3) chemically pure dry sodium hydroxide (NaOH; JSC "Ekos", Moscow, Russia).

The total mass of the chemical reagents required to achieve an alkaline pH and the masses of the sequential portions of the reagents were previously calculated based on the stoichiometry of the main reactions, and on the concentration of the elements in the drainage solution. We thoroughly manually mixed the portion of the reagent with 1 L of Ber-1/0 solution in the polyethylene bottle. Then, the solutions were allowed to settle for 20 min, and the following parameters were determined in situ in the clarified solution: pH, Eh, and electrical conductivity (EC). The pH and Eh values were determined using the Expert 001 pH/ion-meter (JSC "Ekoniks-Expert", Moscow, Russia). The EC values were measured using a Cond 315i/SET device (JSC "Wissenschaftlich Technische Werkstatten GmbH", Weilheim, Germany), with automatic temperature compensation and a TetraCon 325 sensor.

An aliquot of 5 mL of the clarified solution was filtered through a membrane filter with a 0.45 μm pore diameter (CC "Vladipor", Vladimir, Russia) in a plastic tube, and then it was acidified with 100 μL of chemically pure distilled concentrated HNO_3 acid, to determine the elemental composition according to [60]. Concentrations of major, minor, and trace elements in the water samples were determined using inductively coupled plasma mass spectrometry (ELAN-9000 DRC-e, Perkin Elmer, Shelton, USA) in the certified Chemical Analytical Center "Plasma" (Tomsk, Russia). Accuracy and precision were estimated to be 7% or better at the mg/L concentration level and 10% or better at the μg/L concentration level. Reference standard material (RSM) used in the quality control of the ICP-MS analytical results of liquids is certified wastewater—Trace metals solution (CWW-TM-D, High Purity Standards, Charleston, USA). All measurements were conducted in three replicates (n = 3) for each element. After reaching neutral and alkaline pH values, the experiment was completed, the last portion of the solution was sampled, the rest of the solution was decanted, and the precipitates were dried at room temperature.

Next, the dried precipitates were homogenized, and the mineralogical composition was determined using X-ray spectral and X-ray diffraction analyses in the IGM SB RAS, Novosibirsk. Individual grains were selected to determine the mineralogical composition. A study of the morphology and composition of the grains was carried out with the scanning electron microscope MIRA3 LMU (TESCAN ORSAY HOLDING, Brno-Kohoutovice, Czech Republic), with an energy dispersive attachment Inca-Energy 450 XMax 80 (Oxford Instruments-NanoAnalysis, High Wycombe, UK).

The numerical experiment completely simulated the composition of the initial solution and the addition of the corresponding amount of reagents in grams. In the output file, the equilibrium pH values, Eh values and supersaturation of the solution and composition of solid phases were recorded. The equilibria in the heterophase 20-component system H-O-C-S-N-Ca-Mg-Na-K-Fe-Al-Mn-Sr-As-Sb-Cu-Zn-Pb-Ba-P were calculated at 25 °C, under a total pressure of 1 atm and a partial $CO_{2(gas)}$ pressure of $10^{-3.5}$ atm, using the HCh (HydroChemistry) software, based on the principle of minimization of the thermodynamic potential of the system (Gibbs free energy) and the UNITHERM thermodynamic database [61].

We compared the experimental solutions composition and the water chemistry obtained using HCh modeling of each-step solutions supersaturation after adding the reagents. The objective was to determine if the observed concentrations of major cations (Ca, Mg, K, Na), As and metals can be reached by the precipitation of the various minerals. The mineralogy of the precipitates was modeled with the suite of minerals from the extended UNITHERM database. The model is adequate, while small differences in details are due to the fact that (a) there are no thermodynamic data for a number of minerals, for example, chalcophyllite, (b) the experiment solid-solutions system does not always achieve complete thermodynamic equilibrium, (c) the mineral composition was determined only after the experiment, and not during the step-by-step addition of each reagent, as in calculations.

3. Results

3.1. Precipitation Experiment with Ca(OH)$_2$

A change in the pH values during the experiment indicates a large buffer capacity of the solution with the initial pH = 2.1. The pH values increased by only 1 unit when S/R = 33 (30 g of Ca(OH)$_2$ was added, Figure 2).

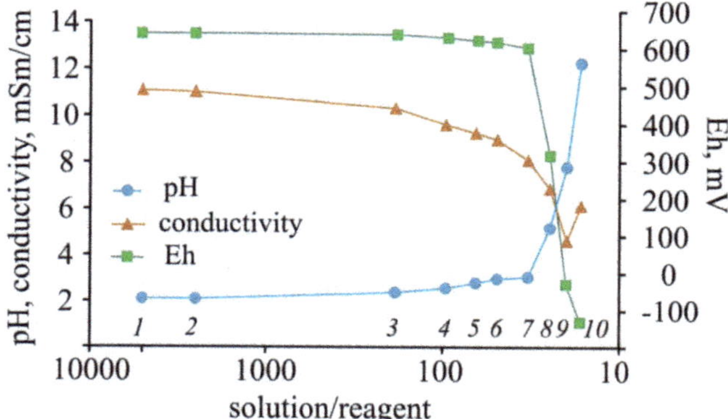

Figure 2. Changes in the physicochemical conditions and EC during the experiment with Ca(OH)$_2$. Hereinafter, the italic numbers above the abscissa axis indicate the steps of the experiment. At the experiment with Ca(OH)$_2$, physico-chemical parameters and element concentrations were measured at solution/reagent ratios (S/R): *1*—5000; *2*—2500; *3*—185; *4*—96; *5*—65; *6*—49; *7*—33; *8*—25; *9*—20; *10*—17.

A noticeable increase in pH to 5.17 occurred when 40 g of reagent was added to the initial solution with a volume of 1 L (S/R = 25). In the next steps of the experiment, the acidity of the solution began to decrease sharply. At S/R = 20, the pH increased to 7.79, and at the end of the experiment, when 60 g of Ca(OH)$_2$ was added to the solution (S/R = 17), the pH increased sharply to 12.22.

The pH value increased during the experiment, whereas the EC, reflecting the total mineralization of the solution, gradually decreased to the ratio S/R = 20 (pH = 7.79), due to the precipitation of the Ca-containing solid phases. This phenomenon is evidenced by the stable Ca concentrations in the solution throughout the experiment, despite the gradual addition of Ca(OH)$_2$. However, at the last step of the experiment (pH = 12.22), the electrical conductivity sharply increased by 1.5 mSm/cm, which indicates the leaching of elements into the solution. This result is not unexpected, since alkaline solutions are very aggressive. The Eh in the first seven steps of the experiment changed from 657 to 619 mV; in step No. 8, the Eh decreased to 333 mV and became negative in the last two subsequent steps (−13 and −116 mV).

The pH value increased during the experiment, whereas the EC, reflecting the total mineralization of the solution, gradually decreased to the ratio S/R = 20 (Ph = 7.79), due to the precipitation of the Ca-containing solid phases. This phenomenon is evidenced by the stable Ca concentrations in the solution throughout the experiment, despite the gradual addition of Ca(OH)$_2$. However, at the last step of the experiment (pH = 12.22), the electrical conductivity sharply increased by 1.5 mSm/cm, which indicates the leaching of elements into the solution. This result is not unexpected, since alkaline solutions are very aggressive. The Eh in the first seven steps of the experiment changed from 657 to 619 mV; in step No. 8, the Eh decreased to 333 mV and became negative in the last two subsequent steps (−13 and −116 mV).

According to the Nernst equation [62], the dependence of Eh-pH suggests such a coherent change. However, we note that, in ponds exposed to the atmosphere, the Eh value is only 657 mV, which indicates that the potential-determining system is most likely the pair Fe^{2+}/Fe(OH)$_{3(s)}$, since the initial concentration of iron is 8600 mg/L (Table 1). Thus, the process can be divided into two main stages: S/R = 33 (a slight increase in pH) and S/R = 25-20-17 (a total of 60.4 g of Ca(OH)$_2$ was added), Equations (1)–(3):

$$Fe^{2+} + 2H_2O = Fe(OH)_2 + 2H^+ \tag{1}$$

$$Fe^{3+} + 3H_2O = Fe(OH)_{3(s)}\downarrow + 3H^+ \tag{2}$$

$$2Fe(OH)_2 + 0.5O_2 + 2H_2O = 2Fe(OH)_3 + H_2O \tag{3}$$

The buffer capacity of the solution was high, and allowed for the stability of pH values in a narrow range of one unit. After reaching S/R = 33, other reactions involving arsenic and metals take place in the system. The content of Fe, Cu, Zn, Pb, and other metals, as well as metalloids (As, Sb, Bi, Sn), slightly decreased while the solution was acidic (to pH < 3.06), which corresponded to S/R = 33 (Figure 3).

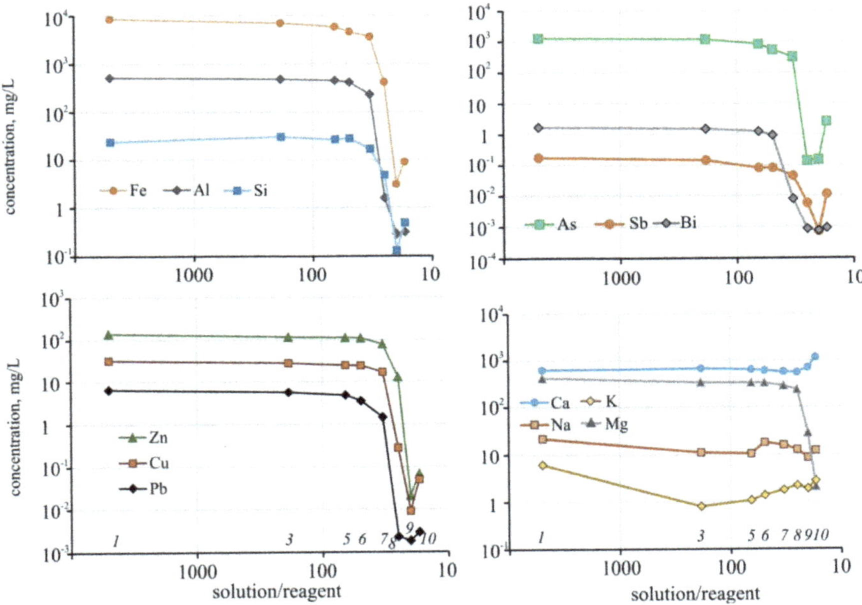

Figure 3. Changes in the elemental concentrations during the experiment with Ca(OH)$_2$.

Then, with a noticeable increase in pH to 5.17 at S/R = 25, the concentrations of almost all the elements decreased sharply and continued to decrease in the next step when the pH reached slightly alkaline values (pH = 7.79, S/R = 20). Adsorption-coprecipitation with hydrous ferric oxides and sorption reactions with kaolinite and montmorillonite play a significant role in mobility of Zn and Pb [63]. However, with a further increase in alkalinity (pH = 12.2, S/R = 17), the concentrations of the elements began to increase. The contents of the alkali metals (Na, K) fluctuated insignificantly throughout the experiment.

The precipitate formed as a result of the interaction of Ca(OH)$_2$ with the solution was a semiamorphous dark gray substance; after drying, it changed to a fine-grained, earthy mass. The XRD analysis data showed the presence of mainly gypsum and the amorphous phase of iron and aluminum hydroxides (plohmite) in the products from the final phase after the experiment. Ettringite, bassanite, alunite, and carbonates of Ca and Fe were formed in smaller amounts. In addition, traces of pickeringite (precipitated Mg) and arsenosiderite Ca$_3$Fe$_4$(AsO$_4$)(OH)$_6$·3H$_2$O were found in the sample

(Table 3). Thus, we have evidence of the formation of aqueous sulfates and complex compositions, hydroxides, and carbonates. The mineral phases of Cu and Zn were not identified.

Table 3. Mineral composition of the precipitates after the experiment with $Ca(OH)_2$ (XRD analysis).

Major		Minor		Trace	
Gypsum	$CaSO_4 \cdot 2H_2O$	Ettringite	$Ca_6Al_2(OH)_{12}(SO_4)_3 \cdot 26H_2O$	Pickeringite	$MgAl_2(SO_4)_4 \cdot 22H_2O$
Amorphous phase	$Fe(OH)_3$	Bassanite	$CaSO_4 \cdot 0.5H_2O$	Plohmite	$AlO(OH)$
		Alunite	$KAl_3(SO_4)_2(OH)_6$	Arsenosiderite	$Ca_3Fe_4(AsO_4)(OH)_6 \cdot 3H_2O$
		Calcite	$CaCO_3$	Butlerite	$Fe(OH)SO_4 \cdot 2H_2O$
		Siderite	$FeCO_3$		

However, as a result of the electron microscopy studies of the precipitate, Cu and Zn impurities were detected in the Fe-Ca minerals. In addition, ultrafine inclusions of the As-containing phases, presumably arsenosiderite, which were identified by XRD analysis (Figure 4), were found in some grains of gypsum and Fe-Ca minerals.

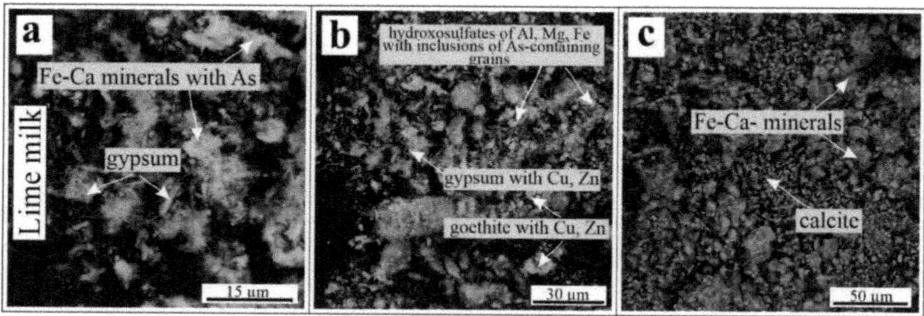

Figure 4. Scanning electron microscopy images of the precipitates after the experiments with $Ca(OH)_2$: (**a**) fine-grained crystals of gypsum in the groundmass of Fe-Ca minerals containing As (possibly arsenosiderite); (**b**) colloform mixture of gypsum, goethite, and hydroxosulfates of Al, Mg, and Fe, with ultrafine inclusions of As-containing grains. Cu and Zn were found as impurities in gypsum and goethite; (**c**) a mixture of Fe-Ca- minerals with calcite.

Numerical Simulation of the $Ca(OH)_2$ Experiment

A theoretical model with one liter of the Fe-As-sulfate solution (Table 1) by milk of lime has shown that, when 2.1 g of $Ca(OH)_2$ is added, the solution becomes supersaturated with, $Fe(OH)_{3(am)}$, gypsum, and anglesite (Table 4).

Due to the formation of suspensions, 2800 mg of iron and 4.6 mg of lead can precipitate from the solution. The Ca content in the solution remains stable because it is controlled by the solubility of gypsum. The amounts of the formed suspension obtained in the numerical simulation are slightly higher than the decrease in the concentration of elements in the experimental solutions at the corresponding stages; however, taking the analysis error (10%) into account at such high concentrations of elements in the solution, the agreement is satisfactory. At pH ~3 (S/R = 49 and 33), the situation changes, such that gibbsite, alunite (aluminum minerals), apatite, and traces of shultenite and conichalcite (arsenates containing heavy metals) appear. Even though only 40 mg of arsenic precipitates from the solution, the precipitation process begins. With optimal S/R ratios of 25 and 20, the precipitation of arsenic occurs in the form of aqueous calcium arsenate, and carbonates may be present in the solid phase (Table 4). The formation of $Ca_3(AsO_4)_2 \cdot 4H_2O$; occurs by many technological schemes in mining plants, although it is a toxic compound for humans [64,65].

Table 4. Model quantity of precipitated minerals, mg (numerical simulation for $Ca(OH)_2$ experiments).

Precipitant	Formula	Mass of Minerals					
	Mass of $Ca(OH)_{2(s)}$, g	2.1	10.4	20.4	40.4	50.4	60.4
	S/R ratio	480	96	49	25	20	17
Tenorite	CuO						40
Hausmannite	$Mn^{2+}Mn^{3+}_2O_4$						19
Zincite	ZnO						170
Gibbsite	$Al(OH)_3$			1400	96		
Portlandite	$Ca(OH)_2$						35,000
Brucite	$Mg(OH)_2$						1007
$Fe(OH)_{3(am)}$		4500	6500	2300			
Calcite	$CaCO_3$				21,000	1790	
Dolomite	$CaMg(CO_3)_2$				3000	93	
Cerussite	$PbCO_3$					7.9	
Rhodochrosite	$MnCO_3$				23	7	
Alunite	$KAl_3(SO4)_2(OH)_6$			60			
Gypsum	$CaSO_4 \cdot 2H_2O$	6030	20,200	14,400	1780		
Anglesite	$PbSO_4$	6.6	6.7				
Apatite-OH	$Ca_5(PO_4)_3(OH)$			170	10		
	$Ca_3(AsO_4)_2 \cdot 4aq$				3310	150	620
Shultenite	$PbHAsO_4$			11			
Austinite	$CaZnAsO_4(OH)$				500	63	
Conichalcite	$CaCuAsO_4$			120			
Ettringite	$Ca_6Al_2(SO_4)_3(OH)_{12} \cdot 26H_2O$						12,100

However, due to the precipitation of $Ca_3(AsO_4)_2 \cdot 4H_2O$, the arsenic concentration equals 0.14 g/L. At the last step of the experiment (pH 12.2, Eh -116 mV), ettringite (an indicator of alkaline solutions and a product of concrete corrosion) and (hydr)oxides (tenorite CuO, hausmannite $Mn^{2+}Mn^{3+}_2O_4$, zincite ZnO) appear. Indeed, the model solution becomes richer in Fe, Cu, and Ca, but does not become richer in arsenic. The increase in the arsenic concentrations at the end of the experiment is because the stability of iron hydroxocomplexes increases at pH 12.2; here, its solid phases dissolve, which releases trace elements into the solution. In the air, calcium arsenate is converted to carbonate, with the release of arsenic into the solution, according to the following reaction (Equation (4)):

$$Ca_3(AsO_4)_2 \cdot 4H_2O + CO_3^{2-} = CaCO_3 + 2CaAsO_4^- + 4H_2O \qquad (4)$$

Portlandite and Brusite were not detected by the microscopy. Not all phases can be taken into account in the model, due to the lack of thermodynamic data. That is why, instead of complex sulfate-arsenate phases, simple hydroxides such as $Ca(OH)_2$ could appear under alkaline conditions. Some supersaturation of the solution in respect to portlandite cannot be ruled out upon its final addition.

Arsenic species in solutions change with increasing pH and mineralization (EC). In the first stages of the experiment, the majority of arsenic in the solution was in the form of positively charged complexes with iron $FeH_2AsO_4^+$, $FeHAsO_4^+$ and $FeH_2AsO_4^{2+}$ and arsenic acid H_3AsO_4 (Figure S1). With a decrease in the S/R ratio to 49, due to the binding of iron in $Fe(OH)_{3(am)}$, $H_2AsO_4^-$ becomes a major species, with a meaningful proportion of arsenate complexes with the metals $MgH_2AsO_4^+$, $CaH_2AsO_4^+$ and $AlHAsO_4^+$. At the next steps of the numerical simulation, at subalkaline and alkaline pH values, arsenic is in the form of a neutral complex of $CaHAsO_4$ and anionic $CaAsO_4^-$, which remains the only compound of arsenic in solution, at a concentration ~0.1 mg/L. Iron also has a variety of species. In acidic solutions, Fe^{2+} (0.01 mole/L), sulfate and arsenite complexes ($FeH_2AsO_4^+$) are the predominant species, with a meaningful proportion of arsenate complexes. At pH 7.79 and 12.22, hydroxocomplexes ($Fe(OH)_4^-$) are the predominant species. The study of the elemental species is important from a scientific (cycle of elements in the biosphere) and practical (technology) perspective. This phenomenon can be traced back to the example of As in the S/R ratio, changing from 25 through 20 to 17. In the first case (S/R = 25), the proportion of the As complexes is low, solutions are supersaturated with respect to minerals and 3310 mg of precipitate are formed (supersaturation is calculated through the product of the uncomplexed ions activity). In the second case (S/R = 20), the proportion of As complexes is almost 50%, and only 150 mg of precipitate are formed; in addition, at S/R = 17, despite the fact that AsO_4^{3-} < 1%, 620 mg of $Ca_3(AsO_4)_2 \cdot 4H_2O$ is precipitated because it is no longer gypsum, and ettringite controls a higher concentration of Ca in the solution (Figure 3).

3.2. Redox Experiment with Na₂S

It is known that sulfur, in the form of sulfide, forms low soluble compounds with heavy metals under reducing conditions [66]. Therefore, in nature, there are massive deposits of hypogenic sulfides and sulfides of the secondary enrichment zone. The highest activity, due to the significant content of sulfur in the form of sulfide, has sodium sulfide (Na_2S), which yields strongly alkaline solutions. In the experiment with Na_2S, the acidity of the solutions decreased, due to a slightly different scheme (Figure 5).

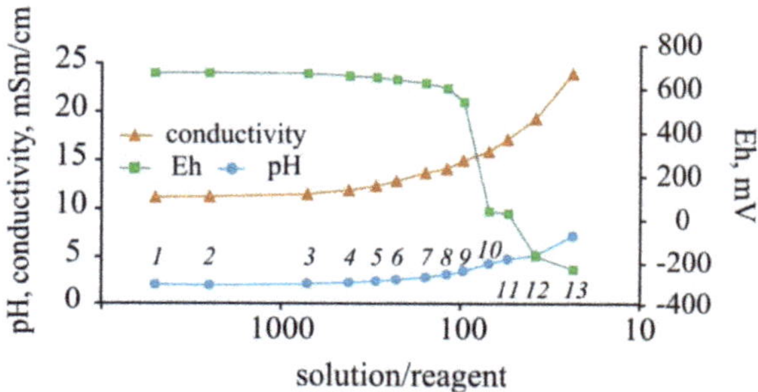

Figure 5. Changes in the physicochemical conditions and EC during the experiment with Na_2S. S/R ratio decreased as follows: *1—5000; 2—2500; 3—714; 4—417; 5—294; 6—227; 7—156; 8—119; 9—96; 10—69; 11—54; 12—38; 13—24.*

Initially, at high S/R ratios, the increase in pH was insignificant (0.5 units), which was similar in the experiment with $Ca(OH)_2$. Moreover, for an identical decrease in acidity, three times less reagent was required than in the experiment with $Ca(OH)_2$ (pH 2.9 requires 6.4 g of Na_2S and 20.4 g of $Ca(OH)_2$). With S/R = 24–25, the pH value in the experiment with Na_2S was 7.29, and in the experiment with $Ca(OH)_2$, it was only 5.17.

It is reasonable that the redox conditions of the solutions changed more significantly, and at the end of the experiment (S/R = 24), Eh was −244 mV, which corresponds to the reducing environment at pH 7.29. The fundamental difference from the experiment with $Ca(OH)_2$ is a change in EC. Due to an increase in the Na concentrations (up to 7.4 g/L), the EC increased. Unlike calcium, Na was deposited in minimal amounts at the last steps of the experiment. The Na concentration, starting from S/R = 700, exceeds the MPC; additionally, at the end of the experiment, the excess concentration was 60 times higher, which is certainly an unfavorable result. The iron concentration during the experiment began to decrease under acidic conditions (S/R = 227, pH 2.69) and then gradually decreased as the conditions changed to S/R = 38 and pH 5.32 (Figure 5).

In the next step of the experiment (S/R = 24, pH 7.29), the Fe content decreased sharply (from 3410 to 9 mg/L), accompanied by a decrease in Eh to 244 mV. The behavior of $Al^{(III)}$, affected only by pH, was similar. A significant decrease in the Al concentrations (to 278 mg/L, almost two-fold) began with the ratio S/R = 69 and pH 4.41. It is noteworthy that arsenic precipitates from the solution, even at higher ratios (S/R = 119, pH 3.25) and under oxidizing conditions (Eh = 587 mV). Thus, the behavior of As is affected by the Eh value, pH value and sulfide sulfur concentration. In addition, at the subsequent steps of the experiment, As concentrations fluctuated and reached a minimum (0.78

mg/L) at S/R = 38 and pH 5.32. The experiment was terminated because it had already reached pH 7.22, and arsenic was released back to the solution at concentrations up to 113 mg/L. We assume that, due to reducing conditions, bound $As^{(V)}$ began to release into the solution in the form of $HAsO_3^{2-}$, which is the highly toxic acid of $As^{(III)}$. Antimonous acid is its analog (green and red lines, Figure 6).

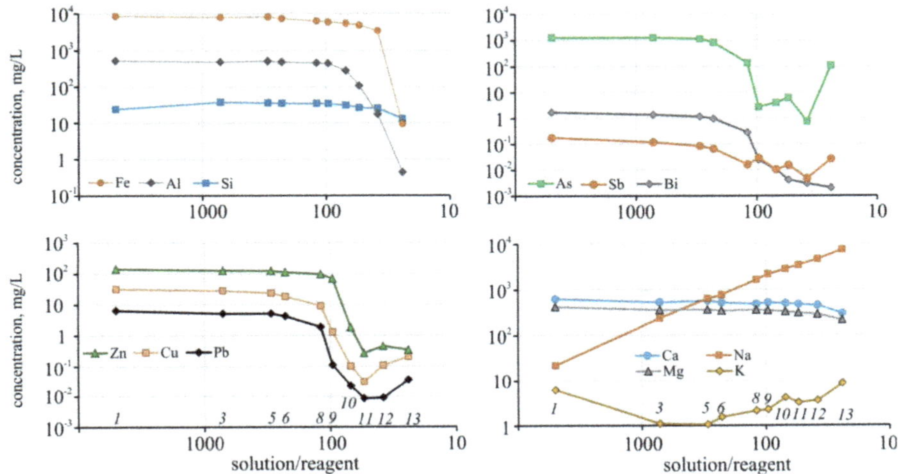

Figure 6. Changes in the elemental concentrations during the experiment with Na_2S.

A decrease in the concentrations of Pb and Cu also began under acidic conditions S/R = 119 and pH 3.25, but the Zn concentration decreased in the next step of the experiment at pH 3.61. The minimum metal concentrations were achieved under slightly acidic conditions (pH 4.87), and then the concentrations began to increase.

The precipitate formed in the experiment was a dark gray, almost black substance; after drying, it was a fine-grained mass. According to XRD analysis, sulfides, sulfates and arsenates formed as a result of the solution reduction (Table 5).

Table 5. Mineral composition of the precipitates after the experiment with Na_2S (XRD analysis).

Major		Minor		Trace	
Pyrite	**FeS$_2$**	**Thenardite**	**Na$_2$SO$_4$**	**Uzonite**	**As$_4$S$_5$**
		Scorodite	FeAsO$_4$·2H$_2$O	Basaluminite	Al$_4$[(OH)$_{10}$SO$_4$]·3.3–5H$_2$O
Amorphous phase	Fe(OH)$_3$	Tennantite	Cu$_{12}$As$_4$S$_{13}$	Mirabilite	Na$_2$SO$_4$·10H$_2$O
				-	Zn$_2$As$_2$O$_7$
		Chalcophyllite	Cu$_{18}$Al$_2$(AsO$_4$)$_4$(SO$_4$)$_3$(OH)$_{24}$·36H$_2$O	-	As$_2$S$_5$
Gypsum	CaSO$_4$·2H$_2$O			-	(Fe,Zn,Cu)SO$_4$·H$_2$O
Sulfur	S				

Pyrite, gypsum, native sulfur, and amorphous Fe-phases are the main phases. The following phases were determined in minor amounts: thenardite, scorodite, tennantite, chalcophyllite; in addition, the following phases were determined in trace amounts: uzonite, basaluminite, mirabilite, As-phases, and Fe, Cu and Zn sulfate (solid solution). We do not exclude the possibility of the formation of soluble Na salts thenardite and mirabilite from residual solutions during the drying of the precipitates.

Using electron microscopy, crystals of native sulfur, gypsum (Figure 7a), Fe and Na sulfates with impurities of As and Zn (Figure 7a–c), thenardite (Figure 7b), pyrite veins and borders with an admixture of As were found.

Figure 7. Scanning electron microscopy images of the precipitates after the experiments with Na_2S: (**a**) crystals of native sulfur in the mass of sulfates; (**b**) fine-grained intergrowths of sulfates; (**c**) veins and borders of pyrite in the mass of sulfates.

Numerical Simulation for the Na_2S Experiment

Thermodynamic modeling was carried out according to the same scheme as for the $Ca(OH)_2$ experiment: titration of the sulfate Fe-As solution with Na_2S. After adding 0.2 g of Na_2S, the solution was supersaturated with respect to silica, $Fe(OH)_{3(am)}$, gypsum, and anglesite, and this condition determined the beginning of a decrease in the concentrations of elements, such as Si, Fe, Ca, and Pb. Here, at pH 2.3 (10.4 g Na_2S, S/R = 96), symplesite $Fe^{(II)}_3(AsO_4)_2 \cdot 8H_2O$ begins to precipitate (Table 6).

Table 6. Model quantity of precipitated minerals, mg (numerical simulation for Na_2S experiments).

Mineral	Formula	Mass of Mineral					
	Mass of $Na_2S_{(s)}$, g	0.2	10.4	14.4	18.4	26.4	42.4
	S/R ratio	5000	96	69	54	38	24
Silica	SiO_2	51					
Pyrite	FeS_2					6421	9029
Galenite	PbS					7.4	
Sphalerite	ZnS					204	4.9
Chalcocite	Cu_2S			40			
Chalcopyrite	$CuFeS_2$					92	
Argentite	Ag_2S			0.01			
Gibbsite	$Al(OH)_3$			466	934	37	
$Fe(OH)_{3(am)}$		2570	7508				
Alunite	$KAl_3(SO_4)_2(OH)_6$			66			
Gypsum	$CaSO_4 \cdot 2H_2O$	162	530	26			
Anglesite	$PbSO_4$	7					
Apatite-OH	$Ca_5(PO_4)_3(OH)$				114	67	2.8
Kaolinite	$Al_2(Si_2O_5)(OH)_4$			81			
Symplesite	$Fe^{2+}_3(AsO_4)_2 \cdot 8H_2O$		4610	362	140	120	100

The binding of arsenic to symplesite explains the sharp decrease in the arsenic concentrations in the numerical simulation from 1300 to 35.8 mg/L, and the decrease in iron concentrations to 668 mg/L. The process can be described by the reaction as follows (Equation (5)):

$$Na_2S + 3Fe(SO_4)^0 + 2AsO_4^{3-} + 12H_2O = Fe^{(II)}_3(AsO_4)_2 \cdot 8H_2O + 4SO_4^{2-} + 2Na^+ + 4H_{2(gas)} \quad (5)$$

We did not see any symplesite $Fe^{2+}_3(AsO_4)_2 \cdot 8H2O$ in our experimental solids. Instead, a mixture of $Fe(OH)_3$ + Scorodite $FeAsO4 \cdot 2H2O$ + Chalcophyllite $Cu_{18}Al_2(AsO_4)_4(SO_4)_3(OH)_{24} \cdot 36H_2O$ was discovered (Table 5). Nevertheless, the precipitation of symplesite at S/R ~100, and then its gradual dissolution, correctly described the variation of As and Fe in according experimental solutions.

Because of the active binding and precipitation of a large amount of As, in the next steps of the modeling, the amount of formed simplesite sharply decreases due to the formation of pyrite, which can be seen in Table 6. The negative result is high sulfate ion concentrations in solution, which is clearly indicated by the calculations. To precipitate the sulfate ion at this stage, $Ca(OH)_2$ should probably be added (using a combined scheme), which would lead to the formation of gypsum. Of the metal sulfides, chalcocite (Cu_2S) first appeared, which in nature marks the zone of secondary sulfide enrichment, and this explains the decrease in the concentration of Cu from 32 to 1.3 mg/L (at 25 times). A subsequent decrease in the Cu concentrations is also determined by the formation of chalcocite to pH ~5.0. In parallel with Cu sulfide, sulfates, such as gypsum, alunite, gibbsite, and kaolinite, are formed. Then, during the transition from weakly oxidizing to reducing conditions, sulfides, such as pyrite, chalcopyrite, sphalerite, galena, and argentite, begin to form in neutral and subalkaline conditions. Arsenopyrite was not recorded in the calculations.

The change in the As species in an acidic solution is similar to the experiment with $Ca(OH)_2$. At high S/R ratios, iron arsenate cationic complexes, $FeH_2AsO_4^+$, $FeHAsO_4^+$, $FeH_2AsO_4^{2+}$, and arsenic acid H_3AsO_4, are the predominant species of As (Figure S2). Arsenate complexes with Al ($AlHAsO_4^+$) and Mg ($MgH_2AsO_4^+$) were formed in a significantly smaller amount. Starting with the S/R = 69, the differences in the experiment with $Ca(OH)_2$ are dramatic, since there is no Ca in the solution; i.e., Ca completely precipitated (gypsum and then apatite). Arsenic in solution is present in the form of highly toxic arsenic acid $H_3As^{(III)}O_3$. In this case, in the model solutions, a clear increase in the As content is observed, which agrees with the experimental results.

3.3. Neutralization Experiment with NaOH

During the experiment with NaOH, the pH values gradually increased to pH = 2.94 (S/R = 156). Subsequently, at S/R = 119, the pH value increased by almost 1 unit (pH 3.86), and then the pH value further increased at each step by 1–1.5 units (Figure 8).

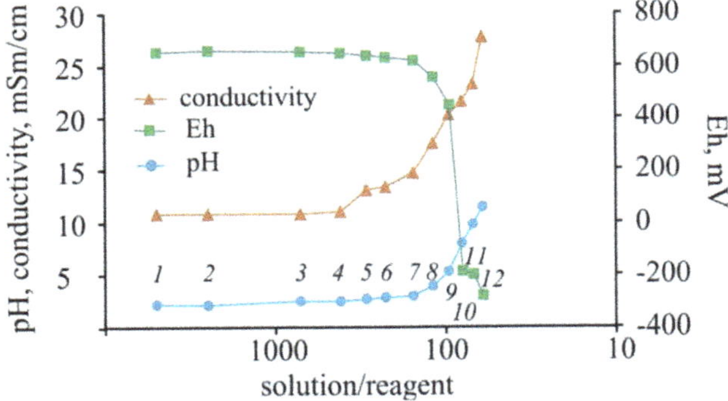

Figure 8. Changes in the physicochemical conditions and EC during the experiment with NaOH. S/R ratio decreased as follows: *1—5000; 2—2500; 3—714; 4—417; 5—294; 6—227; 7—156; 8—119; 9—96; 10—81; 11—69; 12—61.*

The EC due to the release of sodium into the solution remained stable only in the first four steps. When the S/R ratio was decreased, the EC sharply increased. The redox potential began to change markedly when the medium became near-neutral, and then, in the alkaline region, the conditions changed to reducing conditions.

The concentration of Fe decreased by four times with the increase in pH from 2.94 to 3.86, and then continued to decrease until alkaline conditions were established (Figure 9).

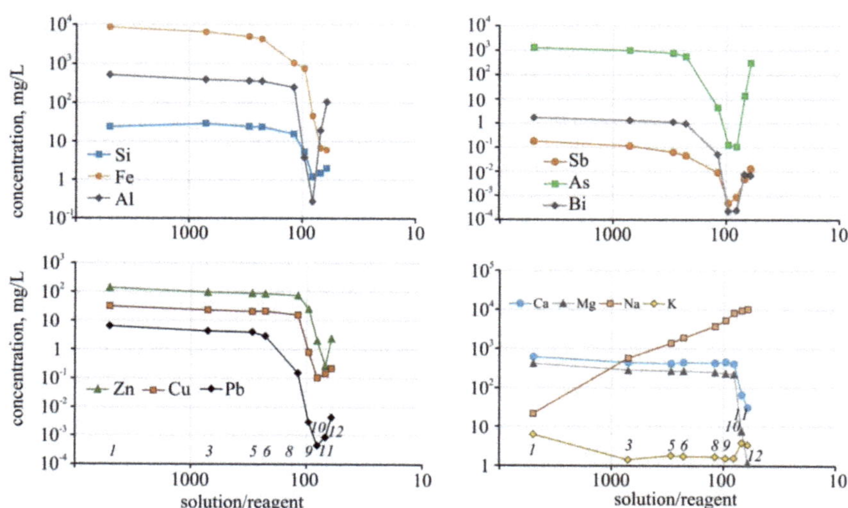

Figure 9. Changes in the element concentrations during the experiment with NaOH.

The Al concentration began to decrease one step later, with a change in pH from 3.86 to 5.29; however, the decrease was sharper than for the iron. In addition, the pH range of its minimum values was even narrower: in an alkaline environment, aluminum actively began releasing into the solution. The behavior of As, Sb and Bi was similar. The minimum As concentration (0.13–0.11 mg/L) was reached in the neutral-slightly alkaline conditions, and As increased sharply (up to 310 mg/L) at pH = 11.45.

Metals (Zn, Cu, and Pb) began to precipitate in neutral conditions; however, as expected under alkaline conditions, they were released into solution. Similar to the experiment with Na_2S, the Na concentration increased linearly during the experiment, and at S/R = 700, the Na concentration was 4.5 times higher than that of the MPC. By the end of the experiment, Na was almost 2 orders of magnitude higher than that of the MPC. The precipitate mainly consists of thenardite, gypsum, and iron hydroxides (Table 7, Figure 10a–c). Calcite, chalcophyllite, scorodite, kaatialaite, and untitled phases (Na and Al sulfate and Mg hydroxo arsenate) were identified as impure minerals.

In the electron microscopy study, As was revealed in the amorphous Fe-phase together with an admixture of Cu (Figure 10a). Impurities of Cu, Zn, Al, and As were revealed in small inclusions of Fe-sulfate in thenardite (Figure 10b). Perhaps these inclusions are the smallest grains of As minerals identified by XRD, and the impurities in them are due to matrix capture. The grains of calcite are surrounded by the druse of tenardite (Figure 10c).

Table 7. Mineral composition of the precipitates after the experiment with NaOH (XRD analysis).

Major		Minor		Trace	
Thenardite	Na_2SO_4	Calcite	$CaCO_3$	Kaatialaite	$FeAs_3O_9 \cdot 8H_2O$
Gypsum	$CaSO_4 \cdot 2H_2O$	Chalcophyllite	$Cu_{18}Al_2(AsO_4)_4(SO_4)_3(OH)_{24} \cdot 36H_2O$	Scorodite	$FeAsO_4 \cdot 2H_2O$
Amorphous phase	$Fe(OH)_3$				$Na_3Al(SO_4)_3$
					$Mg(H_2AsO_4)_2$

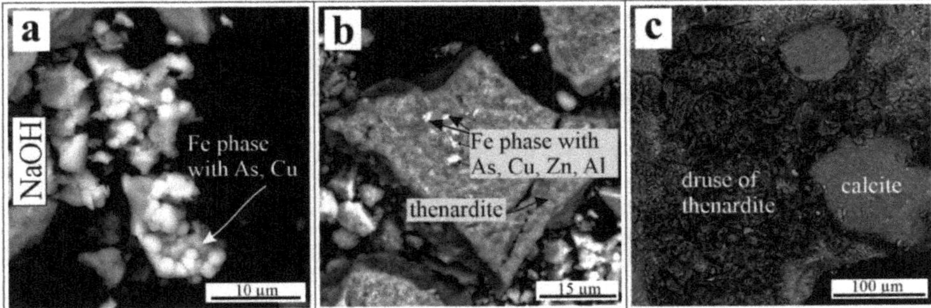

Figure 10. Scanning electron microscopy images of the precipitates after the experiments with NaOH: (**a**) colloforms of Fe minerals containing As (up to 3.2%) and Cu (0.34%); (**b**) fine-grained inclusions of Fe-minerals (presumably scorodite) in thenardite, with impurities of Cu (up to 0.4%), Zn (up to 0.53%), and Al (up to 1.6%); (**c**) calcite grains in a druse of thenardite.

Numerical Simulation for the NaOH Experiment

The acidic solutions are supersaturated with respect to silica, $Fe(OH)_{3(am)}$, gypsum, anglesite and simplesite $Fe_3(AsO_4)_2 \cdot 8H_2O$ (Table 8). The neutralization processes are faster than in the experiment with $Ca(OH)_2$. It was possible to achieve a pH = 5.29 with the addition of only 10.4 g of NaOH. At this moment (Eh 0.44), gibbsite, alunite, and apatite were precipitated, and conichalcite $CaCuAsO_4(OH)$ (Table 8) were formed instead of symplesite and shultenite. At pH 9.8 and 11.6, the removal of As was associated with the formation of calcium arsenate, which was mixed with illite, apatite, and goethite. Therefore, the formation of native copper and austenite $CaZnAsO_4(OH)$ is possible.

Table 8. Model quantity of the precipitated minerals, mg (numerical simulation for NaOH experiments).

Minerals	Formula	Mass of Mineral					
	Mass of $NaOH_s$, g	0.4	3.4	10.4	12.4	14.4	16.4
	S/R Ratio	2500	294	96	81	69	61
Copper	Cu						32
Silica	SiO_2	51					
Chalcocite	Cu_2S				40		
Gibbsite	$Al(OH)_3$			1450	49		
Brucite	$Mg(OH)_2$					187	816
$Fe(OH)_{3(am)}$		11,000	1304	1055		27	
Dolomite	$CaMg(CO_3)_2$				870		
Rhodochrosite	$MnCO_3$				11	25	2
Alunite	$KAl_3(SO_4)_2(OH)_6$			61			
Gypsum	$CaSO_4 \cdot 2H_2O$	2130	323				
Anglesite	$PbSO_4$	7					
Apatite-OH	$Ca_5(PO_4)_3(OH)$			163	21		
Illite	$K_{0.65}Al_{2.0}[Al_{0.65}Si_{3.35}O_{10}](OH)_2$					94	
	$Ca_3(AsO_4)_2 \cdot 4aq$					1450	250
Shultenite	$PbHAsO_4$			11			
Austinite	$CaZnAsO_4(OH)$				420	140	
Conichalcite	$CaCuAsO_4$			122			
Symplesite	$Fe^{2+}_3(AsO_4)_2 \cdot 8H_2O$	1900	1100		1200		
	$Pb(OH)_2$				7		

At the beginning of the experiment, the As species in solution are the same, namely, iron arsenate complexes with a small proportion of magnesium and aluminum complexes, but arsenic acid predominates (H_3AsO_4 and $H_2AsO_4^-$, Figure S3). When the S/R = 96, the As concentration sharply decreases, and the proportion of arsenic complexes with metals also decreased. In general, for the experiment with NaOH, the As complexes with metals are not characteristic in the solution, due to the sharp decrease in the iron concentration in the solution at high S/R ratio, which will be discussed subsequently. $HAsO_4^{2-}$ and AsO_4^{3-} are the predominant species, with a small proportion of calcium arsenate.

So, in experimental settings, abundant Cu, Al, Fe, Mg arsenates precipitation can occur compared to model mineral phases via pH increase Fe arsenate symplesite → Ca, Cu, Zn, Pb arsenates (Conichalcite, Shultenite Austinite) → Ca arsenate $Ca_3(AsO_4)_2 \cdot 4H_2O$. HCh modeling, together with experimental results, indicate that Ca in these solutions is very low due to the low solubility of its arsenates. Most likely, that native copper, quartz, as well as illite require higher reaction time in order to precipitate in the experiment.

4. Discussion

For a comparative assessment of the effectiveness of the reagents used, the changes in the different parameters are summarized in the corresponding diagrams (Figures S4 and S5). A decrease in the acidity of the solution (increasing pH values) most rapidly occurred in the experiment with NaOH, and the slightly acidic conditions (pH 5.29) were recorded at S/R = 96 (10.4 g of reagent was added). The same pH value (5.32) in the experiment with Na_2S was reached at S/R = 38 (26.4 g was added); in addition, in the experiment with $Ca(OH)_2$, a pH of 5.17 was reached only after adding 40.4 g (S/R = 25).

In experiments with NaOH and Na_2S, the mineralization of the solution increased with the decrease in the S/R ratio, due to an increase in the Na concentration. Sodium precipitation in the form of thenardite and mirabilite did not have a significant effect on its concentration in the solution, due to the high solubility of thenardite and mirabilite. However, in the experiment with $Ca(OH)_2$, the mineralization decreased with increasing pH, due to the formation of a large amount of gypsum in the first stages, and then due to the formation of calcite. The element deposition also occurred differently. The most effective deposition of Fe, As, and Zn occurred in the experiment with NaOH and began at earlier stages. However, in the experiment with NaOH, the release of arsenic to 310 mg/L into the solution was observed at earlier stages compared with the other experiments, and this release negated the effectiveness of this reagent. An increase in the As concentrations occurred in the other two experiments, up to 2.6 mg/L in the experiments with $Ca(OH)_2$, and up to 113 mg/L in experiments with Na_2S. Note that Cu began to precipitate earlier in the experiment with Na_2S, which was facilitated by the formation of a chalcocite suspension.

These processes can be explained by the thermodynamic calculation of the amounts and mineral compositions of the suspension formed during the experiments at different S/R ratios. As a result of the interaction of the solution with $Ca(OH)_2$, gypsum and $Fe(OH)_{3(am)}$ are already formed, even at high S/R ratios. Only at S/R = 49 do small amounts of schultenite and conichalcite begin to form, which leads to the removal of 40 mg of arsenic from the solution in the form of As phases. At the same stage (S/R = 49), gibbsite and alunite are formed. In a field experiment with S/R = 49, 760 mg of arsenic was removed from the solution, which we associated with sorption on the surface of amorphous Fe-hydroxides (Figure S6). The effective purification of As-containing solutions using iron sulfate (the formation of amorphous iron hydroxides) has been known for a long time and is widely used [67].

The most significant binding and As removal occurred when the solution was at saturation with respect to the $Ca_3(AsO_4)_2 \cdot 4H_2O$. When the S/R ratio was 25, 990 mg of As precipitated according to the numerical simulation results. In the field experiment, it was at this stage (S/R = 25) that a sharp decrease in As concentrations from 320 to 0.14 mg/L was also recorded.

Unlike $Ca(OH)_2$, the Na_2S at much earlier stages caused the binding of arsenic in the form of symplesite, which began to form under acidic conditions at S/R = 96 (Figure S7). At this stage, the formation of its highest amount took place; in the subsequent steps of the experiment, symplesite remained the only As mineral. However, if the As-containing phases were formed up to the very end in the experiment with $Ca(OH)_2$, then the formation of symplesite in the experiment with Na_2S was reduced up to the end of the experiment. Perhaps this formation of As-containing phases, in addition to desorption from iron hydroxides, explains the sharp increase in the As concentrations at the end of the experiment with Na_2S.

The formation of As-containing pyrite in the last stages of the experiment with Na_2S under alkaline conditions led to a decrease in the Fe concentration (Figure 6).

From the first steps of the experiment with NaOH, due to the neutralization process, a large suspension quantity was formed, mainly $Fe(OH)_{3(am)}$ and symplesite, which immediately removed 240 mg of arsenic (Figure S8). In the subsequent stages, As minerals were formed in greater or lesser amounts. Moreover, if there was only symplesite at the first two steps of the experiments with NaOH, then in the future steps, conichalcite and schultenite were formed, and then again, the symplesite was in a mixture with austenite; at the final stages, the $Ca_3(AsO_4)_2 \cdot 4H_2O$ phase appeared, which led to a sharp decrease in the As concentration. However, although the As-containing minerals in this experiment could be formed until the end of the process, As leaching in the alkaline solutions was also observed, due to desorption and dissolution.

NaOH is the most effective compound for the deposition of As and Fe: the pH value increased to 11.45 at S/R = 60. To achieve the neutral pH value of the solution, it was necessary to add approximately 10 g of NaOH, 30 g of Na_2S, and 50 g of $Ca(OH)_2$. It is advisable to use NaOH in the first stage of the technological treatment. The removal of 99.9% arsenic and other elements from the solution was achieved in a narrow pH range for experiments with all three reagents. When using $Ca(OH)_2$, the best result is obtained at S/R ~20; in the experiment with Na_2S, this S/R ratio can be ~55–50; and with NaOH, the best result is obtained at S/R ~100–80. In the experiments with $Ca(OH)_2$ and NaOH, the minimum As concentrations were 0.14–0.15 mg/L and 0.13–0.11 mg/L, respectively. This result required 40–50 g $Ca(OH)_2$ and 10–12 g NaOH. In the experiment with Na_2S, the minimum As concentration was 0.78 mg/L, which is a high value even for technological effluents. In this case, the presence of Fe has a negative effect. Since Fe is the first element to be reduced in the solution, a competing reaction of iron sulfide formation occurs. For the precipitation of Zn and Cu, $Ca(OH)_2$ was the most effective reagent; the minimum concentrations of Zn and Cu were 0.021 mg/L and 0.009 mg/L, respectively. Lead was best removed in the experiment with NaOH. A serious drawback of the use of Na_2S and NaOH is the high Na mobility. At the end of the experiment with Na_2S, the excess concentration of Na was 60 times higher than the MPC for Na; additionally, in the experiment with NaOH, it was almost two orders of magnitude higher than the MPC for Na.

The correlation analysis of As bonds with other elements in the solution after the interaction with $Ca(OH)_2$, Na_2S, and NaOH and deposition of solid phases was performed (Table 9). High positive correlation coefficients were revealed between arsenic and metals (Cu, Zn, Pb, Cd, Cr, Ni, Ag, V, Mo, Sn, In) and metalloids (Sb, Bi, P) when interacting with all reagents. This means similar behavior of these components in all three experiments and deposition of solid phases of complex composition: 1) arsenates, sulfo- and hydroxo-arsenates, oxides of Ca, Mg, Fe, Zn, Cu, and Pb; 2) colloidal hydroxides of Fe, Mn, and Al with metals and As sorbed on their surface; 3) compounds of unknown composition that do not occur in nature.

Positive correlation between As and Ca, Mg, Al, Fe, Zn, Cu, and Pb in solutions after interaction with $Ca(OH)_2$ is explained by the formation of arsenates, sulfo- and hydroxo-arsenates and other compounds confirmed by thermodynamic modeling and found in sediments using XRD analysis: $Ca_3(AsO_4)_2 \cdot 4aq$, Fe-Ca minerals containing As, arsenosiderite $Ca_3Fe_4(AsO_4)(OH)_6 \cdot 3H_2O$ austinite, conichalcite, shultenite (Tables 3 and 4, Figure 8). The positive correlations of As with Fe, Zn, Cu in the experiment with Na_2S are due to the formation of symplesite, scorodite, Fe-Na sulfates with As and Zn, pyrite with As, chalcophyllite, tennantite, and $Zn_2As_2O_7$ (Tables 5 and 6, Figure 9). Positive correlations of As and Ca, Mg, Pb, Zn, Cu, and Fe, in the experiment with NaOH, are associated with the formation of $Ca_3(AsO_4)_2 \cdot 4aq$, $Mg(H_2AsO_4)_2$, shultenite $PbHAsO_4$, austinite $CaZnAsO_4(OH)$, conichalcite $CaCuAsO_4$, symplesite $Fe^{2+}_3(AsO_4)_2 \cdot 8H_2O$, scorodite $FeAsO_4 \cdot 2H_2O$, chalcophyllite $Cu_{18}Al_2(AsO_4)_4(SO_4)_3(OH)_{24}·36H_2O$, and kaatialaite $FeAs_3O_9·8H_2O$ (Tables 7 and 8, Figure 10).

Table 9. Values of correlation coefficients between As and elements in solution during the experiments with $Ca(OH)_2$, Na_2S and NaOH. Bold indicates values above which the correlation is statistically significant, with a probability of 99%.

	Ca	Mg	Na	K	Al	Fe	Mn	Cr	Co	Ni	Cu	Zn
$Ca(OH)_2$	−0.29	0.81	0.43	0.26	**0.94**	**0.98**	**0.86**	**0.98**	**0.90**	**0.89**	**0.94**	**0.92**
Na_2S	0.64	0.62	−0.72	−0.17	0.69	0.72	0.45	**0.87**	0.65	**0.83**	**0.98**	**0.86**
NaOH	0.49	0.62	−0.76	0.43	**0.90**	**0.96**	0.72	**0.96**	0.78	0.51	**0.86**	**0.84**
	Pb	Cd	Ba	Rb	Sr	Sb	Bi	P	V	Sn	In	Ag
$Ca(OH)_2$	**0.99**	**0.95**	**0.83**	0.00	0.32	**0.98**	**0.97**	**0.98**	**0.98**	**0.97**	**0.96**	**0.97**
Na_2S	**0.98**	**0.93**	−0.02	0.39	0.54	**0.92**	**0.95**	**1.00**	**0.87**	**0.99**	**0.91**	**0.82**
NaOH	**0.98**	**0.81**	0.63	0.59	0.72	**0.96**	**0.97**	**0.99**	**0.97**	**0.98**	**0.95**	**0.99**

The deposition of arsenic due to sorption on colloidal hydroxide iron (III), manganese (IV) and aluminum (III) compounds when the pH shifts to the alkaline values in experiments with $Ca(OH)_2$ and NaOH is indicated by positive significant correlations of the contents of As and Al, Fe, and Mn ($r = 0.8/0.9$). Above, we showed the formation of colloforms of Fe minerals containing As (up to 3.2%) and Cu (0.34%) after the interaction with NaOH (Figure 10) and colloform mixture of gypsum, goethite, and hydroxosulfates of Al, Mg, and Fe, with ultrafine inclusions of As-containing grains after the interaction with $Ca(OH)_2$ using scanning electron microscopy (SEM analysis) (Figure 4). When forming a suspension, not only those phases that were found in precipitation and determined by the thermodynamic modelling, but also multicomponent compounds, which, possibly, have no analogues in nature, including Cr, Ni, Bi, In, V, Mo, Sn, and Ag. The co-precipitation of many elements must be taken into account when developing practical recommendations for the subsequent use of the precipitate.

On the whole, the use of NaOH is advisable in the first steps of the solution neutralization, and the further precipitation of metals and metalloids should be carried out using $Ca(OH)_2$.

5. Conclusions

1. A case study for both acidic solutions neutralization and metal precipitation from highly mineralized solutions was conducted in situ experiments. This study was developed in order to evaluate feasibility and duration prior to the performance of full-scale research and applied projects to remove dissolved arsenic (approximately 80–120 kg) from the brown ponds, on the surface of long-stored tailings.

2. Neutralization of acidic multicomponent solutions of reservoirs on the Berikul tailings is most effective by NaOH (caustic soda) in comparison with $Ca(OH)_2$ and Na_2S. Due to its advantages, at a solution/reagent ratio of ~100, pH increased to 5.3, but using Na_2S to 3.6, and with $Ca(OH)_2$ only to 2.6. Arsenic compounds (symplesite) began to form and precipitate in the first steps of the NaOH experiment, which led to its efficient removal from solution. However, its disadvantage is a higher As concentration at pH 11.45 of 310 mg/L.

3. In terms of efficiency, $Ca(OH)_2$ exhibits the smallest pH-buffer ability, but generates the greatest bulk sediments (100 g), which is due to gypsum and carbonate precipitation. In the grains of gypsum and other Fe-Ca minerals, ultrafine inclusions of the As-containing phases, hypothetically, arsenosiderite and $Ca_3(AsO_4)_2·4H_2O$, were detected. Calcium hydroxide is the preferred method of most facilities due to its low cost. Sodium sulfide Na_2S treatment seems more advantageous than lime, because it can precipitate arsenic faster than the other chemicals at pH = 3.6, due to the sedimentation of simplesite. However, minimum arsenic content in solution was 0.78 mg/L, which is significantly higher than that in the experiments with $Ca(OH)_2$ and NaOH.

4. The obtained findings are quite important, since they allow us to recommend the use of NaOH in the first stages for full-scale experiments in combination with Na_2S and then, starting with the S/R = 100, replace them with $Ca(OH)_2$. None of the reagents alone work well for these acid multicomponent solutions.

Supplementary Materials: The following are available online at http://www.mdpi.com/2075-163X/10/10/867/s1, Figure S1. Changes in the As species in the solutions during the experiments with $Ca(OH)_2$ at a decrease in the solution/reagent ratio. Figure S2. Changing As species in solution during the experiment with Na_2S at a decrease in the water/rock ratio. Figure S3. The change in As species during the experiment with NaOH at a decrease in the solution/reagent ratio. Figure S4. Comparison of the changes in the physicochemical parameters of the solution in the experiments. Figure S5. Comparison of changes in elemental concentrations in experiments. Figure S6. The amount and phase composition of the resulting precipitate in the experiment with $Ca(OH)_2$. Hereinafter: numbers above circles denote the amount of As removed from the solution; Figure S7. The amounts and phase compositions of the resulting suspensions in the experiment with Na_2S. Figure S8. The amounts and phase compositions of the resulting suspensions in the experiment with NaOH.

Author Contributions: S.B.: Conceptualization, Methodology, Investigation, Writing–Original Draft, Visualization. O.G.: Formal analysis, Data Curation, Validation, Writing–Review and Editing, N.Y.: Methodology, Investigation, Resources, Writing–Review and Editing, O.S.: Formal analysis, Resources, Investigation, Validation. N.A.: Formal analysis, Data Curation, Writing–Review and Editing, Visualization. All authors have read and agreed to the published version of the manuscript.

Funding: This work was supported by the Russian Science Foundation [19–17–00134].

Acknowledgments: The authors gratefully thank Academic Editor Carlito Tabelin, and anonymous reviewers for helpful comments and recommendations on this manuscript.

Conflicts of Interest: The authors declare that they have no known competing financial interests or personal relationships that could have appeared to influence the work reported in this paper.

References

1. Gazea, B.; Adam, K.; Kontopoulos, A. A review of passive systems for the treatment of acid mine drainage. *Min. Eng.* **1996**, *9*, 23–42. [CrossRef]
2. Lee, G.; Bigham, J.M.; Faure, G. Removal of trace metals by coprecipitation with Fe, Al and Mn from natural waters contaminated with acid mine drainage in the Ducktown Mining District, Tennessee. *Appl. Geochem.* **2002**, *17*, 569–581. [CrossRef]
3. Tabelin, C.B.; Hashimoto, A.; Igarashi, T.; Yoneda, T. Leaching of boron, arsenic and selenium from sedimentary rocks: II. pH dependence, speciation and mechanisms of release. *Sci. Total Environ.* **2014**, *473–474*, 244–253. [CrossRef] [PubMed]
4. Tamoto, S.; Tabelin, C.B.; Igarashi, T.; Ito, M.; Hiroyoshi, N. Short and long term release mechanisms of arsenic, selenium and boron from a tunnel-excavated sedimentary rock under in situ conditions. *J. Contam. Hydrol.* **2015**, *175–176*, 60–71. [CrossRef] [PubMed]
5. Pozo, G.; Pongy, S.; Keller, J.; Ledezma, P.; Freguia, S. A novel bioelectrochemical system for chemical-free permanent treatment of acid mine drainage. *Water Res.* **2017**, *126*, 411–420. [CrossRef]
6. Tabelin, C.B.; Igarashi, T.; Villacorte-Tabelin, M.; Park, I.; Opiso, E.M.; Ito, M.; Hiroyoshi, N. Arsenic, selenium, boron, lead, cadmium, copper, and zinc in naturally contaminated rocks: A review of their sources, modes of enrichment, mechanisms of release, and mitigation strategies. *Sci. Total Environ.* **2018**, *645*, 1522–1553. [CrossRef]
7. Rodriguez, M.; Manuel Baena-Moreno, F.; Arroyo, F.; Zhang, Z. Remediation of acid mine drainage. *Springer* **2019**, *17*, 1529–1538. [CrossRef]
8. Tomiyama, S.; Igarashi, T.; Tabelin, C.B.; Tangviroon, P.; Ii, H. Acid mine drainage sources and hydrogeochemistry at the Yatani mine, Yamagata, Japan: A geochemical and isotopic study. *J. Contam. Hydrol.* **2019**, *225*, 103502. [CrossRef]
9. dos Santos, K.B.; de Almeida, V.O.; Weiler, J.; Schneider, I.A.H. Removal of Pollutants from an AMD from a Coal Mine by Neutralization/Precipitation Followed by "In Vivo" Biosorption Step with the Microalgae Scenedesmus sp. *Minerals* **2020**, *10*, 711. [CrossRef]
10. Ríos, C.A.; Williams, C.D.; Roberts, C.L. Removal of heavy metals from acid mine drainage (AMD) using coal fly ash, natural clinker and synthetic zeolites. *J. Hazard. Mater.* **2008**, *156*, 23–35. [CrossRef]

11. Makarova, M.; Abrosimova, N.; Rybkina, E.; Fiaizullina, R.; Nikolaeva, I. Experimental investigation of sorption of microelements from mine drainage on zeolite and clay. *SGEM* **2017**, *17*, 447–455.

12. Gaskova, O.L.; Bukaty, M.B. Sorption of different cations onto clay minerals: Modelling approach with ion exchange and surface complexation. *Phys. Chem. Earth Part A/B/C* **2008**, *33*, 1050–1055. [CrossRef]

13. Saeva, O.P.; Yurkevich, N.V.; Kabannik, V.G.; Kolmogorov, Y.P. Determining the effectiveness of natural reactive barriers for acid drainage neutralization using SR-XRF method. *Bull. Russ. Acad. Sci. Phys.* **2013**, *77*. [CrossRef]

14. Farooq, S.H.; Chandrasekharam, D.; Berner, Z.; Norra, S.; Stüben, D. Influence of traditional agricultural practices on mobilization of arsenic from sediments to groundwater in Bengal delta. *Water Res.* **2010**, *44*, 5575–5588. [CrossRef]

15. Kefeni, K.K.; Mamba, B.B. Evaluation of charcoal ash nanoparticles pollutant removal capacity from acid mine drainage rich in iron and sulfate. *J. Clean. Prod.* **2020**, *251*, 119720. [CrossRef]

16. Hua, M.; Zhang, S.; Pan, B.; Zhang, W.; Lv, L.; Zhang, Q. Heavy metal removal from water/wastewater by nanosized metal oxides: A review. *J. Hazard. Mater.* **2012**, *211–212*, 317–331. [CrossRef]

17. Sepehrian, H.; Ahmadi, S.J.; Waqif-Husain, S.; Faghihian, H.; Alighanbari, H. Adsorption Studies of Heavy Metal Ions on Mesoporous Aluminosilicate, Novel Cation Exchanger. *J. Hazard. Mater.* **2010**, *176*, 252–256. [CrossRef]

18. Fu, F.; Wang, Q. Removal of heavy metal ions from wastewaters: A review. *J. Environ. Manag.* **2011**, *92*, 407–418. [CrossRef]

19. Abrosimova, N.; Saeva, O.; Bortnikova, S.B.; Edelev, A.V.; Korneeva, T.V.; Yurkevich, N.V. Metals and Metalloids Removal from Mine Water Using Natural and Modified Heulandite. *Int. J. Environ. Sci. Dev.* **2019**, *9*, 202–205. [CrossRef]

20. Kiventerä, J.; Piekkari, K.; Isteri, V.; Ohenoja, K.; Tanskanen, P.; Illikainen, M. Solidification/stabilization of gold mine tailings using calcium sulfoaluminate-belite cement. *J. Clean. Prod.* **2019**, *239*, 118008. [CrossRef]

21. Wilkin, R.T.; McNeil, M.S. Laboratory evaluation of zero-valent iron to treat water impacted by acid mine drainage. *Chemosphere* **2003**, *53*, 715–725. [CrossRef]

22. Hammarstrom, J.M.; Sibrell, P.L.; Belkin, H.E. Characterization of limestone reacted with acid-mine drainage in a pulsed limestone bed treatment system at the Friendship Hill National Historical Site, Pennsylvania, USA. *Appl. Geochem.* **2003**, *18*, 1705–1721. [CrossRef]

23. Doye, I.; Duchesne, J. Neutralisation of acid mine drainage with alkaline industrial residues: Laboratory investigation using batch-leaching tests. *Appl. Geochem.* **2003**, *18*, 1197–1213. [CrossRef]

24. Park, I.; Tabelin, C.B.; Jeon, S.; Li, X.; Seno, K.; Ito, M.; Hiroyoshi, N. A review of recent strategies for acid mine drainage prevention and mine tailings recycling. *Chemosphere* **2019**, *219*, 588–606. [CrossRef] [PubMed]

25. Jung, H.; Shin, T.; Cho, N.; Kim, T.; Kim, J.; Ryu, T.; Song, K.; Hwang, S.; Ryu, B.; Han, B. Thermochemical study for remediation of highly concentrated acid spill: Computational modeling and experimental validation. *Chemosphere* **2020**, *247*, 126098. [CrossRef]

26. Igarashi, T.; Herrera, P.S.; Uchiyama, H.; Miyamae, H.; Iyatomi, N.; Hashimoto, K.; Tabelin, C.B. The two-step neutralization ferrite-formation process for sustainable acid mine drainage treatment: Removal of copper, zinc and arsenic, and the influence of coexisting ions on ferritization. *Sci. Total Environ.* **2020**, *715*, 136877. [CrossRef]

27. Tabelin, C.B.; Corpuz, R.D.; Igarashi, T.; Villacorte-Tabelin, M.; Ito, M.; Hiroyoshi, N. Hematite-catalysed scorodite formation as a novel arsenic immobilisation strategy under ambient conditions. *Chemosphere* **2019**, *233*, 946–953. [CrossRef]

28. Mulopo, J. Continuous pilot scale assessment of the alkaline barium calcium desalination process for acid mine drainage treatment. *J. Environ. Chem. Eng.* **2015**, *3*, 1295–1302. [CrossRef]

29. Kefeni, K.K.; Msagati, T.A.M.; Mamba, B.B. Acid mine drainage: Prevention, treatment options, and resource recovery: A review. *J. Clean. Prod.* **2017**, *151*, 475–493. [CrossRef]

30. Lewis, A.; Van Hille, R. An exploration into the sulphide precipitation method and its effect on metal sulphide removal. *Hydrometallurgy* **2006**, *81*, 197–204. [CrossRef]

31. Reis, F.D.; Silva, A.M.; Cunha, E.C.; Leão, V.A. Application of sodium- and biogenic sulfide to the precipitation of nickel in a continuous reactor. *Sep. Purif. Technol.* **2013**, *120*, 346–353. [CrossRef]

32. Chen, T.; Yan, B.; Lei, C.; Xiao, X. Pollution control and metal resource recovery for acid mine drainage. *Hydrometallurgy* **2014**, *147–148*, 112–119. [CrossRef]

33. Silwamba, M.; Ito, M.; Hiroyoshi, N.; Tabelin, C.B.; Fukushima, T.; Park, I.; Jeon, S.; Igarashi, T.; Sato, T.; Nyambe, I.; et al. Detoxification of lead-bearing zinc plant leach residues from Kabwe, Zambia by coupled extraction-cementation method. *J. Environ. Chem. Eng.* **2020**, *8*, 104197. [CrossRef]

34. Mondal, P.; Majumder, C.B.; Mohanty, B. Laboratory based approaches for arsenic remediation from contaminated water: Recent developments. *J. Hazard. Mater.* **2006**, *137*, 464–479. [CrossRef] [PubMed]

35. Asta, M.P.; Cama, J.; Martínez, M.; Giménez, J. Arsenic removal by goethite and jarosite in acidic conditions and its environmental implications. *J. Hazard. Mater.* **2009**, *171*, 965–972. [CrossRef]

36. Jelenová, H.; Majzlan, J.; Amoako, F.Y.; Drahota, P. Geochemical and mineralogical characterization of the arsenic-, iron-, and sulfur-rich mining waste dumps near Kaňk, Czech Republic. *Appl. Geochem.* **2018**. [CrossRef]

37. Nazari, A.M.; Radzinski, R.; Ghahreman, A. Review of arsenic metallurgy: Treatment of arsenical minerals and the immobilization of arsenic. *Hydrometallurgy* **2017**, *174*, 258–281. [CrossRef]

38. Asere, T.G.; Stevens, C.V.; Du Laing, G. Use of (modified) natural adsorbents for arsenic remediation: A review. *Sci. Total Environ.* **2019**, *676*, 706–720. [CrossRef]

39. Kim, M.J.; Nriagu, J. Oxidation of arsenite in groundwater using ozone and oxygen. *Sci. Total Environ.* **2000**, *247*, 71–79. [CrossRef]

40. Lièvremont, D.; Bertin, P.N.; Lett, M.C. Arsenic in contaminated waters: Biogeochemical cycle, microbial metabolism and biotreatment processes. *Biochimie* **2009**, *91*, 1229–1237. [CrossRef]

41. Twidwell, L.G.; McCloskey, J.; Lee, M.; Saran, J. Arsenic Removal from Mine and Process Waters by Lime/Phosphate Precipitation. Proceedings of Arsenic Metallurgy: Fundamentals and Applications, Plenary Lecture, TMS, Warrendale, PA, USA, 13–17 February 2005; pp. 71–86.

42. Noël, M.-C.; Demopoulos, G.P. Precipitation of crystalline svabite (Ca5(AsO4)3F) and Svabite/Fluorapatite (Ca5(AsO4)X(PO4)3-XF) Compounds and Evaluation of Their Long-Term Arsenic Fixation Potential. In Proceedings of the Metallurgists Proceedings, Canadian Institute of Mining, Metallurgy and Petroleum, Vancouver, BC, Canada, 11–14 May 2014.

43. Valenzuela, A. Arsenic Management in the Metallurgical industry. Ph.D. Thesis, Université Laval, Québec, QC, Canada, 2000.

44. Filippou, D.; St-Germain, P.; Grammatikopoulos, T. Recovery of metal values from copper—Arsenic minerals and other related resources. *Miner. Process. Extr. Metall. Rev.* **2007**, *28*, 247–298. [CrossRef]

45. De Klerk, R.J.; Jia, Y.; Daenzer, R.; Gomez, M.A.; Demopoulos, G.P. Continuous circuit coprecipitation of arsenic(V) with ferric iron by lime neutralization: Process parameter effects on arsenic removal and precipitate quality. *Hydrometallurgy* **2012**, *111–112*, 65–72. [CrossRef]

46. Demopous, G. Arsenic Immobilization Research Advances: Past, Present and Future. Proceedings of Conference of Metallurgists (COM), Canadian Institute of Mining, Metallurgy and Petroleum, Vancouver, BC, Canada, 28 September–1 October 2014.

47. Duquesne, K.; Lebrun, S.; Casiot, C.; Bruneel, O.; Personné, J.C.; Leblanc, M.; Elbaz-Poulichet, F.; Morin, G.; Bonnefoy, V. Immobilization of Arsenite and Ferric Iron by Acidithiobacillus ferrooxidans and Its Relevance to Acid Mine Drainage. *Appl. Environ. Microbiol.* **2003**, *69*, 6165–6173. [CrossRef] [PubMed]

48. Alam, R.; McPhedran, K. Applications of biological sulfate reduction for remediation of arsenic—A review. *Chemosphere* **2019**, *222*, 932–944. [CrossRef]

49. Yun, H.; Jang, M.; Shin, W.; Choi, J. Remediation of arsenic-contaminated soils via waste-reclaimed treatment agents: Batch and field studies. *Miner. Eng.* **2018**, *127*, 90–97. [CrossRef]

50. Gugushe, A.S.; Nqombolo, A.; Nomngongo, P.N. Application of Response Surface Methodology and Desirability Function in the Optimization of Adsorptive Remediation of Arsenic from Acid Mine Drainage Using Magnetic Nanocomposite: Equilibrium Studies and Application to Real Samples. *Molecules* **2019**, *24*, 1792. [CrossRef]

51. Yuan, Z.; Zhang, G.; Ma, X.; Yu, L.; Wang, X.; Wang, S.; Jia, Y. Rapid abiotic As removal from As-rich acid mine drainage: Effect of pH, Fe/As molar ratio, oxygen, temperature, initial As concentration and neutralization reagent. *Chem. Eng. J.* **2019**, *378*, 122156. [CrossRef]

52. Houngaloune, S.; Kawaai, T.; Hiroyoshi, N.; Ito, M. Study on schwertmannite production from copper heap leach solutions and its efficiency in arsenic removal from acidic sulfate solutions. *Hydrometallurgy* **2014**, *147–148*, 30–40. [CrossRef]

53. Okibe, N.; Koga, M.; Morishita, S.; Tanaka, M.; Heguri, S.; Asano, S.; Sasaki, K.; Hirajima, T. Microbial formation of crystalline scorodite for treatment of As(III)-bearing copper refinery process solution using Acidianus brierleyi. *Hydrometallurgy* **2014**, *143*, 34–41. [CrossRef]

54. Jahromi, F.G.; Ghahreman, A. In-situ oxidative arsenic precipitation as scorodite during carbon catalyzed enargite leaching process. *J. Hazard. Mater.* **2018**, *360*, 631–638. [CrossRef]

55. Tabelin, C.; Sasaki, A.; Igarashi, T.; Tomiyama, S.; Villacorte-Tabelin, M.; Ito, M.; Hiroyoshi, N. Prediction of acid mine drainage formation and zinc migration in the tailings dam of a closed mine, and possible countermeasures. *MATEC Web Conf.* **2019**, *268*, 06003. [CrossRef]

56. Tabelin, C.B.; Silwamba, M.; Paglinawan, F.C.; Mondejar, A.J.S.; Duc, H.G.; Resabal, V.J.; Opiso, E.M.; Igarashi, T.; Tomiyama, S.; Ito, M.; et al. Solid-phase partitioning and release-retention mechanisms of copper, lead, zinc and arsenic in soils impacted by artisanal and small-scale gold mining (ASGM) activities. *Chemosphere* **2020**, *260*, 127574. [CrossRef] [PubMed]

57. Zhang, D.; Yuan, Z.; Wang, S.; Jia, Y.; Demopoulos, G.P. Incorporation of arsenic into gypsum: Relevant to arsenic removal and immobilization process in hydrometallurgical industry. *J. Hazard. Mater.* **2015**, *300*, 272–280. [CrossRef] [PubMed]

58. Majzlan, J.; Drahota, P.; Filippi, M. Parageneses and crystal chemistry of arsenic minerals. *Rev. Mineral. Geochemistry* **2014**, *79*, 17–184. [CrossRef]

59. Bortnikova, S.; Olenchenko, V.; Gaskova, O.; Yurkevich, N.; Abrosimova, N.; Shevko, E.; Edelev, A.; Korneeva, T.; Provornaya, I.; Eder, L. Characterization of a gold extraction plant environment in assessing the hazardous nature of accumulated wastes (Kemerovo region, Russia). *Appl. Geochem.* **2018**, *93*, 145–157. [CrossRef]

60. Horowitz, A.; Demas, C.; Fitzgerald, K.; Miller, T. *US Geological Survey Protocol for the Collection and Processing of Surface-Water Samples for the Subsequent Determination of Inorganic Constituents in Filtered Water (No. 94-539)*; US Geological Survey; USGS Earth Science and Information Center, Open-File Reports Section: Reston, VA, USA, 1994.

61. Shvarov, Y.V. HCh: New potentialities for the thermodynamic simulation of geochemical systems offered by Windows. *Geochem. Inter.* **2008**, *46*, 834–839. [CrossRef]

62. Garrels, R.; Christ, C. *Solutions, Minerals, and Equilibria*; Harper and Row: New York, NY, USA, 1965.

63. Tabelin, C.B.; Corpuz, R.D.; Igarashi, T.; Villacorte-Tabelin, M.; Alorro, R.D.; Yoo, K.; Raval, S.; Ito, M.; Hiroyoshi, N. Acid mine drainage formation and arsenic mobility under strongly acidic conditions: Importance of soluble phases, iron oxyhydroxides/oxides and nature of oxidation layer on pyrite. *J. Hazard. Mater.* **2020**, *399*, 122844. [CrossRef]

64. Bortnikova, S.; Bessonova, E.; Gaskova, O. Geochemistry of arsenic and metals in stored tailings of a Co-Ni arsenide-ore, Khovu-Aksy area, Russia. *Appl. Geochem.* **2012**, *27*, 2238–2250. [CrossRef]

65. Gaskova, O.L.; Bessonova, E.P.; Bortnikova, S.B. Leaching Experiments on Trace Element Release from the Arsenic-Bearing Tailings of Khovu-Aksy (Tuva Republic, Russia). *Appl. Geochem.* **2003**, *18*, 1361–1371. [CrossRef]

66. Andreev, S.Y.; Kochergin, A.C.; Malutina, T.V.; Alekseeva, T.V. *The Use of Precipitating Agents in the Wastewater Treatment of Galvanic Plants*; PH PEGAS: Pensa, Russia, 2016; Volume 116. (In Russian)

67. Gas'kova, O.L.; Bortnikova, S.B.; Airiyants, A.A.; Kolmogorov, Y.P.; Pashkov, M.V. Geochemical features of an anthropogenic impoundment with cyanidation wastes of gold-arsenopyrite-quartz ores. *Geochem. Int.* **2000**, *38*, 281–291.

Article

Forecast of AMD Quantity by a Series Tank Model in Three Stages: Case Studies in Two Closed Japanese Mines

Chiharu Tokoro [1,*] **, Kenichiro Fukaki** [2]**, Masakazu Kadokura** [2] **and Shigeshi Fuchida** [1]

[1] Faculty of Science and Engineering, Waseda University, 3-4-1 Okubo, Shinjuku-ku, Tokyo 169-8555, Japan; sfuchida@aoni.waseda.jp

[2] Graduate School of Creative Science and Engineering, Waseda University, 3-4-1 Okubo, Shinjuku-ku, Tokyo 169-8555, Japan; kf-2219-rikoh@ruri.waseda.jp (K.F.); masa-kado0413@akane.waseda.jp (M.K.)

* Correspondence: tokoro@waseda.jp; Tel.: +81-3-5286-3320

Received: 8 April 2020; Accepted: 9 May 2020; Published: 11 May 2020

Abstract: There are about 100 sites of acid mine drainage (AMD) from abandoned/closed mines in Japan. For their sustainable treatment, future prediction of AMD quantity is crucial. In this study, AMD quantity was predicted for two closed mines in Japan based on a series tank model in three stages. The tank model parameters were determined from the relationship between the observed AMD quantity and the inflow of rainfall and snowmelt by using the Kalman filter and particle swarm optimization methods. The Automated Meteorological Data Acquisition System (AMeDAS) data of rainfall were corrected for elevation and by the statistical daily fluctuation model. The snowmelt was estimated from the AMeDAS data of rainfall, temperature, and sunshine duration by using mass and heat balance of snow. Fitting with one year of daily data was sufficient to obtain the AMD quantity model. Future AMD quantity was predicted by the constructed model using the forecast data of rainfall and temperature proposed by the Max Planck Institute–Earth System Model (MPI–ESM), based on the Intergovernmental Panel on Climate Change (IPCC) representative concentration pathway (RCP) 2.6 and RCP8.5 scenarios. The results showed that global warming causes an increase in the quantity and fluctuation of AMD, especially for large reservoirs and residence time of AMD. There is a concern that for mines with large AMD quantities, AMD treatment will be unstable due to future global warming.

Keywords: acid mine drainage; life cycle simulation; global warming; earth system model; RCP2.6; RCP8.5

1. Introduction

Japan has more than 5000 abandoned/closed mines, and about 100 of their sites produce acid mine drainage (AMD) due to the presence of sulfide mineralization [1]. The general treatment for AMD is neutralization and sedimentation by addition of a neutralizer, such as lime, calcium carbonate, and sodium hydroxide [2], and solid/liquid separation [3] of the produced sludge from the neutralized effluents. In these treatments, all toxic elements are concentrated into the sludge by precipitation [4–6] and adsorption [7–12], and the sludge is controlled in a tailing pond at a mine site or final disposal site. For these last several decades, AMD has been treated properly in Japan and has not caused severe pollution. However, since our results of the statistical calculation (details are shown below) suggested that some mines have required AMD treatment for over 150 years [13,14] and other groups suggested that more than 1000 years of treatment will be necessary [15] in the current situation, more sustainable treatment to reduce both AMD generation [16,17] and treatment cost [18] is needed. To reduce the treatment cost of the addition of chemicals and of sludge generation, for example, a passive treatment that utilizes the natural environment of mines, such as topography, plants, and microorganisms, has attracted attention as a sustainable AMD treatment based on new concepts [19–21]. Several researchers are trying to successfully reuse this sludge as an industrial material [22,23].

Unlike industrial wastewater, AMD quantity and quality differ significantly in mines due to regional, geological, mineralogical, and biological factors. Therefore, it is necessary to customize an appropriate treatment for each mine. To select an optimal treatment method from the various treatment technologies, including those based on both active and passive concepts, an accurate understanding of the current potential for AMD generation [24,25] and the future forecast of AMD quantity and quality are essential.

The objective of this study was to determine a forecast for AMD quantity. To accomplish this, we constructed a model that reproduces the current AMD quantity using previous monitoring data of AMD, and then extrapolated it to the future. To reproduce AMD quantity, there are two ways: one is a hydraulic simulation [26–31] and the other is a tank model. A hydraulic simulation provides detailed information on the origin and distribution of AMD and can be a powerful tool for discussing AMD generation countermeasures, but it requires, in addition to meteorological data, detailed geological, mineralogical, and hydraulic data, which are generally difficult to obtain, especially for abandoned and closed mines. On the other hand, a tank model is a blackbox to determine the relationship between inflow and outflow, but just inflow data of rainfall and snowmelt and outflow of AMD quantity are sufficient for the model [32,33]. In this study, the tank model was selected for AMD quantity modeling, and rainfall and snowmelt were used as inflow. The rainfall data were corrected for elevation and adjusted using the statistical daily fluctuation model to suit each AMD site. Snowmelt was also estimated from rainfall by considering mass and heat balance by using temperature and sunshine duration data. We did not select a hydraulic simulation but chose a statistical model because our target mines are closed and it was difficult to obtain detailed monitoring, geological, and hydraulic data for this study.

For the AMD quality model, we previously reported the geochemical calculation with first-order elution kinetics of sulfide minerals [13,14]. In the model, sulfides that should be the source of AMD were selected from the quality data, and their first-order elution rate and initial AMD generation potential were estimated by fitting to the time change of their elution amount obtained from the AMD quantity and quality data. The AMD quality could be estimated by the coupling of the kinetics for sulfide elution and oxidation, and the geochemical code for the chemical equilibrium calculation of precipitation and adsorption [34–37]. This means that accurate estimation of AMD quantity is crucial for the AMD quality model.

In this study, the AMD quantity model was constructed using two case studies of underground mines: a closed sulfur mine (Mine A) and a closed black-ore copper, lead, and zinc mine (Mine B). Mine A has a large quantity of AMD, averaging 18 m^3 min^{-1} with a small fluctuation, which is opposite to that from Mine B (1.5 m^3 min^{-1}). From these case studies, the parameters of the model were estimated and the future AMD quantity for the next few decades was predicted using forecast data for rainfall and temperature based on the MPI–SEM (Max Planck Institute–Earth System Model) [37]. For this, we selected two kinds of global warming scenarios proposed by IPCC (Intergovernmental Panel on Climate Change): a low-stabilization scenario of RCP (representative concentration pathway) 2.6 and a high-level greenhouse gas emission scenario of RCP 8.5 [37]. We further discuss the effects of global warming on the forecast of AMD quantity stemming from the closed sulfide mines that were examined.

2. Materials and Methods

2.1. AMD Quantity Model

2.1.1. Tank Model

The AMD quantity model was constructed by using a series tank model in three stages, as shown in Figure 1. The first and second stages correspond to the nonpolluted water reservoir on the surface and inside the mine, respectively. The third stage corresponds to the polluted water reservoir in the ore deposit that causes AMD. Inflow, r (mm), is the summation of rainfall, r_w, and snowmelt, r_s. In each tank, a part of the inflow is distributed to the outflow (mm h^{-1}), q_{oi} (i = 1, 2, 3), and seepage flow to the next tank, q_{si} (i = 1, 2), according to the water reservoir height (mm), x_i (i = 1, 2, 3) and outflow height (mm), b_i (i = 1, 2). The water balance in each tank is as follows:

$$\mathrm{d}x_i/\mathrm{d}t = q_{si\text{-}1} - q_{si} - q_{oi}, (i = 1, 2, 3) \tag{1}$$

where t is time, $q_{s0} = r = r_w + r_s$, and $q_{s3} = 0$. The outflow is calculated from:

$$q_{oi} = a_{oi} (x_i - b_i), (i = 1, 2, 3) \tag{2}$$

where a_{oi} (i = 1, 2, 3) is the outflow coefficient and $b_3 = 0$. The seepage flow is also calculated from:

$$q_{si} = a_{si} x_i, (i = 1, 2) \tag{3}$$

where a_{si} (i = 1, 2) is the seepage coefficient.

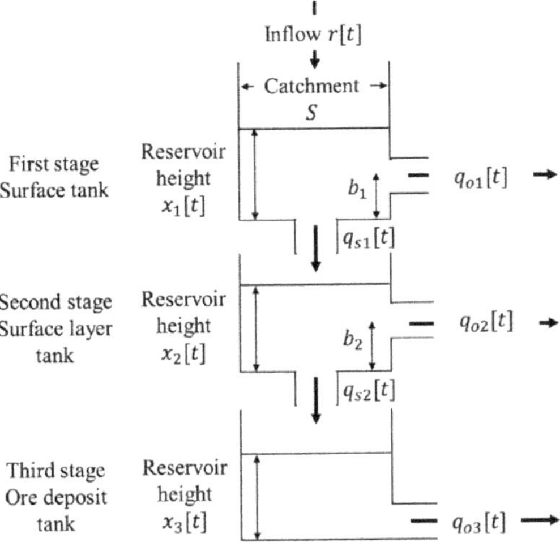

Figure 1. Schematic of the tank model in three stages used for the acid mine drainage (AMD) quantity model in this study.

In this study, q_{o3} corresponded to AMD quantity. The inflow, r, was set using the following procedure. The Kalman filter and particle swarm optimization methods were used for fitting q_{o3} to the observed data of AMD quantity to estimate the x_i, b_i, a_{si}, and a_{oi} parameters [38].

2.1.2. Correction of Rainfall Data and Judgment of Snowfall

Rainfall data near each mine were derived from AMeDAS (Automated Meteorological Data Acquisition System) provided by the Japan Meteorological Agency [39]. To suit each mine situation, the daily data of rainfall and temperature obtained from AMeDAS were corrected for elevation and adjusted by using the statistical daily fluctuation model, as shown in Figure 2. Also, snowfall was estimated according to the corrected temperature, and if the temperature was under 2 °C, it was judged to be snowfall and not rainfall.

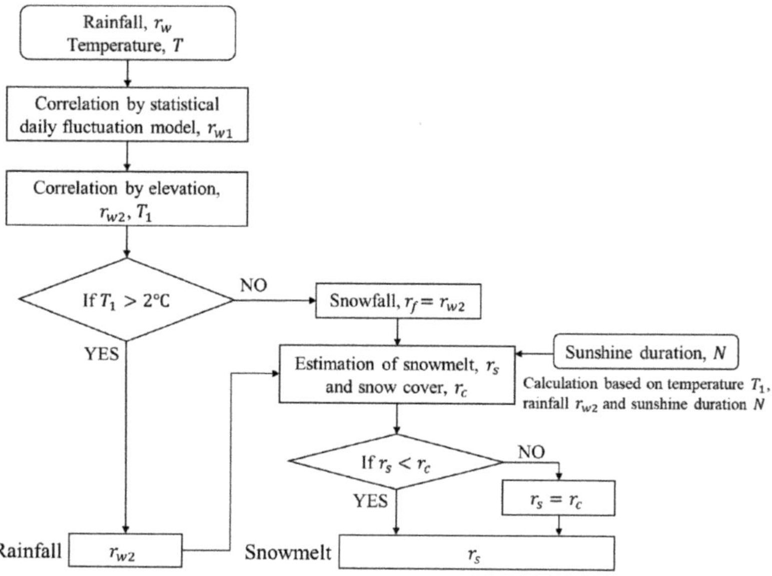

Figure 2. Procedural flow to obtain the inflow for the tank model (rainfall and snowmelt) from the Automated Meteorological Data Acquisition System (AMeDAS) data of rainfall, temperature, and sunshine duration.

Since the capture rate of rainfall particles by the rain gauge decreases as the wind speed increases, known as the Jebons effect, Kondo et al. proposed the following statistical correction for daily rainfall data obtained from AMeDAS [40]:

$$r_{w1} = (1.25 + 0.15 \cos(w(d - 20)))r_{w0} \tag{4}$$

$$w = 2\pi/365 \tag{5}$$

where r_{w0} (mm) is the raw rainfall data, r_{w1} (mm) is the corrected rainfall data by the model, d (days) refers to the days from 1 January. Furthermore, the amount of rainfall near mines depends greatly on elevation due to rapid updraft and adiabatic expansion, and is corrected as follows [40]:

$$r_{w2} = (1 + c(h - h_0))r_{w1} \tag{6}$$

where r_{w2} (mm) is the corrected rainfall data by elevation, h (km) is the elevation of mine site, h_0 (km) is the elevation of the AMeDAS observation point, and c is the coefficient (0.001 km^{-1} for 5 °C or less and 0.00064 km^{-1} for 5 °C or more). AMeDAS data of temperature, T, were also corrected according to following equation:

$$T_1 = -6(h - h_0) + T \tag{7}$$

where T_1 (°C) is the corrected temperature.

As shown in Figure 2, if the temperature at the mine site was above 2 °C, it was assumed that there was no snowfall and the inflow of rainfall was set as r_{w2}. On the other hand, if the temperature was below 2 °C, the rainfall data, r_{w2}, were judged to be equivalent to snowfall, $r_f = r_{w2}$, and $r_{w2} = 0$.

2.1.3. Estimation of Snowmelt and Snow Cover

Daily snowmelt, r_s, and snow cover, r_c, were also estimated by following the mass and heat balance of snow using the AMeDAS data of rainfall, temperature, and sunshine duration, as shown in Figure 2. If the temperature was below 0 °C, no snowmelt was assumed and $r_s = 0$; otherwise, snowmelt was calculated according to the following procedure.

Snowmelt, r_s (mm), was calculated from the ratio of fusion heat, Q (J), and latent heat, L (334 J kg^{-1}), the catchment area, S (mm^2), and the density of water, ρ (9.97 × 10^{-7} kg mm^{-3}):

$$r_s S \rho = Q/L \tag{8}$$

The fusion heat was calculated from the heat balance of snow exposure as follows:

$$Q = Q_1 + Q_2 + Q_3 + Q_4 + Q_5 \tag{9}$$

where Q_1 is the short-wavelength radiation, Q_2 is the long-wavelength radiation, Q_3 is the sensible heat transfer, Q_4 is the latent heat transfer, and Q_5 is the transfer heat due to rainfall. Here, heat changes in the snow layer and heat transfer from the ground were assumed to be negligible [41].

The short-wavelength radiation was calculated from albedo, r, which is the ratio of reflected sunshine radiation to sunshine radiation on the earth's surface and the daily average of solar irradiance, I (W m^{-2}):

$$Q_1 = (1 - r)I \tag{10}$$

The average of solar irradiance was the function of the ratio of sunshine duration, N, and astronomical sunshine duration, N_0 [42]:

$$I/I_0 = 0.179 + 0.550N/N_0, \text{ for } 0 \leq N \leq N_0 \tag{11}$$

$$I/I_0 = 0.114, \text{ for } N = 0 \tag{12}$$

where I_0 is the solar irradiance at the top of atmosphere. The values of solar irradiance at the top of the atmosphere, I_0, astronomical sunshine duration, N_0, and albedo, r, are available from references [39,43], and sunshine duration data, N, are available from AMeDAS.

The long-wavelength radiation was the difference between the radiation from the atmosphere, Q_a, and the radiation from the snow surface, Q_s:

$$Q_2 = Q_a - Q_s \tag{13}$$

$$Q_a = \sigma(T_1 + 273.15)^4(0.605 + 0.048e^{0.5}) \tag{14}$$

$$Q_s = 0.9\sigma(T_s + 273.15)^4 \tag{15}$$

$$e = 6.1078 \times 10^{7.5T/(T+273.3)} \tag{16}$$

where σ is the Stefan–Boltzmann constant (5.67 × 10^{-8} W m^{-2} K^{-4}) and e is amount of saturated water vapor [44]. The temperature of the snow surface, T_s, was calculated from [45]:

$$T_s = 1.13T_1 - 1.67 \tag{17}$$

when $T_1 \leq 1.47$.

The sensible heat was calculated from [46]:

$$Q_3 = K(1 - 0.0065h/(T_1 + 273.15+0.0065h))^{5.257}, \text{ for } T_1 \geq 0, \tag{18}$$

$$Q_3 = 0, \text{ for } T_1 \leq 0 \tag{19}$$

where K is the transfer coefficient of the sensible heat and the latent heat; 3.5 was proposed for the area near the mines that were modeled in this case study [46]. The latent heat was calculated from:

$$Q_4 = 1.53K(e - 6.11), \text{ for } T_1 \geq 7 \tag{20}$$

$$Q_4 = 0, \text{ for } T_1 \leq 7 \tag{21}$$

The transfer heat due to rainfall was calculated from:

$$Q_5 = \rho c_w(273.15 + T_1)r_w S \tag{22}$$

where c_w is the specific heat of the water (4.186 J kg^{-1} K^{-1}).

The snowmelt, r_s, calculated from Equations (8) to (22), should be less than the snow cover, r_c. Snow cover was calculated from following summation of daily mass balance:

$$r_c = \sum r_f - \sum r_s \tag{23}$$

if $r_s > r_c$, then snowmelt should be $r_s = r_c$.

2.2. Forecast Data of Temperature, Rainfall, and Sunshine Duration

In the above-mentioned AMD quantity and quality models, daily data of rainfall, average temperature, and sunshine duration obtained from AMeDAS were used for model construction. Therefore, future forecasts of daily data of rainfall, average temperature, and sunshine duration were also necessary for the forecast of AMD quantity and quality in the future. For the AMD quality model, the geochemical calculation with first-order elution kinetics of sulfide minerals was used as mentioned above; the first-order elution rate and initial AMD generation potential were estimated by fitting to the time change of their elution amount obtained from the AMD quantity and quality data. The daily output data of MPI–ESM (Max Planck Institute–Earth System Model), which is an earth system model proposed by Max Planck Institute, were used in this study. Two kinds of IPCC RCPs for the greenhouse gas (GHG) concentration scenario were selected: RCP2.6 and RCP8.5. The former is the scenario with the lowest GHG emission to keep future temperature rise below 2 °C, and the latter is the scenario with the highest GHG emissions.

From the system, daily forecast data of rainfall, average temperature, and maximum and minimum temperatures were available. The daily forecast of the average of solar irradiance, I, was estimated from [47]:

$$I = 0.76I_0(1 - \exp(-A\Delta T^{2.2})) \tag{24}$$

$$A = 0.036\exp(-0.154\Delta T_{ave}) \tag{25}$$

where ΔT is the difference between the maximum temperature and the minimum temperature, and ΔT_{ave} is the monthly average of ΔT.

2.3. Case Studies in Two Closed Mines

In this study, two closed underground mines in the northern part of Japan were selected as a case study; Mine A has a large quantity of AMD, with small fluctuation of quantity and quality, and Mine B has a small quantity of AMD, with large fluctuation of quantity and quality. Snow is observed in winter at both of the closed mines. The locations of the mines are shown in Figure 3, and the AMD characteristics are shown in Table 1.

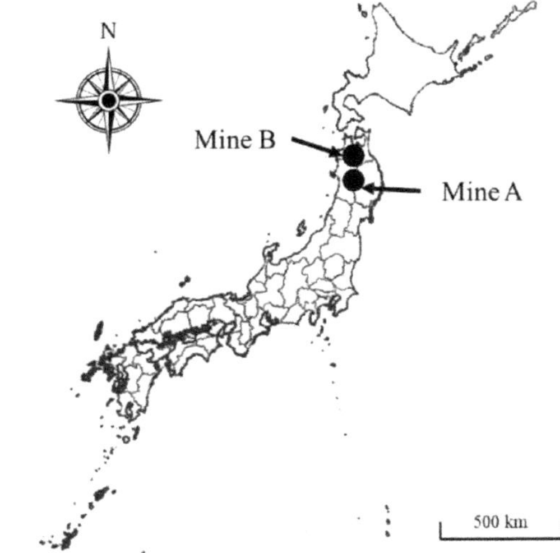

Figure 3. Location of the two closed mines for case study.

Table 1. The AMD characteristics of Mine A and B.

Items	Mine A		Mine B		Effluent
	April 1982	March 2017	April 1972	March 2017	Standard
pH	2.0	2.3	3.2	4.7	5.8–8.6
Fe	547 mg L^{-1}	178 mg L^{-1}	N.D.	N.D.	10 mg L^{-1}
As	3.3 mg L^{-1}	0.93 mg L^{-1}	N.D.	N.D.	0.01 mg L^{-1}
Cd	N.D.	N.D.	0.360 mg L^{-1}	0.024 mg L^{-1}	0.03 mg L^{-1}
Zn	N.D.	N.D.	90.4 mg L^{-1}	14.9 mg L^{-1}	2.0 mg L^{-1}
Pb	N.D.	N.D.	0.700 mg L^{-1}	0.117 mg L^{-1}	0.1 mg L^{-1}
Q	17.6 m^3/min	14.8 m^3 min^{-1}	2.50 m^3 min^{-1}	0.77 m^3 min^{-1}	

N.D.: Not detected; Q: Quantity.

In Mine A, native sulfur and pyrite were mined during its operation. The size of the ore deposit was about 1500 m East–West, about 1500 m North–South, and 25–150 m of thickness, and ore reserves were about 230 million tons per year. The mine produced about 1 million tons of ore and one-third of Japan's sulfur demand, but it closed in 1971 due to the market influence of sulfur recovered from oil refining. The quantity of AMD is about 18 m^3 min^{-1} on average annually, which is one of the largest AMD values in Japan [48]. The AMeDAS point is located 11 km east and 825 m below the mine.

In Mine B, copper, lead, and zinc were mined during its operation. The ore deposit was a black ore type, which has changed from lower to yellow ore, black ore, and quartz band. The mine produced a maximum of about 25,000 tons per year, but closed in 1985 due to ore depletion. The annual average of the AMD quantity was 1.72 m^3 min^{-1} in 2017 and increased from 5 to 7 m^3 min^{-1} during the snowmelt season [49]. The AMeDAS point is located 18 km northwest and 465 m below the mine.

3. Results and Discussion

3.1. AMD Quantity Model Construction

The relation between the input data of AMeDAS rainfall (upper side) and the observed data and the calculated value of AMD quantity (underside) are shown in Figure 4. In this calculation, the daily observed data of AMD quantity in the prior one year were used for fitting, and later one-year data were used for the model validation. The fitting period was also changed from half a year to two years and the correlation coefficients between the observed and calculated values were compared, as shown in Table 2 and Supplementary Figure S1. Of course, the longer the fitting period, the higher the correlation coefficient in the validation period, but a fitting period of one year seemed to be generally sufficient for the reproduction of AMD quantity in the next one year. As shown in Supplementary Figure S2, when the correction for elevation and the statistical daily fluctuation and the snowmelt estimation were not conducted, the reproducibility of AMD quantity became worse, especially for Mine A. This is because Mine A is located at a higher elevation and the effects of elevation correction and snowfall are larger than for Mine B.

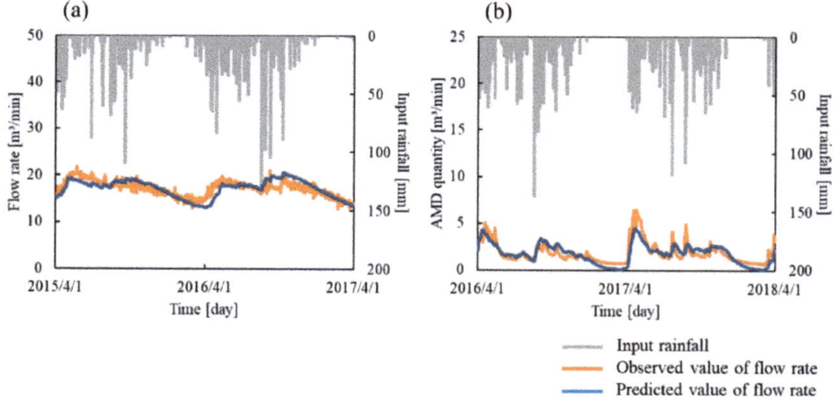

Figure 4. Observed and calculated AMD quantity by the tank model for (**a**) Mine A and (**b**) Mine B.

Table 2. Relationship between the fitting period and the correlation coefficients.

Mine	Fitting Period	Correlation Coefficient in the Fitting Period	Correlation Coefficient in the Validation Period [7]
A	Half-year [1]	0.89	0.65
	One year [2]	0.81	0.69
	Two years [3]	0.80	0.67
B	Half-year [4]	0.86	0.77
	One year [5]	0.86	0.83
	Two years [6]	0.87	0.85

[1] 2015/10/1–2016/3/31; [2] 2015/4/1–2016/3/31; [3] 2014/4/1–2016/3/31; [4] 2016/10/1–2017/4/1; [5] 2016/4/1–2017/3/31; [6] 2015/4/1–2017/3/1; [7] 2016/4/1–2017/3/31 for Mine A; 2017/4/1–2018/3/31 for Mine B.

The parameters obtained for the tank model are shown in Table 3. Mine A has a smaller outflow height and a larger AMD reservoir than Mine B. Additionally, Mine A has the smaller value of outflow coefficient in the third stage, which directly affects AMD generation, compared to Mine B. This trend means that Mine A has the bigger reservoir and the longer residence time of AMD, which resulted in the smaller fluctuation of AMD, compared to Mine B. As we mentioned in the previous section, ore production was 230 million tons per year in Mine A and 25,000 tons per year in Mine B. This difference in scale should directly affect the difference in reservoir and residence time of AMD.

Table 3. Parameters of the tank model obtained for Mines A and B.

Mine	Tank Stage	Outflow Coefficient a_o (day^{-1})	Seepage Coefficient a_s (day^{-1})	Outflow Height b (mm)
	First stage	0.768	0.568	0.00484
A	Second stage	0.936	0.352	0.00140
	Third stage	0.00282		
	First stage	0.738	0.863	0.0358
B	Second stage	0.292	0.607	0.0205
	Third stage	0.0500		

3.2. Forecast of AMD Quantity

The forecast of AMD quantity (underside) is shown in Figure 5, with the forecast of rainfall (upperside) proposed by MPI–ESM. According to MPI–ESM in RCP2.6, the forecast for temperature rise around the mines is about 2 °C by 2050, but remaining at about 1.0–2.5 °C after 2050. On the other hand, in RCP8.5, the forecast for temperature continues to rise and reaches +5.9 °C in 2100.

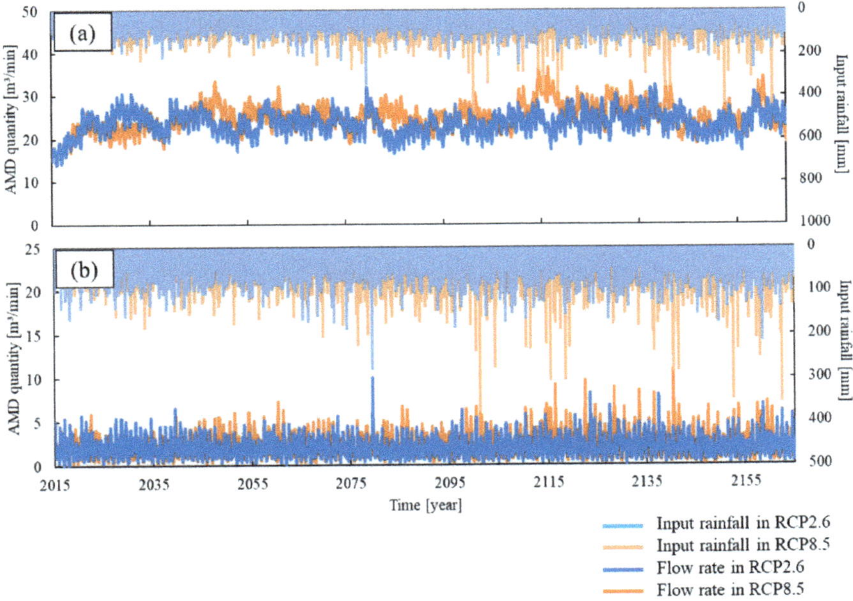

Figure 5. Forecast of AMD quantity calculated from the model until 2165, with the forecast of rainfall proposed by the Max Planck Institute–Earth System Model (MPI–ESM) for (**a**) Mine A and (**b**) Mine B.

In Mine A, the MPI–ESM shows that both rainfall and heavy rain frequency, which is the number of days per year with greater than 50 mm of rainfall, increase due to the temperature rise. In 2100, the forecast of rainfall is +21% for both RCP2.6 and RCP8.5, and forecast of heavy rain frequency increases 4 days/year for RCP2.6 and 8 days/year for RCP8.5, compared to the present. According to these trends, the AMD quantity calculated from the constructed model increases, as shown in Figure 5. The forecast for AMD quantity in 2100 is +27% for RCP2.6 and +31% for RCP8.5.

In Mine B, the MPI–ESM shows that the temperature rise of around 2 °C in the RCP2.6 scenario does not have much effect on rainfall and heavy rain frequency. In 2100, the forecast for rainfall decreases 1.5% and heavy rain frequency decreases 2 days/year, which results in a 0.55% increase for the forecast of AMD quantity, compared to the present. However, the temperature rise of 5.9 °C in the

RCP8.5 scenario affects the forecast of rainfall and heavy rain frequency as much as for Mine A. In 2100, the rainfall forecast increases 22% and heavy rain frequency increases 5.5 days/year, which results in a 25% increase for the forecast of AMD quantity, compared to the present.

The forecast of the standard deviation of AMD quantity is shown in Figure 6. A comparison of the coefficient of variation of AMD quantity between the present and the future is shown in Table 4. Here, the coefficient of variation was calculated for 10 years from 2010 to 2020 for the present and from 2100 to 2110 for the future. The temperature rise due to global warming caused larger fluctuations in the AMD quantity for Mine A. In the case of Mine B, since the AMD reservoir is small and the AMD residence time is short, even if the rainfall fluctuation increases in the future due to global warming, the AMD fluctuation will remain largely as it is now. On the other hand, in the case of Mine A, since the AMD reservoir is larger and the AMD residence time is longer, AMD fluctuation tends to increase gradually in the future, affected by increases in rainfall fluctuation due to global warming. This trend suggests that AMD treatment might be unstable because of global warming in the future, especially for mines with larger AMD quantities.

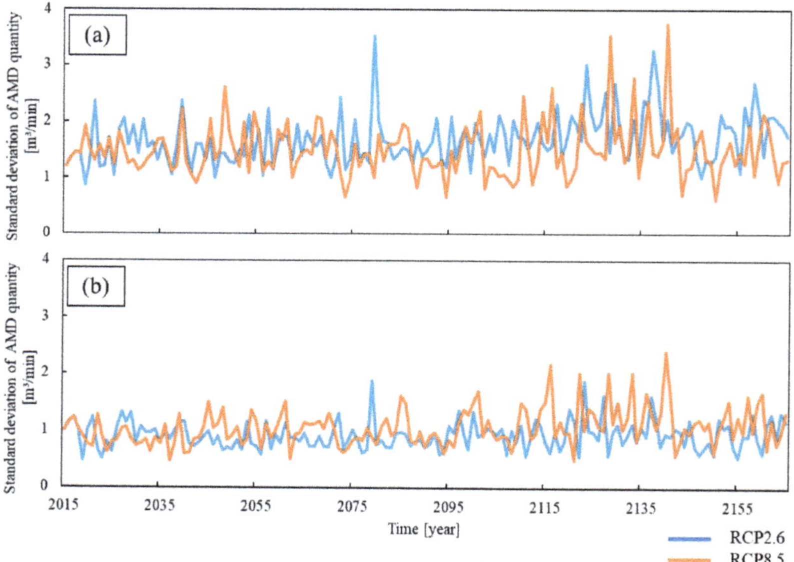

Figure 6. Forecast of the standard deviation of AMD quantity for (**a**) Mine A and (**b**) Mine B.

Table 4. The coefficient of variation for AMD quantity at present and in the future.

Mine Name	Present	Future
Mine A	0.083	RCP2.6: 0.092 RCP8.5: 0.090
Mine B	0.56	RCP2.6: 0.49 RCP8.5: 0.54

In general, AMD quality tends to deteriorate as the AMD quantity increases. This is because when the AMD quantity increases, AMD comes in contact with a new pollution source in the mine. Actually, at present, the fluctuation of AMD quantity is larger in Mine B than in Mine A, and the fluctuation in AMD quality tends to be larger in Mine B as well. This suggests that fluctuations in AMD quantity due to global warming will cause large fluctuations in AMD quality.

4. Conclusions

The AMD quantity model was constructed for two closed mines in Japan. The model was constructed with a series tank model, and fitted by using daily data for one year, which were enough to obtain adequate parameters. The results showed that Mine B has a smaller AMD reservoir and a shorter AMD residence time than Mine A, resulting in a large fluctuation of AMD quantity in Mine B. The forecast of AMD quantity was also estimated based on the forecast of rainfall and temperature proposed by the MPI–ESM with IPCC RCP2.6 and RCP8.5 scenarios. The forecast results showed that temperature rise due to global warming will cause an increase in rainfall, resulting in increased AMD quantity. The fluctuation of rainfall will also increase due to global warming, increasing the fluctuation of AMD quantity in Mine A. The effect of global warming in Mine A will be bigger than in Mine B due to its larger reservoir and longer residence time of AMD.

In this study, it is expected that the quantity and fluctuation of AMD might increase due to global warming. This suggests that fluctuations in AMD quality might also increase. Therefore, when selecting future treatment methods, careful consideration should be given to whether or not the AMD fluctuation can be sufficiently dealt with in the future, especially for passive treatment.

Supplementary Materials: The following are available online at http://www.mdpi.com/2075-163X/10/5/430/s1, Figure S1: Observed and calculated AMD quantity by the tank model for (a) Mine A with half-year fitting, (b) Mine A with two-year fitting, (c) Mine B with half-year fitting, and (d) Mine B with two-year fitting; Figure S2: Observed and calculated AMD quantity by the tank model without correction of rainfall and temperature or considering snowmelt for (a) Mine A and (b) Mine B.

Author Contributions: Conceptualization, C.T.; methodology, K.F.; validation, M.K. and S.F.; investigation, K.F.; data curation, S.F.; writing—original draft preparation, C.T.; writing—review and editing, K.F. and S.F.; supervision, C.T. All authors have read and agreed to the published version of the manuscript.

Funding: This research was partially funded by JOGMEC (Japan Oil, Gas and Metals National Corporation) and the Center for Eco-Mining, Japan.

Acknowledgments: Part of this work was performed as the activities of the Waseda Research Institute for Science and Engineering and Research Organization for Open Innovation Strategy, Waseda University.

Conflicts of Interest: The authors declare no conflict of interest.

References

1. JOGMEC (Japan Oil, Gas and Metals National Corporation). Mine Pollution Control. Available online: http://www.jogmec.go.jp/english/mp_control/index.html (accessed on 22 March 2020).
2. Tokoro, C. Removal mechanism in anionic co-precipitation with hydroxides in acid mine drainage treatment. *Resour. Process.* **2015**, *62*, 3–9. [CrossRef]
3. Tokoro, C. As (V) removal by Fe (III), Al or Pb salts and rapid solid/liquid separation in wastewater containing dilute arsenic—A fundamental study for efficient treatment of wastewater containing dilute arsenic (Part 1). *J. Mmij.* **2005**, *121*, 399–406. [CrossRef]
4. Onoguchi, A.; Granata, G.; Haraguchi, D.; Hayashi, H.; Tokoro, C. Kinetics and mechanism of selenate and selenite removal in solution by green rust-sulfate. *R. Soc. Open Sci.* **2019**, *6*, 182147. [CrossRef] [PubMed]
5. Mamun, A.; Onoguchi, A.; Granata, G.; Tokoro, C. Role of pH in green rust preparation and chromate removal from water. *Appl. Clay Sci.* **2018**, *165*, 205–213. [CrossRef]
6. Tokoro, C.; Suzuki, S.; Haraguchi, D.; Izawa, S. Silicate removal in aluminum hydroxide co-precipitation process. *Materials* **2014**, *7*, 1084–1096. [CrossRef] [PubMed]
7. Tokoro, C.; Kadokura, M.; Kato, T. Mechanism of arsenate coprecipitation at the solid/liquid interface of ferrihydrite: A perspective review. *Adv. Powder Technol.* **2020**, *31*, 859–866. [CrossRef]
8. Mamun, A.; Morita, M.; Matsuoka, M.; Tokoro, C. Sorption mechanisms of chromate with coprecipitated ferrihydrite in aqueous solution. *J. Hazard. Mater.* **2017**, *334*, 142–149. [CrossRef]
9. Haraguchi, D.; Tokoro, C.; Oda, Y.; Owada, S. Sorption mechanisms of arsenate in aqueous solution during coprecipitation with aluminum hydroxide. *J. Chem. Eng. Jpn.* **2012**, *46*, 173–180. [CrossRef]
10. Tokoro, C.; Koga, H.; Oda, Y.; Owada, S.; Takahashi, Y. XAFS investigation for As (V) co-precipitation mechanism with ferrihydrite. *J. Mmij.* **2011**, *127*, 213–218. [CrossRef]

11. Tokoro, C.; Yatsugi, Y.; Koga, H.; Owada, S. Sorption mechanisms of arsenate during coprecipitation with ferrihydrite in aqueous solution. *Environ. Sci. Technol.* **2010**, *44*, 638–643. [CrossRef]

12. Sasaki, K.; Qiu, X.; Moriyama, S.; Tokoro, C.; Ideta, K.; Miyawaki, J. Characteristic sorption of H_3BO_3/B $(OH)_4-$on magnesium oxide. *Mater. Trans.* **2013**, *54*, 1809–1817. [CrossRef]

13. Koide, R.; Tokoro, C.; Murakami, S.; Adachi, T.; Takahashi, A. A model for prediction of neutralizer usage and sludge generation in the treatment of acid mine drainage from abandoned mines: Case studies in Japan. *Mine Water Environ.* **2012**, *31*, 287–296. [CrossRef]

14. Otsuka, H.; Murakami, S.; Yamatomi, J.; Koide, R.; Tokoro, C. A predictive model for the future treatment of acid mine drainage with regression analysis and geochemical modeling. *J. Mmij.* **2014**, *130*, 488–493. [CrossRef]

15. Tabelin, C.; Sasaki, A.; Igarashi, T.; Tomiyama, S.; Tabelin, M.V.; Ito, M.; Hiroyoshi, N. Prediction of acid mine drainage formation and zinc migration in the tailings dam of a closed mine, and possible countermeasures. *Matec. Web Conf.* **2019**, *268*, 06003. [CrossRef]

16. Ogbughalu, O.T.; Gerson, A.R.; Qian, G.; Smart, R.S.C.; Schumann, R.C.; Kawashima, N.; Fan, R.; Li, J.; Short, M.D. Heterotrophic microbial stimulation through biosolids addition for enhanced acid mine drainage control. *Minerals* **2017**, *7*, 105. [CrossRef]

17. Qian, G.; Schumann, R.C.; Li, J.; Short, M.D.; Fan, R.; Li, Y.; Kawashima, N.; Zhou, Y.; Smart, R.S.C.; Gerson, A.R. Strategies for reduced acid and metalliferous drainage by pyrite surface passivation. *Minerals* **2017**, *7*, 42. [CrossRef]

18. Nguyen, H.T.H.; Nguyen, B.Q.; Duong, T.T.; Bui, A.T.K.; Nguyen, H.T.A.; Cao, H.T.; Mai, N.T.; Nguyen, K.M.; Pham, T.T.; Kim, K.-W. Pilot-scale removal of arsenic and heavy metals from mining wastewater using adsorption combined with constructed wetland. *Minerals* **2019**, *9*, 379. [CrossRef]

19. Kato, T.; Fukushima, R.; Granana, G.; Sato, K.; Yamagata, S.; Tokoro, C. Quantitative modeling incorporating surface complexation for zinc removal using leaf mold. *J. Soc. Powder Technol. Jpn.* **2019**, *56*, 136–141. [CrossRef]

20. Lefticariu, L.; Behum, P.T.; Bender, K.S.; Lefticariu, M. Sulfur Isotope fractionation as an indicator of biogeochemical processes in an AMD passive bioremediation system. *Minerals* **2017**, *7*, 41. [CrossRef]

21. Johnson, D.B. Recent developments in microbiological approaches for securing mine wastes and for recovering metals from mine waters. *Minerals* **2014**, *4*, 279–292. [CrossRef]

22. Herrera, P.S.; Uchiyama, H.; Igarashi, T.; Asakura, K.; Ochi, Y.; Ishizuka, F.; Kawada, S. Acid mine drainage treatment through a two-step neutralization ferrite-formation process in northern Japan: Physical and chemical characterization of the sludge. *J. Miner. Eng.* **2007**, *20*, 1309–1314. [CrossRef]

23. Igarashi, T.; Herrera, P.S.; Uchiyama, H.; Miyamae, H.; Iyatomi, N.; Hashimoto, K.; Tabelin, C.B. The two-step neutralization ferrite-formation process for sustainable acid mine drainage treatment: Removal of copper, zinc and arsenic, and the influence of coexisting ions on ferritization. *Sci. Total Environ.* **2020**, *715*, 136877. [CrossRef]

24. Matsumoto, S.; Ishimatsu, H.; Shimada, H.; Sasaoka, T.; Kusuma, G.J. Characterization of mine waste and acid mine drainage prediction by simple testing methods in terms of the effects of sulfate-sulfur and carbonate minerals. *Minerals* **2018**, *8*, 403. [CrossRef]

25. Chopard, A.; Marion, P.; Mermillod-Blondin, R.; Plante, B.; Benzaazoua, M. Environmental impact of mine exploitation: An early predictive methodology based on ore mineralogy and contaminant speciation. *Minerals* **2019**, *9*, 397. [CrossRef]

26. Kitamura, A.; Kurikami, H.; Sakuma, K.; Malins, A.; Okumura, M.; Machida, M.; Mori, K.; Tada, K.; Tawara, Y.; Kobayashi, T.; et al. Redistribution and export of contaminated sediment within eastern Fukushima Prefecture due to typhoon flooding. *Earth Surf. Process Landf.* **2016**, *41*, 1708–1726. [CrossRef]

27. Sakuma, K.; Kitamura, A.; Malins, A.; Kurikami, H.; Machida, M.; Mori, K.; Tada, K.; Kobayashi, T.; Tawara, Y.; Tosaka, H. Characteristics of radio-cesium transport and discharge between different basins near to the Fukushima Dai-ichi Nuclear Power Plant after heavy rainfall events. *J. Environ. Radioact.* **2017**, *169–170*, 137–150. [CrossRef]

28. Sakuma, K.; Malins, A.; Funaki, H.; Kurikami, H.; Niizato, T.; Nakanishi, T.; Mori, K.; Tada, K.; Kobayashi, T.; Kitamura, A.; et al. Evaluation of sediment and 137Cs redistribution in the Oginosawa River catchment near the Fukushima Dai-ichi Nuclear Power Plant using integrated watershed modeling. *J. Environ. Radioact.* **2018**, *182*, 44–51. [CrossRef]

29. Tomiyama, S.; Igarashi, T.; Tabelin, C.B.; Tangviroon, P.; Ii, H. Modeling of the groundwater flow system in excavated areas of an abandoned mine. *J. Contam. Hydrol.* **2020**, *230*, 103617. [CrossRef]

30. Tomiyama, S.; Igarashi, T.; Tabelin, C.B.; Tangviroon, P.; Ii, H. Acid mine drainage sources and hydrogeochemistry at the Yatani mine, Yamagata, Japan: A geochemical and isotopic study. *J. Contam. Hydrol.* **2020**, *225*, 103502. [CrossRef]

31. Kato, T.; Kawasaki, Y.; Kadokura, M.; Suzuki, K.; Tawara, Y.; Ohara, Y.; Tokoro, C. Quantitative modeling of arsenic removal by ferrihydrite coprecipitation in an artificial wetland and pond for chemical reactions coupled GETFLOWS. *Minerals* **2020**. under second review.

32. Ahmad, S.W. Tank Model Application for runoff and infiltration analysis on sub-watersheds in Lalindu River in South East Sulawesi Indonesia. *J. Phys. Conf. Ser.* **2017**, *846*, 012019. [CrossRef]

33. Aqili, S.W.; Hong, N.; Hama, T.; Suenaga, Y.; Kawagoshi, Y. Application of modified tank model to simulate groundwater level fluctuations in Kabul Basin, Afghanistan. *J. Water Enrivon. Technol.* **2016**, *14*, 57–66. [CrossRef]

34. Tokoro, C.; Sakakibara, T.; Suzuki, S. Mechanism investigation and surface complexation modeling of zinc sorption on aluminum hydroxide in adsorption/coprecipitation processes. *Chem. Eng. J.* **2015**, *279*, 86–92. [CrossRef]

35. Tokoro, C.; Yatsugi, Y.; Sasaki, H.; Owada, S. A quantitative modeling of co-precipitation phenomena in wastewater containing dilute anions with ferrihydrite using a surface complexation model. *Resour. Process.* **2008**, *55*, 3–8. [CrossRef]

36. Tokoro, C.; Maruyama, Y.; Badulis, G.C.; Sasaki, H. Application of surface complexation model for dilute As removal in wastewater by Fe (III) or Al (III) salts—A fundamental study for efficient treatment of wastewater containing dilute arsenic (Part 2). *J. Mmij.* **2005**, *121*, 532–537. [CrossRef]

37. Giorgetta, M.A.; Jungclaus, J.; Reick, C.H.; Legutke, S.; Bader, J.; Böttinger, M.; Brovkin, V.; Crueger, T.; Esch, M.; Fieg, K.; et al. Climate and carbon cycle changes from 1850 to 2100 in MPI-ESM simulations for the Coupled Model Intercomparison Project phase 5. *J. Adv. Model. Earth Syst.* **2013**, *5*, 572–597. [CrossRef]

38. Tada, T. Parameter optimization of hydrological model using the PSO algorithm. *J. Jpn. Soc. Hydrol. Water Resour.* **2007**, *20*, 450–461. [CrossRef]

39. Japan Meteorological Agency. Previous Meteorological Data and Download. Available online: https://www.data.jma.go.jp/gmd/risk/obsdl/index.php (accessed on 22 March 2020).

40. Kondo, J.; Motoya, K.; Matsushima, D. A study on annual variations of the soil water content and water equivalent of snow in a watershed, runoff and the river water temperature by use of the new bucket-model. *Meteorol. Soc. Jpn.* **1995**, *42*, 821–831.

41. Kurihara, J.; Yamakoshi, T.; Irasawa, M.; Sasahara, K.; Takahashi, M.; Yoshida, M. Study on the applicability of the simplified snowmelt prediction method to the Imokawa river basin, Niigata prefecture, Japan. *Int. J. Eros. Control Eng.* **2007**, *59*, 47–54.

42. Kondo, J.; Xu, J.; Haginoya, S. Empirical formula for estimating the solar radiation at an upland from the sunshine duration data. *J. Jpn. Soc. Hydrol Water Resour.* **1996**, *9*, 468–472. [CrossRef]

43. Yamazaki, T.; Kondo, J.; Taguchi, B. Estimation of the heat balance in small snowcovered forested catchment basin. *Tenki* **1994**, *41*, 71–77.

44. Hiramatsu, S.; Irasawa, M.; Hongo, K. Study on occurrence of hillside landsides caused by snowmelt. *Int. J. Eros. Control Eng.* **1998**, *51*, 27–34.

45. Hashimoto, T.; Ohta, T.; Ishibashi, H. Estimation of the effects of deciduous forest to the surface snowmelt by a heat balance analysis. *J. Jpn. Soc. Snow Ice* **1992**, *54*, 131–143. [CrossRef]

46. Suizu, S. A snowmelt and water equivalent snow model applicable to an extensive area. *J. Jpn. Soc. Snow Ice* **2002**, *64*, 617–630. [CrossRef]

47. Shinohara, Y.; Komatsu, H.; Otsuki, K. A method for estimating global solar radiation from daily maximum and minimum temperatures: Its applicability to Japan. *J. Jpn. Soc. Hydrol Water Resour.* **2007**, *20*, 462–469. [CrossRef]

48. Fujii, N. On the wastewater processing plant at the closed Matsuo Mine. *J. Clay Sci. Soc. Jpn.* **1994**, *34*, 184–186.
49. Asami, Y.; Nishida, Y.; Kimura, H.; Iwasawa, M. Nurukawa Kozan no tankokaihatsu oyobi sonogo no sogyo jokyo. *J. Mininig Metall. Inst. Jpn.* **1988**, *104*, 185–190. (In Japanese)

Article

Immobilization of Lead and Zinc Leached from Mining Residual Materials in Kabwe, Zambia: Possibility of Chemical Immobilization by Dolomite, Calcined Dolomite, and Magnesium Oxide

Pawit Tangviroon [1,*], Kenta Noto [2], Toshifumi Igarashi [1], Takeshi Kawashima [3], Mayumi Ito [1], Tsutomu Sato [1], Walubita Mufalo [2], Meki Chirwa [4], Imasiku Nyambe [4], Hokuto Nakata [5], Shouta Nakayama [5] and Mayumi Ishizuka [5]

[1] Division of Sustainable Resources Engineering, Faculty of Engineering, Hokkaido University, Sapporo 060-8628, Japan; tosifumi@eng.hokudai.ac.jp (T.I.); itomayu@eng.hokudai.ac.jp (M.I.); tomsato@eng.hokudai.ac.jp (T.S.)

[2] Division of Sustainable Resources Engineering, Graduate School of Engineering, Hokkaido University, Sapporo 060-8628, Japan; kn1855@docon.jp (K.N.); wmufalo@gmail.com (W.M.)

[3] Yoshizawa Lime Industry Co., LTD., Libra Bldg, 3-2 Nihonbashi-Kobuncho, Chuo-Ku, Tokyo 103-0024, Japan; tkawashi@yoshizawa.co.jp

[4] IWRM Centre/Geology Department, School of Mines, The University of Zambia, Lusaka 32379, Zambia; meki.chirwa@gmail.com (M.C.); inyambe@gmail.com (I.N.)

[5] Faculty of Veterinary Medicine, Hokkaido University, Kita 18, Nishi 9, Kita-Ku, Sapporo 060-0818, Japan; hokuto.nakata@vetmed.hokudai.ac.jp (H.N.); shouta-nakayama@vetmed.hokudai.ac.jp (S.N.); ishizum@vetmed.hokudai.ac.jp (M.I.)

* Correspondence: tangviroon.p@eng.hokudai.ac.jp or tangviroon.p@gmail.com; Tel.: +81-90-6217-0676

Received: 22 July 2020; Accepted: 24 August 2020; Published: 28 August 2020

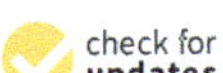

Abstract: Massive amount of highly contaminated mining residual materials (MRM) has been left unattended and has leached heavy metals, particularly lead (Pb) and zinc (Zn) to the surrounding environments. Thus, the performance of three immobilizers, raw dolomite (RD), calcined dolomite (CD), and magnesium oxide (MO), was evaluated using batch experiments to determine their ability to immobilize Pb and Zn, leached from MRM. The addition of immobilizers increased the leachate pH and decreased the amounts of dissolved Pb and Zn to different extents. The performance of immobilizers to immobilize Pb and Zn followed the following trend: MO > CD > RD. pH played an important role in immobilizing Pb and Zn. Dolomite in RD could slightly raise the pH of the MRM leachate. Therefore, the addition of RD immobilized Pb and Zn via adsorption and co-precipitation, and up to 10% of RD addition did not reduce the concentrations of Pb and Zn to be lower than the effluent standards in Zambia. In contrast, the presence of magnesia in CD and MO significantly contributed to the rise of leachate pH to the value where it was sufficient to precipitate hydroxides of Pb and Zn and decrease their leaching concentrations below the regulated values. Even though MO outperformed CD, by considering the local availability of RD to produce CD, CD could be a potential immobilizer to be implemented in Zambia.

Keywords: mine waste; contamination; batch experiments; lead; zinc; immobilization; remediation; Kabwe; Zambia

---

1. Introduction

Kabwe District was one of the most important mining regions in Zambia for almost a century (1902–1994). It was regarded as Southern Africa's principal lead (Pb)-zinc (Zn) producer, producing over 1.8 and 0.8 Mt of Zn and Pb, respectively [1]. While in operation, no pollution laws were enforced to regulate the discharge from wastes of the mine; therefore, operations of the mine have left Kabwe with a massive amount of unattended mining residual materials (MRM), which still contain elevated amounts of heavy metals, particularly Zn, Pb, and iron (Fe). Weathering of MRM causes heavy metals to transport from the contaminated sites to the surrounding environments (groundwater, surface water, and soil) [2]. In particular, the redistribution of heavy metals through solute transport processes has been reported to be one of the most dangerous pathways, which invokes harmful effects on water sources of nearby ecosystems and health-threatening to the nearby residents [3–9]. Therefore, the remediation of heavy metals in and around the mine is necessary.

In our recent studies, several potential remediation techniques have been investigated to remediate the contaminated site in Kabwe. Silwamba et al. (2020) [10,11] have proposed the concurrent dissolution and cementation method. The method shows promising results in terms of Pb removal and recovery. However, Zn could not be effectively recovered from the extraction solution, and further investigation is needed. Biocementation by locally available bacteria has been studied by Mwandira et al. (2019) [12,13]. The results indicate that the biocemented material can be effectively used as a covering layer to prevent airborne contamination of metallic dust and infiltration of water into the waste. In the present study, chemical immobilization is introduced as an alternative and practical method to remediate the site.

Remediation techniques for heavy metals polluted sites can be classified into two main categories, in-situ and ex-situ. In general, ex-situ treatment has high efficiency; however, it is less cost-effective than in-situ remediation due significantly to the costs for excavation and transport of large quantities of contaminated materials. In-situ remediation avoids the excavation and transportation costs because of on-site treatments of contaminants. Various kinds of in-situ remediation techniques have been developed to immobilize or extract the heavy metals in the contaminated sites. Among them, chemical immobilization is cheap, easy to implement, and quick in execution [14,15]. Thus, this is the most promising technique, especially to be applied in one of the developing countries.

In in-situ chemical immobilization, the leaching potential of heavy metals from contaminated soils is reduced via sorption and/or precipitation processes by adding chemical agent (immobilizer) into the contaminated area. The performance of a variety of immobilizers, including carbonates, phosphates, alkaline agents, clay, iron-containing minerals, and organic matters, has been evaluated [14–23]. However, most of the studies have been conducted to remediate contaminated soil samples in which they generally contain much lesser metals contents than those in MRM. Moreover, because of the complex interactions between solutes and immobilizers, the definite efficiency of the immobilizer remains site-specific. In other words, there is no guarantee on the effectiveness of a particular immobilizer implemented on different contaminated sites. Therefore, the objective of this study was to evaluate and compare the performance of selected potential immobilizers (e.g., locally available, low-cost) to reduce the mobility of Pb and Zn leached from the highly contaminated sample (MRM).

It is necessary to apply immobilizers that is low cost and abundant in nature for remediating contaminated areas. Hence, in this study, raw dolomite (RD) was selected as one of the potential candidates because it is naturally available in a large quantity in Zambia [24]. It is a carbonate mineral; therefore, it can increase and buffer pH of MRM, leading to more adsorption and precipitation of cationic heavy metal ions [25–27]. In the present study, calcined dolomite (CD) was also used as an immobilizer. The heat treatment was performed to change the carbonate property of dolomite to be more alkaline [28]. As a result, the immobilizer was expected to strongly increase the pH of MRM, favoring the immobilization of heavy metals by hydroxide precipitation in addition to adsorption and precipitation of other secondary minerals. At the same time, commercially available alkaline-based agent, magnesium oxide (MO), was also tested to compare the ability of RD and CD on immobilizing heavy metals in MRM. The current study will provide meaningful information for the development of chemical immobilization to remediate heavy metals contaminated sites in Zambia.

2. Materials and Methods

2.1. Solid Sample Collection, Preparation, and Characterization

MRM was collected from the dumping site of Pb-Zn mine wastes in Kabwe, Zambia. The sampling was done using shovels at random points within the area shown in Figure 1. This leaching residual was selected as one of the representative wastes because the leaching concentrations of Zn and Pb from the waste were higher compared with the other wastes. Moreover, the storage size of this waste occupies more than 50% of the total dumping area. The sample was stored in vacuum bags and transported to the laboratory in Japan with permission by the Ministry of Agriculture, Forestry, and Fisheries of Japan. In preparation, it was air-dried under ambient conditions, lightly crushed, sieved using a 2 mm aperture screen, and kept in a polypropylene bottle before use. Particle sizes of less than 2 mm were chosen to follow the Japanese standard for the leaching test of contaminated soils [29].

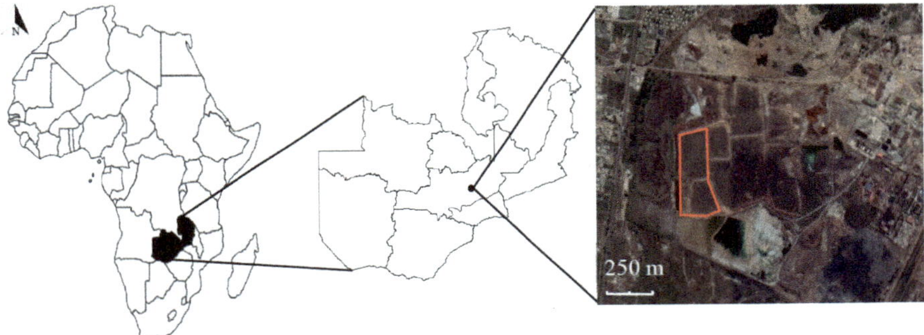

Figure 1. Dumping site in Kabwe; (□) sampling area (top left: 14°27′39″ South, 28°25′42″ East; down right: 14°27′55″ South, 28°26′16″ East).

Three types of immobilizing agents were selected to immobilize Zn and Pb in the waste: raw dolomite, calcined dolomite, and magnesium oxide denoted as RD, CD, and MO, respectively. RD was taken from a dolomite quarry source near the MRM storage site, while CD was prepared by burning RD with particle sizes of less than 2 mm in a furnace at 700 °C for 2 h. MO was commercially available, purchased from Ube Industry, Japan. The same preparation procedure as that for MRM was also applied to these materials.

Chemical and mineralogical properties of all solid samples were characterized on the pressed powder of finely crushed samples (<50 μm) using an X-ray fluorescence spectrometer (XRF) (Spectro Xepos, Rigaku Corporation, Tokyo, Japan) and X-ray diffractometer (XRD) (MultiFlex, Rigaku Corporation, Tokyo, Japan), respectively. Sequential extraction was conducted to evaluate solid-phase heavy metals speciation of MRM. The procedure used in this study was modified from two well-known procedures, Tessier et al. (1979) [30] and Clevenger (1990) [31]. The modification was done by Marumo et al. (2003) [32], and it was widely used to extract the tailings sample, mineral processing wastes, and leaching residues [10,33–36]. The process can divide solid-phase heavy metals bounded to solid into five different phases, including exchangeable, carbonates, Fe-Mn oxides, sulfide/organic matter, and residual. The details of the extraction procedure are summarized in Table 1. The extraction was done on 1 g of the <2 mm MRM sample. Between each step, the extractant solution of the previous step was retrieved by centrifugation of the suspension at 3000 rpm for 40 min to separate the residue out of the leachate. The residual was then washed with 20 mL of deionized water. Finally, the washing and extractant solutions were mixed, diluted to 50 mL, filtrated through 0.45-μm Millex® filters, and kept in polypropylene bottles prior to chemical analysis.

Table 1. Sequential extraction for heavy metals speciation.

Step	Extractant	pH	Liquid to Solid Ratio (mL/g)	Temperature (°C)	Duration (h)	Mixing Speed (rpm)	Extracted Phase
1	1 M $MgCl_2$	7	20/1	25	1	120	Exchangeable
2	1 M CH_3COONa	5	20/1	25	5	120	Carbonates
3	0.04 M $NH_2OH \cdot HCl$ in 25% acetic acid		20/1	50	5	120	Reducible
4	0.04 M $NH_2OH \cdot HCl$ in 25% acetic acid; 30% H_2O_2; 0.02 M HNO_3		36/1	85	5	120	Oxidizable
5	Calculated						Residual

2.2. Batch Leaching Experiments

Batch leaching experiments were performed using 250 mL polypropylene Erlenmeyer flasks with a lateral reciprocating shaker (EYELA Multi Shaker MMS, Tokyo Rikakikai Co., Ltd., Tokyo, Japan). All batches were conducted under ambient conditions by mixing 15 g of solid sample to 150 mL of deionized water (1:10 solid-to-liquid ratio) at 200 rpm for 6 h. Five replications of the leaching tests of MRM were conducted, while a single run of every immobilization experiment was done. The reason is that in immobilization tests, we adjusted the addition of immobilizers and did not control the pH of the suspension. In other words, pH is determined by the complex chemical and physical interactions between immobilizer and MRM. Thus, a variation of pH can be easily observed even though the same mixing ratio between MRM and immobilizer and the conditions is employed. To avoid the uncertainty of the variation of pH at the same immobilizer:MRM mixing ratio, we varied the immobilizer:MRM mixing ratios at 1:100, 3:100, 1:20, and 1:10 to evaluate the performance of immobilizers. After 6 h of shaking, the pH, electrical conductivity (EC), redox potential (ORP), and temperature were immediately measured. The leachates were collected by first centrifuging the mixtures at 3000 rpm for 40 min to separate the suspended particles. The supernatants were then filtered with 0.45-μm Millex® filters (Merck Millipore, Burlington, MA, USA) and kept in air-tight polypropylene bottles prior to chemical analysis.

2.3. Chemical Analysis of Liquid Samples

Inductively coupled plasma atomic emission spectrometer (ICP-AES) (ICPE-9000, Shimadzu Corporation, Kyoto, Japan) and inductively coupled plasma atomic emission mass spectrometry (ICP-MS) (ICAP Qc, Thermo Fisher Scientific, Waltham, MA, USA) were used to quantify the dissolved concentrations of heavy metals and coexisting ions. The analyses were performed on the pretreated liquid samples in which 1% by volume of 60% nitric acid (HNO_3) was added to the liquid samples. The acidification was done to make sure that all target elements were in a dissolved form. The non-acidified samples, on the other hand, were used to determine alkalinity or acid resistivity. This parameter is generally reported as bicarbonate concentration (meq/L), quantified by titration of a known volume of sample with 0.01 M sulfuric acid (H_2SO_4) until pH 4.8. All chemicals used were reagent-grade.

2.4. Geochemical Modeling

An aqueous geochemical modeling program, PHREEQC (Version 3, U.S. Geological Survey, Sunrise Valley Drive Reston, VA, USA) [37], was used to aid in the interpretation of the experimental results. The program can determine the parameters that may affect the mobility of heavy metals from MRM, such as stability of minerals and chemical species. The input data included temperature, pH, ORP, and concentrations of heavy metals and other coexisting ions. Thermodynamic properties were taken from the WATEQ4F database.

3. Results and Discussion

3.1. Properties of Solid Samples

The mineralogical and chemical compositions of MRM and immobilizers, including RD, CD, and MO, are listed in Tables 2 and 3, respectively. MRM was composed of anglesite ($PbSO_4$) as a primary mineral; zinkosite ($ZnSO_4$) and quartz (SiO_2) as the second-highest; and goethite ($FeOOH$), hematite (Fe_2O_3), and gypsum ($CaSO_4 \cdot 2H_2O$) as the miner minerals. Anglesite and zinkosite are commonly found as the weathering products of Pb- and Zn-sulfides under natural oxygenated environments [38,39]. Therefore, the presence of these two minerals indicates that MRM has already been exposed to the surface environment for a long time before the sampling was done. It can also be expected that other than goethite and hematite, MRM also contained amorphous iron-(hydr)oxides and iron-sulfate salts (e.g., melanterite, coquimbite) since goethite and hematite were found as the miner minerals but Fe_2O_3 content was the highest among all compositions detected. The contents of Pb and Zn in MRM were 10.9% and 8.1%, respectively. The values of both metals were extremely high and exceeded the permissible limit in soil, 600 mg/kg for Pb and 1500 mg/kg for Zn [40]. However, this does not guarantee that MRM can release significant amounts of Pb and Zn since their mobility also depends significantly on the chemical speciation. Sequential extraction was then performed, and the result showed that around 40% of the total contents of Pb and Zn were in mobile fractions (exchangeable, carbonates, and oxidizable) under surface environments (Figure 2). This confirms that MRM could be a potential source contaminating the surrounding environment with Pb and Zn.

Table 2. Mineralogical composition of solid samples.

	MRM	RD	CD	MO
Quartz	++	-	-	-
Gypsum	+	-	-	-
Anglesite	+++	-	-	-
Zinkosite	++	-	-	-
Hematite	+	-	-	-
Goethite	+	-	-	-
Dolomite	-	+++	+++	-
Calcite	-	-	+	-
Magnesia	-	-	+	+++

+++: Strong; ++: Moderate; +: Weak; -: None.

Table 3. Chemical composition of solid samples (the unit is in wt%).

	MRM	RD	CD	MO
SiO_2	20.9	0.39	0.37	<0.01
TiO_2	0.35	<0.01	<0.01	<0.01
Al_2O_3	1.91	0.24	0.11	<0.01
Fe_2O_3	45.8	0.52	0.68	<0.01
MnO	1.59	0.46	0.43	<0.01
MgO	<0.01	36.9	33.3	100
CaO	4.64	60.7	62.4	<0.01
Na_2O	<0.01	<0.01	<0.01	<0.01
K_2O	<0.01	0.15	0.16	<0.01
P_2O_5	<0.01	0.14	<0.01	<0.01
SO_3	2.71	0.15	0.2	<0.01
Pb	10.9	<0.01	<0.01	<0.01
Zn	8.1	<0.01	<0.01	<0.01

The most dominant mineral found in RD was dolomite ($CaMg(CO_3)_2$). Magnesium (Mg) and calcium (Ca) oxides contents accounted for more than 95% with a molar ratio of Ca to Mg of 1.2. This indicates that RD adequately consisted of pure dolomite. Burning RD at 700 °C for 2 h generated the new type of immobilizer, CD. Calcite ($CaCO_3$) and magnesia (MgO) were detected in addition to dolomite. With almost the same molar ratio of Ca to Mg in CD compared with that in RD, it clearly indicates that the calcination process transformed dolomite into calcite and magnesia. MO composed only of magnesia with 100% MgO content, which shows pure magnesia.

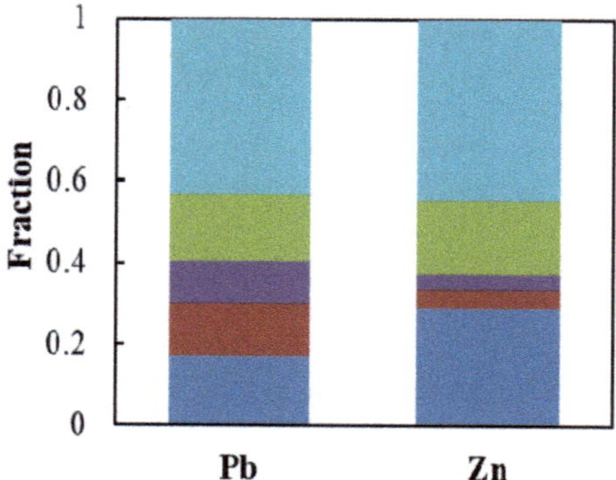

Figure 2. Solid-phase Pb and Zn speciation of mining residual materials (MRM); (□) exchangeable, (□) carbonates, (□) sulfide/organic matter, (□) Fe-Mn oxides, and (□) residual.

3.2. Leaching Characteristic of MRM

Table 4 shows the leaching characteristic of MRM. The experimental values were reproducible with high precision since the standard deviations of all parameters were quite small. Four heavy metals, Pb, Zn, cadmium (Cd), and copper (Cu), were leached at the concentrations falling within the instrument detection limits of ICP-AES and -MS. Among these heavy metals, the leaching concentrations of only Pb (2.1 mg/L) and Zn (365 mg/L) exceeded the effluent standard in Zambia (0.5 mg/L for Pb and 1 mg/L for Zn) [41]. Therefore, this study focused only on the mobility of these two metals. PHREEQC simulation on the saturation indices of all possible Pb- and Zn-minerals showed that with the given conditions and components in MRM leachate, only saturation index of anglesite fell within a common error interval used to indicate saturation equilibrium (±0.2) (Table 5). The result indicates that the low solubility of anglesite restricted the dissolved concentration of Pb, while no restriction by means of precipitation was observed on the leaching of Zn. This can explain why 4.51% of the total Zn content was leached from MRM, while only 0.02% was observed in the case of Pb leaching.

Table 4. Leaching characteristic of MRM (n = 5).

	MRM	**Reg. Value * (mg/L)**
pH	5.26 ± 0.04	-
ORP (mv)	300 ± 36	-
EC (mS/cm)	2.7 ± 0.09	-
Pb (mg/L)	2.1 ± 0.008	0.5
Zn (mg/L)	365 ± 18	1
Cd (mg/L)	0.21 ± 0.009	0.5
Cu (mg/L)	0.08 ± 0.01	1.5
Ca^{2+} (mg/L)	547 ± 48.2	-
Mg^{2+} (mg/L)	27.7 ± 3.1	-
SO_4^{2-} (mg/L)	1907 ± 36.8	-
Si (mg/L)	13 ± 0.5	-

* Regulated value in mg/L specified by the Environment Management Act (2013) [40].

Table 5. Calculated saturation indices of possible Pb- and Zn-minerals in MRM leachate.

Mineral	Saturation Index
Anglesite	−0.2
Cerussite	−2.25
Pb(OH)$_2$	−3.4
Smithsonite	−2.45
Willemite	−3.62
Zn(OH)$_2$	−3.82
Hydrocerussite	−8.41
Hydrozincite	−3.08

Calcium ion (Ca^{2+}) and sulfate ion (SO$_4^{2-}$) were the major ions in the leachate, accounting for more than 85% of the total dissolved ions. These ions were likely to be enriched by the dissolution of soluble phase minerals, such as gypsum and zinkosite [42,43]. However, the dissolution of these two minerals might not only be the sources of SO$_4^{2-}$ since the molar ratio of Ca^{2+} and Zn to SO$_4^{2-}$ was lower than one. The sulfide fractions of both metals in MRM (Figure 2), together with the slightly acidic pH (5.2) and positive ORP (300 mV) of the leachate, suggest that the oxidation of sulfide minerals (e.g., pyrite, galena, sphalerite) and dissolution of iron-sulfate salts (e.g., melanterite, coquimbite) also occurred and attributed to the enrichment of SO$_4^{2-}$, even though they were not detected by XRD. This could also partly contribute to the enrichment of Pb and Zn in the leachate.

3.3. Potential of Immobilizers

3.3.1. Effects of Addition of Immobilizers on pH and Coexisting Ions

Changes in the pH of the leachate as a function of the amounts of addition of RD, CD, and MO are shown in Figure 3. The pH increased from 5.2 to 6.7, 8.2, and 9.8 with increasing RD, CD, and MO addition from 0 to 10%. When the same amount of immobilizer was added, the performance of the immobilizers to increase the leachate pH followed the order: MO > CD > RD. The results clearly showed the improvement of the alkaline property of CD over RD. The variation in pH could mainly be attributed to the liming effect(s) of dolomite (Equation (1)) in RD treatments, of dolomite (Equation (1)), of calcite (Equation (2)), and of magnesia (Equation (3)) in CD treatments, and of magnesia (Equation (3)) in MO treatments [44–47].

$$CaMg(CO_3)_2 + 2\,H^+ \Leftrightarrow Ca^{2+} + Mg^{2+} + 2HCO_3^{-}, \tag{1}$$

$$CaCO_3 + H^+ \Leftrightarrow Ca^{2+} + HCO_3^{-}, \tag{2}$$

$$MgO + H_2O \Leftrightarrow Mg(OH)_2 \Leftrightarrow Mg^{2+} + 2OH^-. \tag{3}$$

Figure 4a–c illustrates the leaching concentrations of major coexisting ions, Ca^{2+}, magnesium (Mg^{2+}), and SO$_4^{2-}$, as a function of the amount of immobilizer added. The dissolved concentrations of Ca^{2+} and Mg^{2+} in RD treatments were higher than those in MRM and increased with the increasing addition of RD, indicating the simultaneous leaching of Ca^{2+} and Mg^{2+} from the dissolution of dolomite (Equation (1)). In CD treatments, Ca^{2+} and Mg^{2+} were also leached at higher concentrations than those in MRM leachate. At the same amount of CD and RD addition, the leaching concentration of Mg^{2+} in CD treatment was higher than that in RD treatment, while almost the same dissolved concentration of Ca^{2+} was observed in both treatments. This, together with the result that CD contained less dolomite and more calcite compared to those in RD, suggest the occurrence of hydration of magnesia (Equation (3)) in addition to the dissolution of carbonate minerals (Equations (1) and (2)) in CD treatments. Moreover, the difference in the leaching concentration of Mg^{2+} between CD and RD treatments became more significant as the addition of immobilizers increased, which indicates that as pH increased, the hydration of magnesia (Equation (3)) played a more important role in controlling pH. In MO treatments, the concentration of Mg^{2+} increased with higher addition of MO, while almost no change in the concentration of Ca^{2+} from that in MRM was observed. This result suggests the occurrence of hydration of magnesia (Equation (3)) in MO treatments.

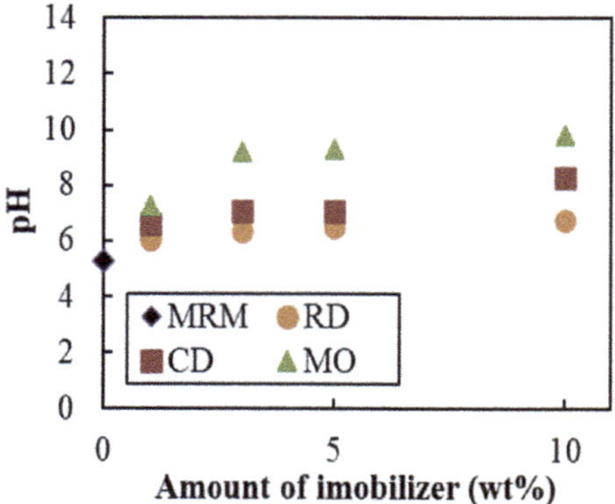

Figure 3. pH of leachate upon addition of immobilizers.

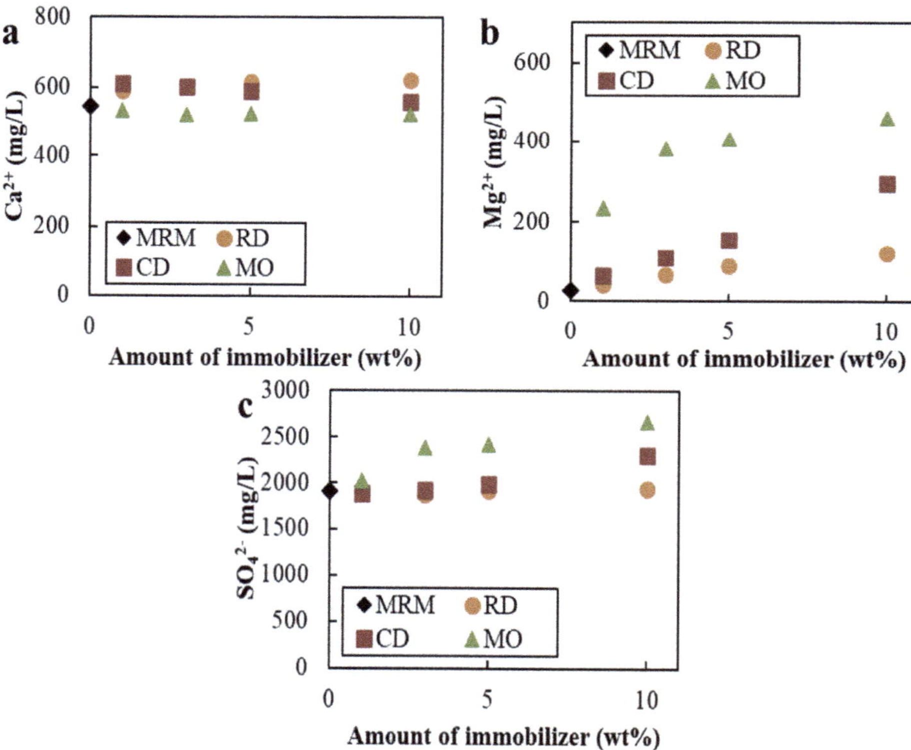

Figure 4. Leaching concentrations of major ions upon addition of immobilizers: (**a**) Ca^{2+}, (**b**) Mg^{2+}, and (**c**) SO_4^{2-}.

At the same amount of immobilizer added, the leaching concentration of SO_4^{2-} was the highest when treating MRM with MO, followed by CD and RD (Figure 4c). In consideration of the trace amount of SO_3 content in all immobilizers, the results suggest that MRM should be the source of SO_4^{2-}, and the leaching of SO_4^{2-} might be caused by the change in the parameter(s) of the leachate, triggered by the addition of immobilizer. Figure 5 illustrates the correlation between the leaching concentration of SO_4^{2-} vs. pH. The SO_4^{2-} concentration exhibited a strong positive correlation with pH (correlation coefficient (r) = 0.92, $p < 0.01$), suggesting that SO_4^{2-} level could mainly be influenced by pH due possibly to the following mechanisms: (1) desorption, (2) production by oxidation of sulfide minerals, (3) common-ion of between calcite, dolomite, and gypsum, and (4) dissolution of anglesite. The pH increase led to a higher negative surface potential of MRM, thereby decreasing the affinity of SO_4^{2-} toward the surface of MRM. However, Tabatabai (1987) [48] reported that since the adsorption of SO_4^{2-} was only flavored under strongly acidic conditions, the amount of adsorbed SO_4^{2-} became almost negligible under weakly acidic pH. This means that the desorption might not be a viable explanation in this study since all leachates' pH ranged from weakly acidic (5.2) to moderately alkaline (9.8). Therefore, the fact that the oxidation rate of sulfide minerals, such as pyrite and galena, increases with pH becomes the potential reason contributing to the higher leaching concentration of SO_4^{2-} [49,50]. However, the enrichment of SO_4^{2-} could be restricted by the solubility of gypsum because the saturation index of gypsum was within the equilibrium condition range of ±0.2 in all leachates. This could explain why the leaching concentration of SO_4^{2-} slightly increased with pH at lower pH region where the dissolution of dolomite containing in RD and calcite and dolomite containing in and CD tended to control the pH (Equations (1) and (2)). In other words, Ca^{2+} produced from the dissolution of calcite and dolomite precipitated with SO_4^{2-} to form gypsum, thereby restricting SO_4^{2-} concentration. Meanwhile, as pH became more alkaline, the concentration of SO_4^{2-} increased rapidly. This could be attributed to the less contribution of calcite and dolomite dissolution (Equations (1) and (2)) to control the leachate pH in the case of CD addition, conjointly in MO treatments, only hydration of magnesia (no production of Ca^{2+}) (Equation (3)) was found to control the pH of the leachates. The pH-dependent solubility of anglesite could also be attributed to the rapid increase of SO_4^{2-} concentration under alkaline conditions since anglesite was originally contained in MRM and is unstable under alkaline pH [51].

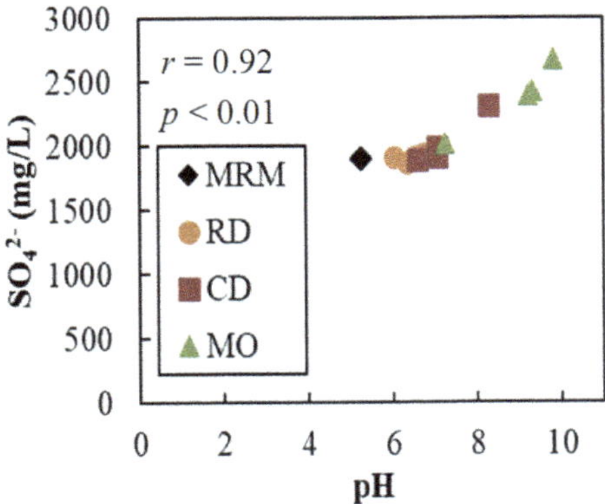

Figure 5. Leaching concentration of SO_4^{2-} vs. pH.

3.3.2. Effects of the Addition of Immobilizers on Mobility of Heavy Metals

In this study, the performance of immobilizers was evaluated based on the solubility of Pb and Zn. Figure 6a,b shows the changes in Pb and Zn concentrations as a function of the amount of addition of immobilizers. In general, the leaching concentrations of Pb and Zn exponentially decreased with an increase in the dose of immobilizers. Figure 7a,b illustrates the correlation between leaching concentrations of Pb and Zn vs. pH. The mobility of heavy metals was strongly influenced by pH, indicated by significant negative correlations of Pb-pH ($r = -0.92$) and Zn-pH ($r = -0.87$) at the 0.01 significance level (2-tailed). Coupled with the results of the leaching concentration of SO_4^{2-} and the characteristic of immobilizers, as well as the leaching condition used in the current study, the major modes of Pb and Zn attenuation could be either one or more of the following mechanisms: precipitations of metal-sulfate, -carbonate, and/or -hydroxide, co-precipitation of metal with iron-(oxy)hydroxides, and metal ion adsorption to immobilizer. The formation of low soluble anglesite under acidic conditions was expected since the leaching concentration of SO_4^{2-} was high and increased with pH (Figure 5) [52,53]. The carbonate property of RD and CD might result in the precipitation of cerussite ($PbCO_3$), hydrocerussite ($Pb_3(CO_3)_2(OH)_2$), smithsonite ($ZnCO_3$), and hydrozincite ($Zn_3(CO_3)_2(OH)_2$) [35,54–56]. The carbonate precipitations of Pb and Zn are also expected to occur in MO treatments since the experiments were done under atmospheric conditions in which carbon dioxide (CO_2) in the atmosphere was freely dissolved [57]. However, the simulation results by PHREEQC showed that except hydrozincite in 3% addition of MO treatment, the precipitations of anglesite, cerussite, hydrocerussite, smithsonite, and hydrozincite were thermodynamically unfavorable (saturation index < −0.2) regardless of the type and amount of immobilizer added (Table 6). Therefore, the possible immobilization mechanisms of Pb and Zn in all types of immobilizers could be narrowed down to the hydroxide precipitation, adsorption, and co-precipitation.

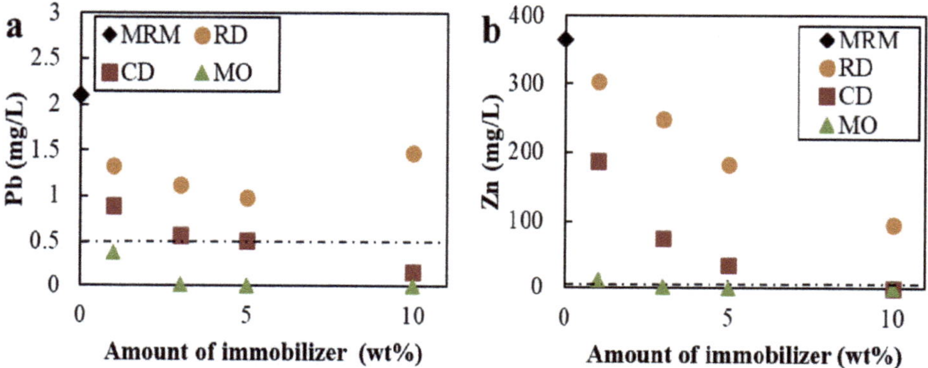

Figure 6. Leaching concentrations of heavy metals upon addition of immobilizers: (**a**) Pb and (**b**) Zn (dashed lines represent the effluent standards of Pb and Zn in Zambia).

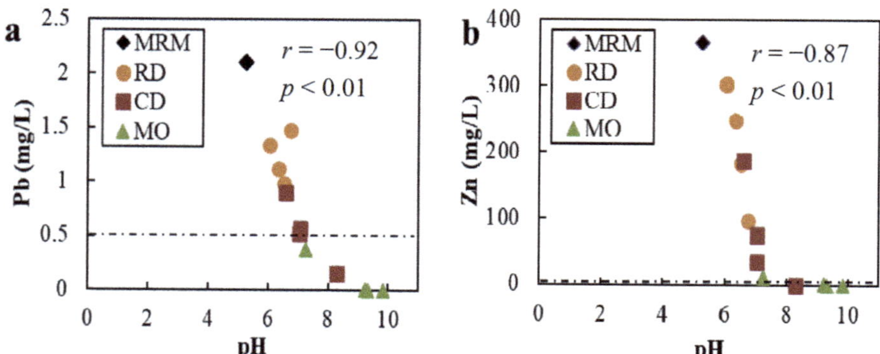

Figure 7. Leaching concentrations of heavy metals vs. pH: (**a**) Pb vs. pH and (**b**) Zn vs. pH (dash lines represent the effluent standards of Pb and Zn in Zambia).

To verify the dominant mechanism(s), pH-dependent solubility diagrams of Pb- and Zn-hydroxides were plotted (Figure 8a,b). Points in the figures represent the relationship between the logarithmic activity of divalent heavy metal and pH in each batch test. The solid lines demonstrate the solubility of heavy metal hydroxides. Therefore, any point located on or close to the line implies the hydroxide precipitation-controlled sequestration process.

Table 6. Calculated saturation indices of anglesite, cerussite, hydrocerussite, smithsonite, and hydrozincite in leachates with the addition of RD, CD, and MO.

Treatment		Saturation Index				
		Anglesite	Cerussite	Smithsonite	Hydrocerussite	Hydrozincite
RD treatments	1%	−0.4	−1.15	−1.63	−5.63	−1.82
	3%	−0.49	−1.13	−1.21	−4.28	−1.35
	5%	−0.55	−0.98	−1.13	−3.68	−1.17
	10%	−0.41	−0.48	−1.01	−2.11	−1.01
CD treatments	1%	−0.6	−0.78	−0.87	−3.19	−0.98
	3%	−0.92	−0.27	−0.44	−1.56	−0.48
	5%	−0.99	−0.23	−0.66	−1.58	−0.79
	10%	−2.39	−0.3	−1.24	−0.64	−0.64
MO treatments	1%	−1.13	−0.38	−1.16	−1.57	−1
	3%	−4.36	−1.6	−1.16	−3.31	0.22
	5%	−4.72	−1.86	−2.19	−4.05	−0.78
	10%	−5.49	−2.52	−2.99	−5.16	−1.02

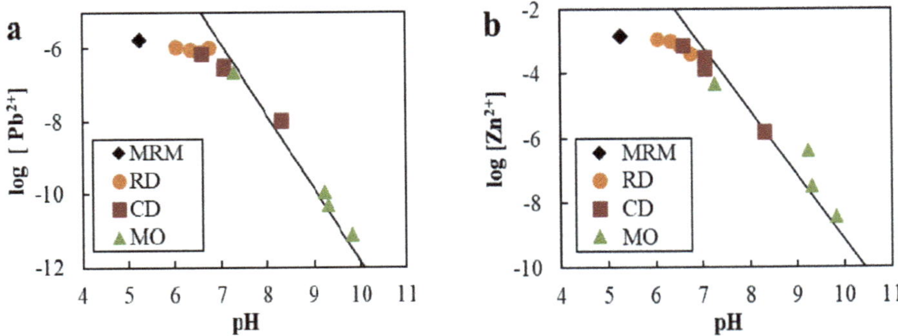

Figure 8. pH-dependent solubility diagrams of (**a**) Pb-hydroxide and (**b**) Zn-hydroxide ([ ] represents activity).

In the case of RD addition, a discrepancy from the equilibrium line for Pb and Zn was observed. This means that at pH from weekly acid to neutral, hydroxide precipitation was not the main mechanism controlling the mobility of Pb and Zn. Therefore, Pb and Zn were suspected to be immobilized by sorption and co-precipitation with iron-(oxy)hydroxides. The sorption was likely to occur since dolomite, major mineral in the immobilizer, can adsorb Pb and Zn [58,59]. Besides, adding more of this immobilizer induced the leachate pH increase. This could change the surface charge of goethite, hematite, and iron-(hydr)oxide compounds in MRM to more negative, thereby increasing their adsorption ability against cationic divalent Pb and Zn [58–63]. During the neutralization process under ambient conditions, iron-(oxy)hydroxides precipitate from the oxidative dissolution of pyrite and dissolution of iron-bearing salts [63–66]. The precipitations of iron-(oxy)hydroxides have been reported by many studies to induce co-precipitation of divalent metals, including Pb(II) and Zn(II) [50,67,68], and thus the co-precipitation of Pb and Zn with iron-(oxy)hydroxides was also expected in RD treatments. From the above explanations, adding more RD should reduce the leaching concentration of both metals. However, leaching concentration of Pb increased from 0.98 mg/L to 1.4 mg/L when RD rose from 5% to 10% (Figure 6a). This could be attributed to the stability of Pb(II) species as a function of pH. Theoretically, as pH increases under acidic region, more of free Pb(II) ion tends to complex with OH⁻ and CO_3^{2-}, generating larger ion with lower charge ($Pb(OH)^+$ and $PbCO_3$), which lowers the affinity of Pb to the surface of the potential adsorbents and inhibits the co-precipitation [69–71].

On the other hand, adding CD and MO made most of the logarithmic leaching activities of Pb^{2+} and Zn^{2+} to approach their solubility product lines. This means that hydroxide precipitation is the dominant mechanism of attenuating Pb and Zn. Regardless of the type of immobilizer, at low pH, the logarithmic activities of both metals were slightly lower than their equilibrium lines and then tended to stay on or be slightly higher than the lines afterward. This probably indicates that at low pH, adsorption and co-precipitation with iron-(oxy)hydroxides also occurred in addition to the precipitation, but as pH got higher, they diminished. There are two probable explanations for this phenomenon as follows: (1) competition with strong competing ion (Mg^{2+}) and (2) change in specification of the dissolved metals. The pH alteration mechanisms of CD and MO appeared to generate Mg^{2+} as a by-product (Equations (1) and (3)). Because of this, the concentration of Mg^{2+} significantly increased with pH with a correlation coefficient of 0.97, $p < 0.01$ (Figure 9). Therefore, as pH increased, high concentration of Mg^{2+} could compete for Pb and Zn for adsorption sites and for co-precipitation with iron-(oxy)hydroxides, attributing to the less contribution of the adsorption and co-precipitation on the immobilization process. Increasing pH could also result in the redistributions of Pb(II) and Zn(II) species. The fraction of free Pb(II) and Zn(II) reduces as pH increases since they are thermodynamically preferable to be hydrolyzed forming $-(OH)^{+}$, $-(OH)_2$, and $-(OH)_3^{-}$ [69–72]. Moreover, since the systems contained high dissolved carbonate, the formation of carbonate complexes of Pb(II) and Zn(II) was also expected [69,73]. Once these complexes are formed, their abilities to get adsorbed and co-precipitated are inhibited by the larger size and lower positive potential they become.

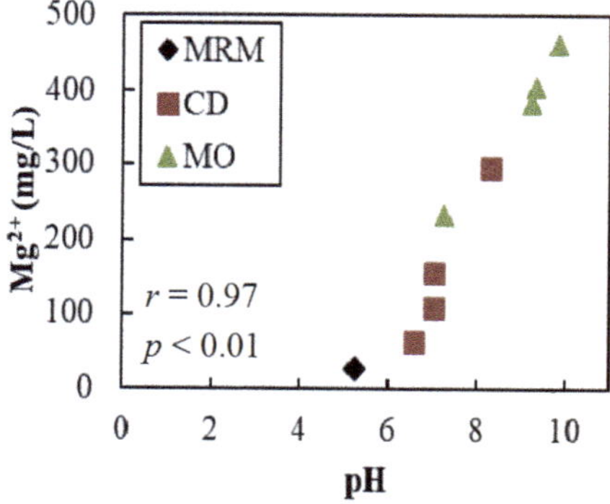

Figure 9. Leaching concentration of Mg^{2+} vs. pH.

3.3.3. Performance of Immobilizers

When the same amount of immobilizer was added, the dissolved concentrations of Pb and Zn were the highest in RD treatment, second highest in CD treatment, and the lowest in MO treatment (Figure 6a,b). As previously mentioned, adsorption, co-precipitation, and hydroxide precipitation were the major sink of Pb and Zn, and their mobilities depended strongly on pH. Because of the carbonate property of dolomite, RD could not raise the pH of MRM leachate to the value favoring the precipitations of Pb- and Zn-hydroxides. Therefore, RD treatments remediated Pb and Zn by adsorption and co-precipitation in which it was insufficient to reduce the leaching concentrations of Pb and Zn down below their regulated values. On the other hand, magnesia in CD and MO played a significant role in increasing the leachate pH of MRM into the alkaline region. Lead and Zn were then mainly immobilized by precipitation as hydroxides. Thus, metal concentrations as high as 368 mg/L for Zn and 2.1 mg/L for Pb released from MRM were reduced to the values below their regulated concentrations. Efficiency-wise, MO was the most effective immobilizer in immobilizing Pb and Zn since it contained the highest MgO content. However, CD could be the immobilizer of choice since it can be produced from the naturally abundant material in Zambia (RD), and its performance was almost the same as that of MO.

4. Conclusions

The leachate of MRM was slightly acidic (pH 5.2) and contained high concentrations of Pb (2.1 mg/L) and Zn (365 mg/L), exceeding those of Zambian regulation. When immobilizers were introduced, the leachate pH increased, and the leaching concentrations of Pb and Zn decreased. Lead and Zn immobilized by RD were interpreted by the adsorption and co-precipitation mechanisms. On the contrary, chemical immobilization using CD and MO suppressed Pb and Zn leaching mainly by the hydroxide precipitation. Of the immobilizers investigated, only CD and MO decreased the dissolved Pb and Zn concentrations to below their regulated values, in which MO had a higher performance than CD. The results show that heat treatment on RD to produce CD drastically improved the immobilizing performance of Pb and Zn. Even though MO provided the highest efficiency, Pb and Zn could also be effectively immobilized by giving an adequate amount of CD. Therefore, by considering the availability of CD in the local area, CD could be the most promising chemical agent to be implemented in Zambia.

Author Contributions: Conceptualization, P.T.; methodology, P.T., K.N., T.I., and T.K.; formal analysis, P.T. and T.I.; writing—original draft preparation, P.T.; writing—review and editing, P.T., T.I., M.I. (Mayumi Ito), T.S., W.M., M.C., and I.N.; supervision, P.T. and T.I.; project administration, M.I. (Mayumi Ishizuka), S.N., and H.N.; funding acquisition, M.I. (Mayumi Ishizuka), S.N., and H.N.; All authors have read and agreed to the published version of the manuscript.

Funding: This research was supported by JST/JICA SATREPS (Science and Technology Research Partnership for Sustainable Development; No. JPMJSA1501) and aXis (Accelerating Social Implementation for SDGs Achievement; No. JPMJAS2001) funded by JST.

Acknowledgments: The authors would like to acknowledge JST/JICA SATREPS (Science and Technology Research Partnership for Sustainable Development; No. JPMJSA1501) and aXis (Accelerating Social Implementation for SDGs Achievement; No. JPMJAS2001) funded by JST for the financial support.

Conflicts of Interest: The authors declare no conflict of interest.

References

1. Kamona, A.F.; Friedrich, G.H. Geology, mineralogy, and stable isotope geochemistry of the Kabwe carbonate-hosted Pb-Zn deposit, Central Zambia. *Ore Geol. Rev.* **2007**, *30*, 217–243. [CrossRef]

2. Kříbek, B.; Nyambe, I.; Majer, V.; Knésl, I.; Mihaljevič, M.; Ettler, V.; Vaněk, A.; Penížek, V.; Sracek, O. Soil contamination near the Kabwe Pb-Zn smelter in Zambia: Environmental impacts and remediation measures proposal. *J. Geochem. Explor.* **2019**, *197*, 159–173. [CrossRef]

3. Fuge, R.; Pearce, M.F.; Pearce, N.J.G.; Perkins, W.T. Geochemistry of Cd in the secondary environment near abandoned metalliferous mines, Wales. *J. Appl. Geochem.* **1993**, *8*, 29–35. [CrossRef]

4. Paulson, A.J. The transport and fate of Fe, Mn, Cu, Zn, Cd, Pb and SO4 in a groundwater plume and in downstream surface waters in the Coeur d'Alene Mining District, Idaho, U.S.A. *J. Appl.* **1997**, *12*, 447–464. [CrossRef]

5. Yabe, J.; Nakayama, S.M.M.; Ikenaka, Y.; Muzandu, K.; Ishizuka, M.; Umemura, T. Uptake of lead, cadmium, and other metals in the liver and kidneys of cattle near a lead-zinc mine in Kabwe, Zambia. *Environ. Toxicol. Chem.* **2011**, *30*, 1892–1897. [CrossRef]

6. Yabe, J.; Nakayama, S.M.M.; Ikenaka, Y.; Muzandu, K.; Choongo, K.; Mainda, G.; Kabeta, M.; Ishizuka, M.; Umemura, T. Metal distribution in tissues of free-range chickens near a lead-zinc mine in Kabwe, Zambia. *Environ. Toxicol. Chem.* **2013**, *32*, 189–192. [CrossRef]

7. Yabe, J.; Nakayama, S.M.M.; Ikenaka, Y.; Yohannes, Y.B.; Bortey-Sam, N.; Oroszlany, B.; Muzandu, K.; Choongo, K.; Kabalo, A.N.; Ntapisha, J.; et al. Lead poisoning in children from townships in the vicinity of a lead-zinc mine in Kabwe, Zambia. *Chemosphere* **2015**, *119*, 941–947. [CrossRef]

8. Yabe, J.; Nakayama, S.M.M.; Ikenaka, Y.; Yohannes, Y.B.; Bortey-Sam, N.; Kabalo, A.N.; Ntapisha, J.; Mizukawa, H.; Umemura, T.; Ishizuka, M. Lead and cadmium excretion in feces and urine of children from polluted townships near a lead-zinc mine in Kabwe, Zambia. *Chemosphere* **2018**, *202*, 48–55. [CrossRef]

9. Basta, N.T.; McGowen, S.L. Evaluation of chemical immobilization treatments for reducing heavy metal transport in a smelter-contaminated soil. *Environ. Pollut.* **2004**, *127*, 73–82. [CrossRef]

10. Silwamba, M.; Ito, M.; Hiroyoshi, N.; Tabelin, C.B.; Hashizume, R.; Fukushima, T.; Park, I.; Jeon, S.; Igarashi, T.; Sato, T.; et al. Recovery of lead and zinc from zinc plant leach residues by concurrent dissolution-cementation using zero-valent aluminum in chloride medium. *Metals* **2020**, *10*, 531. [CrossRef]

11. Silwamba, M.; Ito, M.; Hiroyoshi, N.; Tabelin, C.B.; Fukushima, T.; Park, I.; Jeon, S.; Igarashi, T.; Sato, T.; Nyambe, I.; et al. Detoxification of lead-bearing zinc plant leach residues from Kabwe, Zambia by coupled extraction-cementation method. *J. Environ. Chem. Eng.* **2020**, *8*, 104197. [CrossRef]

12. Mwandira, W.; Nakashima, K.; Kawasaki, S.; Ito, M.; Sato, T.; Igarashi, T.; Banda, K.; Chirwa, M.; Nyambe, I.; Nakayama, S.; et al. Efficacy of biocementation of lead mine waste from the Kabwe Mine site evaluated using Pararhodobacter sp. *Environ. Sci. Pollut. Res.* **2019**, *26*, 15653–15664. [CrossRef] [PubMed]

13. Mwandira, W.; Nakashima, K.; Kawasaki, S.; Ito, M.; Sato, T.; Igarashi, T.; Chirwa, M.; Banda, K.; Nyambe, I.; Nakayama, S.; et al. Solidification of sand by Pb(II)-tolerant bacteria for capping mine waste to control metallic dust: Case of the abandoned Kabwe Mine, Zambia. *Chemosphere* **2019**, *228*, 17–25. [CrossRef] [PubMed]

14. Liu, L.; Li, W.; Song, W.; Guo, M. Remediation techniques for heavy metal-contaminated soils: Principles and applicability. *Sci. Total Environ.* **2018**, *633*, 206–219. [CrossRef]

15. Gray, C.W.; Dunham, S.J.; Dennis, P.G.; Zhao, F.J.; McGrath, S.P. Field evaluation of in situ remediation of a heavy metal contaminated soil using lime and red-mud. *Environ. Pollut.* **2006**, *142*, 530–539. [CrossRef]

16. Kumpiene, J.; Lagerkvist, A.; Maurice, C. Stabilization of As, Cr, Cu, Pb and Zn in soil using amendments—A review. *J. Waste Manag.* **2008**, *28*, 215–225. [CrossRef]

17. Farrell, M.; Jones, D.L. Use of composts in the remediation of heavy metal contaminated soil. *J. Hazard. Mater.* **2010**, *175*, 575–582. [CrossRef]

18. He, M.; Shi, H.; Zhao, X.; Yu, Y.; Qu, B. Immobilization of Pb and Cd in contaminated soil using nano-crystallite hydroxyapatite. *Procedia Environ. Sci.* **2013**, *18*, 657–665. [CrossRef]

19. Bolan, N.; Kunhikrishnan, A.; Thangarajan, R.; Kumpiene, J.; Park, J.; Makino, T.; Kirkham, M.B.; Scheckel, K. Remediation of heavy metal(loid)s contaminated soils—To mobilize or to immobilize? *J. Hazard. Mater.* **2014**, *266*, 141–166. [CrossRef]

20. Mahar, A.; Wang, P.; Li, R.; Zhang, Z. Immobilization of lead and cadmium in contaminated soil using amendments: A review. *Pedosphere* **2015**, *25*, 555–568. [CrossRef]

21. Ali, A.; Guo, D.; Zhang, Y.; Sun, X.; Jiang, S.; Guo, Z.; Huang, H.; Liang, W.; Li, R.; Zhang, Z. Using bamboo biochar with compost for the stabilization and phytotoxicity reduction of heavy metals in mine-contaminated soils of China. *Sci. Rep.* **2017**, *7*, 2690. [PubMed]

22. Seshadri, B.; Bolan, N.S.; Choppala, G.; Kunhikrishnan, A.; Sanderson, P.; Wang, H.; Currie, L.D.; Tsang, D.C.W.; Ok, Y.S.; Kim, G. Potential value of phosphate compounds in enhancing immobilization and reducing bioavailability of mixed heavy metal contaminants in shooting range soil. *Chemosphere* **2017**, *184*, 197–206. [CrossRef]

23. Andrunik, M.; Wołowiec, M.; Wojnarski, D.; Zelek-Pogudz, S.; Bajda, T. Transformation of Pb, Cd, and Zn minerals using phosphates. *Minerals* **2020**, *10*, 342. [CrossRef]

24. Southwood, M.; Cairncross, B.; Rumsey, M.S. Minerals of the Kabwe ("Broken Hill") mine, central province, Zambia. *Rocks Miner.* **2019**, *94*, 114–149. [CrossRef]

25. Trakal, L.; Neuberg, M.; Tlustoš, P.; Száková, J.; Tejnecký, V.; Drábek, O. Dolomite limestone application as a chemical immobilization of metal-contaminated soil. *Plant Soil Environ.* **2011**, *4*, 173–179.

26. Deer, W.A.; Howie, R.A.; Zussman, J. *An Introduction to The Rock Forming Minerals*, 2nd ed.; Longmans: London, UK, 1966; pp. 489–493.

27. Krauskopf, K.B.; Bird, D.K. *Introduction to Geochemistry*, 3rd ed.; McGraw-Hill: New York, NY, USA, 1995.

28. Salameh, Y.; Albadarin, A.B.; Allen, S.; Walker, G.; Ahmad, M.N.M. Arsenic (III,V) adsorption onto charred dolomite: Charring optimization and batch studies. *Chem. Eng. J.* **2015**, *259*, 663–671. [CrossRef]

29. JLT-13. Departmental notification No. 13 on leaching test method for landfill wastes. *Jap. Environ. Agcy.* **1973**. Available online: http://www.env.go.jp/hourei/11/000178.html (accessed on 26 August 2020).

30. Tessier, A.; Campbell, G.C.; Bisson, M. Sequential extraction procedure for the speciation of particulate trace metals. *Anal. Chem.* **1979**, *51*, 844–850.

31. Clevenger, T.E. Use of sequential extraction to evaluate the heavy metals in mining wastes. *Water Air Soil Pollut.* **1990**, *50*, 241–253.

32. Marumo, K.; Ebashi, T.; Ujiie, T. Heavy metal concentrations, leachabilities and lead isotope ratios of Japanese soils. *Shigen Chihsitsu* **2003**, *53*, 125–146. (In Japanese)

33. Dang, Z.; Liu, C.; Haigh, M.J. Mobility of heavy metals associated with the natural weathering of coal mine spoils. *Environ. Pollut.* **2002**, *118*, 419–426. [CrossRef]

34. Anju, M.; Banerjee, D.K. Comparison of two sequential extraction procedures for heavy metal partitioning in mine tailings. *Chemosphere* **2010**, *78*, 1393–1402. [CrossRef] [PubMed]

35. Tabelin, C.B.; Silwamba, M.; Paglinawan, F.C.; Mondejar, A.J.S.; Duc, H.G.; Resabal, N.J.; Opiso, E.M.; Igarashi, T.; Tomiyama, S.; Ito, M.; et al. Solid-phase partitioning and release-retention mechanisms of copper, lead, zinc and arsenic in soils impacted by artisanal and small-scale gold mining (ASGM) activities. *Chemosphere* **2020**, *260*, 127574. [CrossRef] [PubMed]

36. Khoeurn, K.; Sasaki, A.; Tomiyama, S.; Igarashi, T. Distribution of zinc, copper, and iron in the tailings dam of an abandoned mine in Shimokawa, Hokkaido, Japan. *Mine Water Environ.* **2018**, *38*, 119–129. [CrossRef]

37. Parkhurst, D.L.; Appelo, C.A.J. User's guide to PHREEQC (Version 2): A computer program for speciation, batch-reaction, one-dimensional transport, and inverse geochemical calculations. *Water Resour. Investig. Rep.* **1999**, *99*, 312.

38. Blowes, D.W.; Jambor, J.L.; Alpers, C.N. The environmental geochemistry of sulfide mine-wastes. *Mineral. Assoc. Can.* **1994**, *22*, 59–102.

39. Hayes, S.M.; White, S.A.; Thompson, T.L.; Maier, R.M.; Chorover, J. Changes in lead and zinc lability during weathering-induced acidification of desert mine tailings: Coupling chemical and micro-scale analyses. *Appl. Geochem.* **2009**, *42*, 2234–2245. [CrossRef]

40. Wuana, R.A.; Okieimen, F.E. Heavy metals in contaminated soils: A review of sources, chemistry, risks and best available strategies for remediation. *ISRN Ecol.* **2011**, *2011*, 20. [CrossRef]

41. The Environment Management Act. The environment management (licensing) regulations. *SI Govt. Zambia* **2013**, *112*, 737–858.

42. Klimchouk, A.B. The dissolution and conversion of gypsum and anhydrite. *Speleolog* **1996**, *25*, 21–36. [CrossRef]

43. Hester, R.E.; Harrison, R.M. *Contaminated Land and Its Reclamation*, 1st ed.; Royal Society of Chemistry: London, UK, 1997.

44. Lui, Z.; Dreybrodt, W. Dissolution kinetics of calcium carbonate minerals in H_2O-CO_2 solutions in turbulent flow: The role of the diffusion boundary layer and the slow reaction $H_2O + CO_2 \leftrightarrow H^+ + HCO_3^-$. *Geochem. Cosmochim. Acta* **1997**, *61*, 2879–2889.

45. Pokrovsky, O.S.; Schott, J.N. Kinetics and mechanism of dolomite dissolution in neutral to alkaline solutions revisited. *Am. J. Sci.* **2001**, *301*, 597–626. [CrossRef]

46. Fu, J.; He, Q.; Miedziak, P.J.; Brett, G.L.; Huang, X.; Pattisson, S.; Douthwaite, M.; Hutchings, G.J. The role of $Mg(OH)_2$ in the so-called "base-free" oxidation of glycerol with Au Pd catalysts. *Chemistry* **2018**, *24*, 2396–2402. [CrossRef]

47. Xing, Z.; Bai, L.; Ma, Y.; Wang, D.; Li, M. Mechanism of magnesium oxide hydration based on the multi-rate model. *Materials* **2018**, *11*, 1835. [CrossRef] [PubMed]

48. Tabatabai, M.A. Physicochemical fate of sulfate in soils. *JAPCA* **1987**, *37*, 34–38. [CrossRef]

49. Evangelou, V.P. *Pyrite Oxidation and Its Control*, 1st ed.; CRC Press: New York, NY, USA, 1995.

50. Tabelin, C.B.; Igarashi, T.; Tabelin, M.V.; Park, I.; Opiso, E.M.; Ito, M.; Hiroyoshi, N. Arsenic, selenium, boron, lead, cadmium, copper, and zinc in naturally contaminated rocks: A review of their sources, modes of enrichment, mechanisms of release, and mitigation strategies. *Sci. Total Environ.* **2018**, *645*, 1522–1553. [CrossRef]

51. Cappuyns, V.; Alian, V.; Vassilieva, E.; Swennen, R. pH dependent leaching behavior of Zn, Cd, Pb, Cu and as from mining wastes and slags: Kinetics and mineralogical control. *Waste Biomass Valor.* **2014**, *5*, 355–368. [CrossRef]

52. Lindsay, W.L. *Chemical Equilibria in Soils*, 1st ed.; Wiley: Hoboken, NJ, USA, 1979.

53. Tatsuhara, T.; Arima, T.; Igarashi, T.; Tabelin, C.B. Combined neutralization-adsorption system for the disposal of hydrothermally altered excavated rock producing acidic leachate with hazardous elements. *Eng. Geol.* **2012**, *139*, 76–84. [CrossRef]

54. McBride, M.B. Reactions controlling heavy metal solubility in soils. *Adv. Soil Sci.* **1989**, *10*, 1–56.

55. Mench, M.J.; Didier, V.L.; Leoffler, M.; Gomez, A.; Pierre, M. A mimicked in-situ remediation study of metal-contaminated soils with emphasis on cadmium and lead. *J. Environ. Qual.* **1994**, *23*, 58–63. [CrossRef]

56. Chlopecka, A.; Adriano, D.C. Mimicked in-situ stabilization of metals in a cropped soil: Bioavailability and chemical form of zinc. *Environ. Sci. Technol.* **1996**, *30*, 3294–3303. [CrossRef]

57. Carbonate Equilibria in Natural Waters. Available online: https://www.fkit.unizg.hr/_download/repository/CO2_ravnoteza%5B1%5D.pdf (accessed on 20 July 2020).

58. Tozsin, G. Inhibition of acid mine drainage and immobilization of heavy metals from copper flotation tailings using marble cutting waste. *Int. J. Min. Met. Mater.* **2016**, *23*, 1–6. [CrossRef]

59. Gruszecka-Kosowska, A.; Baran, P.; Wdowin, M.; Franus, W. Waste dolomite powder as an waste dolomite powder as an adsorbent of Cd, Pb(II), and Zn from aqueous solutions. *Environ. Earth. Sci.* **2017**, *79*, 521. [CrossRef]

60. Gadde, R.R.; Laitinen, H.A. Studies of heavy metal adsorption by hydrous iron and manganese oxides. *Anal. Chem.* **1974**, *46*, 2022–2026. [CrossRef]

61. Millward, G.E.; Moore, R.M. The adsorption of Cu, Mn and Zn by iron oxyhydroxide in model estuarine solutions. *Water Res.* **1982**, *16*, 981–985. [CrossRef]

62. Tangviroon, P.; Endo, Y.; Fujinaga, R.; Kobayashi, M.; Igarashi, T.; Yamamoto, T. Change in arsenic leaching from silty soil by adding slag cement. *Water Air Soil Pollut.* **2020**, *231*, 259. [CrossRef]

63. Marove, C.A.; Tangviroon, P.; Tabelin, C.B.; Igarashi, T. Leaching of hazardous elements from Mozambican coal and coal ash. *J. Afr. Earth Sci.* **2020**, *168*, 103861. [CrossRef]

64. Tangviroon, P.; Hayashi, R.; Igarashi, T. Effects of additional layer(s) on the mobility of arsenic from hydrothermally altered rock in laboratory column experiments. *Water Air Soil Pollut.* **2017**, *228*, 191. [CrossRef]

65. Tangviroon, P.; Igarashi, T. Modeling and evaluating the performance of river sediment on immobilizing arsenic from hydrothermally altered rock in laboratory column experiments with Hydrus-1D. *Water Air Soil Pollut.* **2017**, *228*, 465. [CrossRef]

66. Huyen, D.T.; Tabelin, C.B.; Thuan, H.M.; Dang, D.H.; Truong, P.T.; Vongphuthone, B.; Kobayashi, M.; Igarashi, T. The solid-phase partitioning of arsenic in unconsolidated sediments of the Mekong Delta, Vietnam and its modes of release under various conditions. *Chemosphere* **2019**, *233*, 512–523. [CrossRef]

67. Igarashi, T.; Herrera, P.S.; Uchiyama, H.; Miyamae, H.; Iyatomi, N.; Hashimoto, K.; Tabelin, C.B. The two-step neutralization ferrite-formation process for sustainable acid mine drainage treatment: Removal of copper, zinc and arsenic, and the influence of coexisting ions on ferritization. *Sci. Total Environ.* **2020**, *715*, 136877. [CrossRef] [PubMed]

68. HoungAloune, S.; Kawaai, T.; Hiroyoshi, N.; Ito, M. Study on schwertmannite production from copper heap leach solutions and its efficiency in arsenic removal from acidic sulfate solutions. *Hydrometallurgy* **2014**, *147*, 30–40. [CrossRef]

69. Powell, K.J.; Brown, P.L.; Byrne, R.H.; Gajda, T.; Hefter, G.; Leuz, A.K.; Sjöberg, S.; Wanner, H. Chemical speciation of environmentally significant metals with inorganic ligands, Part 3. The Pb^{2+}, OH^-, Cl^-, CO_3^{2-}, SO_4^{2-}, and PO_4^{3-} systems. IUPAC Technical Report. *Pure Appl. Chem.* **2009**, *81*, 2425–2476. [CrossRef]

70. Badawy, S.H.; Helal, M.I.D.; Chaudri, A.M.; Lawlor, K.; McGrath, S.P. Heavy metals in the environment. Soil solid-phase controls lead activity in soil solution. *J. Environ. Qual.* **2002**, *31*, 162–167. [CrossRef]

71. Namieśnik, J.; Rabajczyk, A. The speciation and physicochemical forms of metals in surface waters and sediments. *Chem. Speciat. Bioavailab.* **2010**, *22*, 1–24. [CrossRef]

72. Masliy, A.N.; Shapnik, M.S.; Kuznetsov, A.M. Quantum-chemical investigation of electrochemical processes. Part I. Investigation of the mechanism of Zn(II) complex electroreduction from alkaline water solutions. Quantum chemical simulation, chemistry, and computational simulation. *Butlerov. Commun.* **2000**, *3*, 1–6.

73. Fouillac, C.; Criaud, A. Carbonate and bicarbonate trace metal complexes: Critical reevaluation of stability constants. *Geochem. J.* **1984**, *18*, 297–303. [CrossRef]

minerals

Article

Genomic Analysis of a Newly Isolated *Acidithiobacillus ferridurans* JAGS Strain Reveals Its Adaptation to Acid Mine Drainage

Jinjin Chen [1,†] , Yilan Liu [1,†], Patrick Diep [1] and Radhakrishnan Mahadevan [1,2,*]

1 Department of Chemical Engineering and Applied Chemistry, University of Toronto, Toronto, ON M5S 3E5, Canada; jinjin.chen@utoronto.ca (J.C.); liu.yilan@utoronto.ca (Y.L.); patrick.diep@mail.utoronto.ca (P.D.)
2 Institute of Biomedical Engineering, University of Toronto, Toronto, ON M5S 3G9, Canada
* Correspondence: krishna.mahadevan@utoronto.ca
† Jinjin Chen and Yilan Liu contribute equally to this work.

Abstract: *Acidithiobacillus ferridurans* JAGS is a newly isolated acidophile from an acid mine drainage (AMD). The genome of isolate JAGS was sequenced and compared with eight other published genomes of *Acidithiobacillus*. The pairwise mutation distance (Mash) and average nucleotide identity (ANI) revealed that isolate JAGS had a close evolutionary relationship with *A. ferridurans* JCM18981, but whole-genome alignment showed that it had higher similarity in genomic structure with *A. ferrooxidans* species. Pan-genome analysis revealed that nine genomes were comprised of 4601 protein coding sequences, of which 43% were core genes (1982) and 23% were unique genes (1064). *A. ferridurans* species had more unique genes (205–246) than *A. ferrooxidans* species (21–234). Functional gene categorizations showed that *A. ferridurans* strains had a higher portion of genes involved in energy production and conversion while *A. ferrooxidans* had more for inorganic ion transport and metabolism. A high abundance of *kdp*, *mer* and *ars* genes, as well as mobile genetic elements, was found in isolate JAGS, which might contribute to its resistance to harsh environments. These findings expand our understanding of the evolutionary adaptation of *Acidithiobacillus* and indicate that *A. ferridurans* JAGS is a promising candidate for biomining and AMD biotreatment applications.

Keywords: *Acidithiobacillus*; acid mine drainage; biomining; comparative genomics

Citation: Chen, J.; Liu, Y.; Diep, P.; Mahadevan, R. Genomic Analysis of a Newly Isolated *Acidithiobacillus ferridurans* JAGS Strain Reveals Its Adaptation to Acid Mine Drainage. *Minerals* **2021**, *11*, 74. https://doi.org/10.3390/min11010074

Received: 20 November 2020
Accepted: 11 January 2021
Published: 13 January 2021

Publisher's Note: MDPI stays neutral with regard to jurisdictional claims in published maps and institutional affiliations.

1. Introduction

With continually increasing concerns about acid mine drainage (AMD) contamination and the depletion of high-grade ores, innovative and sustainable methods to recover heavy metals from tailings and AMD as well as to treat AMD pollution are urgently needed [1]. Pyrometallurgical and hydrometallurgical routes are the conventional methods for metal recovery, but they are environmentally unsustainable, with a high cost in terms of operating on low-grade ores [2,3]. Though many techniques have been applied for AMD management, such as neutralization, adsorption, oxygen barriers, bactericides and so on, most of those options are unsustainable and unaffordable [4]. Compared with conventional and other emerging reprocessing techniques, bioleaching is considered as a simple, highly efficient, safe, low-cost, more easily managed and eco-friendly technique to facilitate sustainable mining and prevent AMD [3]. Bioleaching facilitates metal mobilization from solid metal sulfides into their water-soluble forms by different microorganisms via direct and indirect bioleaching [5]. Direct bioleaching can be summarized as:

$$MeS + 2O_2 \rightarrow MeSO_4 \tag{1}$$

while indirect bioleaching can be described as:

$$MeS + Fe_2(SO_4)_3 \rightarrow MeSO_4 + 2Fe_2SO_4 + S^0 \tag{2}$$

Microorganisms take part in and accelerate the oxidation of mineral sulfide to sulfate or the reoxidation of ferrous iron to ferric iron [6]. Although many factors affect the bioleaching process such as temperature, pH, dissolved oxygen, redox potential and formation of secondary minerals [7–9], indigenous bacteria play a crucial role in effective bioleaching [10]. Therefore, an increasing research effort has been placed on discovering and characterizing indigenous robust microbes that are resistant to high metal concentrations in extremely acidic environments [11,12].

Acidithiobacillus is a group of Gram-negative, chemoautotrophic, acidophilic aerobes which dominates in all types of extremely acidic habitats, suggesting its tremendous potential in the bioleaching process [13]. Genome sequencing and analysis of this genus, revealing its adaption to harsh environments, will help us to better understand its mechanisms and provide insights for future genetic engineering strategies to make bioleaching more efficient and versatile. Li et al. [14] analyzed and validated genomic information from this genus, including seven species and 37 strains, and revealed that *Acidithiobacillus* spp. recruited and consolidated novel functionalities via horizontal gene transfer, gene duplication and purifying selection to cope with challenging environments. Presently, there are 10 species reported in this genus: *A. ferrooxidans*, *A. ferridurans*, *A. ferrivorans*, *A. ferrianus*, *A. ferriphilus*, *A. albertensis*, *A. caldus*, *A. thiooxidans*, *A. sulfuriphilus* and *A. cuprithermicus*. The first five species were reported to generate energy by oxidizing ferrous iron, sulfur and hydrogen. Since soluble ferric iron produced from this ferrous iron oxidation can serve as a powerful oxidant to accelerate the dissolution of sulfidic minerals to release target metals, *Acidithiobacillus* species such as *A. ferrooxidans* have drawn focused attention [7]. Since the 1940s, more than 500 isolates of *A. ferrooxidans* have been reported and whole-genome sequencing has been performed for nine isolates. The strain *A. ferrooxidans* ATCC23270 was chosen as the model strain in this genus because of its extensive description in the literature and its wide usage [15]. It can oxidize ferrous iron, reduced inorganic sulfur compounds (RISCs) and hydrogen to generate energy as ATP. Additionally, it can fix atmospheric carbon dioxide and nitrogen as nutrition sources [16]. *A. ferrooxidans* has been successfully applied to recover metals such as copper, nickel, zinc, arsenic and uranium from low-grade ores, sewage sludge and contaminated sediments [13].

A. ferridurans is a species reclassified from *A. ferrooxidans* by Hedrich and Johnson [17] in 2013 because the species' DNA–DNA hybridization (63%) was lower than the threshold value (70%) used to delineate species. Moreover, when comparing *A. ferrooxidans* ATCC23270 with *A. ferridurans* JCM18981 (formerly called *A. ferrooxidans* ATCC33020), the latter showed better resistance to lower pH and higher concentrations of Fe^{2+}, Ni^{2+} and Mg^{2+} [17,18]. Thus, the genus *A. ferridurans* might be a good candidate chassis for industrial applications in the field of biomining and bioremediation. For instance, a newly isolated *A. ferridurans* SBU-SH2 was used for flask and column bioleaching from low-grade uranium ore, which generated 96% and 95.5% uranium extraction in 7 and 26 days, respectively [19,20]. However, at present, only one whole-genome sequence of *A. ferridurans* species, *A. ferridurans* JCM18981 isolated from uranium drainage water in Japan, is available [21]. The limited genome information of *A. ferridurans* species hinders our understanding of the mechanism and evolutionary history underpinning its unique metal and acid resistance.

In our previous study, a dominant strain belonging to the *Acidithiobacillus* genus was found based on 16S rRNA gene sequence analysis, which took up 92.6% of the enriched culture from acid mine drainage (AMD) in Sudbury, Canada [22]. After whole-genome sequencing and assembly, we reclassified and named it as *A. ferridurans* JAGS based on simple 16S rRNA gene and ANIb analyses and announced its genome [23]. To better understand this strain and provide useful data for future research, a detailed genomic analysis was further performed in this study. First, a side-by-side comparison of iron, nickel and low-pH tolerance between *A. ferridurans* JAGS and *A. ferrooxidans* ATCC23270 was conducted. To reveal the genetic traits associated with heavy metal and acid resistance in isolate JAGS, its genomic data were compared with the reported genomes of *Acidithiobacillus* strains. A pan-genome analysis was further conducted on these genomes to explore the metabolic features leading to the diversity of physico-biochemical traits. Functional genes and pathways responsible for heavy metal and acid resistance were analyzed and compared. A mobile genetic element analysis further suggested that gene transfers among these strains likely enabled adaptation to challenging environments. The insights gained in this study enhanced our understanding of the mechanism and evolutionary history of heavy metal and acid resistance in *A. ferridurans* and we suggest possible approaches for engineering *A. ferridurans* as a microbial chassis for biomining processes.

2. Materials and Methods

2.1. Culture Media, Phenotypic and Growth Observations

The strain *A. ferridurans* JAGS was isolated from acidic mine drainage in our previous study [22]. It was the dominant species and made up 92.6% of the enriched culture, based on 16S rRNA gene sequence analysis. The strain *A. ferrooxidans* ATCC23270 was purchased from American Type Culture Collection (ATCC). The phenotypic features of isolated JAGS were observed on a light microscope (Nikon Eclipse E400, Nikon, Melville, NY, USA) and a scanning electron microscope (FEI XL30 SEM, Philips, Eindhoven, Holland), separately. The strain JAGS was cultured with either 9K-Fe^{2+} (160 mM ferrous iron, pH 2.0) or 9K-S^0 (0.5% elemental sulfur, pH 3.0) at 30 °C with shaking at 180 rpm or with 2:2 solid medium in an incubator at 30 °C [24].

The abilities to tolerate elevated concentrations of ferrous iron and nickel and low pH were tested in a side-by-side comparison between *A. ferridurans* JAGS and *A. ferrooxidans* ATCC23270. Cultures grown in the 9K-Fe^{2+} medium were inoculated at a ratio of 20% into the same medium with Fe^{2+} (200 or 320 mM), Ni^{2+} (100 or 200 mM) or at pH 1.5 and then incubated at 30 °C for 22 h. The Fe^{2+} concentration was tested by the colorimetric ferrozine-based assay [25] and the ferrous iron oxidation rate was calculated by using consumed Fe^{2+} divided by its initial concentration, as in a previous study [26].

The growth features of *A. ferridurans* JAGS in 9K-Fe^{2+} or 9K-S^0 medium were further monitored by detecting pH, iron or sulfate concentrations and cell numbers during incubation by removing samples at intervals. The pH value was detected using a pH meter (Thermo Scientific®, Orion Star A211, Waltham, MA, USA). Ferrous and ferric iron concentrations were examined by the colorimetric ferrozine-based assay [25]. Sulfate was detected using a turbidimetric method [27]. Three cell-counting methods were tested: direct cell counting, optical density (OD_{600}) measurements and plate counting. For the direct cell-counting method, samples were taken from media and cell numbers were estimated using a hemocytometer (Hausser Scientific, Horsham, PA, USA). For the OD_{600} method, cells were harvested, washed twice with a basal salt buffer (4.5 g/L $(NH_4)_2SO_4$, 0.15 g/L KCl and 0.75 g/L $MgSO_4 \cdot 7H_2O$) and resuspended in 6% betaine prior to measurements. The plate counting was carried out by spreading proper diluted samples on 2:2 solid plates and colonies were counted after 7–10 days.

2.2. Comparative Genomics

Details of *A. ferridurans* JAGS genomic DNA extraction, sequencing, assembly and annotation are described in our previous paper [23]. The complete genome sequence of the isolate JAGS contains 2,933,811 bp with a GC (guanine-cytosine) content of 58.6%. The Similar Genome Finder service on the PATRIC website was used with the default parameters to find the other similar *Acidithiobacillus* genomes published and to calculate their Mash/MinHash distances with isolate JAGS [28]. For these genomes, average nucleotide identities based on BLAST (ANIb) and MUMmer (ANIm) were calculated in JSpeciesWS [29]. Genome alignment among four whole-genome sequences was achieved using progressiveMauve within PATRIC [30]. Genes related to acid stress and metal resistance were analyzed using PATRIC and created with BioRender (https://biorender.com).

2.3. Pan-Genome Analysis

NCBI PGAP [31] was used to predict coding sequences for *A. ferridurans* JAGS and 8 other *Acidithiobacillus* genomes, and these amino acid sequences were used as the input for the Bacterial Pan-genome Analysis tool (BPGA ver. 1.2) to estimate core and pan genomes using the USEARCH program (ver. 9.0) available in BPGA, with a 50% cut-off of sequence identity [32]. The empirical power law equation $f(n) = a \times n^{\alpha}$ and the exponential equation $f1(n) = c \times e^{(d.n)}$ were used for extrapolation of the pan and core genome curves, respectively. Core, accessory and unique genes defined in USEARCH were mapped into various cluster of orthologous group (COG) categories and Kyoto Encyclopedia of Genes and Genomes (KEGG) pathways. The EggNOG (ver. 5.0.0) program [33] with default parameters was further used to cluster genes into functionally related groups and to analyze metabolic pathways.

2.4. Prediction of Mobile Genetic Elements

Insertion sequences (ISs) and transposases (Tn) distributed over the 9 *Acidithiobacillus* genomes were predicted and classified using the ISFinder platform with manual inspection of search hits (E-value $\leq 10^{-5}$) [34]. IslandViewer (ver. 4), which has integrated the three most accurate and complementary genomic islands (GIs) prediction tools, IslandPath-DIMOB, SIGI-HMM and IslandPick [35], was applied for the computational identification of putative GIs. In addition, the web tool CRISPRFinder was mainly used to identify the Clustered Regularly Interspaced Short Palindromic Repeats-Cas protein (CRISPR-Cas) array.

3. Results and Discussion

3.1. Phenotypic and Growth Features

The genus of *Acidithiobacillus* is widely distributed in natural environments such as acid mine drainage (AMD) settings. We isolated an *Acidithiobacillus* strain from an AMD sample collected from Sudbury, Canada, and named it *A. ferridurans* JAGS in our previous study [23]. Orange-brown colonies of isolate JAGS formed on solid media, taking the shape of dots after 10 days of incubation (Figure 1A). The cells of isolate JAGS collected from 9K-Fe^{2+} liquid media showed single and paired rods, approximately 0.5–1.5 μm long and 0.3 μm wide (Figure 1B), which is slightly smaller than the reported *A. ferrooxidans* that is 1–2 μm long and 0.3–0.6 μm wide.

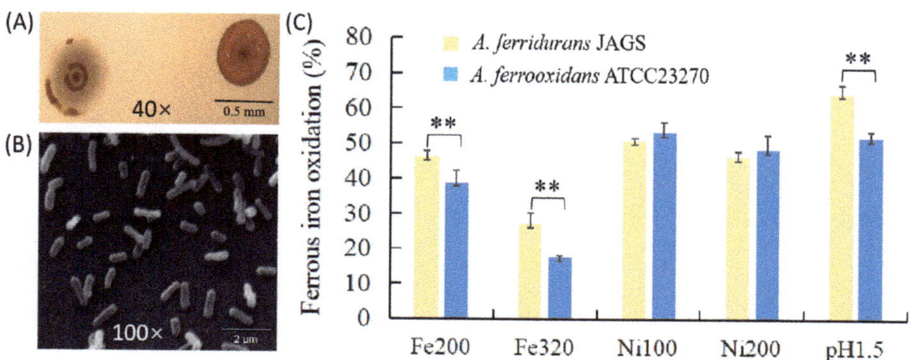

Figure 1. Differentiation of *Acidithiobacillus ferridurans* JAGS from closely related species. (**A**) Colony morphologies of *A. ferridurans* JAGS observed under 40× optical microscopy for both top and bottom images. (**B**) Cell morphology observed under SEM. (**C**) Ferrous iron oxidation rates of *A. ferridurans* JAGS and *A. ferrooxidans* ATCC23270 under different pressures of Fe200 (200 mM Fe^{2+}), Fe320 (320 mM Fe^{2+}), Ni100 (100 mM Ni^{2+}), Ni200 (200 mM Ni^{2+}) and pH 1.5. ** indicates $p < 0.01$.

The species of *A. ferridurans* was reported to have a notably higher tolerance to many metals such as Fe^{2+}, Ni^{2+} and Mg^{2+} and lower pH when compared with other *Acidithiobacillus* species [17,18]. In addition, there is 0.5–1% Ni existing in the pyrrhotite tailings of Sudbury [36]. Therefore, we compared the Fe^{2+} oxidation rate (%) between *A. ferridurans* JAGS and *A. ferrooxidans* ATCC23270 under elevated concentrations of Fe^{2+} and Ni^{2+} and low pH pressures (Figure 1C). The results showed that isolate JAGS had higher Fe^{2+} oxidation rates compared with ATCC23270 under high concentrations of Fe^{2+} and low pH values but similar levels of Fe^{2+} oxidation rates under high concentrations of Ni^{2+}. These results indicated the adaptation of *A. ferridurans* JAGS to the acid mine drainage in Sudbury and suggested that it might be a great candidate as the ferrous oxidizer in low-pH bioleaching.

The growth features of *A. ferridurans* JAGS in culture media with 9K-Fe^{2+} and 9K-S^0 were investigated (Supplementary Material Figure S1). Figure S1A,B present a standard curve of OD_{600} versus the cell count obtained by plate counting. When OD_{600} = 1, we estimated that there were 8.8×10^9 cells/mL of *A. ferridurans* JAGS, which is slightly higher than the reported number of *A. ferrooxidans* (8.3×10^9 cells/mL) [37]. This may be due to the smaller cell size of isolate JAGS that causes it to absorb less light in the cuvette. We noted that counting cells of isolate JAGS was difficult due to its very low cell density in the lag phase and the interference of precipitate formation in the exponential phase. Therefore, monitoring growth required indirect tracking via changes in the pH and electron donor concentrations, but this was corroborated with data from OD_{600} measurements and the plate count method. The growth behavior of isolate JAGS in 9K-Fe^{2+} (Figure S1C) and 9K-S0 culture media (Figure S1D) shared similarities with what was reported for *A. ferrooxidans* ATCC 23270 [38,39], which indicates a close relationship between the two species. However, the cell numbers examined by OD_{600} were much lower than those obtained by the plate count method, which is likely due to the cells lost during the process of precipitate removal prior to OD_{600} measurements.

3.2. Genomic Features

To better understand the isolated strain *A. ferridurans* JAGS, its genome was sequenced (GenBank: CP044411) and analyzed. The genome of *A. ferridurans* JAGS is a single circular chromosome comprising 2,933,811 bases with a GC content of 58.56%, which contains 3001 protein-coding sequences (CDSs), 46 tRNAs and 6 rRNAs [23]. The genomic features of *A. ferridurans* JAGS are quite similar to those of *A. ferridurans* JCM18981, which are 2,921,399 bases with 58.4% GC content, containing 3026 CDSs, 47 tRNAs and 6 rRNAs.

To explore the relationship of *A. ferridurans* JAGS with other *Acidithiobacillus* species, the Similar Genome Finder service from PATRIC was used to find similar genomes with isolate JAGS as the reference. There were eight *Acidithiobacillus* genomes found (Table 1). These strains were collected from different environments but mainly from acid mine waters. Their genomes varied in size from 2.7 to 3.2 Mb, with total CDS numbers ranging from 2634 to 3179. For the Mash/MinHash distances, *A. ferridurans* JCM18981 and *A. ferrooxidans* IO-2C showed the closest distance with *A. ferridurans* JAGS compared to the other six strains, which suggests that the IO-2C strain might be incorrectly classified. Based on the average nucleotide identity (ANI) relatedness analysis, it appears that the strains JAGS, JCM18981 and IO-2C are all *A. ferridurans* species as they shared ANI values >98% with each other (Table 1 and Table S1), which is larger than the reported threshold of ≥96% for classification [14].

Genome alignment among the complete-genome sequences of *A. ferridurans* JAGS, *A. ferridurans* JCM18981, *A. ferrooxidans* ATCC53993 and *A. ferrooxidans* ATCC23270 was performed using progressiveMauve [24] and is shown in Figure 2. Surprisingly, the genomic arrangement of *A. ferridurans* JAGS had better co-linearity with *A. ferrooxidans* ATCC53993 and *A. ferrooxidans* ATCC23270 than with *A. ferridurans* JCM18981. The same-color blocks suggest high conservation of gene orders among multiple genomes that are likely inherited through vertical transfer, while *A. ferridurans* JCM18981 had a large number of gene rearrangements, insertions and/or deletions. This result indicates multiple recombination events and a rich evolutionary history of *A. ferridurans* species.

Table 1. General features and genomic comparison (pairwise mutation (Mash) distance, average nucleotide identity (ANI)) between *A. ferridurans* JAGS and selected representatives.

Genome Name	Geographic Origin	Contigs	Genome Size	GC%	No. of CDS	BioProject Accession	Mash Distance	ANIb	ANIm
A. ferridurans JAGS	Acid mine drainage, Canada	1	2,933,811	58.6	3001	PRJNA573091	-	-	-
A. ferridurans JCM18981	Uranium mine drainage water, Japan	1	2,921,399	58.4	3026	PRJDB7175	0.0090	99.13	99.66
A. ferrooxidans IO-2C	Acid seep soil, USA	23	2,716,894	58.7	2634	PRJNA432283	0.0136	98.69	99.23
A. ferrooxidans ATCC53993	-	1	2,885,038	58.9	2826	PRJNA16689	0.0446	94.98	95.51
A. ferrooxidans RVS1	Andacollo gold mining area, Argentina	49	2,826,311	58.8	2705	PRJNA499028	0.0463	94.68	95.33
A. ferrooxidans CCM4253	Mine waters, Czech Republic	15	3,196,562	58.6	3059	PRJNA475418	0.0480	94.70	95.32
A. ferrooxidans YQH-1	Wudalianchi volcano water, China	96	3,111,222	58.6	3089	PRJNA294114	0.0482	94.71	95.34
A. ferrooxidans Hel18	Flue dust	123	3,109,160	58.6	3179	PRJNA308169	0.0484	94.69	95.34
A. ferrooxidans ATCC23270	bituminous coal mine effluent	1	2,982,397	58.8	3147	PRJNA53	0.0493	94.76	95.37

Note: Average nucleotide identity (ANI) based on BLAST+ (ANIb) and MUMmer (ANIm).

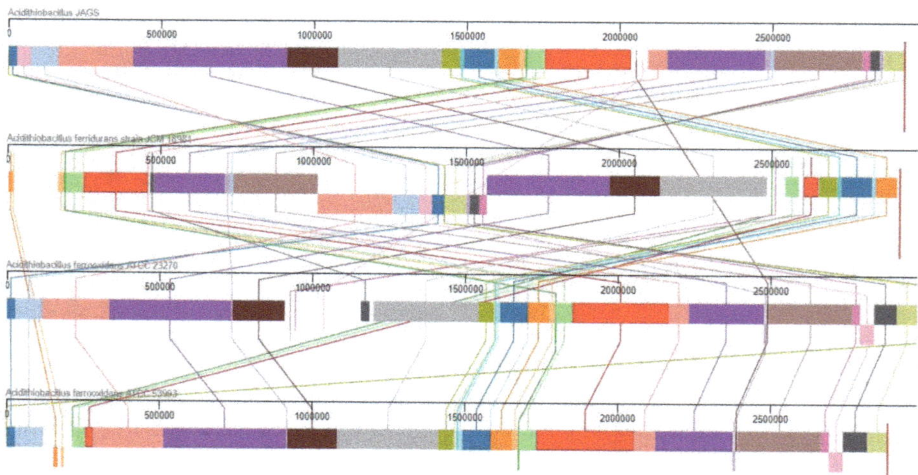

Figure 2. Whole-genome alignment of *A. ferridurans* JAGS, *A. ferridurans* JCM18981, *A. ferrooxidans* ATCC53993 and *A. ferrooxidans* ATCC23270. Locally collinear blocks (LCBs) identified by Mauve are color-coded; links between LCBs are indicated by the thin colored lines.

3.3. Pan-Genome and Functional Gene Analysis

Pan-genome analysis was carried out using the Bacterial Pan-genome Analysis (BPGA) tool to provide insights regarding genomic features, diversity and evolution [32]. It is well accepted that more than five genomes in a pan-genome analysis could provide sufficient data for extrapolation of the information for species [40]. In this study, genomes of *A. ferridurans* JAGS and eight other strains of *A. ferridurans* and *A. ferrooxidans* were used for the pan-genome analysis since we wanted to investigate the genetic diversity and ecological adaption of these two species. As shown in Figure 3A,B, according to the Heaps' Law function (f(n) = $2819.13n^{0.22}$), the pan genome is open as the γ was calculated as 0.22, which means that the addition of new genomes will provide novel genes and it indicates evolutionary changes in these genomes [41]. The pan genome contains a total of 4601 genes, of which 1982 genes are in the core genome and 1064 genes are in the unique genome. The richness of unique genes in *A. ferridurans* JAGS, JCM18981, IO-2C and *A. ferrooxidans* ATCC23270 suggests that they may actively exchange genes with other genera to adapt to different environmental conditions.

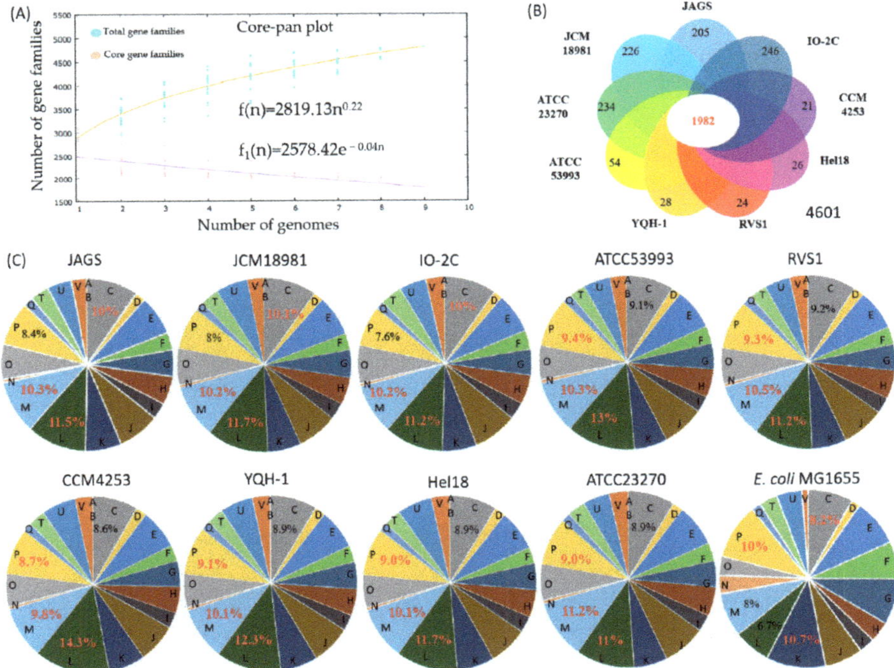

Figure 3. Pan-genome and EggNOG analysis. (**A**) Core-pan plot of studied *Acidithiobacillus* genomes; (**B**) Venn diagram of the pan genome; (**C**) pie charts of cluster of orthologous groups (COGs) of studied *A. ferridurans* and *A. ferrooxidans* strains and *E. coli* MG1655 (reference). Note: In Figure 3C, the numbers represent the percentage of each category. A, RNA processing and modification; B, Chromatin structure and dynamics; C, Energy production and conversion; D, Cell cycle control and mitosis; E, Amino acid metabolism and transport; F, Nucleotide metabolism and transport; G, Carbohydrate metabolism and transport; H, Coenzyme metabolism; I, Lipid metabolism; J, Translation; K, Transcription; L, Replication and repair; M, Cell wall/membrane/envelope biogenesis; N, Cell motility; O, Post-translational modification, protein turnover, chaperone functions; P, Inorganic ion transport and metabolism; Q, Secondary structure; T, Signal transduction; U, Intracellular trafficking and secretion; V, Defense mechanisms. The categories of "R, General functional prediction only" and "S, Function unknown" were omitted.

The cluster of orthologous group (COG) distributions of the annotated genes for each studied *Acidithiobacillus* strain are illustrated in Figure 3C. These COGs fell into 20 COG classes, not including "General functional prediction only" and "Function unknown". *Escherichia coli* MG1655 was used as a reference, which showed that *E. coli* had the highest portion of genes corresponding to (K) Transcription, while *Acidithiobacillus* strains showed higher portions of functional genes related to (L) Replication, recombination and repair

(10.7–14%) and (M) Cell wall/membrane/envelope biogenesis (9.4–10.9%). This was not surprising given that these categories of proteins have been reported to be necessary for acid and heavy metal resistance and, likely, long-term adaptation mechanisms to extreme environments [42]. When the two *Acidithiobacillus* species were compared, *A. ferridurans* had more genes associated with functions supporting (C) Energy production and conversion (9.7–9.9%), while *A. ferrooxidans* had more genes related to the function of (P) Inorganic ion transport and metabolism (8.5–9.2%).

3.4. Genetic Mechanisms of Acid Stress and Metal Resistance.

In response to acidic heavy metal stress, acidophiles have developed different genetic mechanisms to survive and they are described, which can be very complex [43,44]. The metabolic diversity and adaptive mechanisms of *Acidithiobacillus* spp. responding to extremely acidic environments have been reviewed [45] and are beyond the scope of this study. Here, we focused on five major mechanisms (Figure 4A) for acid and heavy metal resistance in representative *Acidithiobacillus* strains by analyzing related gene clusters: (1) a membrane barrier created by outer membrane proteins (Omp40) and hopanoids; (2) maintenance of a membrane potential by influx of potassium and sequestration of metal ions intra-/extracellularly; (3) active removal by antiporters or exporters; (4) decarboxylation and detoxification; and (5) DNA and protein repair systems.

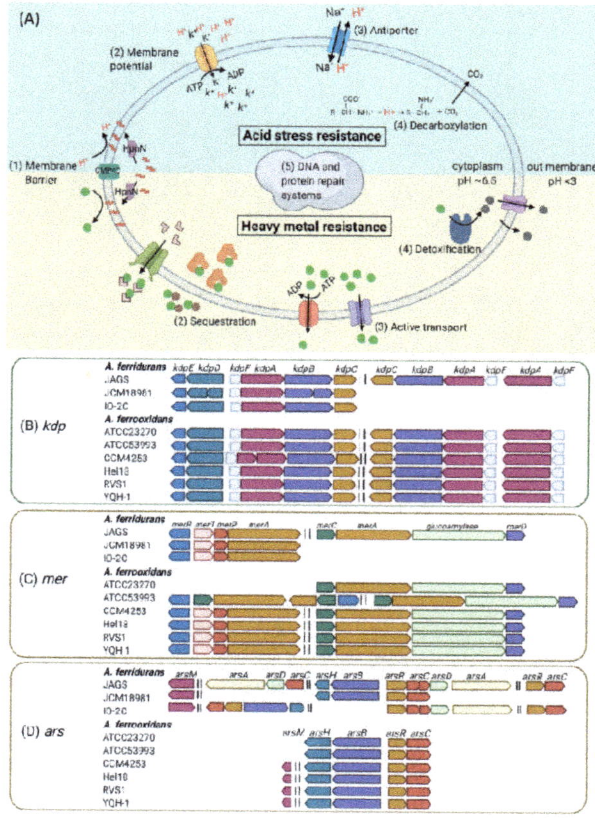

Figure 4. Overview of adaptive strategies for acid and heavy metal resistance. Potential resistance mechanisms (**A**) and comparisons of operons for *kdp* (**B**), *mer* (**C**) and *ars* (**D**) genes among *Acidithiobacillus* strains. *Kdp*, a high-affinity K$^+$ transport system; *mer*, Hg^{2+}-resistant genes; *ars*, As$^{2+/3+}$-resistant genes.

The mechanisms responsible for acid stress resistance are complex. We analyzed relevant genes in *A. ferridurans* JAGS and listed them in Table S2 (Supplementary Materials). OMP40 (Gene ID: F6A13_00370) was discovered and it was previously reported as an anionic porin in the outer membrane to restrict the influx of protons in *A. ferrooxidans* [46]. A number of genes coding for hopanoid-synthesis proteins were identified in the isolate JAGS genome, including a glycosyltransferase (HpnB, Gene ID: F6A13_01630) and a cluster for hopanoid-associated proteins, a squalene-hopene cyclase and a squalene synthase (HpnMHNKJIAG-Sch-Sqs, Gene IDs: F6A13_09105–09160). Hopanoid is an important type of bacterial lipid that can alter membrane fluidity and permeability to restrict H^+ influx. It was reported to be synthesized from squalene by SHC, HpnG and other proteins enriched in the outer membrane by a transporter HpnN, although the functions of some factors are still unknown [47].

The Na^+/H^+ antiporter (Gene IDs: F6A13_09475, 04755) can export excess protons by coupling the uptake of Na^+, while decarboxylases (SpeA, PanD and Psd; Gene IDs: F6A13_06090, 14595, 04545) will consume excess protons in the cytoplasm. For most of the proteins, the interspecies identities ranged from 94% to 98%, while intraspecies identities were 100%. Interestingly, we found the *A. ferridurans* JAGS genome to process three *kdp* clusters (*kdpEDFABC*, Gene IDs: F6A13_09590-09605; *kdpFABC*, Gene IDs: 11020–11005; kdpFA, GeneIDs: 11030–11025), similar to other *A. ferrooxidans* strains. However, *A. ferridurans* JCM18981 and IO-2C only had one *kdp* cluster (Figure 4B). The *kdp* clusters code for a high-affinity K^+ transport system, which could generate a reversed membrane potential through the active influx of K^+ to cope with acid resistance [48]. In Figure S2 (Supplementary Material), a neighbor-joining (NJ) phylogenetic tree is constructed based on the kdpA protein sequences. Even though *A. ferridurans* JAGS had a similar pattern of *kdp* clusters to *A. ferrooxidans* species (Figure 4B), its *kdpA* (I) showed higher sequence identity with the other *A. ferridurans* strains, suggesting that *kdp* clusters can be acquired more than once in these genomes and have redundancy.

We also tried to find genes involved in heavy metal resistance pathways in *A. ferridurans* JAGS (Figure 4 and Table S2). In total, eight genes (Gene IDs: F6A13_04860, 04905, 04925, 04945, 08625, 10865, 10890, 11740) were predicted to code for p-type ATPases to transport Pb^{2+}, Cd^{2+}, Zn^{2+}, Hg^{2+} and Cu^{2+} [49]. Similar numbers of these genes were detected in ATCC53993 (8) and JCM18981 (9), while only 5–6 genes were found in other strains. Some of these genes are highly similar, such as the genes F6A13_10865 and F6A13_04945, which indicates gene duplication within a genome. There are several other clusters belonging to the Resistance-nodulation-division (RND) transporter system, in the *czc* or *znu* families, which corresponds to Ni^{2+}, Mn^{2+}, $Fe^{2+/3+}$, Mo^{2+}, Co^{2+}, Cd^{2+} and Zn^{2+} and CorAC for Mg^{2+}/Co^{2+} export. We also noticed that these *Acidithiobacillus* strains possessed two types and several copies of *czcABC* clusters located at different sites, suggesting that *czc* clusters may be acquired more than once from different origins. Overall, this pattern of redundancy of resistance clusters in *Acidithiobacillus* suggests that it is part of the adaptive strategy for survival in acidic heavy metal conditions.

We also found genes coding for proteins related to mercury (Hg^{2+}) and arsenic ($As^{2+/3+}$) resistance. The operons we found for Hg^{2+} resistance in *Acidithiobacillus* strains are summarized in Figure 4C. There were two subgroups of *mer* clusters detected, *merRTPA* and *merCAD*. *A. ferridurans* JAGS and *A. ferrooxidans* CCM4253, Hel18, RVS1 and YQH-1 possessed both *mer* clusters, while *A. ferridurans* JCM18981 and IO-2C lacked the *merCAD* cluster. An intact *merRTPA* cluster was not detected in *A. ferrooxidans* ATCC23270 and ATCC53993, although ATCC53993 had two copies of *merR*. The *merA* in the *merCAD* cluster exhibited 100% identity in all selected strains, suggesting that *merCAD* might come from the same donor. Therefore, we further investigated the genetic context of the cluster *merCAD* and found it adjacent to *kdpCBAFAF* clusters. The DNA sequences of the *kdpCBAFAF–merCAD* clusters (11,906 bp) in *A. ferridurans* JAGS share 100% identity and 100% coverage with *A. ferrooxidans* ATCC53993, ATCC23270, RVS1 and CCM4253 and 100% identity with less coverage for *A. ferrooxidans* Hel18 and YQH-1, possibly due to incomplete sequencing.

This indicates that the *kdpCBAFAF–merCAD* clusters might come from a same donor via a horizontal gene transfer (HGT) event [14]. Two predicted mobile element proteins (Gene IDs: F6A13_10980, 10985) were found upstream of the *kdpCBAFAF–merCAD* clusters in isolate JAGS, supporting the HGT hypothesis.

Annotation of arsenic resistance clusters included *arsHBRCDA, arsHRBC, arsCDA, arsRC* and *arsM* in the studied genomes. *A. ferridurans* JAGS possessed all of these clusters, while the *arsCDA* and *arsRC* clusters were absent in *A. ferridurans* JCM18981 and the studied *A. ferrooxidans* strains. The largest abundance of *ars* clusters in isolate JAGS compared with other strains might contribute to its dominance in our metal-rich mine drainage sample. Besides, the analysis of *arsC* protein sequences further suggested that the *ars* clusters *arsHBRCDA, arsHBRC* and *arsRC* are likely acquired from different donors during evolution. Since it was also reported that gene copy number alterations can benefit microorganisms' survival under selective pressure [50], we hypothesize that *A. ferridurans* JAGS might have gained *ars* clusters from other species during adaption to the metal-rich environments.

In summary, compared to the other *Acidithiobacillus* strains that we studied, *A. ferridurans* JAGS had several genes and considerable redundancy that likely contributes to its acid and heavy metal resistance, which highlights its strong potential for usage in biomining processes, especially for cinnabar (HgS) or arsenopyrite (FeAsS) tailings. Additionally, we speculate that *A. ferridurans* JAGS might be an intermediate species between *A. ferrooxidans* and *A. ferridurans* based on the evidence of the gene cluster types and genomic structure (Figure 4) and the genome alignment result (Figure 2).

3.5. Mobile Genetic Elements Analysis

Mobile genetic elements (MGEs) play a great role in genome plasticity and evolution, shaping both genes and genomes to respond to drastic changes in environmental conditions [51]. MGEs, including insertion sequences (ISs) and genomic islands (GIs), are listed in Table 2. The number of ISs per strain ranged from 38 (*A. ferridurans* IO-2C) to 78 (*A. ferridurans* JAGS and *A. ferridurans* JCM18981), which might be due to the genome assembly level. High similarity regarding IS type was observed in all studied strains: IS1595, IS21, IS3, ISL3 and Tn families, which were the most common IS families. IS3 was the most abundant family. However, closer inspection demonstrated several differences. For instance, *A. ferridurans* JAGS has a much higher number of IS1595 when compared with other species. The nine genomes harbored 19–26 GIs ranging from 4 to 158 kb in size, representing many versatile gene pools. Several ISs, such as IS110 and IS66, were presented in the predicted GIs, suggesting that these putative GIs were likely acquired by horizontal gene transfer. In addition, GIs carrying mercury resistance genes (merRTPA) were found in all *A. ferridurans* species but not in the model strains *A. ferrooxidans* ATCC23270 and *A. ferrooxidans* ATCC53993, which might provide *A. ferridurans* with an adaptive advantage in mercury-rich environments.

Furthermore, we examined the clustered regularly interspaced short palindromic repeats (CRISPR) systems in all studied genomes using the CRISPRCasFinder [52]. Interestingly, *A. ferrooxidans* ATCC23270 was the only strain with a predicted CRISPR system. One unique type IV Cas cluster (csf4-1-2-3) and five spacers presented in the vicinity were found and presumed to function in conjunction with other CRISPR arrays [53].

Table 2. The prediction of mobile genetic elements including insertion sequences (ISs) and genomic islands (GIs) of the *Acidithiobacillus* strains studied.

A. The Putative Insertion Sequences							
IS Family	**JAGS**	**JCM18981**	**IO-2C**	**ATCC53993**	**CCM4253**	**ATCC23270**	**RVS1**
IS110	3	3	1	4	4	3	1
IS1182	0	1	0	1	0	0	-
IS1595	15	1	3	6	0	1	-
IS1634	2	6	4	2	1	6	-
IS200/IS605	3	2	0	1	0	2	-
IS21	7	17	3	9	6	17	1
IS256	7	7	0	2	2	7	-
IS3	21	23	14	21	22	23	-
IS481	1	0	0	1	1	0	-
IS5	1	1	2	12	2	1	-
IS66	2	2	2	2	2	2	-
IS7	1	1	1	1	1	1	-
ISL3	8	8	7	7	8	8	-
Tn	7	6	1	5	3	6	-
In total	78	78	38	74	52	77	-
B. The Predicted Genomic Islands							
GI No.	22	22	20	21	26	21	19
Size range (Kb)	4–42	4–46	4–28	4–158	4–63	4–25	4–35

Note: Since RVS1, Hel18 and YQH-1 have many contigs, ISFinder and IslandViewer cannot predict ISs and GIs exactly, respectively.

4. Conclusions

In the present study, we provide the growth characteristics and genomic insights of a newly isolated strain, *A. ferridurans* JAGS. The growth features of isolate JAGS in 9K-Fe^{2+} and 9K-S^0 liquid media are similar to the *A. ferrooxidans* type strain ATCC23270, but it shows a higher oxidation rate under elevated concentrations of Fe^{2+} and low pH. Genomic comparison and pan-genome analysis among nine strains of two species of *A. ferridurans* and *A. ferrooxidans* revealed obvious genetic differences between the two species. *A. ferridurans* JAGS showed a closer evolutionary relationship with other *A. ferridurans* species but a higher similarity of genomic structure with the *A. ferrooxidans* strains. This suggests that *A. ferridurans* JAGS might be an intermediary strain. Investigations of gene clusters (*kdp*, *mer* and *ars*) and mobile genetic elements indicated that there have been frequent gene transfers between their genomes during evolution. The high abundance of acid and metal resistance genes found in *A. ferridurans* JAGS points to its unique abilities to survive in harsh mining environments, which highlights its strong potential for applications in bio-mining processes. Further transcriptomic and proteomic analyses are required to find the exact genes, proteins and possible mechanisms that lead to the increased resistance of the isolate JAGS strain.

Supplementary Materials: The following are available online at https://www.mdpi.com/2075-163X/11/1/74/s1, Figure S1: Growth features of *A. ferridurans* JAGS. (**A**) Standard curve of optical density (OD$_{600}$) versus cell numbers obtained by plate count method. (**B**) Colonies on 2:2 solid medium for cell count. (**C**) Growth on 9K-Fe^{2+} medium, (**C1**) pH value; (**C2**) Fe oxidation; (**C3**) cell numbers by OD$_{600}$; (**C4**) cell numbers by plate count method. (**D**) Growth on 9K-S^0 medium, (**D1**) pH value; (**D2**) sulfur oxidation; (**D3**) cell count by OD$_{600}$; (**D4**) cell numbers by plate count method. Figure S2: Neighbor-joining (NJ) phylogenetic tree of the kdpA protein sequences derived from nine *Acidithiobacillus* strains. Bootstrap values indicated at each node are based on a total of 500 bootstrap replicates. Table S1: Average nucleotide identity (ANI) (%) based on whole-genome alignments among *Acidithiobacillus* strains by JSpeciesWS. Table S2: Genes predicted to be involved in acid and heavy metal tolerance in *A. ferridurans* JAGS.

Author Contributions: Conceptualization, J.C. and R.M.; methodology, J.C. and Y.L.; software, formal analysis and writing—original draft preparation, J.C. and Y.L.; writing—review and editing, P.D. and R.M.; supervision, R.M.; project administration, R.M.; funding acquisition, R.M. All authors have read and agreed to the published version of the manuscript.

Funding: This work was funded through the Elements of Biomining Grant from the Province of Ontario through the ORF Research Excellence funding program.

Institutional Review Board Statement: Not applicable.

Informed Consent Statement: Not applicable.

Data Availability Statement: Not applicable.

Acknowledgments: Vale is acknowledged for providing access to sample tailings at their Sudbury, ON, Canada, operations.

Conflicts of Interest: The authors declare no conflict of interest.

References

1. Igarashi, T.; Herrera, P.S.; Uchiyama, H.; Miyamae, H.; Iyatomi, N.; Hashimoto, K.; Tabelin, C.B. The two-step neutralization ferrite-formation process for sustainable acid mine drainage treatment: Removal of copper, zinc and arsenic, and the influence of coexisting ions on ferritization. *Sci. Total Environ.* **2020**, *715*, 136877. [CrossRef]
2. Thao, N.T.; Tsuji, S.; Jeon, S.; Park, I.; Tabelin, C.B.; Ito, M.; Hiroyoshi, N. Redox potential-dependent chalcopyrite leaching in acidic ferric chloride solutions: Leaching experiments. *Hydrometallurgy* **2020**, *194*, 105299. [CrossRef]
3. Baniasadi, M.; Vakilchap, F.; Bahaloo-Horeh, N.; Mousavi, S.M.; Farnaud, S. Advances in bioleaching as a sustainable method for metal recovery from e-waste: A review. *J. Ind. Eng. Chem.* **2019**, *76*, 75–90. [CrossRef]
4. Park, I.; Tabelin, C.B.; Jeon, S.; Li, X.; Seno, K.; Ito, M.; Hiroyoshi, N. A review of recent strategies for acid mine drainage prevention and mine tailings recycling. *Chemosphere* **2019**, *219*, 588–606. [CrossRef] [PubMed]
5. Vera, M.; Schippers, A.; Sand, W. Progress in bioleaching: Fundamentals and mechanisms of bacterial metal sulfide oxidation—part A. *Appl. Microbiol. Biotechnol.* **2013**, *97*, 7529–7541. [CrossRef] [PubMed]
6. Bosecker, K. Bioleaching: Metal solubilization by microorganisms. *FEMS Microbiol. Rev.* **1997**, *20*, 591–604. [CrossRef]
7. Tabelin, C.B.; Corpuz, R.D.; Igarashi, T.; Villacorte-Tabelin, M.; Alorro, R.D.; Yoo, K.; Raval, S.; Ito, M.; Hiroyoshi, N. Acid mine drainage formation and arsenic mobility under strongly acidic conditions: Importance of soluble phases, iron oxyhydroxides/oxides and nature of oxidation layer on pyrite. *J. Hazard. Mater.* **2020**, *399*, 122844. [CrossRef]
8. Rasoulnia, P.; Barthen, R.; Lakaniemi, A.M. A critical review of bioleaching of rare earth elements: The mechanisms and effect of process parameters. *Crit. Rev. Environ. Sci. Technol.* **2020**, 1–50. [CrossRef]
9. Park, I.; Tabelin, C.B.; Magaribuchi, K.; Seno, K.; Ito, M.; Hiroyoshi, N. Suppression of the release of arsenic from arsenopyrite by carrier-microencapsulation using Ti-catechol complex. *J. Hazard. Mater.* **2018**, *344*, 322–332. [CrossRef]
10. Wu, C.; Jiang, M.; Hsieh, L.; Cai, Y.; Shen, Y.; Wang, H.; Lin, Q.; Shen, C.; Hu, B.; Lou, L. Feasibility of bioleaching of heavy metals from sediment with indigenous bacteria using agricultural sulfur soil conditioners. *Sci. Total Environ.* **2020**, *703*, 134812. [CrossRef]
11. Camargo, F.P.; do Prado, P.F.; Tonello, P.S.; Dos Santos, A.C.A.; Duarte, I.C.S. Bioleaching of toxic metals from sewage sludge by co-inoculation of *Acidithiobacillus* and the biosurfactant-producing yeast *Meyerozyma guilliermondii*. *J. Environ. Manag.* **2018**, *211*, 28–35. [CrossRef] [PubMed]
12. Giese, E.C.; Carpen, H.L.; Bertolino, L.C.; Schneider, C.L. Characterization and bioleaching of nickel laterite ore using *Bacillus subtilis* strain. *Biotechnol. Prog.* **2019**, *35*, e2860. [CrossRef] [PubMed]
13. Zhang, S.; Yan, L.; Xing, W.; Chen, P.; Zhang, Y.; Wang, W. *Acidithiobacillus ferrooxidans* and its potential application. *Extremophiles* **2018**, *22*, 563–579. [CrossRef] [PubMed]
14. Li, L.; Liu, Z.; Meng, D.; Liu, X.; Li, X.; Zhang, M.; Tao, J.; Gu, Y.; Zhong, S.; Yin, H. Comparative genomic analysis reveals the distribution, organization, and evolution of metal resistance genes in the Genus *Acidithiobacillus*. *Appl. Environ. Microbiol.* **2019**, *85*, e02153-18. [CrossRef] [PubMed]
15. Valdés, J.; Pedroso, I.; Quatrini, R.; Dodson, R.J.; Tettelin, H.; Blake, R.; Eisen, J.A.; Holmes, D.S. *Acidithiobacillus ferrooxidans* metabolism: From genome sequence to industrial applications. *BMC Genom.* **2008**, *9*, 597. [CrossRef] [PubMed]
16. Campodonico, M.A.; Vaisman, D.; Castro, J.F.; Razmilic, V.; Mercado, F.; Andrews, B.A.; Feist, A.M.; Asenjo, J.A. *Acidithiobacillus ferrooxidans*'s comprehensive model driven analysis of the electron transfer metabolism and synthetic strain design for biomining applications. *Metab. Eng. Commun.* **2016**, *3*, 84–96. [CrossRef]
17. Hedrich, S.; Johnson, D.B. *Acidithiobacillus ferridurans* sp. nov., an acidophilic iron-, sulfur-and hydrogen-metabolizing chemolithotrophic gammaproteobacterium. *Int. J. Syst. Evol. Microbiol.* **2013**, *63*, 4018–4025.
18. Falagán, C.; Moya-Beltrán, A.; Castro, M.; Quatrini, R.; Johnson, D.B. *Acidithiobacillus sulfuriphilus* sp. nov.: An extremely acidophilic sulfur-oxidizing chemolithotroph isolated from a neutral pH environment. *Int. J. Syst. Evol. Microbiol.* **2019**, *69*, 2907–2913.

19. Jalali, F.; Fakhari, J.; Zolfaghari, A. Response surface modeling for lab-scale column bioleaching of low-grade uranium ore using a new isolated strain of *Acidithiobacillus Ferridurans*. *Hydrometallurgy* **2019**, *185*, 194–203. [CrossRef]

20. Jalali, F.; Fakhar, J.; Zolfaghari, A. On using a new strain of *Acidithiobacillus ferridurans* for bioleaching of low-grade uranium. *Sep. Sci. Technol.* **2020**, *55*, 994–1004. [CrossRef]

21. Miyauchi, T.; Kouzuma, A.; Abe, T.; Watanabe, K. Complete genome sequence of *Acidithiobacillus ferridurans* JCM 18981. *Microbiol. Resour Announc* **2018**, *7*, e01028-18. [CrossRef]

22. Garg, S. Abiotic and Biotic Leaching Characteristics of Pyrrhotite Tailings from the Sudbury, Ontario Area. Ph.D. Thesis, University of Toronto, Toronto, ON, Canada, 2017.

23. Chen, J.; Liu, Y.; Diep, P.; Jo, A.; Nesbø, C.; Edwards, E.; Papangelakis, V.; Mahadevan, R. Complete genome sequence of *Acidithiobacillus ferridurans* JAGS, isolated from acidic mine drainage. *Microbiol. Resour. Announc.* **2020**, *9*, 9. [CrossRef]

24. Wang, H.; Liu, X.; Liu, S.; Yu, Y.; Lin, J.; Lin, J.; Pang, X.; Zhao, J. Development of a markerless gene replacement system for *Acidithiobacillus ferrooxidans* and construction of a *pfkB* mutant. *Appl. Environ. Microbiol.* **2012**, *78*, 1826–1835. [CrossRef] [PubMed]

25. Stookey, L.L. Ferrozine—A new spectrophotometric reagent for iron. *Anal. Chem.* **1970**, *42*, 779–781. [CrossRef]

26. Daoud, J.; Karamanev, D. Formation of jarosite during Fe^{2+} oxidation by *Acidithiobacillus ferrooxidans*. *Miner. Eng.* **2006**, *19*, 960–967. [CrossRef]

27. Kolmert, Å.; Wikström, P.; Hallberg, K.B. A fast and simple turbidimetric method for the determination of sulfate in sulfate-reducing bacterial cultures. *J. Microbiol. Methods* **2000**, *41*, 179–184. [CrossRef]

28. Wattam, A.R.; Davis, J.J.; Assaf, R.; Boisvert, S.; Brettin, T.; Bun, C.; Conrad, N.; Dietrich, E.M.; Disz, T.; Gabbard, J.L. Improvements to PATRIC, the all-bacterial bioinformatics database and analysis resource center. *Nucleic Acids Res.* **2017**, *45*, D535–D542. [CrossRef] [PubMed]

29. Richter, M.; Rosselló-Móra, R.; Oliver Glöckner, F.; Peplies, J. JSpeciesWS: A web server for prokaryotic species circumscription based on pairwise genome comparison. *Bioinformatics* **2016**, *32*, 929–931. [CrossRef]

30. Darling, A.E.; Mau, B.; Perna, N.T. progressiveMauve: Multiple genome alignment with gene gain, loss and rearrangement. *PloS ONE* **2010**, *5*, e11147. [CrossRef]

31. Tatusova, T.; DiCuccio, M.; Badretdin, A.; Chetvernin, V.; Nawrocki, E.P.; Zaslavsky, L.; Lomsadze, A.; Pruitt, K.D.; Borodovsky, M.; Ostell, J. NCBI prokaryotic genome annotation pipeline. *Nucleic Acids Res.* **2016**, *44*, 6614–6624. [CrossRef]

32. Chaudhari, N.M.; Gupta, V.K.; Dutta, C. BPGA-an ultra-fast pan-genome analysis pipeline. *Sci. Rep.* **2016**, *6*, 1–10. [CrossRef]

33. Huerta-Cepas, J.; Szklarczyk, D.; Heller, D.; Hernández-Plaza, A.; Forslund, S.K.; Cook, H.; Mende, D.R.; Letunic, I.; Rattei, T.; Jensen, L.J. eggNOG 5.0: A hierarchical, functionally and phylogenetically annotated orthology resource based on 5090 organisms and 2502 viruses. *Nucleic Acids Res.* **2019**, *47*, D309–D314. [CrossRef]

34. Siguier, P.; Pérochon, J.; Lestrade, L.; Mahillon, J.; Chandler, M. ISfinder: The reference centre for bacterial insertion sequences. *Nucleic Acids Res.* **2006**, *34*, D32–D36. [CrossRef] [PubMed]

35. Bertelli, C.; Laird, M.R.; Williams, K.P.; Group, S.F.U.R.C.; Lau, B.Y.; Hoad, G.; Winsor, G.L.; Brinkman, F.S. IslandViewer 4: Expanded prediction of genomic islands for larger-scale datasets. *Nucleic Acids Res.* **2017**, *45*, W30–W35. [CrossRef] [PubMed]

36. Rezaei, S.; Liu, F.; Marcuson, S.; Muinonen, M.; Lakshmanan, V.; Sridhar, R.; Barati, M. Canadian pyrrhotite treatment: The history, inventory and potential for tailings processing. *Can. Metall. Q.* **2017**, *56*, 410–417. [CrossRef]

37. Kernan, T.; Majumdar, S.; Li, X.; Guan, J.; West, A.C.; Banta, S. Engineering the iron-oxidizing chemolithoautotroph *Acidithiobacillus ferrooxidans* for biochemical production. *Biotechnol. Bioeng.* **2016**, *113*, 189–197. [CrossRef]

38. Beard, S.D.; Paradela, A.D.; Albar, J.P.D.; Jerez, C.A.D. Growth of *Acidithiobacillus ferrooxidans* ATCC 23270 in thiosulfate under oxygen-limiting conditions generates extracellular sulfur globules by means of a secreted tetrathionate hydrolase. *Front. Microbiol.* **2011**, *2*, 79. [CrossRef]

39. Shen, X.; Hu, H.; Peng, H.; Wang, W.; Zhang, X. Comparative genomic analysis of four representative plant growth-promoting rhizobacteria in *Pseudomonas*. *BMC Genom.* **2013**, *14*, 271. [CrossRef]

40. Vernikos, G.; Medini, D.; Riley, D.R.; Tettelin, H. Ten years of pan-genome analyses. *Curr. Opin. Microbiol.* **2015**, *23*, 148–154. [CrossRef]

41. Hinger, I.; Ansorge, R.; Mussmann, M.; Romano, S. Phylogenomic analyses of members of the widespread marine heterotrophic genus *Pseudovibrio* suggest distinct evolutionary trajectories and a novel genus, *Polycladidibacter* gen. nov. *Appl. Environ. Microbiol.* **2020**, *86*, 86. [CrossRef]

42. Mi, S.; Song, J.; Lin, J.; Che, Y.; Zheng, H.; Lin, J. Complete genome of *Leptospirillum ferriphilum* ML-04 provides insight into its physiology and environmental adaptation. *J. Microbiol.* **2011**, *49*, 890–901. [CrossRef]

43. Mirete, S.; Morgante, V.; González-Pastor, J.E. Acidophiles: Diversity and mechanisms of adaptation to acidic environments. In *Adaption of Microbial Life to Environmental Extremes*; Springer: Cham, Switzerland, 2017; pp. 227–251.

44. Vergara, E.; Neira, G.; González, C.; Cortez, D.; Dopson, M.; Holmes, D.S. Evolution of predicted acid resistance mechanisms in the extremely acidophilic *Leptospirillum* genus. *Genes* **2020**, *11*, 389. [CrossRef]

45. Zhang, X.; Liu, X.; Liang, Y.; Fan, F.; Zhang, X.; Yin, H. Metabolic diversity and adaptive mechanisms of iron-and/or sulfur-oxidizing autotrophic acidophiles in extremely acidic environments. *Environ. Microbiol. Rep.* **2016**, *8*, 738–751. [CrossRef] [PubMed]

46. Guiliani, N.; Jerez, C.A. Molecular Cloning, Sequencing, and Expression of *omp-40*, the gene coding for the major outer membrane protein from the acidophilic bacterium *Thiobacillus ferrooxidans*. *Appl. Environ. Microbiol.* **2000**, *66*, 2318–2324. [CrossRef] [PubMed]

47. Belin, B.J.; Busset, N.; Giraud, E.; Molinaro, A.; Silipo, A.; Newman, D.K. Hopanoid lipids: From membranes to plant–bacteria interactions. *Nat. Rev. Microbiol.* **2018**, *16*, 304. [CrossRef] [PubMed]
48. Nanatani, K.; Shijuku, T.; Takano, Y.; Zulkifli, L.; Yamazaki, T.; Tominaga, A.; Souma, S.; Onai, K.; Morishita, M.; Ishiura, M. Comparative analysis of kdp and ktr mutants reveals distinct roles of the potassium transporters in the model cyanobacterium *Synechocystis* sp. strain PCC 6803. *J. Bacteriol.* **2015**, *197*, 676–687. [CrossRef] [PubMed]
49. Nies, D.H. Efflux-mediated heavy metal resistance in prokaryotes. *FEMS Microbiol. Rev.* **2003**, *27*, 313–339. [CrossRef]
50. Tang, Y.-C.; Amon, A. Gene copy-number alterations: A cost-benefit analysis. *Cell* **2013**, *152*, 394–405. [CrossRef]
51. Shapiro, J. *Mobile Genetic Elements*; Elsevier: Amsterdam, The Netherlands, 2012.
52. Couvin, D.; Bernheim, A.; Toffano-Nioche, C.; Touchon, M.; Michalik, J.; Néron, B.; Rocha, E.P.; Vergnaud, G.; Gautheret, D.; Pourcel, C. CRISPRCasFinder, an update of CRISRFinder, includes a portable version, enhanced performance and integrates search for Cas proteins. *Nucleic Acids Res.* **2018**, *46*, W246–W251. [CrossRef]
53. Makarova, K.S.; Haft, D.H.; Barrangou, R.; Brouns, S.J.; Charpentier, E.; Horvath, P.; Moineau, S.; Mojica, F.J.; Wolf, Y.I.; Yakunin, A.F. Evolution and classification of the CRISPR–Cas systems. *Nat. Rev. Microbiol.* **2011**, *9*, 467. [CrossRef]

Article

The Role of Mineral Assemblages in The Environmental Impact of Cu-Sulfide Deposits: A Case Study from Norway

Yulia Mun [1],*, Sabina Strmić Palinkaš [1] and Kåre Kullerud [2]

1 Department of Geosciences, UiT The Arctic University of Norway, Hansine Hansens veg 18, 9019 Tromsø, Norway; sabina.s.palinkas@uit.no
2 Norwegian Mining Museum, Hyttegata 3, 3616 Kongsberg, Norway; kk@bvm.no
* Correspondence: Yulia.mun@uit.no; Tel.: +47-483-30-892

Abstract: Metallic mineral deposits represent natural geochemical anomalies of economically valuable commodities but, at the same time, their weathering may have negative environmental implications. Cu-sulfide mineral deposits have been recognized as deposits with a particularly large environmental footprint. However, different Cu deposits may result in significantly different environmental impacts, mostly depending on weathering conditions, but also on geological characteristics (mineralogy, geochemistry, host-rock lithology) of the Cu mineralization. This study presents new mineral and geochemical data from the Repparfjord Tectonic Window sediment-hosted Cu deposits and the Caledonian volcanogenic massive sulfides (VMS) deposits. The deposits share similar mineral features, with chalcopyrite and bornite as the main ore minerals, but they differ according to their trace element composition, gangue mineralogy, and host lithology. The studied sediment-hosted Cu deposits are depleted in most toxic metals and metalloids like Zn, As, Cd, and Hg, whereas the Røros Caledonian VMS mineralization brings elevated concentrations of Zn, Cd, In, Bi, As, and Cd. The conducted leaching experiments were set to simulate on-land and submarine weathering conditions. A high redox potential was confirmed as the main driving force in the destabilization of Cu-sulfides. Galvanic reactions may also contribute to the destabilization of minerals with low rest potentials, like sphalerite and pyrrhotite, even under near-neutral or slightly alkaline conditions. In addition, the presence of carbonates under near-neutral to slightly alkaline conditions may increase the reactivity of Cu sulfides and mobilize Cu, most likely as $CuCO_3$ (aq).

Keywords: Cu-sulfide ore; Nussir; Ulveryggen; Røros VMS deposit; leaching tests; submarine weathering conditions; on-land weathering conditions

Citation: Mun, Y.; Strmić Palinkaš, S.; Kullerud, K. The Role of Mineral Assemblages in The Environmental Impact of Cu-Sulfide Deposits: A Case Study from Norway. *Minerals* **2021**, *11*, 627. https://doi.org/10.3390/min11060627

Academic Editors: Carlito Tabelin, Kyoungkeun Yoo and Jining Li

Received: 31 May 2021
Accepted: 9 June 2021
Published: 12 June 2021

Publisher's Note: MDPI stays neutral with regard to jurisdictional claims in published maps and institutional affiliations.

1. Introduction

Copper is one of the most widely used mineral commodities in modern society, with a particular importance to electronics, electrical power generation, and the renewable energy sector, as well as in electric vehicle technologies [1–3]. In nature, Cu can be found in various types of mineral deposits, but in addition to the Cu-porphyry type (e.g., Chuquicamata, Chile [4]; El Teniente, Chile [5]; Ok Tedi, Papua New Guinea [6]), deposits of volcanogenic massive sulfides (VMS) and Cu-sediment hosted types represent the most important sources of Cu. Worldwide, 20 million tons (Mt) of copper was the total mine production in 2020 [7]. This number decreased from 24.5 Mt in 2019 due to COVID-19 lockdowns in April and May [8]. Chile remains the major copper producer (5.7 Mt) followed by Peru (2.2 Mt), China (1.7 Mt), DR Congo (1.3 Mt), and the US (1.2 Mt) [9]. A range from 60% to 75% of copper is mined from porphyry-copper deposits [10], 20% from sediment-hosted Cu deposits [11], and around 6% of Cu is mined from VMS deposits [12].

In this study, sediment-hosted Cu mineralization is represented by samples from the Nussir and Ulveryggen Cu deposits, from the Repparfjord Tectonic Window, while the Røros deposit, located within the Upper Allochthon of Scandinavian Caledonides, was

273

selected as representative of VMS mineralization (Figure 1). All three deposits are characterized by chalcopyrite and bornite as the main ore minerals, but they significantly differ in trace element composition, gangue mineralogy, and host lithology. The Ulveryggen Cu sediment-hosted deposit was mined in the period from 1972 to 1978/79, and tailings were deposited subaqueously in Repparfjorden. The Nussir deposit has not been mined yet, but there are plans for start-up mining of both the Nussir and Ulveryggen deposits in the near future [13]. The mine tailings from this operation are designated to be disposed of subaqueously in Repparfjorden as well. The Røros VMS deposit was mined from 1644 to 1977 [14], and similar to other historic VMS mines along the Scandinavian Caledonides, the mine waste material was disposed of on land, and still represents a significant environmental threat [15].

Mining activities may result in negative environmental impacts due to the accumulation of large quantities of mine waste material, generation of acid mine drainage (AMD), and dispersion of heavy metals in aquifers, streams, and marine sediments and soils. Copper has been recognized as a commodity with a particularly large environmental footprint (e.g., [16–18]); the environmental impact of Cu mines mostly depends on tailings disposal site conditions and the geological features of Cu mineralization, including mineral, geochemical, and host lithology characteristics (e.g., [19–22]).

AMD is a major problem associated with mineral deposits, in which Cu occurs in the form of sulfides (e.g., chalcopyrite, bornite, chalcocite, covellite) or if barren sulfides (e.g., pyrite, marcasite, pyrrhotite) represent asignificant component in the ore mineral assemblage (e.g., [23,24]). The consequences are often severe, leading to a lowering of the pH of contaminated aquifers, and release of metals and metalloids into the environment (e.g., [25–29]). Traditionally, tailings have been deposited in subaerial conditions, but several countries including Norway, practice subaqueous deposition [30]. Since subaqueous deposition in particular raises environmental concerns, we have tested the stability of representative Cu mineral assemblages from two of the most common types of Cu-sulfide deposits in Norway (sediment-hosted Cu deposits and VMS deposit) in a set of experiments that simulated subaerial and subaquatic weathering site conditions.

Kinetic leaching tests represent a powerful and relatively inexpensive tool to predict generation of AMD (e.g., [31]). They are designed to simulate sulfide-weathering processes in different physicochemical conditions. Kinetic leaching tests can be industrial or performed in the laboratory. Industrial tests are run in leaching columns, heaps, tanks, vats, dumps, large bins or drums [32]. They are placed in the field and subjected to meteoric waters, oxygen from the atmosphere, and changing temperature depending on the season. These tests can be conducted for several months to several years and are infrequently sampled for concentrations of dissolved metals and metalloids, sulfate, and changing pH and Eh parameters [33]. The tests can be accelerated by adding additional water [32].

However, more often the leaching tests are performed in miniature versions and run in laboratory size equipment—batch reactors, leaching columns, and humidity cells (e.g., [34–42]). The results are later extrapolated or mathematically modelled for larger volumes [33]. The tests are well-controlled and parameters such as water pH and Eh, metals and metalloids concentrations are continuously measured. The tests are often accelerated by increased temperature or the addition of hydrogen peroxide (e.g., [43]). The laboratory leaching tests also allow determination of an acid neutralizing capacity of gangue minerals and the acid producing potential of sulfides as well as to test remediation mechanisms, (e.g., [34,44–46]). However, many authors (e.g., [47]) argue that laboratory tests cannot be simply extrapolated to the field conditions. For example, a faster oxidation of pyrite and chalcopyrite from the Aitik site in Northern Sweden was observed in the laboratory compared to the field conditions [47].

The importance of characterization of ore parageneses and their host rocks was recognized as an important tool in the prediction of leaching of heavy metals from naturally contaminated rocks during anthropogenic activities e.g., underground constructions [31]. The tests are designed around primary ore mineralization to observe the oxidation potential of major ore minerals, and aim to extrapolate the results of the study to apply to produced mine tailings. The reactivity of the tailings with the surrounding environment will be significantly higher due to particle size and an increase in surface energy.

Norway has a long shoreline, and the ore-bearing rocks are often subjected to weathering by seawater. In addition, Norway is one of few countries where subaquatic mine tailings deposition is permitted. In both cases, it is important to understand the role of salinity on weathering of sulfides. Therefore, during the experiments seawater was used together with meteoric water.

The aim of this study is to evaluate the potential environmental impact of the studied Cu mineralization, considering geological characteristics including mineralogy, geochemistry, as well as the main physicochemical features of subaerial and subaquatic disposal sites.

2. Geological Settings

2.1. Sediment-Hosted Cu Deposits of Nussir and Ulveryggen, Repparfjord Tectonic Window

The Repparfjord Tectonic Window, Northern Norway, is composed of mafic metavolcanics and carbonate-siliciclastic sequences that were compressed in a SE-NW direction during the Svecofennian Orogeny at ca. 1.84 Ga [48,49]. The rocks are metamorphosed under greenschist to lower amphibolite facies conditions, and [50] determined the age of host volcanics to be about 2.1 Ga, with Nussir mineralization around 1.765 Ga. The Repparfjord Tectonic Window contains numerous sites with Cu mineralization (e.g., [14,51]), of which the Nussir (26.7 Mt at 1.13% Cu) and the Ulveryggen (7.7 Mt at 0.81% Cu) deposits have the greatest implications for the local environment [14,52]. The Nussir deposit is hosted by a thin (no more than 5 m thick) metadolostone layer that can be traced for several kilometres (Figure 1A,C), intercalated with metasandstone, metasiltsone and metapelites. The metasedimentary complex is overlain by a several hundred meters thick metavolcanic sequence [49,50,53]. Despite the close geographical occurrence of the Nussir and Ulveryggen deposits, they have different lithologies. The Ulveryggen deposit is hosted by arkosic metasandstones, metasiltstones, and metaconglomerates with low carbonate content. Mineralization can be traced for more than 1 km in the northeast direction and is structurally confined to tight folds [48,49,54,55]. Chalcopyrite, bornite, sphalerite, and minor pyrite are the main ore minerals at both deposits.

The Nussir mineralization is confined to a thin (up to 5 m thick) dolomitic marble layer intercalated with metasiltstone, metapelites and metasandstones, localizing the mineralization within quartz-carbonate veins as well as disseminated in mafic metavolcanics [49,50,54,55]. The major ore minerals at the Nussir and Ulveryggen deposits are chalcopyrite, bornite, and chalcocite [49,50,54].

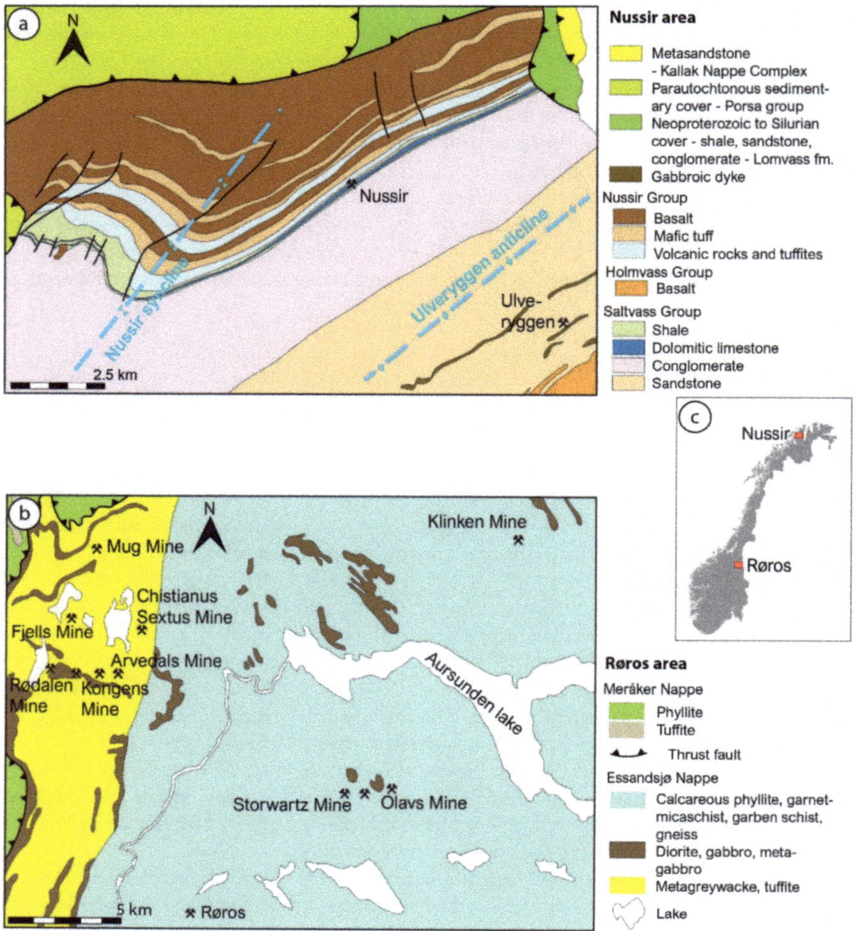

Figure 1. Geological maps showing the locations of the mines in (**A**) the Repparfjord Tectonic Window: Nussir and Ulveryggen sediment-hosted Cu deposits, and (**B**) the Røros area, drawn based on interactive online maps at [56] as well as modified after [49,55]. The map of Norway in (**C**) shows the locations of the Nussir, and Røros areas.

2.2. Volcanogenic Massive Sulfide (VMS) Deposit Røros, the Upper Allochthon of Scandinavian Caledonides

The Røros VMS mining area includes several hundreds of mineralizations in the southeastern part of the Upper Allochthon of the Scandinavian Caledonides [57], including the Røros deposit studied herein. The mineralizations are characterized by predominately chalcopyrite, sphalerite, pyrite, and galena hosted by interbedded metatuffite, metagraywacke, and gabbroic sills and dykes [14,58,59].

The Upper Allochthon of Scandinavian Caledonides extends for about 1500 km, from the Stavanger region in southern Norway to the Barents Sea region in northern Norway. This first-order tectonostratigraphic unit is dominated by sedimentary and magmatic rocks derived from the Iapetus Ocean, including ophiolite and island-arc complexes usually associated with VMS type mineralization (Figure 1B,C; [60–62]).

The Røros mining area in Trøndelag County, south-eastern Norway, hosts numerous occurrences of the VMS type (Figure 1B). Among the largest occurrences are found at the Storwartz and Olav mines in the eastern sector, and the Kongens mine in the north-western sector (Figure 1), with average Cu and Zn contents of about 2.7% and 4.2–5%, respectively [59]. The mineralization is hosted by metagraywacke interbedded with metatuffites, metabasalts and gabbroic sills and dykes [14,58,59], with major ore minerals: chalcopyrite, pyrrhotite, sphalerite, and pyrite.

3. Materials and Methods

For this study, representative samples of the Cu sulfide mineralization were selected from the Nussir and Ulveryggen sediment-hosted Cu deposits, and from the Røros VMS deposit. Two main types of samples were analyzed: (1) Mineral assemblages composed of ore and gangue minerals and (2) Individual Cu-sulfide minerals.

Twenty-seven polished thin sections of representative mineral assemblages were prepared at the Department of Geosciences at UiT The Arctic University of Norway. Three thin sections represented reference samples for the respective deposits. The reference samples were studied under a reflective polarizing light microscope and a Zeiss Merlin Compact Field Emission Scanning Electron Microscope (FE-SEM) equipped with an energy-dispersive X-ray spectrometer (EDS) at UiT The Arctic University of Norway, to determine mineral and geochemical characteristics of the ore mineral assemblages prior to and after the leaching tests. In order to investigate the primary ore, the SEM was run in a high vacuum regime at 20 kV accelerating voltage, 20 s counting time, and with an aperture of 60 μm.

In order to simulate weathering processes corresponding to the tailings disposal site conditions, a set of leaching experiments were performed on the polished thin sections (Figure 2). The experiments were designed to simulate (Figure 3): (1) Subaquatic vs. on-land disposal site conditions; (2) Oxidative vs. reductive conditions; (3) Carbonate buffered vs. carbonate free systems; and (4) Seawater infiltrated vs. meteoric water infiltrated sites (Table 1). Each thin section was placed in an individual beaker of 400 mL with a height of 10.5 cm and diameter of 8.5 cm. The thin sections were then covered with a 4 cm thick layer of quartz sand (200 g) for the simulations of on-land conditions, whereas natural marine sediments (200 g) were used for the simulations of submarine conditions.

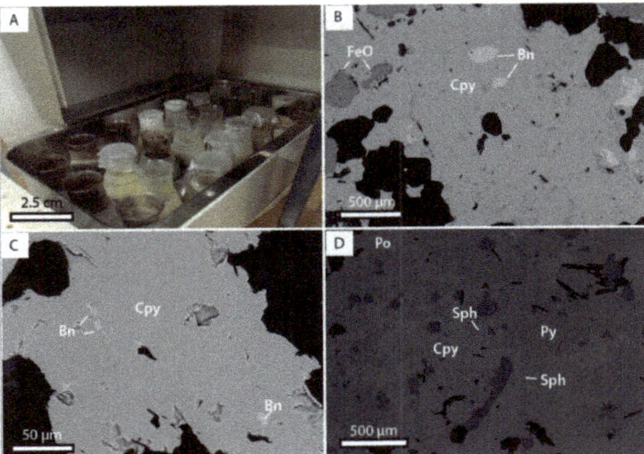

Figure 2. (**A**) Photograph of the test setup; (**B**) backscattered electron image of chalcopyrite (Cpy) in assemblage with bornite (Bn) and inclusions of iron oxides (FeO) from unaltered reference sample E-N-1 (Nussir); (**C**) backscattered electron image of chalcopyrite (Cpy) with small inclusions of bornite (Bn) from unaltered reference sample E-U-1 (Ulveryggen); (**D**) backscattered electron image of the Røros reference sample E-R-1 showing the mineral assemblage of chalcopyrite (Cpy), sphalerite (Sph), pyrite (Py), and pyrrhotite (Po).

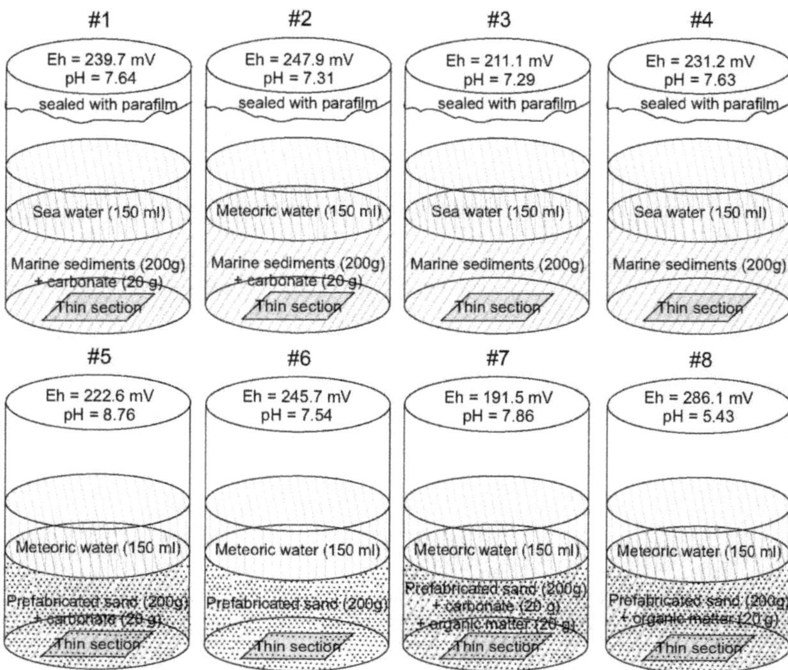

Figure 3. Schematic presentation of test settings #1–#8 conditions. Eh and pH are measured in sediments before the tests. See text for explanation.

The average organic content in the natural marine sediments was 0.82 wt.% (Supplementary Table S1), whereas the quartz sand was free of organic matter. To test the influence of redox potential, half of the beakers with quartz sand were doped with ~10 wt.% of organic matter and sealed with parafilm tape to prevent oxidation reactions. The other half of the beakers were left open during the entire experiment and refilled with circa 150 mL of meteoric water once per day to ensure oxidative conditions. Pure calcium carbonate was used to buffer relevant solutions. The experiments were run for three months, and in order to accelerate the reactions, the beakers were kept in a water bath at 50 °C.

At the end of the experiments, the samples were carefully removed and investigated under a reflective light microscope. The formation of secondary minerals was studied by Raman spectroscopy, conducted at UiT The Arctic University of Norway in Tromsø. A Renishaw inVia confocal Raman microscope equipped with a 532 nm (green) diode laser was used to identify the mineral phases in the studied ore samples, as well as the degree of weathering after simulation of weathering conditions under on-land and subaquatic conditions. The identification was based on Raman spectra published in the literature [63].

Individual grains of Cu sulfides were handpicked under a binocular microscope, washed in an ultrasonic bath and pulverized in an agate mortar. An amount of 0.5 g was analyzed for bulk trace element composition at the AcmeLabs (Vancouver, B.C. Canada), on an ICP MS instrument following the internal LF202 analysis code.

Table 1. Experimental setups. The cross mark corresponds to the ingredient added to the composition of the mixture. Sample last digits correspond to the experiment setup number.

Sample #	Condition #	Condition Description	Marine Sediment	Prefabricated Sand	Sea Water	Meteoric Water	Organic Matter	Carbonate
Env-[1]NS-1		Marine sediments, TOC = 0.82 wt.%.	+		+			+
Env-[2]Ulv-1	1	Carbonate	+		+			+
Env-[3]RS-1		buffered Seawater	+		+			+
Env-NS-2		Marine sediments, TOC = 0.82 wt.%.	+			+		+
Env-Ulv-2	2	Carbonate	+			+		+
Env-RS-2		buffered Meteoric water	+			+		+
Env-NS-3		Marine sediments,	+		+			
Env-Ulv-3	3	TOC = 0.82 wt%,	+		+			
Env-RS-3		non-buffered. Seawater	+		+			
Env-NS-4		Marine sediments,	+			+		
Env-Ulv-4	4	TOC = 0.82 wt.%,	+			+		
Env-RS-4		non-buffered. Meteoric water	+			+		
Env-NS-5		Quartz sand, TOC=0 wt.%,		+		+		+
Env-Ulv-5	5	carbonate-buffered.		+		+		+
Env-RS-5		Meteoric water		+		+		+
Env-NS-6		Quartz sand, TOC=0 wt.%,		+		+		
Env-Ulv-6	6	non-buffered. Meteoric		+		+		
Env-RS-6		water		+		+		
Env-NS-7		Quartz sand, TOC=10 wt.%,		+		+	+	+
Env-Ulv-7	7	carbonate-buffered.		+		+	+	+
Env-RS-7		Meteoric water		+		+	+	+
Env-NS-8		Quartz sand, TOC=10 wt.%,		+		+	+	
Env-Ulv-8	8	non-buffered.		+		+	+	
Env-RS-8		Meteoric water		+		+	+	

[1] NS—Nussir; [2] Ulv—Ulveryggen; [3] RS—Røros, "+"—present in the experiment

4. Results

4.1. Mineral Analyses

4.1.1. Nussir and Ulveryggen

Mineral analyses show that typical ore assemblages from the Nussir and Ulveryggen sediment hosted Cu deposits consist of chalcopyrite, bornite, and chalcocite (Figure 2B,C). The Røros VMS mineralization is represented by massive sulfide bodies predominantly composed of chalcopyrite, sphalerite, pyrite and pyrrhotite (Figure 2D).

The Nussir dolomitic marble contains rhomboidal-shaped fragments of carbonates that are up to 5 mm in diameter. The mineralization is confined to crosscutting quartz-carbonate veins with euhedral to subhedral grains of vein carbonate, which are up to 0.1–0.3 mm in diameter. Quartz grains are anhedral and about 0.1 mm in diameter. Chalcopyrite is the dominant Cu mineral together with bornite; the minerals are intergrown and contain abundant inclusions of pyrite and sphalerite (Figure 2B).

The Ulveryggen arkosic metasandstone contains fragments of quartz and feldspar up to 0.2 mm in size. The ore minerals are disseminated and they have grown interstitially between fragments of quartz and feldspar together with muscovite. The main ore minerals are bornite and chalcopyrite (Figure 2C), with minor amounts of pyrite and chalcopyrite.

4.1.2. Røros

Since the Røros samples were prepared from pieces of massive ore, only ore minerals were observed under the microscope. The main ore minerals are pyrrhotite, pyrite, chalcopyrite and sphalerite, which show various intergrowth textures. Chalcopyrite, which is the most abundant mineral, occurs as individual grains that are several centimetres in diameter. Pyrite and pyrrhotite crystals also show large grain-sizes (several centimetres in diameter), while sphalerite forms small inclusions of less than 0.1 mm in diameter (Figure 2D).

4.2. Leaching Tests

Experimental conditions #1 (Marine sediments, TOC = 0.82 wt.%; carbonate buffered; infiltrated with seawater (Figure 3); Eh_{sed} = 239.7 mV, pH_{sed} = 7.64, where Eh_{sed} and pH_{sed} are values of redox potential and pH for pore water in sediments measured after initial stabilization of conditions, i.e. 60 h after the experiment started) did not affect stability of mineral assemblages from the Nussir and Ulveryggen deposits (Figures 4A and 5A). However, for the sample from the Røros VMS deposit, small grains of sphalerite were weathered while some pyrite grains remained well-preserved (Figure 6A; Supplementary Figures S3H and S4A,B).

Experimental conditions #2 (Marine sediments, TOC = 0.82 wt.%; carbonate buffered; infiltrated with meteoric water; Eh_{sed} = 247.9 mV, pH_{sed} = 7.31; Figure 3) resulted in partial oxidation of sulfides from the Nussir and Ulveryggen deposits (Figures 4 and 5B; Supplementary Figures S1A–C and S2B). As for the experimental conditions #1, small grains of sphalerite from the Røros VMS deposit were affected (Figure 6B; Supplementary Figure S4D). Pyrite was partly oxidized along rims, while other parts remained unaltered (Figure 6B; Supplementary Figure S4C–E). Pyrrhotite was weathered significantly (Supplementary Figure S4D), and chalcocite was weathered while chalcopyrite remained unaltered (Supplementary Figure S4C,E).

Experimental conditions #3 (Marine sediments, TOC = 0.82 wt.%; no added carbonates; infiltrated with seawater; Eh_{sed} = 211.1 mV, pH_{sed} = 7.29; Figure 3) did not affect the samples from the Nussir and Ulveryggen deposits (Figures 4C and 5C; Supplementary Figure S2D), but sulfides from the Røros VMS deposit went through extensive oxidation reactions (Figure 6C; Supplementary Figure S4F,G). Pyrrhotite was oxidized significantly, however chalcopyrite crosscutting pyrrhotite remained well-preserved (Figure 6C). Pyrite was partly oxidized along the rims. Covellite was observed along cracks in the pyrite.

Experimental conditions #4 (Marine sediments, TOC = 0.82 wt.%; no added carbonates; infiltrated with meteoric water; Eh_{sed} = 231.2 mV, pH_{sed} = 7.63; Figure 3) partly affected the Nussir and Ulveryggen samples: fine-grained fragments were significantly oxidized (Figures 4D and 5D; Supplementary Figures S1D–F and S2E), while larger grains were partly oxidized. Chalcopyrite obtained oxidized rims while bornite remained well-preserved (Supplementary Figures S1D–F and S2E). Chalcocite from the Nussir deposit was partly oxidized, while chalcocite in contact with pyrite from Ulveryggen remained well-preserved (Supplementary Figure S2E). The Røros samples were weathered significantly (Figure 6D; Supplementary Figure S4H). Notably, bornite was weathered to a higher degree than chalcopyrite. The latter was found in veins within bornite and remained well-preserved (Supplementary Figure S4H). Oxidized rims were formed around pyrite grains, whereas pyrrhotite and sphalerite had undergone significant weathering (Figure 6F).

Experimental conditions #5 (quartz sand, TOC = 0 wt.%; carbonate buffered, infiltrated with meteoric water; Eh_{sed} = 222.6 mV, pH_{sed} = 8.76; Figure 3) affected the sulfides from all three studied deposits to different degrees. Chalcopyrite and bornite from Nussir were extensively weathered (Figure 4E; Supplementary Figure S1G), whereas for the Ulveryggen sample, chalcopyrite and chalcocite were partly weathered while most of the pyrite and bornite remained well-preserved (Figure 5E; Supplementary Figures S2F–H and S3A). Sulfides from Røros were more oxidized in comparison to sulfides from Nussir and Ulveryggen. In contrast to pyrite and chalcopyrite, which remained well-preserved or only partly oxidized, pyrrhotite and sphalerite were significantly weathered (Figure 6E; Supplementary Figure S5A–C).

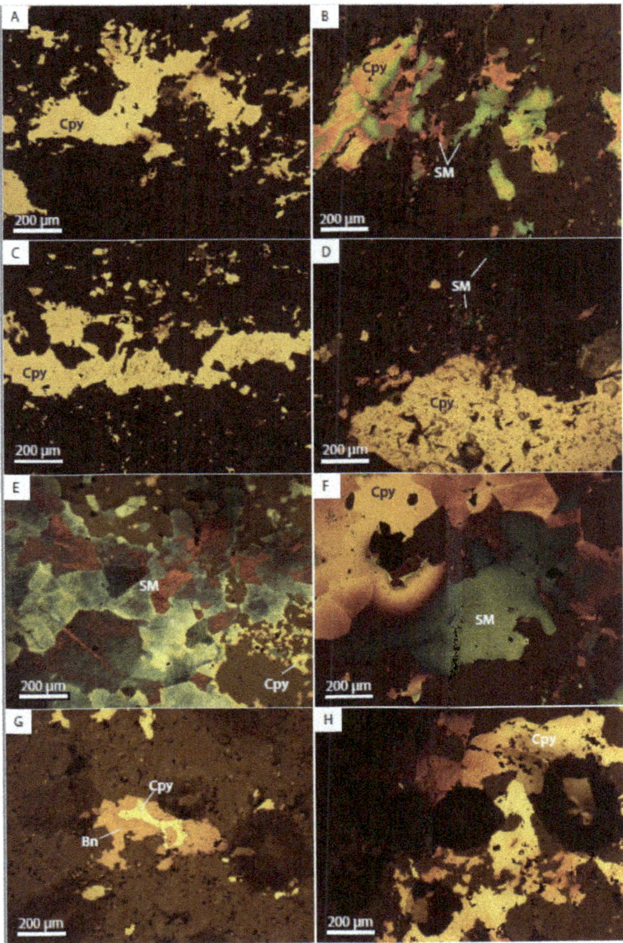

Figure 4. Microphotographs of samples under the reflected light microscope from the Nussir test runs after 90 days of the weathering experiment. The photographs are taken in crossed polars and correspond to test setups from 1 to 8 in Table 1. (**A**) reductive condition #1: well-preserved chalcopyrite (Cpy) grain (sample ENV-NS-1); (**B**) reductive condition #2: secondary minerals (SM) on top of chalcopyrite (Cpy) (sample ENV-NS-2d); (**C**) reductive condition #3: well-preserved chalcopyrite, secondary minerals are possibly forming in the cavities in the grain (Sample ENV-NS-3); (**D**) reductive condition #4: secondary minerals formed on the surface of chalcopyrite (Cpy, sample ENV-NS-4); (**E**) oxidative condition: secondary minerals on the surface of Cpy (ENV-NS-5); (**F**) oxidative condition #6: Cpy grain is partly oxidized (ENV-NS-6); (**G**) oxidative condition #7: well-preserved assemblage of Cpy and bornite (Bn) (ENV-NS-7); (**H**) oxidative condition #8: intensively oxidized chalcopyrite grain (#ENV-NS-8).

Experimental conditions #6 (quartz sand, TOC = 0 wt.%; no added carbonates; infiltrated with meteoric water; Eh_{sed} = 245.7 mV, pH_{sed} = 7.54; Figure 3) resulted in extensive oxidation of the sulfides from the Nussir and Ulveryggen deposits. Chalcopyrite, bornite and chalcocite obtained a weathered rim around the grains (Figures 4F and 5F; Supplementary Figures S1H and S3B,C). Some grains of chalcopyrite were entirely covered with a thin film of weathering products (Figure 4F). Pyrite from Ulveryggen remained well-preserved, but might have accelerated the oxidation of bornite (Supplementary Figure S3C). Pyrrhotite and sphalerite from the Røros deposit were significantly weathered, whereas pyrite remained relatively well-preserved with insignificant formation of iron oxides (Figure 6F; Supplementary Figure S5D). A weathered Cu-containing mineral was also observed within the pyrite (Supplementary Figure S5D).

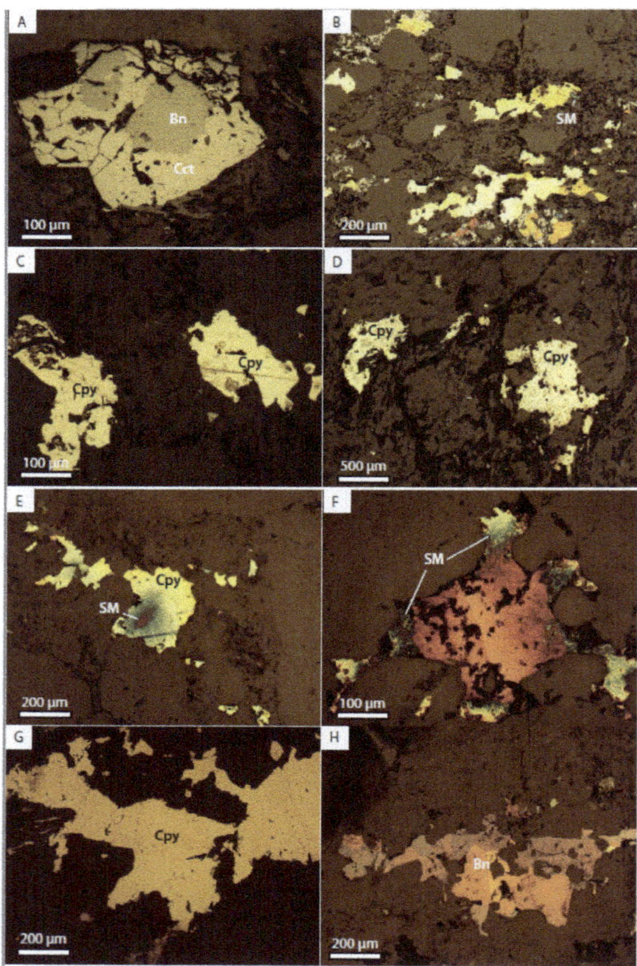

Figure 5. Microphotographs of Ulveryggen samples after 90 days of experiments. The photographs are taken under a reflected light microscope, with crossed polarizers. Setup description is given in Table 1, (**A–H**) microphotographs correspond to #1–8 conditions. (**A**) Intergrowth of chalcocite (Cct) with bornite (Bn, ENV-Ulv-1); (**B**) well-preserved chalcopyrite (Cpy) with minor oxidation of fine grains (ENV-Ulv-2); (**C**) well-preserved chalcopyrite grains (ENV-Ulv-3); (**D**) unaltered chalcopyrite grains (ENV-Ulv-4); (**E**) partly oxidized chalcopyrite grain with secondary mineral and oxidation cover on the surface; (**F**) oxidized Cu sulfide, likely chalcopyrite with secondary minerals formed on the lateral parts; (**G**) well-preserved chalcopyrite grain; and (**H**) micro-assemblage of well-preserved bornite (Bn) with chalcopyrite.

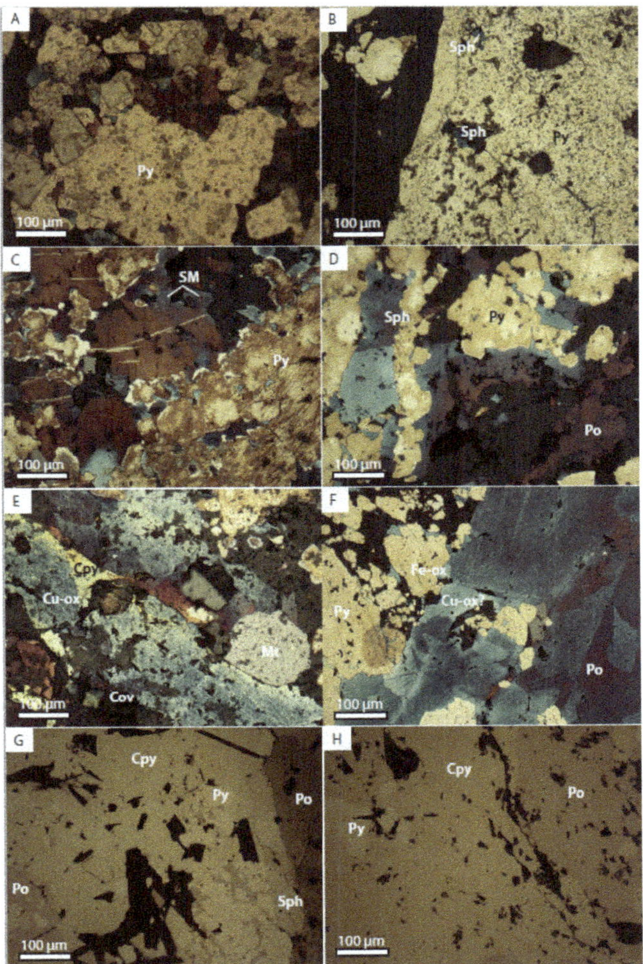

Figure 6. Microphotographs of polished thin sections from the Røros deposit after the 90-day experiment. The photographs (**A–H**) are taken under crossed poles, reflected light microscope and correspond to test conditions #1–8 (Table 1). (**A,B**) well-preserved pyrite grains (Py) with partly oxidized sphalerite (Sph) triangles (blue); (**C**) partly oxidized pyrite with secondary minerals formed on the rim of Cu-sulfides; (**D**) oxidation of sulfide mineral assemblages: pyrite (Py), sphalerite (Sph), pyrrhotite (Po), and secondary minerals likely formed on Cu-sulfides; (**E**) Intensive oxidation and the formation of secondary minerals (Cu-ox) of chalcopyrite (Cpy); (**F**) relatively well-preserved pyrite and intensely oxidized sphalerite, with minor amount of iron oxides (Fe-ox); (**G,H**) micro-assemblages of well-preserved sphalerite (Sph), chalcopyrite (Cpy), pyrite (Py), and pyrrhotite (Po).

Experimental conditions #7 (quartz sand, TOC $\approx$ 10 wt.%, carbonate buffered, infiltrated with meteoric water; initial Eh_{sed} = 191.5 mV, pH_{sed} = 7.86; Figure 3) did not affect the Nussir and Ulveryggen sulfides. The minerals remained well-preserved (Figure 4G and Figure 5G; Supplementary Figure S3D,E). The Røros ore minerals also remained almost unaffected, however some pyrite grains were slightly tarnished with bright blue secondary covellite and brownish iron hydroxides (Figure 6G; Supplementary Figure S5E,F).

Experimental conditions #8 (quartz sand, TOC ≈ 10 wt.%, no added carbonates, infiltrated with meteoric water; Eh_{sed} = 286.1 mV, pH_{sed} = 5.43; Figure 3) resulted in well-preserved sulfides from the Ulveryggen and Røros deposits (Figures 5H and 6H; Supplementary Figures S3F,G and S5G). Chalcopyrite from Nussir was observed both as relatively well-preserved grains (Figure 4H) and significantly weathered grains (Supplementary Figure S2B).

4.3. Raman Spectroscopy

Raman spectra (Figure 7) obtained from chalcopyrite and pyrite from the Nussir, Ulveryggen, and Røros deposits suggest formation of new peaks 450 and 500 cm^{-1} after exposing samples from the Nussir and Ulveryggen to experimental conditions #2 and #5, respectively (Figure 7A–D).

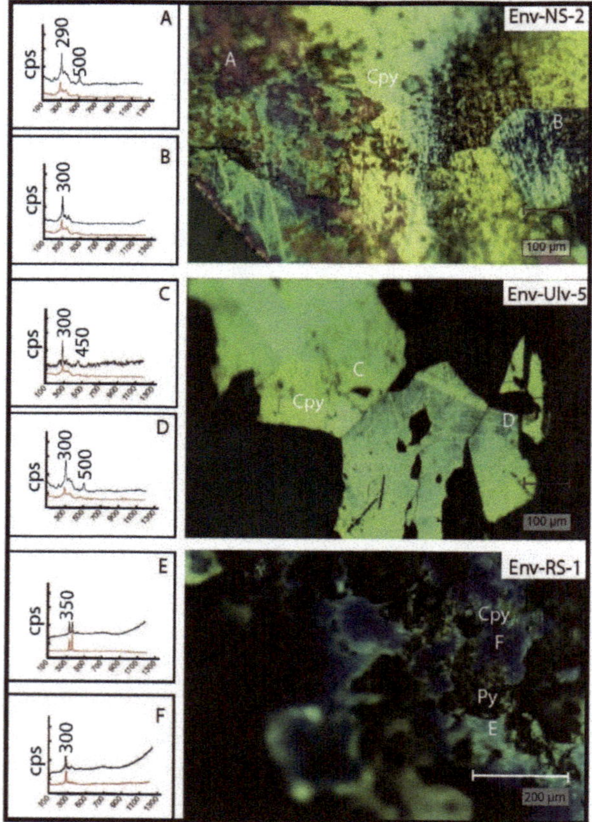

Figure 7. Raman spectrometry of chalcopyrite (Cpy) and pyrite (Py) from Nussir, Ulveryggen and Røros after 90-days of testing in selected conditions. Sample numbers correspond to condition numbers and can be found in Table 1. Cps—counts per second. **A–E**: right images are microphotographs of samples under reflected light microscope; left diagrams relate to Raman spot analyses. **A,B**—chalcopyrite (Cpy) from the Nussir deposit (sample Env-NS-2, experimental condition #2); **C,D**—chalcopyrite (Cpy) from the Ulveryggen deposit (sample Env-Ulv-5, experimental condition #5); **E**—pyrite (Py) and **F**—chalcopyrite (Cpy) from the Røros deposit (sample Env-RS-1, experimental condition #1).

4.4. Mineral Chemistry

In addition, to determine the concentration of potentially toxic elements including Cu, Zn, Ni, Hg, Cd, and As (e.g., [64]), the bulk chemical compositions of hand-picked Nussir, Røros chalcopyrite as well as Ulveryggen bornite were analyzed (Table 2). The Nussir and Ulveryggen results are taken from [53].

Table 2. Lithogeochemistry of hand-picked chalcopyrite from the Nussir deposit (NS-35-ccp), bornite from the Ulveryggen deposit (Ulv-2-bn) and chalcopyrite Røros Mine (RSL-ccp).

Element, ppm	LD	[1] NS-35-ccp	[2] Ulv-2-bn	[3] RSL-ccp
As	5	LLD	LLD	226.1
Ba	3	12	527	18
Bi	0.1	0.3	0.1	9.9
Cd	0.1	LLD	LLD	30
Co	1	85	2	535.9
Cu	10	>10,000	>10,000	>10,000
Hg	0.01	LLD	LLD	4.59
Mo	2	LLD	5	6.5
Ni	0.1	100	LLD	7.8
Pb	5	LLD	LLD	196.1
Rb	1	LLD	24	2.5
Sb	0.2	LLD	LLD	1.8
Sn	1	LLD	LLD	8
Tl	0.05	3.69	0.96	2
V	5	5	27	LLD
W	0.5	0.6	LLD	0.7
Zn	30	310	LLD	2942

[1] NS—Nussir, [2] Ulv—Ulveryggen, [3] RSL—Røros; Bn—bornite, ccp—chalcopyrite, LD—limit of detection, LLD—lower than limit of detection, ND—no data.

4.4.1. Nussir and Ulveryggen

The Nussir chalcopyrite contains 100 ppm of Ni, 85 ppm of Co and 310 ppm of Zn. The content of Bi is 0.3 ppm, while As, Mo, Cd, Sb, Pb, and Hg contents are minor or below the detection limit. The Nussir chalcopyrite also contains Ba (12 ppm), most likely in the form of nearly insoluble inclusions of barite.

Ulveryggen bornite contains 2 ppm of Co and 5 ppm of Mo. Nickel, Zn, As, Sn, Sb, Cd, Pb, and Hg contents are minor or below the detection limit, while the Rb content is 24 ppm. The content of Ba in the Ulveryggen bornite is 527 ppm.

4.4.2. Røros

Chalcopyrite was picked from the crushed Røros sample. The Co content is 535.9 ppm. The chalcopyrite also contains: Ni 7.8 ppm and Zn 2942 ppm. As content is 226.1 ppm, and Sn and Sb contents are 8 and 1.8 ppm respectively. Chalcopyrite contains 30 ppm of Cd and 18 ppm of Ba. The Bi content is 9.9 ppm and Hg content is 4.59 ppm.

5. Discussion

The Cu mineralization found in the Nussir and Ulveryggen sediment-hosted Cu deposits is characterized by predomination of chalcopyrite, bornite and chalcocite. Mine tailings from these Cu-sulfide deposits are associated with a high risk for the generation of acid mine drainage (AMD) because of the high Fe^{2+}/Fe^{3+} and S^{2-}/SO_4^{2-} ratios in their mineral assemblages. Furthermore, the Ulveryggen deposit has a low carbonate content that additionally increases the risk. In contrast, a low content of potentially toxic elements such as As, Cd, Hg, and Zn in both, reduces their environmental threat [53].

The Røros Cu-Zn VMS deposit, similar to other VMS deposits worldwide [65,66], is characterized by a polymetallic composition (Table 2). The main ore minerals are sulfides, including chalcopyrite, bornite, pyrite, and pyrrhotite. High Fe^{2+} and S^{2-} contents together with an absence of carbonates from its mineral assemblages point to a high risk for the generation of AMD in this deposit (e.g., [23,27,46,67,68]). In addition, the enrichment in a wide spectrum of potentially toxic metals and metalloids (e.g., As, Bi, Cd, In, and Zn) magnifies the environmental risk associated with mining activities and/or processes of natural weathering in this type of ore deposits (e.g., Rio Tinto VMS deposits, the Iberian Pyrite Belt in Spain [26,28,69–71] and the Britannia Creek VMS deposit, Canada [48]).

Leaching tests are recognized as powerful tools to predict the behavior of sulfides in different conditions (e.g., [20,32,37,39,40]). Such tests play an important role in the initial phases of mining planning, while deciding the potential placement of tailings for future storage. The tests are usually performed in batch reactors specially equipped with sensors controlling temperature, pH, and the amount of dissolved oxygen (e.g., [39,47,72]). Otherwise, the tests can be performed in leaching columns (e.g., [32,73]) or in static conditions (e.g., [46,74]) for periods of several days to several years. The leaching experiments in this study were designed to test the stability of ore mineral parageneses from three different Cu deposits under diverse physicochemical conditions (Figure 3, Table 1), and predict the behavior of ore-bearing mineral assemblages disposed in on-land and submarine conditions. As expected, the high redox potential was the main driving factor in destabilization of sulfides (e.g., [75]):

$$4\,FeS_2 + 15\,O_2 + 14\,H_2O \leftrightarrow 4\,Fe(OH)_3 + 16\,H{+} + 8\,SO_4{}^{2-} \tag{1}$$

$$4\,FeS + 10\,O_2 + 9\,H_2O \leftrightarrow 4\,Fe(OH)_3 + 8\,H{+} + 4\,SO_4{}^{2-} \tag{2}$$

$$4\,CuFeS_2 + 15\,O_2 + 14\,H_2O \leftrightarrow 4\,Cu^{2+} + 4\,Fe(OH)_3 + 16\,H^+ + 8\,SO_4{}^{2-} \tag{3}$$

The organic matter content of the natural marine sediments that were used (0.82 wt.%) was sufficient to prevent oxidation of the sulfides in experimental conditions in which the sulfide parageneses were exposed to seawater. In the setups with meteoric water, a different scenario was observed. A lower solubility of oxygen in seawater (4.6 mg/L at 50 °C) compared to meteoric water (5.6 mg/L at 50 °C) was probably one of the controlling factors [76]. When sediments were doped with an additional 10 wt.% of organic matter, the differences between seawater and meteoric water influence were not recorded.

The carbonate buffered conditions were mostly less reactive due to a slightly alkaline pH value of the infiltrating aqueous solutions. However, the samples from the Røros VMS deposit revealed that sphalerite can enter galvanic reactions and be extensively dissolved even in alkaline or near-neutral conditions. The prerequisite is that sphalerite (-0.24 V; [77]) occurs in direct contact with sulfides with greater rest potentials like pyrite (0.63 V; [78]) or chalcopyrite (0.54 V; [79]). In interactions with oxygen-rich solutions, sphalerite will act as an anode:

$$ZnS \leftrightarrow Zn^{2+} + S^0 + 2e^- \tag{4}$$

and prevent oxidation of pyrite, and Cu-sulfides reacting with oxygen adsorbed on their surface:

$$\frac{1}{2}\,O_2\,(aq) + 2H^+ + 2e^- \leftrightarrow H_2O \tag{5}$$

This reaction will not affect the pH of the aquifer, but it will promote leaching of Zn.

The galvanic reaction may represent a particularly large environmental issue in mineral deposits in which pyrrhotite (-0.24 V; [77]) is intergrown or occurs in direct contact with Cu-sulfides and/or pyrite:

$$FeS \leftrightarrow Fe^{2+} + S^0 + 2e^- \tag{6a}$$

$$Fe^{2+} + 3H_2O \leftrightarrow Fe(OH)_3 + 3H^+ + e^- \tag{6b}$$

The anode reaction will result in dissolution of pyrrhotite, oxidation of ferrous to ferric ions, and consequently in acidification of the system.

In the experimental setups buffered with carbonates, the Cu sulfides showed an increased reactivity (Figure 4B,E, Figure 5B,E and Figure 6E) whereas pyrite did not show significant changes under these conditions (Figure 6B). Carbonates are often used for prevention of acid mine drainage (e.g., [80]), but the results of this study show that in near-neutral to slightly alkaline conditions Cu can be mobilized from sulfides, most likely in the form of $CuCO_3$ (aq) [53,81]. This reaction is more intensive in solutions with a higher redox potential (Figures 4E, 5E and 6E).

In addition, gangue mineralogy might also play a role in the rate and degree of oxidation. The Nussir mineralogy is hosted by dolomitic marble, while Ulveryggen Cu sulfides are hosted by arkosic sandstone, which retards the oxidation of sulfides [53]. Røros sulfides, studied here, were sampled from a massive ore with an absence of gangue minerals. Therefore, galvanic interaction was favoured to higher degree of oxidation and was not prevented or retarded by gangue mineralogy. Weathering in mine tailings will be accelerated due to higher surface area, however this is also true for gangue mineralogy which in the case of Nussir and Ulveryggen might play a buffering role. On the other hand Røros mineralization being hosted by massive mafic volcanic rocks, which are extremely soluble, will most probably not retard the oxidation reaction or buffer it to limited degree.

Raman spectroscopy confirmed the slight distortion of crystal lattices of pyrite and chalcopyrite (Figure 7), however, the signal was low. Optically, even when the blue tinge was observed in the Nussir and Ulveryggen samples, Raman analyses did not record any changes in crystal structure of sulfides from these two deposits. This is attributed to oxidation occurring in a thin layer of secondary minerals, which tends to be amorphous and is characterized by the absence of a detectable lattice. In addition, the signal from primary minerals is significantly higher.

6. Conclusions

The mineral assemblages from all three studied deposits point to a high risk for generation of acid mine drainage due their high Fe^{2+}/Fe^{3+} and S^{2-}/SO_4^{2-} ratios. However, different ore-forming conditions of sediment hosted Cu deposits (Nussir and Ulveryggen) from conditions related to the formation of VMS deposits (Røros) resulted in contrasting behavior of the trace elements. As a consequence, the Nussir and Ulveryggen deposits are depleted in most potentially toxic elements, such as As, Cd, Hg, and Zn, whereas the Røros VMS mineralization is enriched in a wide spectrum of potentially toxic metals and metalloids, including As, Bi, Cd, In, and Zn.

The leaching experiments revealed the redox potential as the main factor that controls the stabilities of Cu-sulfides for both on-land as well as submarine conditions. Galvanic reactions may contribute to the destabilization of minerals with low rest potentials, like sphalerite and pyrrhotite, even under near-neutral or slightly alkaline conditions. The destabilization of pyrrhotite can have particularly negative environmental consequences due to the release of ferrous ions to an aquifer and acidification of the system as a result of oxidation of ferrous to ferric ions.

Although carbonates are often used for prevention of acid mine drainage, the presence of carbonates under near-neutral to slightly alkaline conditions may increase the reactivity of Cu sulfides and mobilize Cu, most likely in the form of $CuCO_3$ (aq).

More complex ore minerals assemblages lead to deeper weathering in given conditions. Thus, Røros chalcopyrite were notably more altered than chalcopyrite from the Ulveryggen and Nussir deposits.

Supplementary Materials: The following are available online at https://www.mdpi.com/article/10.3390/min11060627/s1, Figure S1: Microphotographs demonstrating the Nussir sulfides reaction after 90-day tests; Figure S2: Microphotographs taken under a reflected light microscope; Figure S3: Microphotographs taken under reflected light; Figure S4: Microphotographs of Røros samples under reflected light; Figure S5: Microphotographs of Røros samples under the reflected light microscope after 90-days of experimental tests; Table S1: Total organic carbon (TOC) of gravity core HH-12-002-MF-GC obtained from Repparfjord.

Author Contributions: Conceptualization, Y.M., S.S.P.; methodology, Y.M. and S.S.P.; validation, Y.M. and S.S.P.; formal analysis, Y.M. and S.S.P.; investigation, Y.M. and S.S.P.; resources, Y.M., S.S.P. and K.K.; data curation, Y.M., S.S.P. and K.K.; writing—original draft preparation, Y.M.; writing—review and editing, Y.M., S.S.P. and K.K.; visualization, Y.M. and K.K.; supervision, S.S.P. and K.K.; project administration, S.S.P. and K.K.; funding acquisition, K.K. All authors have read and agreed to the published version of the manuscript.

Funding: This research was funded by Tromsø Fylkes Kommune and SINTEF through the PhD project of the first author, grant number RDA12/167. The publication charges for this article have been funded by a grant from the publication fund of UiT The Arctic University of Norway.

Data Availability Statement: Not applicable.

Acknowledgments: We are thankful to Carlito Tabelin, the Academic Editor of the special issue "Novel and Emerging Strategies for Sustainable Mine Tailings and Acid Mine Drainage Management" for constructive comments and suggestions that significantly improved the text. The authors are thankful to Kai Neufeld for the assistance with the SEM work. The laboratory staff of the Department of Geosciences, UiT, Trine Merete Dahl, Karina Monsen, Ingvild Hald are acknowledged for the efficient work and assistance in samples preparation and TOC analysis. Nussir ASA and in particular Øystein Rushfeldt is thanked for providing this research with study material. The authors thank Calvin Shackleton for correction of text.

Conflicts of Interest: The authors declare no conflict of interest.

References

1. Gordon, R.B.; Bertram, M.; Graedel, T.E. Metal Stocks and Sustainability. *Proc. Natl. Acad. Sci. USA* **2006**, *103*, 1209–1214. [CrossRef]
2. Schlesinger, M.E.; King, M.J.; Sole, K.C.; Davenport, W.G. *Extractive Metallurgy of Copper*, 5th ed.; Elsevier: Oxford, UK, 2011; p. 472.
3. Elshkaki, A.; Graedel, T.E.; Ciacci, L.; Reck, B.K. Copper demand, supply, and associated energy use to 2050. *Global Environ. Chang.* **2016**, *39*, 305–315. [CrossRef]
4. Ossandón, C.G.; Fréraut, C.R.; Gustafson, L.B.; Lindsay, D.D.; Zentilli, M. Geology of the Chuquicamata mine: A progress report. *Econ. Geol.* **2001**, *96*, 249–270. [CrossRef]
5. Cannell, J.; Cooke, D.R.; Walshe, J.L.; Stein, H. Geology, mineralization, alteration, and structural evolution of the El Teniente porphyry Cu-Mo deposit. *Econ. Geol.* **2005**, *100*, 979–1003. [CrossRef]
6. Large, S.J.; Quadt, A.V.; Wotzlaw, J.F.; Guillong, M.; Heinrich, C.A. Magma evolution leading to porphyry Au-Cu mineralization at the Ok Tedi deposit, Papua New Guinea: Trace element geochemistry and high-precision geochronology of igneous zircon. *Econ. Geol.* **2018**, *113*, 39–61. [CrossRef]
7. Garside, M. Copper–Statistics & Facts. Available online: https://www.statista.com/topics/1409/copper/ (accessed on 7 June 2021).
8. Flanagan, D.M. *Mineral Commodity Summaries*; U.S. Geological Survey: Reston, VA, USA, 2021.
9. Garside, M. Major Countries in Copper Mine Production Worldwide from 2010 to 2020. Available online: https://www.statista.com/statistics/264626/copper-production-by-country/ (accessed on 7 June 2021).
10. Mudd, G.M.; Jowitt, S.M. Growing global copper resources, reserves and production: Discovery is not the only control on supply. *Econ. Geol.* **2018**, *113*, 1235–1267. [CrossRef]
11. Pietrzyk, S.; Tora, B. Trends in global copper mining—A review. *Miner. Eng. Conf.* **2018**, *427*, 012002. [CrossRef]
12. Lepan, N. Everything You Need to Know about VMS Deposits. *Vis. Capital.* **2019**. Available online: https://www.mining.com/web/everything-need-know-vms-deposits/ (accessed on 7 June 2021).
13. Nussir ASA. Available online: www.nussir.no (accessed on 30 November 2020).
14. Sandstad, J.S.; Bjerkgård, T.; Boyd, R.; Ihlen, P.; Korneliussen, A.; Nilsson, L.P.; Often, M.; Eilu, P.; Hallberg, A. Metallogenic areas in Norway. In *Mineral Deposits and Metallogeny of Fennoscandia*; Geological Survey of Finland: Espoo, Finland, 2012; Volume 53, pp. 35–138.
15. Banks, D.; Younger, P.L.; Arnesen, R.T.; Iversen, E.R.; Banks, S.B. Mine-water chemistry: The good, the bad and the ugly. *Environ. Geol.* **1997**, *32*, 157–174. [CrossRef]
16. Dudka, S.; Adriano, D.C. Environmental impacts of metal ore mining and processing: A review. *J. Environ. Qual.* **1997**, *26*, 590–602. [CrossRef]
17. Bridge, G. The social regulation of resource access and environmental impact: Production, nature and contradiction in the US copper industry. *Geoforum* **2000**, *31*, 237–256. [CrossRef]
18. Berrill, P.; Arvesen, A.; Scholz, Y.; Gils, H.C.; Hertwich, e.g., Environmental impacts of high penetration renewable energy scenarios for Europe. *Environ. Res. Lett.* **2016**, *11*, 014012. [CrossRef]
19. Seal, R.R., II; Foley, N.K. (Eds.) *Progress on Geoenvironmental Models for Selected Mineral Deposit Types*; U.S. Geological Survey: Reston, VA, USA, 2002; pp. 2002–2195.

20. Parbhakar-Fox, A.K.; Edraki, M.; Walters, S.; Bradshaw, D. Development of a textural index for the prediction of acid rock drainage. *Miner. Eng.* **2011**, *24*, 1277–1287. [CrossRef]
21. Dold, B.; Weibel, L. Biogeometallurgical pre-mining characterization of ore deposits: An approach to increase sustainability in the mining process. Environ. *Sci. Pollut. R.* **2013**, *20*, 7777–7786. [CrossRef]
22. Anawar, H.M. Sustainable rehabilitation of mining waste and acid mine drainage using geochemistry, mine type, mineralogy, texture, ore extraction and climate knowledge. *J. Environ. Manag.* **2015**, *158*, 111–121. [CrossRef]
23. Bussière, B. Acid mine drainage from abandoned mine sites: Problematic and reclamation approaches. In Proceedings of the International Symposium on Geoenvironmental Engineering, ISGE 2009, Hangzhou, China, 8 September 2009; pp. 111–125.
24. Simate, G.S.; Ndlovu, S. Acid mine drainage: Challenges and opportunities. *J. Environ. Chem. Eng.* **2014**, *2*, 1785–1803. [CrossRef]
25. Braungardt, C.B.; Achterberg, E.P.; Elbaz-Poulichet, F.; Morley, N.H. Metal geochemistry in a mine-polluted estuarine system in Spain. *Appl. Geochem.* **2003**, *18*, 1757–1771. [CrossRef]
26. Enspaña, J.S.; López Pamo, E.; Esther, S.; Aduvire, O.; Reyes, J.; Barettino, D. Acid mine drainage in the Iberian Pyrite Belt (Odiel river watershed, Huelva, SW Spain): Geochemistry, mineralogy and environmental implications. *Appl. Geochem.* **2005**, *20*, 1320–1356.
27. Akcil, A.; Koldas, S. Acid mine drainage (AMD): Causes, treatment and case studies. *J. Clean. Prod.* **2006**, *14*, 1139–1145. [CrossRef]
28. Ayora, C.; Caraballo, M.A.; Macias, F.; Rötting, T.S.; Carrera, J.; Nieto, J.-M. Acid mine drainage in the Iberian Pyrite Belt: 2. Lessons learned from recent passive remediation experiences. *Environ. Sci. Pollut. Res.* **2013**, *20*, 7837–7853. [CrossRef] [PubMed]
29. Kefeni, K.K.; Msagati, T.A.M.; Mamba, B.B. Acid mine drainage: Prevention, treatment options, and resource recovery: A review. *J. Clean. Prod.* **2007**, *151*, 475–493. [CrossRef]
30. Dold, B. Submarine tailings disposal (STD)—A review. *Minerals* **2014**, *4*, 642–666. [CrossRef]
31. Tabelin, C.B.; Igarashi, T.; Villacorte-Tabelin, M.; Park, I.; Einstin, M.O.; Ito, M.; Hiroyoshi, N. Arsenic, selenium, boron, lead, cadmium, copper, and zink in naturally contaminated rocks: A review of their sources, modes of enrichment, mechanisms of release, and mitigation strategies. *Sci. Total Environ.* **2018**, *645*, 1522–1553. [CrossRef]
32. Lottermoser, B. *Mine Wastes Characterization, Treatment and Environmental Impacts*, 3rd ed.; Springer: Berlin/Heidelberg, Germany, 2010; p. 400.
33. De Andrade Lima, L.R.P. A mathematical model for isothermal heap and column leaching. *Braz. J. Chem. Eng.* **2004**, *21*, 435–447. [CrossRef]
34. Gottschalk, V.H.; Buehler, H.A. Oxidation of sulphides. *Econ. Geol.* **1912**, *7*, 15–34. [CrossRef]
35. Cheng, C.Y.; Lawson, F. The kinetics of leaching covellite in acidic oxygenated sulphate-chloride solutions. *Hydrometallurgy* **1991**, *27*, 269–284. [CrossRef]
36. Falk, H.; Lavergren, U.; Bergbäck, B. Metal mobility in alum shale from Öland, Sweden. *J. Geochem. Explor.* **2006**, *90*, 157–165. [CrossRef]
37. Rzepka, P.; Walder, I.F.; Aagaard, P.; Bożecki, P.; Rzepa, G. Sub-sea tailings deposition leach modelling. *Geol. Geophys. Environ.* **2014**, *40*, 123–124.
38. Tabelin, C.B.; Veerawattananun, S.; Ito, M.; Hiroyoshi, N.; Igarashi, T. Pyrite oxidation in the presence of hematite and alumina: I. Batch leaching experiments and kinetic modeling calculations. *Sci. Total Environ.* **2017**, *580*, 687–698. [CrossRef] [PubMed]
39. Embile, R.F., Jr.; Walder, I.F. Galena non-oxidative dissolution kinetics in seawater. *Aquat. Geochem.* **2018**, *24*, 107–119. [CrossRef]
40. Embile, R.F., Jr.; Walder, I.F.; Schuh, C.; Donatelli, J.L. Cu, Pb and Fe release from sulphide-containing tailings in seawater: Results from laboratory simulation of submarine tailings disposal. *Mar. Pollut. Bull.* **2018**, *137*, 582–592. [CrossRef] [PubMed]
41. Huyen, D.T.; Tabelin, C.B.; Thuan, H.M.; Dang, H.D.; Truong, P.T.; Vongphuthone, B.; Kobayashi, M.; Igarashi, T. The solid-phase partitioning of arsenic in unconsolidated sediments of the Mekong Delta, Vietnam and its release under various conditions. *Chemosphere* **2019**, *233*, 512–523. [CrossRef]
42. Tomiyama, S.; Igarashi, T.; Tabelin, C.B.; Tangviroon, P. Acid mine drainage sources and hydrogeochemistry at the Yatani mine, Yamagata, Japan: A geochemical and isotopic study. *J. Contam. Hydrol.* **2019**, *225*, 103502. [CrossRef]
43. Silva, L.F.O.; Querol, X.; da Boit, K.M.; Fdez-Ortiz de Vallejuelo, S.; Madariaga, J.M. Brazilian coal mining residues and sulphide oxidation by Fenton's reaction: An accelerated weathering procedure to evaluate possible environmental impact. *J. Hazard. Mater.* **2011**, *186*, 516–525. [CrossRef] [PubMed]
44. Jambor, J.L.; Dutrizac, J.E.; Groat, L.A.; Raudsepp, M. Static tests of neutralization potentials of silicate and aluminosilicate minerals. *Environ. Geol.* **2002**, *43*, 1–17.
45. Ruan, R.; Zhou, E.; Xingyu, L.; Biao, W.; Guiying, Z.; Jiankang, W. Comparison on the leaching kinetics of chalcocite and pyrite with or without barteria. *Rare Metals.* **2010**, *29*, 552–556. [CrossRef]
46. Plante, B.; Bussière, B.; Benzaazoua, M. Static tests response on 5 Canadian hard rock mine tailings with low net acid-generating potentials. *J. Geochem. Explor.* **2012**, *114*, 57–69. [CrossRef]
47. Banwart, S.A.; Destouni, G.; Malmström, M. Assessing mine water pollution: From laboratory to field scale. In *Ground Water Quality: Remediation and Protection, Proceeding of the GQ'98 Conference*; Tübingen, Germany, 21–25 September 1998, IAHS Publications: Wallingford, UK, 1998; Volume 250, pp. 307–311.
48. Pharaoh, T.C.; MacIntyre, R.M.; Ramsay, D.M. K–Ar age determination on the Raipas suite in the Komagfjord Window, northern Norway. *Nor. Geol. Tidsskr.* **1982**, *62*, 51–57.

49. Torgersen, E.; Viola, G.; Sandstad, J.S. Revised structure and stratigraphy of the northwestern Repparfjord Tectonic Window, Northern Norway. *Norw. J. Geol.* **2015**, *95*, 397–421. [CrossRef]

50. Perelló, J.; Clifford, J.A.; Creaser, R.A.; Valencia, V.A. An example of synorogenic sediment-hosted copper Mineralization: Geologic and geochronologic evidence from the Paleoproterozoic Nussir deposit, Finnmark, Arctic Norway. *Econ. Geol.* **2015**, *110*, 677–689. [CrossRef]

51. Torgersen, E.; Viola, G.; Sandstad, J.S.; Stein, H.; Zwingmann, H.; Hannah, J. Effects of frictional–viscous oscillations and fluid flow events on the structural evolution and Re–Os pyrite–chalcopyrite systematics of Cu-rich carbonate veins in northern Norway. *Tectonophysics* **2015**, *659*, 70–90. [CrossRef]

52. Torgersen, E.; Viola, G.; Sandstad, J.S.; Stein, H. Structural constraints on the formation of Cu-rich mesothermal vein deposits in the Repparfjord Tectonic Window, northern Norway. In Proceedings of the 12th SGA Biennial Meeting, Uppsala, Sweden, 12–15 August 2013.

53. Mun, Y.; Strmić Palinkaš, S.; Forwick, M.; Junttila, J.; Pedersen, K.B.; Sternal, B.; Neufeld, K.; Tibljaš, D.; Kullerud, K. Stability of Cu-sulfides in submarine tailing disposals: A case study from Repparfjorden, northern Norway. *Minerals* **2020**, *10*, 169. [CrossRef]

54. Stribrny, B. The conglomerate-hosted Repparfjord copper ore deposit, Finnmark, Norway. In *Monograph Series on Mineral Deposits*; Gebrüder Borntraeger: Berlin, Germany; Stuttgart, Germany, 1985; Volume 24, pp. 1–75.

55. Mun, Y.; Strmić Palinkaš, S.; Kullerud, K.; Nilsen, K.S.; Neufeld, K.; Bekker, A. Evolution of metal-bearing fluids at the Nussir and Ulveryggen sediment-hosted Cu deposits, Repparfjord Tectonic Window, Northern Norway. *Norw. J. Geol.* **2020**, *100*. [CrossRef]

56. Norwegian Geological Survey. Available online: www.ngu.no (accessed on 12 October 2020).

57. Ettner, D. Passive mine water treatment in Norway. In *Water in Mining Environments*; Cidu, R., Frau, F., Eds.; IMWA Symposium: Cagliari, Italy, 2007.

58. Rui, I.J.; Bakke, I. Stratabound sulphide mineralization in the Kjøli Area, Røros District, Norwegian Caledonides. *Nor. Geol. Tidsskr.* **1975**, *55*, 51–75.

59. Bjerkgård, T.; Sandstad, J.S.; Sturt, B.A. Massive sulphide deposits in the south-eastern Trondheim region Caledonides, Norway: A review. In *Mineral Deposits: Processes to Processing*; Stanley, C.J., Ed.; Taylor & Francis: London, UK, 1999; pp. 935–938.

60. Grenne, T.; Ihlen, P.M.; Vokes, F.M. Scandinavian Caledonide metallogeny in a plate tectonic perspective. *Miner. Deposita* **1999**, *34*, 422–471. [CrossRef]

61. Gee, D.G.; Fossen, H.; Henriksen, N.; Higgins, A.K. From the early Paleozoic platforms of Baltica and Laurentia to the Caledonide Orogen of Scandinavia and Greenland. *Episodes* **2008**, *31*, 44–51. [CrossRef]

62. Corfu, F.; Andersen, T.B.; Gasser, D. The Scandinavian Caledonides: Main features, conceptual advances and critical questions. *Geol. Soc. Lond. Spec. Publ.* **2014**, *390*, 9–43. [CrossRef]

63. RRUF Project. Available online: https://rruff.info/ (accessed on 2 February 2021).

64. Miljødirektoratet. *Quality Standards for Water, Sediment and Biota*; M-608; Norwegian Environment Agency (NEA): Oslo, Norway, 2016; p. 24. (In Norwegian)

65. Barrie, C.T.; Hannington, M.D. Classification of volcanic-associated massive sulfide deposits based on host-rock composition. In *Volcanic-Associated Massive Sulfide Deposits: Processes and Examples in Modern and Ancient Settings*; Barrie, C.T., Ed.; Society of Economic Geologists: Littleton, CO, USA, 1999; pp. 1–11.

66. Galley, A.G.; Hannington, M.D.; Jonasson, I.R. Volcanogenic massive sulphide deposits. In *Mineral Deposits of Canada: A Synthesis of Major Deposit-Types, District Metallogeny, the Evolution of Geological Provinces, and Exploration Methods*; Goodfellow, W.D., Ed.; Geological Association of Canada, Mineral Deposits Division, Special Publication: St. John's, NL, Canada, 2007; Volume 5, pp. 141–161.

67. Nordstrom, D.K.; Alpers, C.N. Geochemistry of acid mine waters. In *The Environmental Geochemistry of Mineral Deposits, Part A. Processes, Techniques, and Health Issues*; Plumlee, G.S., Longsdon, M.J., Eds.; Reviews in Economic Geology; Society of Economic Geologists: Littleton, CO, USA, 1999; Volume 6A, pp. 133–160.

68. Hudson-Edwards, K.A.; Jamieson, H.E.; Lottermoser, B.G. Mine wastes: Past, present, future. *Elements* **2011**, *7*, 375–380. [CrossRef]

69. Grimalt, J.O.; Ferrer, M.; Macpherson, E. The mine tailing accident in Aznacollar. *Sci. Total Environ.* **1999**, *242*, 3–11. [CrossRef]

70. Levings, C.D.; Barry, K.L.; Grout, J.A.; Piercey, G.E.; Marsden, A.D.; Coombs, A.P.; Mossop, B. Effects of acid mine drainage on the estuarine food web, Britannia Beach, Howe Sound, British Columbia, Canada. *Hydrobiologia* **2004**, *525*, 185–202. [CrossRef]

71. Parbhakar-Fox, A.K.; Lottermoser, B.; Bradshaw, D. Evaluating waste rock mineralogy and microtexture during kinetic testing for improved acid rock drainage prediction. *Miner. Eng.* **2013**, *52*, 111–124. [CrossRef]

72. McKibben, M.A.; Tallant, B.A.; del Angel, J.K. Kinetics of inorganic arsenopyrite oxidation in acidic aqueous solutions. *Appl. Geochem.* **2008**, *23*, 121–135. [CrossRef]

73. Temminghoff, E.J.M.; Van Der Zee, S.E.A.T.M.; De Haan, F.A.M. Copper mobility in a copper-contaminated sandy soil as affected by pH and solid and dissolved organic matter. *Environ. Sci. Technol.* **1997**, *31*, 1109–1115. [CrossRef]

74. Balci, N.; Demirel, C. Prediction of acid mine drainage (AMD) and metal release sources at the Küre copper mine site, Kastamonu, NW Turkey. *Mine Water Environ.* **2017**, *37*, 56–74. [CrossRef]

75. Pareuil, P.; Pénilla, S.; Ozkan, N.; Bordas, F.; Bollinger, J.-C. Influence of reducing conditions on metallic elements released from various contaminated soil samples. *Environ. Sci. Technol.* **2008**, *42*, 7615–7621. [CrossRef] [PubMed]

76. Engineering ToolBox. Oxygen—Solubility in Fresh Water and Seawater. 2005. Available online: https://www.engineeringtoolbox. com/oxygen-solubility-water-d_841.html (accessed on 4 May 2021).

77. Chizhikov, D.M.; Kovylina, V.N. Investigation of potentials and anodic polarization of the sulfides and their alloys. In *Trudy Chetvertogo Soveshchaniya po Electrokhimii Akademii Nauk SSSR, Proceedings of the 4th Conference on Electrochemistry, Moscow, Russia, 1–6 October 1956*; Frumkin, A.S., Ed.; Akademii Nauk SSSR: Moscow, Russia, 1956; pp. 715–719.
78. Biegler, T.; Swift, D.A. Anodic behaviour of pyrite in acid solutions. *Electrochem. Acta* **1979**, *24*, 415–422. [CrossRef]
79. Warren, G.W. The Electrochemical Oxidation of CuFeS2. Ph.D. Thesis, University of Utah, Salt Lake City, UT, USA, 1978.
80. Ziemkiewicz, P.; Skousen, J.G. Open limestone channels for treating acid mine drainage: A new look at an old idea. *Green Lands* **1994**, *24*, 36–41.
81. Selvarajan, P.; Chandra, G.; Bhattacharya, S.; Sil, S.; Vinu, A.; Umapathy, S. Potential of Raman spectroscopy towards understanding structures of carbon-based materials and perovskites. *Emergent Mater.* **2019**, *2*, 417–439. [CrossRef]

Article

A Methodology Based on Magnetic Susceptibility to Characterize Copper Mine Tailings

Elizabeth J. Lam [1,*], Rodrigo Carle [2], Rodrigo González [2], Ítalo L. Montofré [3,4], Eugenio A. Veloso [5], Antonio Bernardo [6], Manuel Cánovas [4] and Fernando A. Álvarez [3,7]

1   Chemical Engineering Department, Universidad Católica del Norte, Antofagasta CP 1270709, Chile
2   Geological Sciences Department, Universidad Católica del Norte, Antofagasta CP 1270709, Chile; carlerodrigo@gmail.com (R.C.); r_gonzalez@ucn.cl (R.G.)
3   Mining Business School, ENM, Universidad Católica del Norte, Antofagasta CP 1270709, Chile; imontofre@ucn.cl (Í.L.M.); falvarez@ucn.cl (F.A.Á.)
4   Mining and Metallurgical Engineering Department, Universidad Católica del Norte, Antofagasta CP 1270709, Chile; manuel.canovas@ucn.cl
5   Andean Geothermal Center for Excellence (CEGA), Departamento de Ingeniería Estructural y Geotécnica, Pontificia Universidad Católica de Chile, Santiago 7560873, Chile; anveloso@ing.puc.cl
6   Department of Mining Technology, Topography and Structures, Universidad de León, CP 24071 León, Spain; antonio.bernardo@unileon.es
7   Administration Department, Universidad Católica del Norte, Antofagasta CP 1270709, Chile
*   Correspondence: elam@ucn.cl

Received: 28 August 2020; Accepted: 19 October 2020; Published: 22 October 2020

Abstract: This paper intends to validate the application of magnetic techniques, particularly magnetic susceptibility, as sampling tools on a copper tailings terrace, by correlating them analytically. Magnetic susceptibility was measured in both the field and laboratory. Data obtained allowed for designing spatial magnetic susceptibility distribution maps, showing the horizontal variation of the tailings. In addition, boxplots were used to show the variation of magnetic susceptibility and the concentration of the elements analyzed at different depths of the copper tailings terrace. The degree of correlation between magnetic and chemical variables was defined with coefficient R2. The horizontal and vertical variations of magnetic susceptibility, the concentration of elements, and the significant correlations between them show a relationship between magnetic susceptibility and the chemical processes occurring in the tailing management facility, such as pyrite oxidation. Thus, the correlation functions obtained could be used as semiquantitative tools to characterize tailings or other mining residues.

Keywords: copper mine tailings; magnetic susceptibility; sampling; metals

1. Introduction

The treatment of extensively distributed solid wastes generated by metallurgical processes is a challenge for the environmental sustainability of mining regions [1,2]. Soil characterization is a relevant process for subsequent environmental treatments such as decontamination, stabilization, or remediation [3,4]. The physical (pH, density) and the chemical properties of soils are a forcing factor for the growth and pollutant metabolism of plants during phytoremediation, particularly heavy metal content in the sediments at the (superficial) upper levels [5,6]. In this context, heavy metal concentration is commonly determined by a chemical analysis involving a high cost and a long period of analysis [7–9]; however, a low-cost and fast-application method is necessary for effective sediment characterization prior to remediation.

For a reliable tailing storage facility characterization, a large number of sampling points is required so that results can be representative, a fact that has been limited due to the extensive economic resources needed. Sampling quality is essential for estimating the extent of contamination on-site and, therefore, establishing intervention requirements to protect human health and the ecosystem integrity [10]. In this regard, sampling design plays a fundamental role in the tailing characterization stage and must be based on the spatial contamination distribution hypothesis formulated from the results of the exploratory phase of the study. This allows for making an appropriate assessment of the site and its management [11,12]. It is necessary to design a sampling methodology to establish their presence in the mineral species contained in the tailings [13,14]. There are no general rules for soil sampling because each site requires a particular strategy. Therefore, it is important to design a scheme appropriate for each of the tailing storage facilities, considering the optimal location of the sampling points. This scheme must be flexible enough to make adjustments during field activities, owing to, for example, lack of access to preselected sites, unforeseen soil formations, and climatic conditions [15]. The characterization stage involves sampling and the analysis of physical and chemical properties to determine the nature, extent, and extension of contamination. These data are essential for developing, projecting, analyzing, and selecting cost effective technologies to mitigate contamination [16].

Magnetic techniques and methodologies are notable for characterizing soils since they can be validated with a small number of fast chemical analyses at a low cost [17,18]. Although the magnetic phenomenon was recognized in the early 19th century, its application in the different fields of science and technology has increased in the last few decades [19–21]. The magnetic characterization and mapping of the magnetic susceptibility of soils have been widely applied as a proxy to characterize heavy metals and pollutants in soils in urban and industrial areas [22–29]. This type of study allows for determining the origin and extension of contaminant agents and their effect on natural soils. Magnetic properties, particularly magnetic susceptibility, are useful tools for identifying and describing ferromagnetic elements (Fe, Ni, Cr). They allow an indirect characterization of the study area because a small number of chemical analyses are needed later.

Magnetic susceptibility is the ability of a material to magnetize itself under the effect of an induced magnetic field, which has a range of values characteristic of the different ferromagnetic elements. It is possible to measure this property in the field by using easy-to-use portable susceptometers at a low cost [30]. Despite the wide use of magnetic techniques for natural soil contamination, they have been scarcely used for detecting and determining the concentration of metals present in tailing deposits [31]. Therefore, the application of magnetic techniques is a good alternative for the chemical characterization of tailing storage facilities due to their high resolution and low cost. In addition, these techniques have important advantages such as short measurement time and the repetition of analyses at a low cost.

Several factors take part in the control of the retention and mobility of heavy metals in the soil, mineralogy playing an important role among them [32]. The bioavailability of most of the elements, particularly heavy metals, is determined by adsorption–desorption, complexation, precipitation, and ion-change processes. The most important surfaces involved in soil metal adsorption are active inorganic colloids such as clayey minerals, metal oxides and hydroxides, metal carbonates and phosphates, and organic colloids [33]. Regarding texture, clayey soils retain more metals by adsorption or in the exchange complex of the clayey minerals; on the contrary, sandy soils lack this fixing capacity. In particular, each clay mineral is characterized by a specific surface area value and a degree of electrical decompensation, which influence its ability to adsorb or exchange metals [34].

Chilean mining is paramount for the country's development and has become part of its identity. Chile has become a world's copper production leader, with mining being the productive activity contributing most to GDP [35]. However, it also holds a negative aspect due to the large output of residues and toxic wastes resulting from different operations and processes. Copper sulfide ore processing produces residues called tailings, which contain high heavy metal concentrations [36]. Many solid tailing deposits extend for kilometers, their characterization currently being made via chemical analyses at a high cost [9]. There are 742 mine tailing disposal sites in Chile; some of them are abandoned and require an urgent management plan [37]. To propose remediation plans in order to reduce potential risks associated with tailings, their physicochemical and mineralogical characterization is a priority. These data are not available or are rather scarce, as in the case of geochemical data reported by National Geology and Mining Service (SERNAGEOMIN, for its acronym in Spanish) that determined the geochemical characterization of a number of tailing disposal facilities, based on samples from one to four sampling points [10].

The aim of this work is to discard or validate magnetic susceptibility measurement as a technique for determining the contaminants, concentrations, and/or mobility of the elements in tailing storage facilities. This study, conducted by examining copper mine tailings, intends to validate the techniques measuring magnetic properties as sampling tools, correlating them with a limited number of physicochemical analyses. The objective of this study is to validate the measurement of magnetic susceptibility in the tailing terrace of a copper porphyry-type ore deposit. To attain the objective, heavy metal concentration values in the tailings are correlated with magnetic susceptibility to determine if this technique can be used as a fast and cost effective tool for identifying contaminated areas. Our results are relevant and offer a low cost and effective method to characterize solid wastes generated by copper metallurgical processes prior to their management and treatment in an extensive mining region like northern Chile.

2. Methodology

2.1. Site

This study was conducted in the Atacama Desert, located in Antofagasta Region, Chile. This region is characterized by high solar radiation, high saline soil concentrations, extremely high daytime temperatures, and a wide day–night temperature range. Some areas of the Atacama Desert show zero recorded rainfall in 400 years. In general, rainfall occurs every 100 years. All these conditions produce scarce vegetation. Samples were collected from a copper mine tailings site located at latitude 24°9′58.33″ S and longitude 69°2′33.36″ W, at 3200 masl The terraces where the tailings are disposed show high concentrations of heavy metals such as Cu, Cd, Fe, Zn, Mn, and Pb because the tailings are generated in the sulfide concentration process. The total area of the terrace is about 10,000 m² (135 m × 77 m). This site was used as a tailings dump from 1995 to 2006. Figure 1A,B shows the location of the tailings called CMZ.

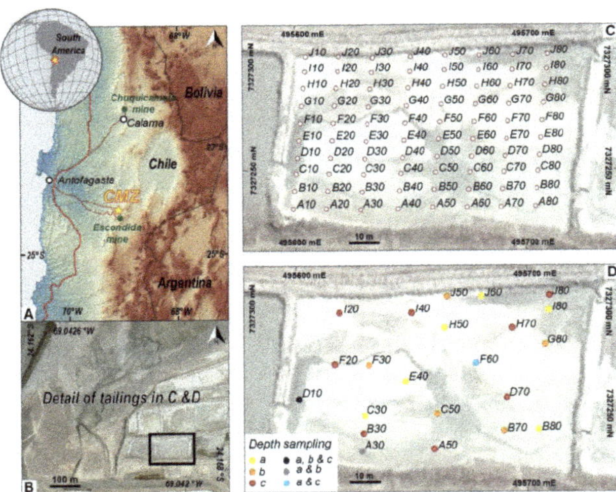

Figure 1. (**A,B**) CMZ tailings location, (**C**) regular sampling grid with 80 sites on CMZ tailings, (**D**) spatial distribution of tailings samples for chemical analyses. Green: depth a; blue: depth b; red: depth c; purple: three depths; yellow: depths a and b; light blue: depths a and c. The black box represents the study area within the tailing terrace.

CMZ is a copper porphyry with low-grade hypogenic mineralization, where a long continuous supergene enrichment process favored by the predominating tectonic regime during the Upper Oligocene–Lower Miocene [38,39] resulted in the formation of economically exploitable ore deposits. The later decrease of the erosion rate and the increase of regional aridity since the Middle Miocene allowed for the preservation of these ore deposits [40]. Intrusive Oligocene, genetically related to CMZ, includes an early phase of rhyolitic porphyries associated with mineralization and a more intense hypogenic alteration. The later phase corresponds to dioritic porphyries.

According to its mineralization, the ore deposit consists of an upper leached zone associated with quartz–sericitic and copper oxide traces; an oxidation area with brochantite and antlerite; and an extremely sericitized secondary rich zone consisting of pyrite, chalcocine, bornite, covelline, and chalcopyrite. The hypogenic mineralization associated with sericitic alteration and a weak silicification consists mainly of pyrite with small amounts of bornite, chalcopyrite, and molybdenite.

2.2. Sampling In Situ

To collect the samples, the terrace was divided into a grid of 15 m × 7 m, covering an internal area of 6615 m^2, to create 80 equidistant sampling points, as shown in Figure 1C. Due to the rough terrain and to facilitate sample collection, the profiles corresponding to the external points of the terrace were not considered. For the 80 internal points, magnetic susceptibility intensity was measured at three different depths: depth a (0–10 cm deep); depth b (10–20 cm deep); and depth c (20–30 cm deep). This property was used to estimate the points with larger metal concentrations [41]. According to this criterion, 33 sample points were selected.

The samples were collected at the first 30 cm, considering characterization as the preliminary analysis of a phytoremediation system. Plant growth and trace element absorption generally occurred at a 10–30 cm depth [42–44].

2.2.1. Sampling Grid Design

The tailings terrace studied is 160 m long and 80 m wide (Figure 1C). For the sampling to include the whole study area, a regular grid was designed, with a 7 m support in the direction of the tailings walls (N05E), while in the W–E direction, they are at a 15 m distance from each other, heading N85E. This area was chosen according to the terrace dimensions so that the sampling could be representative in both directions (nonrandom sampling).

Distances were measured with a 60 m measuring tape, while the direction was obtained with a Brunton compass. The sampling grid was integrated into a geographic data system over a rectified Quickbird image from Google Earth, using ArcGis 10 software.

2.2.2. Sampling

From each tailings sediment site, 1 kg was obtained. The total number of samples amounted to 240. The sediment was extracted with a plastic shovel to avoid altering magnetic measurements. Later, the material collected was stored in clean sealed polyethylene bags, labeled according to site and depth. Table 1 exemplifies the spatial identification of each tailings sample.

Table 1. Example of sample identification.

Type of Sample	Depth (cm)	Sampling Site	Depth Identification	Final Identification
Disaggregated material	0–10	A10	a	A10 a
	10–20	A10	b	A10 b
	20–30	A10	c	A10 c

2.3. Measurement Methodology

2.3.1. Magnetic Susceptibility Measurement

Magnetic susceptibility measurements were made with an SM-30 portable susceptometer (Heritage Geophysics, Littleton, CO, USA) to make quick measurements in the field with a 1×10^{-7} magnetic susceptibility (SI) precision. The measurements were made on dry samples, using the mode 517 of the device at a 43.8 cm^3 fixed volume. Magnetic susceptibility measurements were made at the three depths of the 80 sampling points. At each sampling depth, magnetic susceptibility was measured three times at different points. It was possible to check the validity of the susceptometer data by repeating measurements by horizon. Next, the average value was obtained.

2.3.2. pH Measurement

pH was measured in situ with a pH/°C measuring kit, Model HI 99121 (Hanna Instruments, Woonsocket, RI, USA), which is used for measuring both soil pH and a solution prepared with a soil sample directly.

2.3.3. Heavy Metal Concentration Determination

A total of 33 sampling points were selected for chemical analysis. To do this, the following criteria were considered:

(1)   Tailings pH variation; this property conditions metal bioavailability.
(2)   Magnetic susceptibility variation in tailings sediments; samples were selected to make a representative subsampling of all the values observed in the distribution of this property.
(3)   The three depths were established so as to obtain representative results of the whole tailings volume.

Substrate samples, properly labeled and packed in polyethylene bags, were collected. They were oven-dried at 40 °C until reaching a constant weight [45,46]. Gravel-sized rocks (>2 mm) were removed and the remaining particles reduced in size with mortar and pestle. Particles were then screened with a 2-mm sieve (US N° 10 mesh), which is the standard particle size for most soil testing methods [45,47].

Bioavailable Fe, Mn, Zn, Cr, Cu, Cd, and Pb contents were measured with an atomic absorption spectrophotometer (AAS) after extraction by using a diethylenetriaminepentaacetic acid (DTPA) solution [48]. These metals were collected by shaking 0.01 kg of oven-dried soil for 2 h in 20 mL of 0.005 M DTPA. The filtrate was analyzed for Fe, Mn, Zn, Cr, Cu, Cd, and Pb by AAS. Cadmium, copper, chromium, lead, zinc, and iron were measured by ASS, with detection limits of 0.05 mg kg^{-1} for Cd, Cr and Zn, 4.3 mg kg^{-1} for Pb, 7.5 mg kg^{-1} for Cu, and 24 mg kg^{-1} for Fe. As analysis was conducted separately by hydride generation–atomic absorption spectrometry (HGAAS). Hydride was generated by using a Perkin-Elmer 100 FIAS FIA 100 apparatus (0.005 mg kg^{-1} detection limit). All solutions were filtered with Whatman GF/C fiberglass filter paper [9].

3. Results

Data selected according to the three criteria above (Section 2.3.3) are shown in Table 2. The spatial distribution of the sampling points for the three depths are shown in Figure 1C.

Tables 3–5 show the results of the concentration analyses of eight heavy metals (As, Cd, Cr, Cu, Fe, Ni, Pb, and Zn) and the magnetic susceptibility (MS) values for the three points measured in the field. The denomination is based on the names of the wells and the sampling depth, as shown in Table 1.

Table 2. Samples selected according to pH classes and magnetic susceptibility variability.

Depth	pH Range	Sample	pH	Soil K (μSI)
	pH < 5	A30	3.65	389.3
	5 < pH < 6.5	B80	5.45	532.7
		C30	5.32	655.3
		D70	5.43	672.0
Depth a 0–10 cm		H50	5.65	619.7
		I20	5.12	735.0
		J60	5.97	334.3
	6.5 < pH < 7.5	E40	6.54	520.0
		F60	7.20	655.3
		I80	6.97	679.0
	pH > 7.5	D10	7.99	488.0
	5 < pH < 6.5	A30	5.79	618.3
		B70	5.58	1090.0
		D70	5.92	991.0
		E40	6.12	297.0
Depth b 10–20 cm		E70	6.18	256.8
		G80	6.35	826.0
		I20	6.17	583.0
	6.5 < pH < 7.5	C50	6.58	393.0
		F30	6.89	637.0
		J50	7.02	490.3
	pH > 7.5	D10	7.97	326.7
	5 < pH < 6.5	I40	6.35	272.7
	6.5 < pH < 7.5	A50	7.50	186.0
		D70	6.82	434.3
		E40	6.95	160.0
Depth c 20–30 cm		F20	7.30	816.0
		F60	7.12	682.7
		H70	6.93	1159.0
		I20	6.98	369.3
		J80	7.07	246.3
	pH > 7.5	B30	7.59	364.0
		D10	7.61	479.0

Table 3. Results of chemical analyses for the first stratum at a 0–10 cm depth.

Site	Cd	As	Zn	Cu	Fe	Ni	Pb	Cr	Soil K
					mg·kg^{-1}				μSI
A30 a	2.09	26.3	426.7	12,118	33,050	58.3	412.3	27.2	389.3
B80 a	2.10	33.6	446.4	10,395	36,913	96.2	452.3	29.3	532.7
C30 a	1.90	28.3	473.2	12,090	34,653	77.3	376.6	41.2	655.3
D10 a	2.60	21.8	398.8	14,426	31,278	68.4	420.2	17.6	488.0
D70 a	3.20	21.9	306.6	9323	41,923	67.2	289.5	14.8	672.0
E40 a	2.80	32.1	287.2	8225	33,607	88.2	322.2	32.4	520.0
F60 a	4.10	27.3	327.6	8454	34,756	59.6	254.2	29.5	655.3
H50 a	3.70	21.8	333.6	8857	35,656	72.3	322.1	19.3	619.7
I20 a	2.40	32.5	328.8	12,160	36,675	49.4	312.2	22.4	735.0
I80 a	4.20	20.6	425.3	9519	38,137	88.2	453.8	26.7	679.0
J60 a	2.80	23.4	399.5	10,712	33,716	77.3	408.1	33.9	334.3

Table 4. Results of chemical analyses for the second stratum at a 10–20 cm depth.

Site	Cd	As	Zn	Cu	Fe	Ni	Pb	Cr	Soil K
				mg·kg^{-1}					µSI
A30 b	0.848	21.2	373.7	10,728	34,470	32.5	221.2	17.3	618.3
B70 b	1.187	18.3	375.3	8231	33,219	34.6	245.2	12.8	1090.0
C50 b	0.699	15.6	207.7	10,321	30,846	28.3	282.8	21.4	393.0
D10 b	0.899	18.2	450.8	12,658	31,950	41.2	198.4	19.4	326.7
D70 b	0.599	22.4	280.3	16,296	34,781	19.6	187.5	18.6	991.0
E40 b	0.900	23.9	240.6	14,533	31,733	22.2	201.4	22.6	297.0
E70 b	0.352	21.2	200.1	14,701	27,625	28.3	167.4	21.9	256.8
F30 b	0.798	28.3	500.6	9003	37,297	32.2	135.3	18.2	637.0
G80 b	2.298	12.4	271.2	17,203	36,731	31.4	194.9	19.6	826.0
I20 b	1.797	15.6	300.9	11,096	37,095	17.8	217.8	14.3	583.0
J50 b	2.298	17.8	299.5	12,398	38,489	28.2	175.9	18.8	490.3

Table 5. Results of chemical analyses for the third stratum at a 20–30 cm depth.

Site	Cd	As	Zn	Cu	Fe	Ni	Pb	Cr	Soil K
				mg·kg^{-1}					µSI
A50 c	0.698	10.8	225.4	9186	30,433	12.8	222.7	13.6	186.0
B30 c	0.897	17.3	288.3	9798	35,816	23.9	198.6	17.2	364.0
D10 c	1.800	15.4	458.5	12,207	37,001	21.8	178.5	17.4	479.0
D70 c	0.797	14.1	190.9	8181	23,457	17.5	218.7	9.6	434.3
E40 c	0.900	18.3	200.5	8234	30,839	19.8	218.4	22.5	160.0
F 20c	2.791	28.3	495.4	9452	41,652	21.4	198.4	18.4	816.0
F60 c	2.342	17.9	394.4	10,247	36,012	31.2	219.2	28.3	682.7
H70 c	2.021	15.8	313.5	13,939	33,667	18.9	178.9	12.4	1159.0
I20 c	2.396	13.8	483.2	12,591	39,608	24.8	228.9	17.2	369.3
I40 c	1.879	21.2	393.5	13,872	39,540	22.9	178.4	21.9	272.7
J80 c	2.394	17.4	328.4	14,171	34,704	17.9	189.6	19.4	246.3

3.1. Relationship between Magnetic Susceptibility and Depth

Considering the 240 samples, there is a statistical correlation between sampling depth and magnetic susceptibility. For the three depths, the standard deviation and the variance show high values, indicating a high dispersion of the magnetic susceptibility data measured. The mean value of magnetic susceptibility for the three horizons tends to 411–518 µSI, showing a decrease in depth, as illustrated in Figures 2 and 3, showing a boxplot and an interval plot, respectively. Given a certain relationship between depth and magnetic susceptibility, a hypothesis test was conducted.

The relationship between MS and depth was analyzed with ANOVA. This test shows the influence of one or more factors, in this case depth, over the mean of a continuous variable, in this case MS. Table 6 shows the mean, standard deviation, and 95% CI for the mean of each profile. The ANOVA test was conducted with a 95% CI. Results reveal statistically significant differences between at least two groups (df = 2; F = 7.85; p-value = 0.001). According to the results of the Tukey's post hoc test [49], the group under 30 cm present statistically significant differences in the magnetic susceptibility mean when compared with the other two.

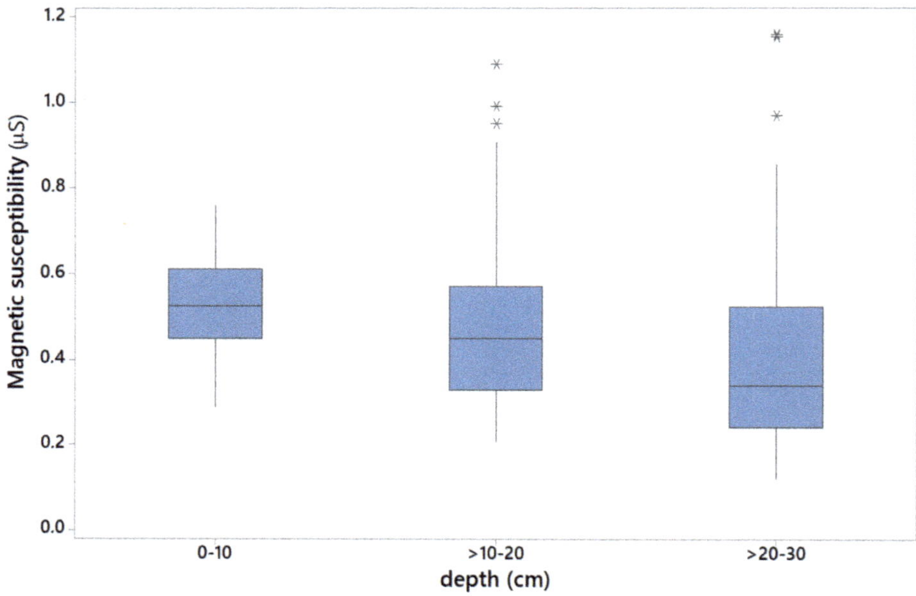

Figure 2. Boxplots for magnetic susceptibility (MS) values at three sampling depths.

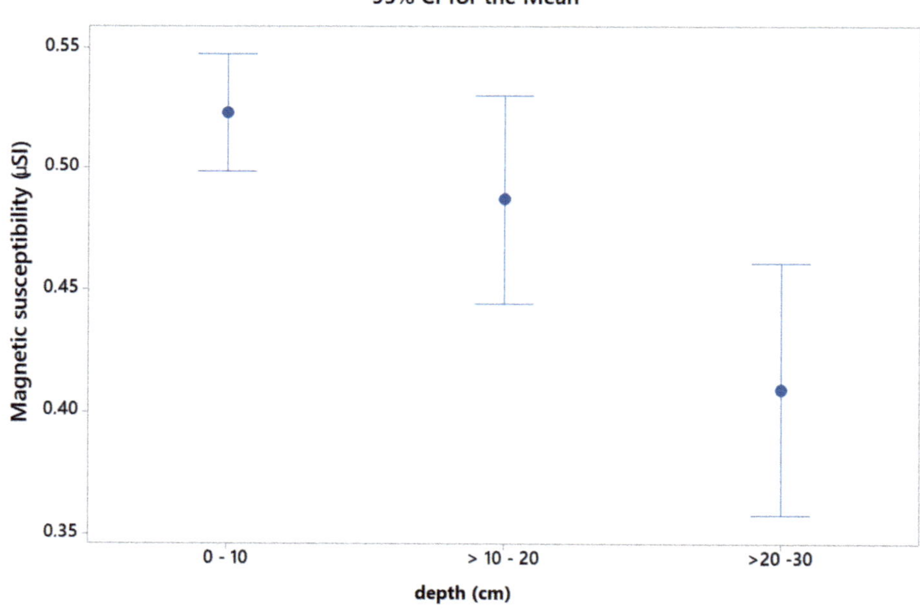

Figure 3. MS values at three sampling depths for a 95% confidence interval (CI).

Table 6. Mean value, standard deviation, and confidence interval per sample depth.

Profile	Mean	SD	95% CI
0–10 cm	0.5230	0.1100	(0.4822; 0.5638)
10–20 cm	0.4872	0.1934	(0.4464; 0.5280)
20–30 cm	0.4095	0.2311	(0.3687; 0.4503)

3.2. Relationship between Magnetic Susceptibility and pH

MS decreases as tailing depth increases. Kapikca et al. [50], Hoffmann et al. [51], Boyko et al. [52], and Magiera et al. [53] observed this tendency, associated with the enrichment of anthropogenic particles on the most superficial layers of the soil from nearby industries or plants. The main difference between this study and those mentioned above is the type of soil where samples are collected, since samples are taken from the tailings mass in this study. For this reason, MS decrease is related to chemical processes occurring within the tailings. MS decrease with depth may be explained by the most important reaction in the tailings, i.e., pyrite oxidation (Equations (1)–(3)). As explained by Dold and Fontboté [54], atmospheric oxygen eruption into the system begins sulfide oxidation, in this case pyrite, the main gangue in the ore deposit.

$$FeS_2 + \frac{7}{2}\,O_2 + H_2O \rightarrow Fe^{+2} + 2\,SO_4^{2-} + 2\,H^+ \tag{1}$$

$$Fe^{+2} + \frac{1}{4}\,O_2 + H^+ \rightarrow Fe^{3+} + \frac{1}{2}\,H_2O \tag{2}$$

$$Fe^{3+} + 3\,H_2O \rightarrow Fe(OH)_3 + 3\,H^+ \tag{3}$$

Pyrite oxidation is the main acidity producer of the system (Equation (1)). The acidity produced by these processes may result in a pH decrease [55–59]. Tailings from copper porphyry-type ore deposits show 1–3% pyrite concentrations [54]. Therefore, ions such as Fe^{3+} precipitate at pH > 3.5. The ferric ion, when not involved in sulfide oxidation, precipitates as secondary mineral such as goethite. Dold and Fontboté [60] conducted chemical, mineralogical, and microbiological analyses on three Chilean copper porphyry ore deposit tailings in different climatic contexts (arid, semiarid, and humid). These tailings were classified as having low sulfidization and carbonate content. The comparison of the behavior of these three tailings shows that the climatic factor controls the direction in which the elements move, free from the chemical reactions in the tailings. For example, in a humid climate, the chemical elements in the tailings move from an oxidizing environment on the upper part to a reducing environment in the direction of the phreatic level. Meanwhile, in regions with an arid climate such as the Atacama Desert, transport is expected to occur in the inverse direction, which is toward a more oxidizing environment due to capillarity, thus forming secondary sulfates on the surface. This effect has also been observed in arid and semiarid environments, favoring oxidation processes and allowing efflorescent mineral formation (hydrated sulfates) on the tailing surface [61,62].

In the tailings studied, horizons a and b show a slightly acid pH, while at depth c, the pH is neutral. This suggests:

(i) As horizon a is more superficial, pyrite oxidizes, resulting in greater acidity (average pH = 6.1).

(ii) There is an enrichment of diamagnetic minerals such as silica toward horizon c; i.e., there is a greater neutralization potential (pH = 7.05), which is supported by Si concentration increase with depth.

(iii) The formation of iron hydroxide should occur at the initial stage (pseudominerals) because, if minerals such as goethite were present in horizon c, the signal would increase owing to the high susceptibility of this mineral, which is not correlated with the MS tendency.

Horizontally, MS shows the highest concentration values at the ends and the lowest concentration values in the middle part of the terrace. Dold [63] indicates that, although tailings particle size is relatively homogeneous, there is a deposition controlled by sedimentological processes resulting in coarse granulometry near the deposition point (sulfides are heavier than silicates). According to this analysis, MS distribution maps show a certain coherence. High MS values may involve the presence of pyrite with positive susceptibility, while low values would be associated with diamagnetic materials such as silica or carbonates.

3.3. Depth Variation of Chemical Elements

As to the variation of chemical elements in the terrace, two factors control element mobility: solubility and pH [64,65]. Solubility is a function of the chemical element concentration of the metal. In lower concentration systems, the elements are more mobile than in those with a higher concentration. Thus, mineralogical composition determines what chemical environment predominates and what elements are liberated and can mobilize [66–68]. In the terrace, Cu and Cd increase their concentration towards horizon c, similarly to the pH. These behaviors are consistent, as reported by Dold and Fontboté [54], who determined that in tailings out of operation, the oxidation process occurs in its initial stage. Metals such as Cu, Pb, Cd, Ni, and Ca are quite mobile at a low pH and are adsorbed when pH increases, explaining the lower concentration in horizon a, as compared with horizons b and c. The opposite is observed for Zn behavior, which mainly concentrates in horizon a. Dold and Fontboté [60] observed the same tendency in Zn and Mg in El Salvador tailings, which concentrate on the most superficial layer, corresponding to an evaporitic horizon with a high pH.

3.4. Relationship between Magnetic Susceptibility and Heavy Metal Concentration

This relationship was analyzed with a MARS (multivariate adaptive regression splines) model suitable for this data structure. MARS was calculated by using the earth [69] library of the statistical package R [70], while the other statistical calculations were made with Minitab. Using data from Tables 3–5, the relationship between MS and the concentrations of eight heavy metals was analyzed by using MARS, as in other engineering applications [71–75] because it fits these data better than other models [76,77].

The model is a MARS analysis, i.e., a type of regression introduced by Jerome Friedman [78]. It is a nonparametric regression technique that may be understood as a linear model extension automatically modeling nonlinear relationships and interactions between variables. In other words, it automatizes prediction model construction by selecting relevant variables, transforming predictive variables, treating missing values, and avoiding overfitting by means of an autotest. MARS is similar to a linear regression without splines.

It is mainly used for predicting a continuous variable $\vec{y}(nx1)$, here MS, from a set of explanatory variables $\vec{X}(nxp)$, here the concentration of heavy metals. So, the MARS model may be represented by the expression $\vec{y}(nx1) = f(\vec{X}) + \vec{e}$, where $\vec{e}$ is an $(nx1)$ error vector.

MARS analysis results in both a linear and a second-grade linear model, without higher-grade models.

In the linear model, a correlation is established between MS and the concentrations of four metals, Cu, Fe, Ni, and Cr, and no correlation obtained with the other heavy metals Cd, As, Zn, and Pb, meaning that the latter are not relevant for MS.

The correlation in this model is given by Equation (4):

$$MS = 0.55 + 7.2\,e^{-05} \times \text{pmax}\,(0, Cu - 12207) - 5.8\,e^{-05} \times \text{pmax}\,(0, 34756 - Fe) - \\ 0.02 \times \text{pmax}\,(0, 31 - Ni) + 0.076 \times \text{pmax}\,(0, 19 - Cr) \tag{4}$$

where pmax is 0, if the other value is not positive. For example, pmax (0, 19-Cr) becomes 0; i.e., Cr does not influence the correlation until Cr concentration does not exceed 0.19 mg·kg^{-1}.

MARS model parameters GCV = 0.047, RSS = 0.82, GRSq = 0.26, and RSq = 0.58 show the model goodness of fit, RSq = 0.58, is not low for spread data.

The linear model, represented by pmax functions, shows that as Fe and Ni concentration increases, MS increases up to a certain value and then remains constant. For Cu concentration, MS remains constant up to a certain value and then increases. As for Cr, MS decreases to a certain concentration value and then remains constant (Figure 4).

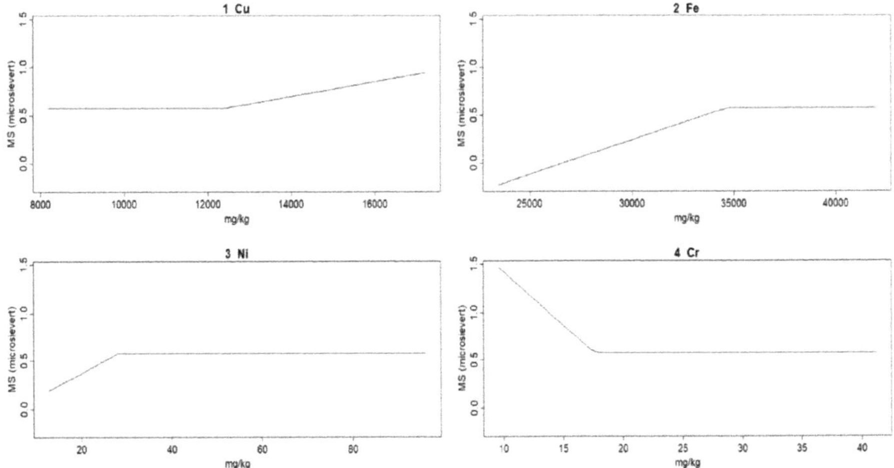

Figure 4. Relationship between metal concentration and magnetic susceptibility, discriminated by pmax functions.

In the second-grade model, the correlation is established between MS and the concentrations of three metals, Cu, Zn, and Cr, the latter only as a second-order term. No correlation is obtained for the other heavy metals. This means that they are not relevant for MS. The fit is better than the one for the first-order model, with RSq slightly over 0.67. The correlation of the second-grade model is given by Equation (5):

$$MS = 0.43682171 + 0.00981946 \times \text{pmax}\,(0,\, Zn{-}287.2) + 0.00981946 \times \text{pmax}\,(0,\, Cu{-}0.9323) +$$
$$\text{pmax}\,(0,\, 287.2{-}Zn) \times Cu - 0.00000040 \times \text{pmax}\,(0,\, Zn{-}287.2) \times Cu - \tag{5}$$
$$0.00000089\, Zn \times \text{pmax}\,(0,\, 17.3{-}Cr) \times 0.00022912$$

MARS model parameters GCV = 0.05631133, RSS = 0.6488601, GRSq = 0.112011, and RSq = 0.670256 show that the model goodness of fit, RSq = 0.67, is better than in the linear model. Figure 5 shows Zn and Cu linear relationships and Cr second-order relationship.

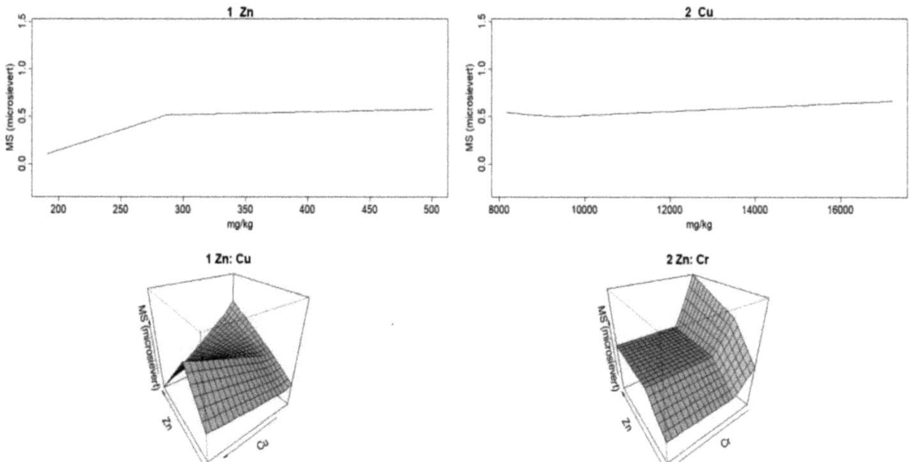

Figure 5. Relationship between metal concentration and magnetic susceptibility.

3.5. MS Advantages over Metal Concentration

The great advantage of susceptibility measurements is their promptness and low cost [79–81]. MS was directly measured in the substrate at 0–10, 10–20, and 20–30 cm depths. For each level, MS was measured at least three times at three different points. Then, the average was determined. The sediment was extracted with a plastic shovel to avoid changing magnetic measurements [82]. The technique results in the lowest cost, as compared with every other type of test and is little invasive for the environment. The chemical analysis was more complex, involving soil homogenization, clod disaggregation, and the removal of larger stones and residues. This was followed by clay content drying and sample sieving. Pretreatment for chemical analysis took about 3 days per sample [5,6]. So, preparing the sample for magnetic susceptibility is much simpler than preparing it for a chemical analysis. In addition, time must also be considered. As to MS, it is possible to measure 6 samples in 1 h on an average, considering the pretreatment stage. So, 40 h are required for the 240 sampling points. Thus, by working 8 h/day, 5 days are needed to obtain results. As to concentrations, pretreatment is more demanding and takes longer. For a further description, 6 samples can be collected each hour, the samples requiring at least a 3-day treatment. So, if 48 samples are collected daily, considering 8 h of work/day and 3 additional days for pretreatment, 4 days are required for collecting and treating 48 samples. Hence, for the 240 sampling points 20 days are required. In brief, the MS measurement of 240 points takes 5 days, while collecting and pretreating the 240 samples for measuring concentration takes 20 days. In addition, 3 months are necessary to measure the samples in the laboratory. So, for the time ratio required, concentrations:MS = 110:5 = 22, i.e., 2100% extra time for the characterization process. Table 7 shows the time necessary for measuring magnetic susceptibility and the chemical analysis of 240 sampling points. The chemical analysis considers the determination of eight heavy metals (As, Cd, Cu, Fe, Hg, Pb, and Zn).

Table 7. Time for determining MS and the concentration of 8 heavy metals in 240 samples.

No. of Samples	MS Measurement Time	Time for the Concentration Measurement of 8 Heavy Metals		
	Sample Collection and Measurement in Situ	Sample Collection	Pretreatment	Measurement
6	1 h	1 h	3 days	45 days
48	1 day	1 day	3 days	45 days
240	5 days	5 days	15 days	90 days
Time for MS measurement of 240 points	5 days	Time for collection, pretreatment, and concentration measurement of 240 points		110 days

The cost of measuring each MS is about 3 USD, while the cost of the chemical analysis of a sample is about 150 USD. For the cost required relationship, chemical analysis:MS = 1:50 USD, i.e., the cost of chemical analyses would be about 4900% higher than the cost of measuring an MS point.

Although it is not possible to measure all the sampling points only with MS, the number of points may decrease if MS can be correlated with the concentration. This renders significant savings, as in this study, where only 33 out of the 240 points were measured, producing savings of more than 30,000 USD.

4. Conclusions

MS measurements may provide further data about soil pollution to estimate the environmental situation in a study area; i.e., magnetic properties show depth variations, which reflect concentration changes, depth being an environmental soil pollution indicator.

A positive linear correlation of magnetic susceptibility with Cr, Fe, Ni, and Cu was observed in the tailings studied, while in the second-grade model a correlation was found for Cu, Zn, and Cr. These heavy metals, well-known as the most hazardous elements, are easily extracted by plants from the soils in the area studied [9]. In addition to traditional geochemical mapping, magnetic susceptibility could be successfully used for determining heavy metal soil pollution in the neighborhood of the site under study.

In the second-grade model, R2 is 0.67. For this kind of problem with spread data, this is not a low value and, therefore, indicates a value correlation between heavy metal concentration and magnetic susceptibility. This correlation is important because the cost of measuring MS is much lower than making chemical analyses for heavy metal concentration. Therefore, an indirect method can be used to assess the presence of heavy metals in tailings without conducting chemical analyses, or perhaps just making a few analyses that may serve as a pattern to calibrate magnetic susceptibility, which has been shown to vary with depth.

The concentration and mobility of chemical elements in the tailings is mainly controlled by pH. The interpretation of this and magnetic data enables detecting the chemical processes in the tailings, such as pyrite oxidation. MS is a good indicator because, as it decreases with depth, it is possible to interpret a greater concentration of diamagnetic minerals toward horizon c, which is complemented by the increasing depth of Cu concentrations.

Tailings sediments correspond to paramagnetic rather than diamagnetic arrangements, where ferromagnetic materials are present in small amounts within the paramagnetic matrix. The grain size of the tailings sediments does not allow a macroscopic recognition. So, for interpreting and understanding the magnetic signal, a detailed mineralogy control is necessary.

The correlation functions obtained can be used as semiquantitative tools for detecting toxic substance formations resulting from chemical reactions.

Magnetic methodologies, along with a small number of chemical analyses on representative samples, make it possible to develop sampling grids with a high spatial resolution at a low cost, thus decreasing costs associated with characterization.

Moreover, the potential use of these measurements to assess the metallic values contained in disposal facilities makes up an issue for further studies.

Author Contributions: Conceptualization, E.J.L., R.C., R.G., Í.L.M., and E.E.V.; methodology, E.J.L., R.C., R.G., Í.L.M., and E.A.V.; validation, E.J.L. and R.G.; formal analysis, A.B., F.A.Á., and M.C.; investigation, E.J.L., R.C., R.G., Í.L.M., and E.A.V.; data curation, A.B., F.A.Á., and M.C.; writing—original draft preparation, E.J.L.; writing—review and editing, E.J.L.; visualization, E.J.L., R.G., Í.L.M., and M.C.; project administration, E.J.L. All authors have read and agreed to the published version of the manuscript.

Funding: This research was funded by CORFO-INNOVA project (08CM01-05), titled "Integrated development of magneto-chemical technologies and phytotechnologies applied to the remediation of heavy metals in the development of mining environmental liabilities".

Acknowledgments: The authors are thankful to CORFO-INNOVA project (08CM01-05), titled "Integrated development of magneto-chemical technologies and phytotechnologies applied to the remediation of heavy metals in the development of mining environmental liabilities", for the financial support of this project.

Conflicts of Interest: The authors declare no conflict of interest.

References

1. Park, I.; Tabelin, C.B.; Jeon, S.; Li, X.; Seno, K.; Ito, M.; Hiroyoshi, N. A review of recent strategies for acid mine drainage prevention and mine tailings recycling. *Chemosphere* **2019**, *219*, 588–606. [CrossRef] [PubMed]

2. Tabelin, C.B.; Silwamba, M.; Paglinawan, F.C.; Mondejar, A.J.S.; Duc, H.G.; Resabal, V.J.; Opiso, E.M.; Igarashi, T.; Tomiyama, S.; Ito, M.; et al. Solid-phase partitioning and release-retention mechanisms of copper, lead, zinc and arsenic in soils impacted by artisanal and small-scale gold mining (ASGM) activities. *Chemosphere* **2020**, *260*, 127574. [CrossRef] [PubMed]

3. Huyen, D.T.; Tabelin, C.B.; Thuan, H.M.; Dang, D.H.; Truong, P.T.; Vongphuthone, B.; Kobayashi, M.; Igarashi, T. The solid-phase partitioning of arsenic in unconsolidated sediments of the Mekong Delta, Vietnam and its modes of release under various conditions. *Chemosphere* **2019**, *233*, 512–523. [CrossRef] [PubMed]

4. Huyen, D.T.; Tabelin, C.B.; Thuan, H.M.; Dang, D.H.; Truong, P.T.; Vongphuthone, B.; Kobayashi, M.; Igarashi, T. Geological and geochemical characterizations of sediments in six borehole cores from the arsenic-contaminated aquifer of the Mekong Delta, Vietnam. *Data Brief.* **2019**, *25*, 104230. [CrossRef] [PubMed]

5. Lam, E.J.; Cánovas, M.; Gálvez, M.E.; Montofré, Í.L.; Keith, B.F.; Faz, Á. Evaluation of the phytoremediation potential of native plants growing on a copper mine tailing in northern Chile. *J. Geochem. Explor.* **2017**, *182*, 210–217. [CrossRef]

6. Lam, E.J.; Gálvez, M.E.; Cánovas, M.; Montofré, Í.L.; Keith, B.F. Assessment of the adaptive capacity of plant species in copper mine tailings in arid and semiarid environments. *J. Soils Sedim.* **2017**, *18*, 2203–2216. [CrossRef]

7. Acosta, J.A.; Faz, A.; Martínez-Martínez, S.; Zornoza, R.; Carmona, D.; Kabas, S. Multivariate statistical and GIS-based approach to evaluate heavy metals behavior in mine sites for future reclamation. *J. Geochem. Explor.* **2011**, *109*, 8–17. [CrossRef]

8. Martin-Crespo, T.; Gomez-Ortiz, D.; Martín-Velázquez, S.; Martínez-Pagán, P.; De Ignacio, C.; Lillo, J.; Faz, Á. Geoenvironmental characterization of unstable abandoned mine tailings combining geophysical and geochemical methods (Cartagena-La Union district, Spain). *Eng. Geol.* **2018**, *232*, 135–146. [CrossRef]

9. Lam, E.J.; Gálvez, M.E.; Cánovas, M.; Montofré, I.L.; Rivero, D.; Faz, A. Evaluation of metal mobility from copper mine tailings in northern Chile. *Environ. Sci. Pollut. Res.* **2016**, *23*, 11901–11915. [CrossRef]

10. Lam, E.J.; Montofré, I.L.; Álvarez, F.A.; Gaete, N.F.; Poblete, D.A.; Rojas, R.J. Methodology to Prioritize Chilean Tailings Selection, According to Their Potential Risks. *Int. J. Environ. Res. Pub. Heal.* **2020**, *17*, 3948. [CrossRef]

11. Lark, M.; Hamilton, E.; Kaninga, B.; Maseka, K.K.; Mutondo, M.; Sakala, G.M.; Watts, M.J. Nested sampling and spatial analysis for reconnaissance investigations of soil: An example from agricultural land near mine tailings in Zambia. *Eur. J. Soil Sci.* **2017**, *68*, 605–620. [CrossRef]

12. Constantinescu, P.; Neagoe, A.; Nicoara, A.; Grawunder, A.; Ion, S.; Onete, M.; Iordache, V. Implications of spatial heterogeneity of tailing material and time scale of vegetation growth processes for the design of phytostabilization. *Sci. Total Environ.* **2019**, *692*, 1057–1069. [CrossRef] [PubMed]

13. Corriveau, M.; Jamieson, H.E.; Parsons, M.B.; Campbell, J.; Lanzirotti, T. Direct characterization of airborne particles associated with arsenic-rich mine tailings: Particle size, mineralogy and texture. *Appl. Geochem.* **2011**, *26*, 1639–1648. [CrossRef]

14. Khelfaoui, M.; Medjram, M.S.; Kabir, A.; Zouied, D.; Mehri, K.; Chikha, O.; Trabelsi, M.A. Chemical and mineralogical characterization of weathering products in mine wastes, soil, and sediment from the abandoned Pb/Zn mine in Skikda, Algeria. *Environ. Earth Sci.* **2020**, *79*, 1–15. [CrossRef]

15. Valencia, I.E.; Hernández, B.A. *Muestreo de Suelos, Preparación de Muestras y Guía de Campo*; Universidad Autónoma de México: Cuautitlán, México, 2002.

16. Eang, K.E.; Igarashi, T.; Kondo, M.; Nakatani, T.; Tabelin, C.B.; Fujinaga, R. Groundwater monitoring of an open-pit limestone quarry: Water-rock interaction and mixing estimation within the rock layers by geochemical and statistical analyses. *Int. J. Min. Sci. Technol.* **2018**, *28*, 849–857. [CrossRef]

17. Siqueira, D.S.; Marques, J.; Pereira, G.T.; Teixeira, D.D.B.; Vasconcelos, V.; De Carvalho, J.O.A.; Martins, E. Detailed mapping unit design based on soil–landscape relation and spatial variability of magnetic susceptibility and soil color. *Catena* **2015**, *135*, 149–162. [CrossRef]

18. Jordanova, D.; Jordanova, N. Thermomagnetic behavior of magnetic susceptibility—Heating rate and sample size effects. *Front. Earth Sci.* **2016**, *3*, 90. [CrossRef]

19. Lecoanet, H.; Lévêque, F.; Segura, S. Magnetic susceptibility in environmental applications: Comparison of field probes. *Phys. Earth Planet. Inter.* **1999**, *115*, 191–204. [CrossRef]

20. Dalan, R.A. A review of the role of magnetic susceptibility in archaeogeophysical studies in the USA: Recent developments and prospects. *Archaeol. Prospect.* **2008**, *15*, 1–31. [CrossRef]

21. Luo, L.; Nguyen, A.V. A review of principles and applications of magnetic flocculation to separate ultrafine magnetic particles. *Sep. Purif. Technol.* **2017**, *172*, 85–99. [CrossRef]

22. Heller, F.; Strzyszcz, Z.; Magiera, T. Magnetic record of industrial pollution in forest soils of Upper Silesia, Poland. *J. Geophys. Res. Space Phys.* **1998**, *103*, 17767–17774. [CrossRef]

23. Strzyszcz, Z.; Magiera, T. Magnetic susceptibility and heavy metals contamination in soils of Southern Poland. *Phys. Chem. Earth* **1998**, *23*, 1127–1131. [CrossRef]

24. Ďurža, O. Heavy metals contamination and magnetic susceptibility in soils around metallurgical plant. *Phys. Chem. Earth A Solid Earth Geod.* **1999**, *24*, 541–543. [CrossRef]

25. Lecoanet, H.; Lévêque, F.; Ambrosi, J.-P. Combination of magnetic parameters: An efficient way to discriminate soil-contamination sources (South France). *Environ. Pollut.* **2003**, *122*, 229–234. [CrossRef]

26. Yang, T.; Liu, Q.; Chan, L.; Cao, G. Magnetic investigation of heavy metals contamination in urban topsoils around the East Lake, Wuhan, China. *Geophys. J. Int.* **2007**, *171*, 603–612. [CrossRef]

27. Rachwał, M.; Magiera, T.; Wawer, M. Coke industry and steel metallurgy as the source of soil contamination by technogenic magnetic particles, heavy metals and polycyclic aromatic hydrocarbons. *Chemosphere* **2015**, *138*, 863–873. [CrossRef]

28. Rachwał, M.; Kardel, K.; Magiera, T.; Bens, O. Application of magnetic susceptibility in assessment of heavy metal contamination of Saxonian soil (Germany) caused by industrial dust deposition. *Geoderma* **2017**, *295*, 10–21. [CrossRef]

29. Wang, B.; Xia, D.; Yu, Y.; Chen, H.; Jia, J. Source apportionment of soil-contamination in Baotou City (North China) based on a combined magnetic and geochemical approach. *Sci. Total Environ.* **2018**, *642*, 95–104. [CrossRef]

30. Gómez-García, C.; Martín-Hernández, F.; López-García, J.Á.; Martínez-Pagán, P.; Manteca, J.I.; Carmona, C. Rock magnetic characterization of the mine tailings in Portman Bay (Murcia, Spain) and its contribution to the understanding of the bay infilling process. *J. Appl. Geophys.* **2015**, *120*, 48–59. [CrossRef]

31. Jordanova, D.; Goddu, S.R.; Kotsev, T.; Jordanova, N. contamination of alluvial soils near Fe–Pb mining site revealed by magnetic and geochemical studies. *Geoderma* **2013**, *192*, 237–248. [CrossRef]

32. De Matos, M.P.; Fontes, M.P.F.; Da Costa, L.M.; Martinez, M. Mobility of heavy metals as related to soil chemical and mineralogical characteristics of Brazilian soils. *Environ. Pollut.* **2001**, *111*, 429–435. [CrossRef]

33. Sarkar, S.; Sarkar, B.; Basak, B.B.; Mandal, S.; Biswas, B.; Srivastava, P. Soil Mineralogical Perspective on Immobilization/Mobilization of Heavy Metals. In *Adaptive Soil Management: From Theory to Practices*; Springer Science and Business Media LLC: Berlin/Heidelberg, Germany, 2017; pp. 89–102.

34. Esquenazi, E.L.; Norambuena, B.K.; Bacigalupo, Í.M.; Estay, M. Gof soil intervention values in mine tailings in northern Chile. *PeerJ* **2018**, *6*, e5879. [CrossRef] [PubMed]

35. Ghorbani, Y.; Kuan, S.H. A review of sustainable development in the Chilean mining sector: Past, present and future. *Int. J. Min. Reclam. Environ.* **2016**, *31*, 137–165. [CrossRef]

36. Araya, N.; Kraslawski, A.; Cisternas, L.A. Towards mine tailings valorization: Recovery of critical materials from Chilean mine tailings. *J. Clean. Prod.* **2020**, *263*, 121555. [CrossRef]

37. Lam, E.J.; Keith, B.F.; Montofré, Í.L.; Gálvez, M.E. Necessity of Intervention Policies For Tailings Identified In The Antofagasta Region, Chile. *Rev. Int. Contam. Ambient.* **2019**, *35*, 515–539. [CrossRef]

38. Alpers, C.N.; Brimhall, G.H. Middle Miocene climatic change in the Atacama Desert, northern Chile: Evidence from supergene mineralization at La Escondida. *GSA Bull.* **1988**, *100*, 1640–1656. [CrossRef]

39. Alpers, C.N.; Brimhall, G.H. Paleohydrologic evolution and geochemical dynamics of cumulative supergene metal enrichment at La Escondida, Atacama Desert, northern Chile. *Econ. Geol.* **1989**, *84*, 229–255. [CrossRef]

40. Maturana, M.R.; Saric, N. Geologia y minerallzacion del yacimiento tipo porfido cuprifero Zaldivar, en los Andes del norte de Chile. *Andean Geol.* **1991**, *18*, 109–120.

41. Durza, O.; Gregor, T.; Antalova, S. The effect of the heavy metals soil contamination on magnetic susceptibility. *Acta Univ. Carol. Geol.* **1993**, *37*, 135–143.

42. Vamerali, T.; Bandiera, M.; Mosca, G. In situ phytoremediation of arsenic- and metal-polluted pyrite waste with field crops: Effects of soil management. *Chemosphere* **2011**, *83*, 1241–1248. [CrossRef]

43. Rashed, M.N.; Awadallah, R.M. Trace elements in faba bean (*Vicia faba* L.) plant and soil as determined by atomic absorption spectroscopy and ion selective electrode. *J. Sci. Food Agric.* **1998**, *77*, 18–24. [CrossRef]

44. Berti, W.R.; Jacobs, L.W. Distribution of Trace Elements in Soil from Repeated Sewage Sludge Applications. *J. Environ. Qual.* **1998**, *27*, 1280–1286. [CrossRef]

45. Fellet, G.; Marchiol, L.; Perosa, D.; Zerbi, G. The application of phytoremediation technology in a soil contaminated by pyrite cinders. *Ecol. Eng.* **2007**, *31*, 207–214. [CrossRef]

46. Marchiol, L.; Fellet, G.; Perosa, D.; Zerbi, G. Removal of trace metals by Sorghum bicolor and Helianthus annuus in a site polluted by industrial wastes: A field experience. *Plant. Physiol. Biochem.* **2007**, *45*, 379–387. [CrossRef] [PubMed]

47. Clemente, R.; Dickinson, N.; Lepp, N.W. Mobility of metals and metalloids in a multi-element contaminated soil 20years after cessation of the pollution source activity. *Environ. Pollut.* **2008**, *155*, 254–261. [CrossRef]

48. Lindsay, W.L.; Norvell, W.A. Development of a DTPA Soil Test for Zinc, Iron, Manganese, and Copper. *Soil Sci. Soc. Am. J.* **1978**, *42*, 421–428. [CrossRef]

49. Tukey, J.W. Comparing Individual Means in the Analysis of Variance. *Biometrics* **1949**, *5*, 99. [CrossRef]

50. Kapicka, A.; Petrovský, E.; Ejordanova, D.; Podrázský, V. Magnetic parameters of forest top soils in Krkonoše mountains, Czech Republic. *Phys. Chem. Earth A Solid Earth Geod.* **2001**, *26*, 917. [CrossRef]

51. Hoffmann, V.; Knab, M.; Appel, E. Magnetic susceptibility mapping of roadside pollution. *J. Geochem. Explor.* **1999**, *66*, 313–326. [CrossRef]

52. Boyko, T.; Scholger, R.; Stanjek, H.; Team, M. Topsoil magnetic susceptibility mapping as a tool for pollution monitoring: Repeatability of in situ measurements. *J. Appl. Geophys.* **2004**, *55*, 249–259. [CrossRef]

53. Magiera, T.; Strzyszcz, Z.; Kapicka, A.; Petrovsky, E. Discrimination of lithogenic and anthropogenic influences on topsoil magnetic susceptibility in Central Europe. *Geoderma* **2006**, *130*, 299–311. [CrossRef]

54. Dold, B.S.; Fontboté, L. A mineralogical and geochemical study of element mobility in sulfide mine tailings of Fe oxide Cu–Au deposits from the Punta del Cobre belt, northern Chile. *Chem. Geol.* **2002**, *189*, 135–163. [CrossRef]

55. Tabelin, C.B.; Veerawattananun, S.; Ito, M.; Hiroyoshi, N.; Igarashi, T. Pyrite oxidation in the presence of hematite and alumina: I. Batch leaching experiments and kinetic modeling calculations. *Sci. Total Environ.* **2017**, *580*, 687–698. [CrossRef] [PubMed]

56. Tabelin, C.B.; Veerawattananun, S.; Ito, M.; Hiroyoshi, N.; Igarashi, T. Pyrite oxidation in the presence of hematite and alumina: II. Effects on the cathodic and anodic half-cell reactions. *Sci. Total Environ.* **2017**, *581*, 126–135. [CrossRef] [PubMed]

57. Li, X.; Gao, M.; Hiroyoshi, N.; Tabelin, C.B.; Taketsugu, T.; Ito, M. Suppression of pyrite oxidation by ferric-catecholate complexes: An electrochemical study. *Miner. Eng.* **2019**, *138*, 226–237. [CrossRef]

58. Tomiyama, S.; Igarashi, T.; Tabelin, C.B.; Tangviroon, P.; Ii, H. Acid mine drainage sources and hydrogeochemistry at the Yatani mine, Yamagata, Japan: A geochemical and isotopic study. *J. Contam. Hydrol.* **2019**, *225*, 103502. [CrossRef]

59. Igarashi, T.; Herrera, P.S.; Uchiyama, H.; Miyamae, H.; Iyatomi, N.; Hashimoto, K.; Tabelin, C.B. The two-step neutralization ferrite-formation process for sustainable acid mine drainage treatment: Removal of copper, zinc and arsenic, and the influence of coexisting ions on fertilization. *Sci. Total Environ.* **2020**, *715*, 136877. [CrossRef]

60. Dold, B.S.; Fontboté, L. Element cycling and secondary mineralogy in porphyry copper tailings as a function of climate, primary mineralogy, and mineral processing. *J. Geochem. Explor.* **2001**, *74*, 3–55. [CrossRef]

61. Del Rio-Salas, R.; Ayala-Ramírez, Y.; Loredo-Portales, R.; Romero, F.; Molina-Freaner, F.; Minjarez-Osorio, C.; Pi-Puig, T.; Ochoa–Landín, L.; Moreno-Rodríguez, V. Mineralogy and Geochemistry of Rural Road Dust and Nearby Mine Tailings: A Case of Ignored Pollution Hazard from an Abandoned Mining Site in Semi-arid Zone. *Nat. Resour. Res.* **2019**, *28*, 1485–1503. [CrossRef]

62. Dold, B. Evolution of Acid Mine Drainage Formation in Sulphidic Mine Tailings. *Minerals* **2014**, *4*, 621–641. [CrossRef]

63. Dold, B.S. Speciation of the most soluble phases in a sequential extraction procedure adapted for geochemical studies of copper sulfide mine waste. *J. Geochem. Explor.* **2003**, *80*, 55–68. [CrossRef]

64. Shu, W.; Ye, Z.; Lan, C.; Zhang, Z.; Wong, M. Acidification of lead/zinc mine tailings and its effect on heavy metal mobility. *Environ. Int.* **2001**, *26*, 389–394. [CrossRef]

65. Kovács, E.; Dubbin, W.E.; Tamás, J. Influence of hydrology on heavy metal speciation and mobility in a Pb–Zn mine tailing. *Environ. Pollut.* **2006**, *141*, 310–320. [CrossRef]

66. Haffert, L.; Craw, D. Mineralogical controls on environmental mobility of arsenic from historic mine processing residues, New Zealand. *Appl. Geochem.* **2008**, *23*, 1467–1483. [CrossRef]

67. Tabelin, C.B.; Sasaki, R.; Igarashi, T.; Park, I.; Tamoto, S.; Arima, T.; Ito, M.; Hiroyoshi, N. Simultaneous leaching of arsenite, arsenate, selenite and selenate, and their migration in tunnel-excavated sedimentary rocks: I. Column experiments under intermittent and unsaturated flow. *Chemosphere* **2017**, *186*, 558–569. [CrossRef] [PubMed]

68. Tabelin, C.B.; Igarashi, T.; Villacorte-Tabelin, M.; Park, I.; Opiso, E.M.; Ito, M.; Hiroyoshi, N. Arsenic, selenium, boron, lead, cadmium, copper, and zinc in naturally contaminated rocks: A review of their sources, modes of enrichment, mechanisms of release, and mitigation strategies. *Sci. Total Environ.* **2018**, *645*, 1522–1553. [CrossRef] [PubMed]

69. Milborrow, S. Notes on the Earth Package. Available online: https://www.milbo.org/doc/earth-varmod.pdf (accessed on 23 June 2017).

70. R Core Team. R: A Language and Environment for Statistical Computing. 2018. Available online: https://www.r-project.org/ (accessed on 13 February 2012).

71. Nieto, P.G.; Lasheras, F.S.; Juez, F.D.C.; Fernández, J.A. Study of cyanotoxins presence from experimental cyanobacteria concentrations using a new data mining methodology based on multivariate adaptive regression splines in Trasona reservoir (Northern Spain). *J. Hazard. Mater.* **2011**, *195*, 414–421. [CrossRef] [PubMed]

72. Presno-Vélez, Á.; Bernardo-Sánchez, A.; Menéndez-Fernández, M.; Fernández-Muñiz, Z. Multivariate Analysis to Relate CTOD Values with Material Properties in Steel Welded Joints for the Offshore Wind Power Industry. *Energies* **2019**, *12*, 4001. [CrossRef]

73. Karimi, R.; Ayoubi, S.; Jalalian, A.; Sheikh-Hosseini, A.R.; Afyuni, M. Relationships between magnetic susceptibility and heavy metals in urban topsoils in the arid region of Isfahan, central Iran. *J. Appl. Geophys.* **2011**, *74*, 1–7. [CrossRef]

74. Harikrishnan, N.; Chandrasekaran, A.; Ravisankar, R.; Alagarsamy, R. Statistical assessment to magnetic susceptibility and heavy metal data for characterizing the coastal sediment of East coast of Tamilnadu, India. *Appl. Radiat. Isot.* **2018**, *135*, 177–183. [CrossRef]

75. Wang, X.S.; Qin, Y. Use of multivariate statistical analysis to determine the relationship between the magnetic properties of urban topsoil and its metal, S, and Br content. *Environ. Earth Sci.* **2006**, *51*, 509–516. [CrossRef]

76. Juez, F.J.D.C.; Lasheras, F.S.; Roqueñí, N.; Osborn, J. An ANN-Based Smart Tomographic Reconstructor in a Dynamic Environment. *Sensors* **2012**, *12*, 8895–8911. [CrossRef] [PubMed]

77. Ordóñez, C.; Lasheras, F.S.; Roca-Pardiñas, J.; Juez, F.J.D.C. A hybrid ARIMA–SVM model for the study of the remaining useful life of aircraft engines. *J. Comput. Appl. Math.* **2019**, *346*, 184–191. [CrossRef]

78. Friedman, J.H. Multivariate Adaptive Regression Splines. *Ann. Stat.* **1991**, *19*, 1–67. [CrossRef]

79. Hanesch, M.; Scholger, R. Mapping of heavy metal loadings in soils by means of magnetic susceptibility measurements. *Environ. Earth Sci.* **2002**, *42*, 857–870. [CrossRef]

80. Siqueira, D.S.; Marques, J., Jr.; Matias, S.S.R.; Barrón, V.; Torrent, J.; Baffa, O.; Oliveira, L.C. Correlation of properties of Brazilian Haplustalfs with magnetic susceptibility measurements. *Soil Use Manag.* **2010**, *26*, 425–431. [CrossRef]

81. Ayoubi, S.; Adman, V.; Yousefifard, M. Use of magnetic susceptibility to assess metals concentration in soils developed on a range of parent materials. *Ecotoxicol. Environ. Saf.* **2019**, *168*, 138–145. [CrossRef]

82. Wawer, M.; Rachwał, M.; Kowalska, J. Impact of noise barriers on the dispersal of solid pollutants from car emissions and their deposition in soil. *Soil Sci. Annu.* **2017**, *68*, 19–26. [CrossRef]

Publisher's Note: MDPI stays neutral with regard to jurisdictional claims in published maps and institutional affiliations.

 *minerals*

Article

Assessment of Native and Endemic Chilean Plants for Removal of Cu, Mo and Pb from Mine Tailings

Pamela Lazo [1,]* and **Andrea Lazo** [2]

[1] Instituto de Química y Bioquímica, Facultad de Ciencias, Universidad de Valparaíso, Avenida Gran Bretaña 1111, Playa Ancha, 2360102 Valparaíso, Chile

[2] Departamento de Ingeniería Química y Ambiental, Universidad Técnica Federico Santa María, Avenida España 1680, 2390123 Valparaíso, Chile; andrea.lazo@usm.cl

* Correspondence: pamela.lazo@uv.cl

Received: 13 October 2020; Accepted: 10 November 2020; Published: 17 November 2020

Abstract: In Chile, 85% of tailings impoundments are inactive or abandoned and many of them do not have a program of treatment or afforestation. The phytoremediation of tailings with *Oxalis gigantea*, *Cistanthe grandiflora*, *Puya berteroniana* and *Solidago chilensis* have been tested in order to find plants with ornamental value and low water requirements, which enable reductions in molybdenum (Mo), copper (Cu) or lead (Pb) concentrations creating an environmentally friendly surrounding. Ex-situ phytoremediation experiments were carried out for seven months and Mo, Cu and Pb were measured at the beginning and at the end of the growth period. The capacity of these species to phyto-remedy was evaluated using the bioconcentration and translocation factors, along with assessing removal efficiency. *Solidago chilensis* showed the ability to phytoextract Mo while *Puya berteroniana* showed potential for Cu and Mo stabilization. The highest removal efficiencies were obtained for Mo, followed by Cu and Pb. The maximum values of removal efficiency for Mo, Cu and Pb were 28.7% with *Solidago chilensis*, 15.6% with *Puya berteroniana* and 8.8% with *Cistanthe grandiflora*, respectively. Therefore, the most noticeable results were obtained with *Solidago chilensis* for phytoextraction of Mo.

Keywords: phytoremediation; heavy metals; mine tailings; endemic species; native species

1. Introduction

Tailings are a mixture of water and heavy metal-bearing fine-grained minerals [1,2]. In Chile, there exists 757 tailings storage facilities (TSF) of which 173 are abandoned, 111 active, 468 inactive and 5 of them are under construction, according to the last record of mine tailings published on August 10th, 2020 by the National Geology and Mining Agency of Chile [3].

Soil contamination by heavy metals can be particularly hazardous due to the properties of these elements [4]. Central Chile presents climatic conditions that favor the dispersion of particles and the occurrence of metal lixiviation [5].

In Chile, mining from porphyry copper and molybdenum deposits occurs and it is common to find—in the areas surrounding mining activities—high concentrations of As, Cd, Cu, Zn, Pb, and Mo, and thus, soil pollution by potentially toxic elements contained in mining tailings is a latent problem that can cause important environmental damage [6,7].

Lead (Pb) is one of the most toxic metals and it has a significant influence on plant growth and development [8]. Under normal environmental conditions, the mobility of Pb is low but it is increased when more acidic conditions prevail [9]. The toxicity and adverse effects of Pb on plant species have been found to occur at very low concentrations, even at micromolar levels [10]. A consensus exists that the Pb taken up by plants from soils remains in the roots [11,12]. Pb may be translocated from roots to the aerial parts of the plant, however, in the majority of plants (>95%) Pb is accumulated in the roots and only a small portion is translocated to the parts above the ground [9]. The threshold level of Pb for plants is around 2 mg·kg^{-1} [13].

Copper (Cu) is an essential metal for plants; however, it is toxic at high concentrations. Normal values of Cu in plants are between 4 and 15 mg Cu·kg^{-1} dry matter and the critical values in roots are in the range of 100 to 400 mg Cu·kg^{-1} dry matter [14]. Oorts, et al. (2013) indicated the onset of Cu toxicity in shoots and leaves between 5 and 40 mg Cu·kg^{-1} dry matter, while Marschner (2000) specified a concentration higher than 20 or 30 mg Cu·kg^{-1}, depending on plant species [14,15].

In the case of molybdenum (Mo), it can be mobile and bioavailable as MoO_4^{2-} [9]. Only small quantities of this element are required by plants, with the normal range for most plant tissues being between 0.3 mg·kg^{-1} and 1.5 mg·kg^{-1}. Moreover, toxicity levels of Mo in plants differ according the species, where values of toxic Mo concentrations have been reported in the range from 100 to 1000 mg·g^{-1} dry matter [16].

The large number of abandoned tailings makes it necessary to find a cost-effective solution and, therefore, to mitigate the negative effects of heavy metals in soils, several methods such as membrane filtration, electrodialysis, and soil washing, among others, have been explored, however, they are expensive and environmentally unfriendly [17]. Among the remediation technologies, several studies have proven the usefulness of phytoremediation as an efficient and environmentally friendly method for removing organic and inorganic contaminants, moreover, it is a cheaper method compared to chemical remediation, biopiles and bioventing, which incorporates the use of plants to remove contaminants from water and soil [18].

Marques et al. (2009) highlighted the three major phytoremediation techniques: phytoextraction, stabilization and volatilization [19]. Additionally, Lam et al. (2018) distinguished two strategies of phytoextraction: the use of plants with a large ability of accumulation in shoots and low biomass, and the use of plant species with high biomass and low ability of extraction [20].

The potential use of certain species for phytoremediation can be evaluated by using the bioconcentration factor (BCF) and translocation factor (TF). BCF is described as the ability of plants for elemental accumulation from the substrate, and the ratio between the concentration of metal present in the plant and the total final metal concentration in soil is considered as an index of bioavailability [21], while TF is used to assess the plant's potential to translocate contaminants [22,23]. BCF values higher than one are indicative of potential success of a certain plant species for phytoremediation, while a TF greater than one indicates the ability to translocate the metal to aerial parts [21]. On the other hand, the consideration of a species as a stabilizer of heavy metals is based on a BCF ≥ 1 and a TF ≤ 1 [24].

Tailings are a poor medium for promoting natural plant growth, they normally have low field capacity, high salinity, high concentration of contaminants such as heavy metals and a lack of organic matter [18]. In order to improve the characteristic of the substrate and to achieve self-sustaining growth of the plants over time, the addition of nutrients, and amendments and/or organic matter are essential for phytoremediation to remediate tailings [25,26].

Prosopis tamarugo, Schinus molle and *Artiplex nummularia,* all of them Chilean native species, have been studied for in-situ phytoremediation of tailings in the region of Antofagasta, Chile, with the addition of an organic compost and water for irrigation [27]. All species showed BCF < 1 with different treatments, but *S. molle* has shown features as an accumulator for Cu, Mn, Pb and Zn, and *P. tamarugo* for Mn, Zn and Cd, with TF > 1. *A. nummularia* was the most promising of these species, it showed an accumulator behavior for Mn, Pb and Zn [27]. Lam et al. (2018) evaluated the potential of *Adesmia atacamensis* in the phytoremediation of mine tailings. The results of TF and BCF allowed for the classification of the plant as a Cu hyperaccumulator [20].

Alfonso et al. (2020) obtained auspicious results with the use of indigenous plants for the in-situ phytoremediation of tailings from the Camaquã Mine (Southern Brazil). Eleven different species of spontaneous occurrence in the mine site were assessed. The translocation factor and bioconcentration factor were calculated. Seven of the studied species showed phytoextraction potential for Pb and four species showed some ability for the phytostabilization of Cu [28].

The aim of this study was to determine the potential of Chilean native or endemic plant species, to phyto-remedy mine tailings. Four species from northern Chile: *Oxalis gigantea*, *Cistanthe grandiflora*, *Puya berteroniana* and *Solidago chilensis*, were chosen according to their low water requirements and ornamental value. The potential of these species for phytostabilization or phytoextraction of Mo, Cu and/or Pb in mine tailings was assessed through ex-situ pot experiments.

2. Materials and Methods

2.1. Characterization and Preparation of Mine Tailing

Paste tailing from Compañía Minera Las Cenizas located in Cabildo, Valparaíso Region, Chile was used. The mine company processed copper sulfide and oxide minerals. The sampling location is presented in Figure 1.

Figure 1. Tailing storage facility (32°28′16.1″ S, 71°05′00.2″ W).

Before the phytoremediation experiments, tailings were dried at 105 °C until achieving constant mass, ground in a ball mill, sieved through an ASTM mesh 19 mm and homogenized [29]. The main properties of the tailings are presented in Table 1. Table 2 shows the initial concentrations of Mo, Cu and Pb measured by ICP-OES.

Table 1. Main geochemical properties of tailing.

Parameter	Value
Specific gravity	2.82
Solid concentration in weight %	83
Granulometry d_{50} micrometers	0.046
Granulometry d_{20} micrometers	0.005
Granulometry d_{80} micrometers	0.240

Table 2. Initial concentration of Mo, Cu and Pb in dry tailing ± confidence interval (IC).

Element	Concentration mg·kg^{-1} Dry Tailing ± IC
Cu	1582.22 ± 78.31
Mo	3.86 ± 0.17
Pb	228.15 ± 2.79
Zn	86.98 ± 3.15
Ni	9.46 ± 0.25
Cd	Under detection limit
Cr	15.46 ± 0.54

2.2. Plants Species

Four different plant species were used for the phytoremediation studies: *Oxalis gigantea, Cistanthe grandiflora, Puya berteroniana* and *Solidago chilensis.*

Oxalis gigantea Barnéoud (Churqui or Churco) is a very common endemic Chilean plant which belongs to the *Oxalidaceae* family. It grows in northern Chile from the Antofagasta to Coquimbo regions and is hardy to USDA Zone 10 and 11. *Cistanthe grandiflora*, frequently called Doquilla or Pata de guanaco is an endemic Chilean plant of the *Portulacaceae* family, which can be found between the Antofagasta and Ñuble regions. It is hardy to USDA Zone 9. *Puya berteroniana* is an endemic Chilean plant of the *Bromeliaceae* family, commonly called Chagual, Cardón or Magüey and has an excellent ornamental value. This plant grows from the Coquimbo to Maule regions and is hardy to USDA Zone 9. Finally, *Solidago chilensis*, or commonly called Fulel, is a native Chilean plant that can be found between the Arica and Parinacota and Los Lagos regions. This plant belongs to the *Astaraceae* family, *Solidago chilensis* is hardy to USDA Zone 9 and Los Lagos [30,31].

2.3. Potted Experiments

Plants with an initial height of 10 cm were placed into pots with 1440 g of dry tailing. The pots were left outdoors over a seven-month period, under similar environmental conditions to those where the mine tailings impoundment is located.

For each plant species, three specimens were placed in tailing. Potable water and biofertilizer were provided weekly and monthly, respectively. The characterization of foliar organic stimulant is presented in Table 3.

Table 3. Foliar organic stimulant composition (based on marine algae *Ascophyllum nodosum*).

Element	Concentration
Nitrogen	0.1% *w/w*
Phosphorous	0.0% *w/w*
Potassium	3.0% *w/w*
Arsenic	<0.5 mg·kg^{-1}
Cadmium	<0.5 mg·kg^{-1}
Lead	<1 mg·kg^{-1}
Mercury	<0.5 mg·kg^{-1}

2.4. Sample Preparation and ICP-OES Measurements

Upon the expiry of the growth period, leaves and stems (aerial part) and roots were divided with a knife and carefully washed with abundant potable water, distilled water and deionized water to remove tailing particles adhering to them and any other type of dirt. Both parts of the plants were cut to reduce their size and placed into waxed paper envelopes, afterwards they were dried at 45 °C until constant mass was achieved, ground and homogenized.

Tailing was carefully cleaned, dried at 105 °C until constant mass was achieved, grounded, sieved through an ASTM mesh N°18 and homogenized.

For ICP-OES measurements, digestion procedure was carried out with 0.200 g of dry sample, which were placed in a Teflon vial for microwave and 8 mL of concentrated HNO_3 and 2.0 mL of concentrated H_2O_2 were added. The vials were covered with parafilm tape and were left to pre-digest for 4 h before the digestion in the microwave. When the samples were at room temperature, they were placed in 25 mL volumetric flasks which were then filled with deionized water to the calibration line. All reagents were of analytical grade.

All samples were prepared in duplicate and digested twice in a microwave Ethos Easy. The temperature program consisted of three segments: the first from 0 to 10 min with an increase in temperature until 180 °C, a second period of 10 min with a constant temperature of 180 °C and the last corresponding to a cool down period of 10 min.

2.5. Heavy Metal Determination

The concentrations of Mo, Cu and Pb were determined in tailing and plants (roots and stems + leaves = aerial part). The metal concentrations in plants and tailings samples were determined by inductively coupled atomic emission spectroscopy (Perkin Elmer), directly from digested solutions at the Institute of Chemistry and Biochemistry, Faculty of Science of Universidad de Valparaíso, Chile.

For the present study, the bioaccumulation factor (BCF) and translocation factor (TF) were calculated with Equations (1)–(3) [4,32–34].

$$TF = \frac{\text{Metal concentration (Stems + Leaves)}}{\text{Metal concentration in roots}} \tag{1}$$

$$BCF_{roots} = \frac{\text{Metal concentration in roots}}{\text{Initial concentration of metal in tailing}} \tag{2}$$

$$BCF_{aerial} = \frac{\text{Metal concentration in aerial parts}}{\text{Initial concentration of metal in tailing}} \tag{3}$$

The removal efficiency (RE) was calculated with the Equation (4).

$$RE = \frac{(C_i - C_f)}{C_i} \times 100\% \tag{4}$$

where C_i and C_f are the initial and final concentration of the element in the tailing.

3. Results

The final concentrations of Mo, Pb and Cu in each plant species, divided into roots and aerial parts, were determined after the growth period. For each plant, three samples of roots and three samples of aerial parts were taken in duplicate, the final mean concentration of each duplicate is presented in Figure 2.

All plant species showed a decreasing trend of Pb and Cu concentrations from tailing to aerial parts (tailings → roots → leaves and stems) but in the case of Mo this decreasing trend is only observed in the case of *Cistanthe grandiflora*.

Oxalis gigantea presented a Mo concentration in aerial parts slightly higher than in roots, while *Puya berteroniana* exhibited a concentration of Mo in roots higher than what was found in the final tailing. *Solidago chilensis* showed the reverse trend with a decreasing concentration of Mo from aerial parts to tailing.

The ability of *Solidago chilensis* and *Puya berteroniana* to accumulate Pb in their roots is notorious, far exceeding the normal threshold levels. Additionally, the same species showed the ability of Cu accumulation in roots.

To evaluate the ability of all species to translocate or stabilize the studied metals, TF, BCF and removal efficiency were calculated; in the case of BCF, this factor was obtained for roots and for aerial parts. The results are presented in Table 4.

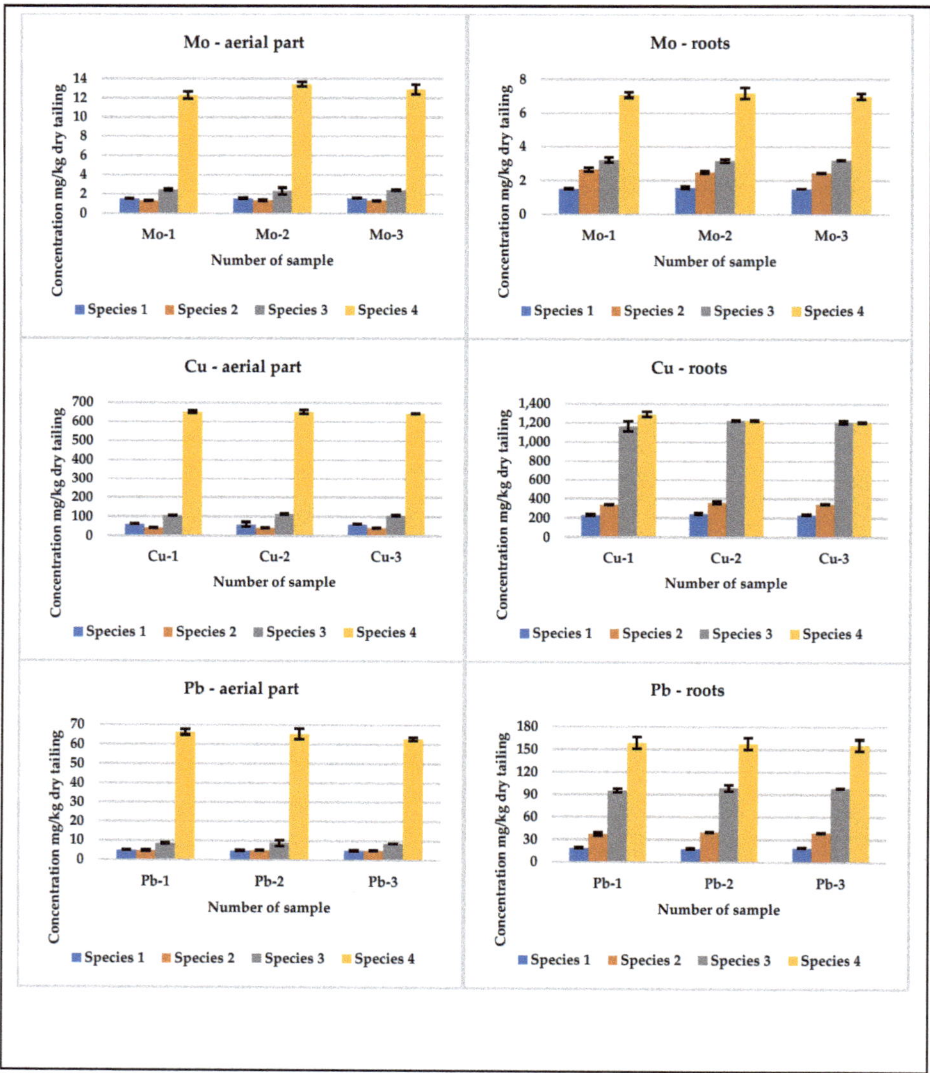

Figure 2. Mean concentration in duplicate samples, aerial part and roots, where species 1: *Oxalis gigantea*, species 2: *Cistanthe grandiflora*, Species 3: *Puya berteroniana*, species 4: *Solidago chilensis*.

Table 4. Translocation factor (TF), bioconcentration factor (BCF) and removal efficiency after the growth period.

Element	Plant Species	*Oxalis gigantea*	*Cistanthe grandiflora*	*Puya berteroniana*	*Solidago chilensis*
Mo	TF	1.03 ± 0.03	0.53 ± 0.03	0.75 ± 0.02	1.82 ± 0.06
	BCF_{roots}	0.47 ± 0.01	0.78 ± 0.03	1.03 ± 0.01	2.57 ± 0.03
	BCF_{aerial}	0.49 ± 0.00	0.42 ± 0.01	0.73 ± 0.02	4.68 ± 0.17
	% RE	16.38 ± 0.75	15.86 ± 0.34	19.48 ± 0.53	28.70 ± 1.57
Pb	TF	0.27 ± 0.01	0.13 ± 0.01	0.09 ± 0.01	0.41 ± 0.01
	BCF_{roots}	0.09 ± 0.01	0.19 ± 0.01	0.46 ± 0.01	0.72 ± 0.01
	BCF_{aerial}	0.02 ± 0.00	0.02 ± 0.00	0.04 ± 0.00	0.30 ± 0.01
	% RE	7.01 ± 0.87	8.78 ± 0.56	7.62 ± 0.45	4.41 ± 0.22
Cu	TF	0.26 ± 0.01	0.12 ± 0.00	0.09 ± 0.00	0.52 ± 0.01
	BCF_{roots}	0.16 ± 0.01	0.24 ± 0.01	0.90 ± 0.02	0.92 ± 0.03
	BCF_{aerial}	0.04 ± 0.01	0.03 ± 0.00	0.08 ± 0.01	0.48 ± 0.02
	% RE	8.63 ± 0.67	8.72 ± 0.78	15.59 ± 1.03	14.91± 0.98

According to the results shown in Table 4, all studied species presented poor ability to translocate Pb and Cu with a TF < 1. In the case of Mo, *Oxalis gigantea* and *Solidago chilensis* are good candidates for Mo phytoextraction with a TF > 1, where the second appears more promising due to its bioconcentration factor values for roots and aerial parts.

The analysis of the values for BCF highlight the potential use of *Puya berteroniana* for Mo phytostabilization. In the case of Cu, *Puya berteroniana* and *Solidago chilensis* showed a potential for phytostabilization with a BCF close to one. These factors could be improved through the study of the use of nanoparticles and/or chemical solutions, also, the mixture of mine tailings with compost or fertilizers could be considered.

The maximum removal efficiencies were obtained for Mo with all studied species, among which, *Solidago chilensis* showed a value close to 30%, followed by *Puya berteroniana* with a 19.5% removal efficiency. In the case of Pb removal, efficiencies were lower than 9%, *Cistanthe grandiflora* presented the best results with a removal efficiency of 8.8%. For Cu, the maximum values of removal efficiency—close to 15%—were obtained with *Puya berteroniana* and *Solidago chilensis*.

4. Discussion

Figure 3 shows the mean concentration ± IC (confidence interval) for each species and each metal after the experimental period. The Mo accumulated in roots decreases as follows: *Solidago chilensis* > *Puya berteroniana* > *Cistanthe grandiflora* > *Oxalis gigantea*, there is little variation in this trend in the case of aerial parts where the Mo concentration decreases as follows: *Solidago chilensis* > *Puya berteroniana* > *Oxalis gigantea* > *Cistanthe grandiflora*. All species outweighed the normal values for most plant tissues and the accumulation of Mo in aerial parts and roots of *Cistanthe grandiflora* by unit of dry matter is the highest in the group of the studied species.

The concentrations of Cu in the roots of *Oxalis gigantea* and *Cistanthe grandiflora* are in the range of critical values indicated by Oorts et al. (2013), in the case of *Puya berteroniana* and *Solidago chilensis* these values are outweighed. In the case of the aerial parts, all species exceeded the toxic levels indicated by Marschner (2000) and Oorts et al. (2013) [14,15].

In the case of Pb, the threshold level of Pb for plants is clearly surpassed. In terms of mass of Pb by unit of dry matter, the increase in concentration in the roots is as follows: *Oxalis gigantea* < *Cistanthe grandiflora* < *Puya berteroniana* < *Solidago chilensis*, with a slight change in the case of aerial parts where the concentration showed by *Oxalis gigantea* is similar to that of *Cistanthe grandiflora*.

It is important to mention that Chile lacks regulations for soil pollutants including heavy metals, and therefore, is not possible to compare with the Chilean norm.

The high ability of *Solidago Chilensis* to accumulate Mo, Pb and Cu, in respect to the other species, is also shown in Figure 3.

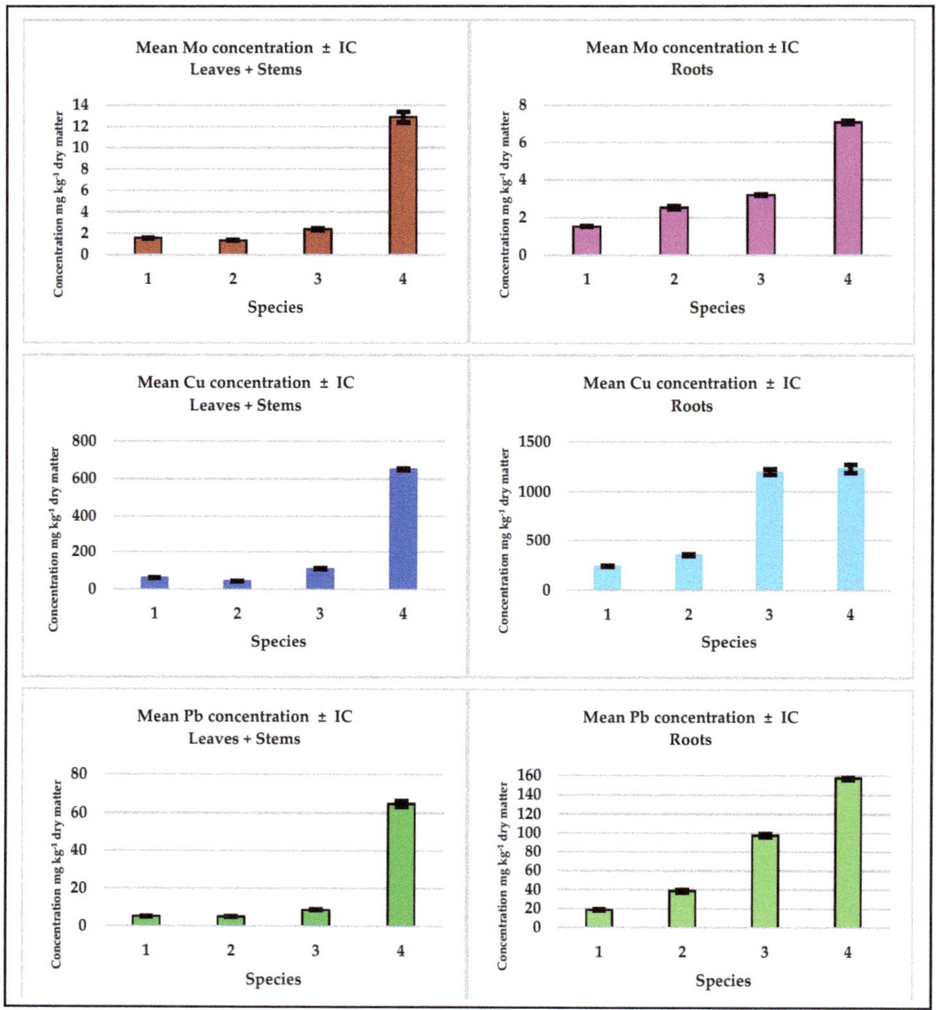

Figure 3. Mean concentration ± IC of Mo, Cu and Pb in roots and aerial parts, where, 1: *Oxalis gigantea*, 2: *Cistanthe grandiflora*, 3: *Puya berteroniana*, 4: *Solidago chilensis*.

The plant species used in the present research have not been studied before for phytoremediation. Some native and endemic species of plants have been previously used but not with the consideration of water requirements.

Lam et al. (2017) studied native Chilean species for the phytoremediation of tailings, among which *Schinus molle* showed the ability to translocate Cu and Pb with TF = 2.78 and 1.33, respectively, and a BCF < 1 in both cases in tailings without amendments [27]. In the same study, *Atriplex nummularia* presented a TF = 1.33 and a BCF < 1 under the same conditions.

In a later study, Lam et al., (2018) established the potential of *Adesmia atacamensis* (TF = 2.47 and BCF = 0.05) to accumulate Pb in aerial part in tailings without treatment [20].

For comparison, the Pb concentrations of *Adesmia atacamensis* reported by Lam et al. (2018) were 4.7 mg·kg^{-1} in roots and 11.6 mg·kg^{-1} in aerial parts. In the case of this study, *Oxalis Gigantea* and *Cistanthe grandiflora* showed concentrations of 5.05 ± 0.07 mg·kg^{-1} and 4.95 ± 0.08 mg·kg^{-1} in the aerial parts and 18.39 ± 0.53 mg·kg^{-1} and 38.56 ± 1.15 mg·kg^{-1} in the roots, respectively. Although *Cistanthe grandiflora* is capable of accumulating higher concentrations of Pb in roots than *Adesmia atcamensis*, it lack the ability to translocate it.

The same behavior for Pb is observed in the case of *Puya berteroniana* and *Solidago chilensis* (8.66 ± 0.10 mg·kg^{-1} and 64.89 ±1.51 mg·kg^{-1} in the aerial parts and 97.39 ± 1.80 mg·kg^{-1} and 157.58 ± 1.34 mg·kg^{-1} in the roots), where both species showed higher concentrations than *Adesmia atacamensis*.

Ortiz-Calderón et al. (2008) analyzed the concentration of Cu in leaves and roots of several species, among them two Chilean native species: *Schinus polygamous* and *Atriplex deserticota* presented a Cu concentration in leaves of 1.213 and 1.358 mg·kg^{-1} dry mass, respectively, while in roots the concentration was 260 and 2160 mg·kg^{-1} dry mass, respectively [35]. Among them *Schinus polygamous* showed a clear ability to translocate Cu and therefore, to extract Cu. In the case of the present study *Puya berteroniana* and *Solidago chilensis* presented a TF close to one, which should be improved in order to increase the ability of these species to translocate Cu.

While this research was carried out ex-situ, it provides substantive information about the potential ability of the studied species to phyto-remedy Cu, Mo and Pb in mine tailings. Future work must be undertaken in order to improve this ability, for example, using joint implementation with another technology. Additionally, experiments in-situ must be performed accompanied by a sequential extraction procedure of the mine tailings for each studied element.

5. Conclusions

This study covers the potential ability of three endemic Chilean plant species and one native plant species, all of them from northern Chile, for the phytoremediation of Mo, Cu and Pb in mine tailings.

The ability of *Solidago chilensis* for the phytoextraction of Mo is highlighted, as is—to a lesser extent—the ability of *Oxalis* gigantea. In the case of Cu, *Puya berteroniana* and *Solidago chilensis* showed potential for phytostabilization which could be increased with the addition of chemicals or via joint implementation of another technique of remediation, which will be the subject of future studies.

It is important to mention that all these species have ornamental value, therefore, the phytoremediation with them, not only serves to decrease the concentration of the studied elements, but also provides a pleasant environment to the community. In addition to the above, the low water requirements of these species allow for their growth and development in water shortage scenarios.

Finally, the most noticeable results were obtained in the case of Mo, where *Solidago chilensis* should be the chosen species for Mo phytoextraction, while with *Puya berteroniana*, high removal efficiencies for Cu and Pb were obtained.

Author Contributions: Conceptualization, A.L.; formal analysis, A.L. and P.L.; investigation, A.L. and P.L.; methodology, A.L. and P.L.; project administration, A.L.; resources, A.L. and P.L.; writing—original draft, P.L.; writing—review and editing, A.L and P.L. All authors have read and agreed to the published version of the manuscript.

Funding: This research was funded by UNIVERSIDAD TÉCNICA FEDERICO SANTA MARÍA, grant number PI_L_17_05.

Acknowledgments: The authors acknowledge to Knud Henrik Hansen for permission to work at the laboratory of electrochemistry at Chemical and Environmental Engineering Department.

Conflicts of Interest: The authors declare no conflict of interest.

References

1. Wang, P.; Sun, Z.; Hu, Y.; Cheng, H. Leaching of heavy metals from abandoned mine tailings brought by precipitation and the associated environmental impact. *Sci. Total Environ.* **2019**, *695*, 133893. [CrossRef] [PubMed]
2. Lottermoser, B. *Mine Wastes: Characterization, Treatment and Environmental Impacts*, 2nd ed.; Springer Science & Business Media: Berlin/Heidelberg, Germany, 2007.
3. Sernageomin. Available online: https://www.sernageomin.cl/datos-publicos-deposito-de-relaves/ (accessed on 10 August 2020).
4. Radziemska, M.; Versova, M.D.; Baryla, A. Phytostabilization—Management Strategy for Stabilizing Trace Elements in Contaminated Soils. *Int. J. Environ. Res. Public Health* **2017**, *14*, 958. [CrossRef] [PubMed]
5. Ochard, C.; León-Lobos, O.; Ginocchio, R. Phytostabilization of massive mine wastes with native phytogenic resources: Potential for sustainable use and conservation of the native flora in north-central Chile. *Cienc. Inv. Agric.* **2009**, *35*, 3. [CrossRef]
6. Frascoli, F.; Hudson-Edwards, K.A. Geochemistry, Mineralogy and Microbiology of Molybdenum in Mining-Affected Environments. *Minerals* **2018**, *8*, 42. [CrossRef]
7. Tapia, J.; Valdés, J.; Orrego, R.; Tchernitchin, A.; Dorador, C.; Bolados, A.; Harrod, C. Geologic and anthropogenic sources of contamination in settled dust of a historic mining port city in northern Chile: Health risk implications. *PeerJ* **2018**, *6*, e4699. [CrossRef]
8. Zulfiqar, U.; Farooq, M.; Hussain, S.; Maqsood, M.; Hussain, M.; Ishfaq, M.; Ahmad, M.; Zohaib Anjum, M. Lead toxicity in plants: Impacts and remediation. *J. Environ. Manag.* **2019**, *250*, 109557. [CrossRef]
9. David, A.J.; Leventhal, J.S. Chapter 2: Bioavailability of metals. In *Book Preliminary Compilation of Descriptive Geoenvironmental Mineral Deposit Models*; Open-File Report; Department of the Interior, U.S. Geological Survey: Reston, VA, USA, 1995; pp. 95–831.
10. Pourrut, B.; Shahid, M.; Dumat, C.; Winterton, P.; Pinelli, E. Lead Uptake, Toxicity, and Detoxification in Plants. *Rev. Environ. Contam. Toxicol.* **2011**, *213*, 113–136. [CrossRef]
11. Kumar, P.B.A.N.; Dushenkov, V.; Motoo, H.; Raskin, I. Phytoextraction: The use of plants to remove heavy metals from soils. *Environ. Sci. Technol.* **1995**, *29*, 1232–1238. [CrossRef]
12. Sharma, P.; Dubey, R. Lead Toxicity in Plants. *Braz. J. Plant Physiol.* **2005**, *17*. [CrossRef]
13. WHO. *Permissible Limits of Heavy Metals in Soil and Plants*; World Health Organization: Geneva, Switzerland, 1996.
14. Oorts, K. Copper. In *Heavy Metals in Soils. Environmental Pollution*; Alloway, B., Ed.; Springer: Dordrecht, The Netherlands, 2013; Volume 22. [CrossRef]
15. Marschner, H. *Mineral Nutrition of Higher Plants*; Academic Press: London, UK, 1995; ISBN 9780124735439.
16. Gupta, U. Deficient, Sufficient, and Toxic Concentrations of Molybdenum in Crops. In *Molybdenum in Agriculture*; Gupta, U., Ed.; Cambridge University Press: Cambridge, UK, 1997; pp. 150–159. [CrossRef]
17. Anning, A.K.; Akoto, R. Assisted phytoremediation of heavy metal contaminated soil from a mined site with *Typha latifolia* and *Chrysopogon zizanioides*. *Ecotoxicol. Environ. Saf.* **2018**, *148*, 97–104. [CrossRef]
18. Wang, L.; Yuehua, B.J.; Hu, Y.; Liu, R.; Sun, W. A review on in situ phytoremediation of mine tailings. *Chemosphere* **2017**, *184*, 594–600. [CrossRef] [PubMed]
19. Marques, A.P.; Rangel, A.; Castro, P. Remediation of heavy metal contaminated soils: Phytoremediation as a potentially promising clean-up technology. *Crit. Rev. Environ. Sci. Technol.* **2009**, *39*, 622–654. [CrossRef]
20. Lam, E.; Keith, B.; Montofré, I. Copper Uptake by Adesmia atacamensis in a Mine Tailing in an Arid Environment. *Air Soil Water Res.* **2018**, *11*. [CrossRef]
21. Misra, V.; Tiwari, A.; Shukla, B.; Seth, C.S. Effects of soil amendments on the bioavailability of heavy metals from zinc mine tailings. *Environ. Monit. Assess.* **2009**, *155*, 467–475. [CrossRef] [PubMed]
22. Mishra, T.; Pandey, V. Phytoremediation of Red Mud Deposits through Natural Succession, Chapter 16. In *Phytomanagement of Polluted Sites*; Elsevier: Amsterdam, The Netherlands, 2019; pp. 409–424. [CrossRef]
23. Shen, Z.; Wang, Y.; Chen, Y.; Zhang, Z. Transfer of Heavy Metals from the polluted Rhizosphere Soil to *Celosia argentea* L., in Copper Mine Tailings. *Hortic. Environ. Biotechnol.* **2017**, *58*, 93–100. [CrossRef]
24. Nirola, R.; Megharaj, M.; Palanisami, T.; Aryal, R.; Venkateswarlu, K.; Naidu, R. Evaluation of metal uptake factors of native trees colonizing an abandoned copper mine—A quest for phytostabilization. *J. Sustain. Min.* **2015**, *14*, 115–123. [CrossRef]

25. Alcantara, H.J.P.; Doronila, A.I.; Nicolas, M.; Ebbs, S.D.; Kolev, S.D. Growth of selected plants species in biosolids-amended mine tailings. *Miner. Eng.* **2015**, *80*, 25–32. [CrossRef]

26. Puga, A.P.; Abreu, C.A.; Melo, I.C.A.; Paz-Ferreiro, J.; Beesley, L. Cadmiun, lead, and zinc mobility and plant uptake in a mine soil amended with sugarcane straw biochar. *Environ. Sci. Pollut. Res.* **2015**, *22*, 17606–17614. [CrossRef]

27. Lam, E.J.; Cánovas, M.; Gálvez, M.E.; Montofré, Í.L.; Keith, B.F.; Faz, Á. Evaluation of the phytoremediation potential of native plants growing on a copper mine tailing in northern Chile. *J. Geochem. Explor.* **2017**, *182*, 210–217. [CrossRef]

28. Afonso, T.F.; Demarco, C.F.; Pieniz, S.; Quadro, M.S.; Camargo, F.A.O.; Andreazza, R. Bioprospection of indigenous flora grown in copper mining tailing area for phytoremediation of metals. *J. Environ. Manag.* **2019**, *256*, 109953. [CrossRef]

29. ASTM E11—20. Standard Specification for Woven Wire Test Sieve Cloth and Test Sieves. In *American Society for Testing and Materials Standard*; ASTM International: West Conshohocken, PA, USA, 2020.

30. United Stated Department of Agriculture (USDA). Forest Service. Available online: https://www.fs.fed.us/wildflowers/Native_Plant_Materials/Native_Gardening/hardinesszones.shtml (accessed on 21 October 2020).

31. Chileflora. Available online: http://www.chileflora.com/Florachilena/FloraSpanish/PIC_NORTHERN_PLANTS_0.php (accessed on 21 October 2020).

32. Mellen, J.J.; Baijnath, H.; Odhav, B. Translocation and accumulation of Cr, Hg, As, Pb, Cu and Ni by *Amaranthus dubius* (*Amaranthaceae*) from contaminated sites. *J. Environ. Sci. Health A* **2009**, *44*, 568–575. [CrossRef] [PubMed]

33. Embrandiri, A.; Rupani, P.F.; Shahadat, M.; Singh, R.P.; Ismail, S.A.; Ibrahim, M.H.; Kadir Abd, M.O. The phytoextraction potential of selected vegetable plants from soil amended with oil palm decanter cake. *Int. J. Recycl. Org. Waste Agric.* **2017**, *6*, 37–45. [CrossRef]

34. Li, X.; Zhang, L.; Wang, X.; Cui, Z. Phytoremediation of multi-metal contaminated mine tailings with *Solanum nigrum* L. and biochar/attapulgite amendments. *Ecotoxicol. Environ. Saf.* **2019**, *180*, 517–525. [CrossRef] [PubMed]

35. Ortiz-Calderón, C.; Alcaide, O.; Li Kao, J. Copper distribution in leaves and roots of plants growing on a copper mine-tailing storage facility in northern Chile. *Rev. Chil. Hist. Nat.* **2008**, *81*, 489–499. [CrossRef]

MDPI

St. Alban-Anlage 66

4052 Basel

Switzerland

Tel. +41 61 683 77 34

Fax +41 61 302 89 18

www.mdpi.com

Minerals Editorial Office

E-mail: minerals@mdpi.com

www.mdpi.com/journal/minerals